Robinson's

BASIC NUTRITION and DIET THERAPY

8TH EDITION

EMMA S. WEIGLEY
Ph.D., R.D., F.A.D.A., School of Nursing, University of Pennsylvania, Philadelphia

DONNA H. MUELLER
Ph.D., R.D., F.A.D.A., Department of Bioscience and Biotechnology,
Drexel University, Philadelphia, Pennsylvania

CORINNE H. ROBINSON
M.S., D.Sc. (Hon.), Professor of Nutrition Emeritus, Formerly Head,
Department of Nutrition and Food, Drexel University, Philadelphia, Pennsylvania

Merrill,
an imprint of Prentice Hall
Upper Saddle River, New Jersey   *Columbus, Ohio*

Library of Congress Cataloging-in-Publication Data

Weigley, Emma Seifrit.
    Robinson's basic nutrition and diet therapy / Emma S. Weigley,
Donna H. Mueller, Corinne H. Robinson. — 8th ed.
        p. cm.
    Previously published: Robinson, Corinne H. Basic nutrition and diet therapy. c1993.
    Includes bibliographical references and index.
    ISBN 0-13-577016-5
      1. Diet therapy. 2. Nutrition.   I. Mueller, Donna H.
II. Robinson, Corinne H. (Corinne Hogden)   III. Robinson, Corinne H.
(Corinne Hogden). Basic nutrition and diet therapy. IV. Title.
V. Title: Basic nutrition and diet therapy
    RM216.R652 1997
    615.8′54—dc20

96-31615
CIP
r96

Cover art: © Paul Cézanne (1839–1906)/Superstock
Editor: Kevin M. Davis
Developmental Editor: Carol S. Sykes
Production Editor: Linda Hillis Bayma
Photo Coordinator: Dawn Garrott
Design Coordinator: Jill E. Bonar
Text Designer: Linda Robertson
Cover Designer: Brian Deep
Production Manager: Pamela D. Bennett
Director of Marketing: Kevin Flanagan
Advertising/Marketing Coordinator: Julie Shough

This book was set in Garamond Book by Carlisle Communications, Ltd. and was printed and
bound by Von Hoffmann Press, Inc. The cover was printed by Von Hoffmann Press, Inc.

 © 1997 by Prentice-Hall, Inc.
Simon & Schuster/A Viacom Company
Upper Saddle River, New Jersey 07458

Earlier editions, entitled *Basic Nutrition and Diet Therapy*, © 1993, 1989, 1984 by Macmillan
Publishing Company, a division of Macmillan, Inc., and © 1980, 1975, 1970, and 1965 by
Corinne H. Robinson.

Photo credits: Pp. xx and 214 by Barbara Schwartz; p. 276 by Barbara Schwartz/Merrill.

Printed in the United States of America

10 9 8 7 6 5 4 3 2 1

ISBN: 0-13-577016-5

Prentice-Hall International (UK) Limited, *London*
Prentice-Hall of Australia Pty. Limited, *Sydney*
Prentice-Hall of Canada, Inc., *Toronto*
Prentice-Hall Hispanoamericana, S. A., *Mexico*
Prentice-Hall of India Private Limited, *New Delhi*
Prentice-Hall of Japan, Inc., *Tokyo*
Simon & Schuster Asia Pte. Ltd., *Singapore*
Editora Prentice-Hall do Brasil, Ltda., *Rio de Janeiro*

PART I

Food and Nutrition Foundations in Health and Disease

LIST OF CASE STUDIES

LIST OF DIETS

Contents

Brief Contents

unstinting support. Donna Mueller and Emma Weigley would especially like to express their gratitude and profound respect to their beloved mentor, Corinne Robinson, who provided the opportunity to "come aboard" and assist with the continuation of the publication of this book.

EMMA S. WEIGLEY
DONNA H. MUELLER

FEATURES

To reinforce learning, each chapter contains a case study or personal study, which is introduced at the beginning of the chapter. There are new figures and tables. Photos highlight health care team members providing nutritional care. Each chapter contains a list, "For Additional Reading," of recently published articles from readily available publications. A Basic Diet is presented in Chapter 2 and followed throughout the book. Each chapter is a "stand alone" with extensive cross-referencing, enabling the instructor to present material in any order desired.

ORGANIZATION OF THE TEXT

The text is arranged in three parts. Part I, "Food and Nutrition Foundations in Health and Disease," emphasizes the role of nutrition in health maintenance. Chapter 3 presents "Nutrition Assessment and Dietary Counseling," because the authors believe assessment is an integral component of nutrition for wellness and should not be considered only for individuals who are ill. Basic nutrients, both macro and micro, are covered here as well as energy balance, fitness, and weight management. Hydration, an especially important concern of health care providers, is highlighted. Every effort is made to present each chapter as a discrete unit so the instructor can organize topics in whatever order best suits his or her needs and those of the students.

Part II, "Food and Nutrition Applications Throughout Life," spans the life cycle. There is a chapter on young and middle-aged adults, a period covering the largest number of years in the life span for most people, yet often given little attention in nutrition texts.

Part III, "Food and Nutrition Therapies," introduces students to nutrition in the clinical setting, be it inpatient, outpatient, or home. These chapters discuss diet modifications, medications and their possible interactions with nutrients, the immune function, stress, and nutrition as it relates to various body systems.

The Appendices have been substantially updated with much new material. Their order has been changed to follow the customary assessment process.

ACKNOWLEDGMENTS

There are many individuals to be thanked—first, Kevin Davis, our editor; Carol Sykes, our developmental editor; Linda Bayma, our production editor; Sheryl Rose, our copy editor; and Dawn Garrott, our photo editor. We appreciated and utilized the suggestions of reviewers of both the seventh edition of this text and of our manuscript for this book: A. J. Clark, Auburn University; Ruth Kessel, Shawnee Community College; Debra Krummel, West Virginia University; Marcia Mitson, Lutheran College of Health Professions; Anne O'Donnell, Santa Rosa Junior College; Judith Radcliff, Fairmont State College; Elaine Rogers, Saint Paul Technical College; and Fariba K. Roughead, University of North Dakota.

We also thank Qun Wang, Lois Pearson, and Peter Bartscherer, who assisted with the electronic preparation of this manuscript. The authors wish to express appreciation to their students in the School of Nursing of the University of Pennsylvania and the Didactic Program in Dietetics of Drexel University who used previous editions of this text and gave invaluable feedback and suggestions.

Finally, many thanks to our families—Katherine Lochel, Russell, Jared, and Catherine Weigley—for sharing our highs and lows while always providing

Preface

Health care in the United States is a major economic, political, and personal issue that affects every individual. Rapid changes are taking place, yet the future direction is still uncertain. Nutrition can, will, and must play an important role. Appropriate diet can foster wellness and prevent chronic disease. Nutritional therapy is based on the principles of normal nutrition and applying these principles to our lives. Excerpts from *Dietary Guidelines for Americans* (1995) and *Healthy People 2000* (1991) help to guide us and are highlighted throughout the text.

This book is intended for students preparing for careers in health care—nursing, dietetic technology, dental hygiene, food service management, sports, fitness, health education, and other health-related fields. It is pragmatic and practice-oriented.

PHILOSOPHY

This text presents practical information on the "how to" of the nutritional aspects of health care, always accompanied by the underlying rationale. It integrates normal and therapeutic nutrition, with nutrition therapy related to a basic diet. Diet therapy focuses on body systems, an approach commonly followed in health care.

NEW AND EXPANDED CONTENT

There is a new chapter on HIV/AIDS, more than in previous editions, emphasizing the increasing importance of this widely studied condition. The 1995 *Dietary Guidelines for Americans* and the 1995 *Exchange Lists for Meal Planning* are featured. Chapters 1 and 2 distill and condense materials covered in several chapters in previous editions to enable students to begin the study of nutrients early in the course. Newer, more comprehensive nutrient analysis tables are included in the Appendices, which have been reordered to follow the flow of nutrition assessment and counseling. The nutrition history form has been revised. The glossary and list of abbreviations have been expanded.

CORINNE H. ROBINSON, B.S., M.S., D.Sc. (Hon.), has a degree in nutrition from the University of Wisconsin and in biochemistry from the Graduate School of the University of Cincinnati. She began her career in nutrition and biochemical research at the Children's Hospital Research Foundation, Cincinnati. Following this she was a supervising therapeutic dietitian at the Presbyterian Hospital in New York City. She is Professor of Nutrition, Emeritus, from Drexel University, where she was Head of the Department of Nutrition and Food. Her extensive experience as an educator has included teaching students at Temple University Medical School, undergraduate and graduate students in dietetics and nutrition at Drexel University, and more than 2,000 students in schools of nursing at Columbia-Presbyterian Medical Center, New York City, and at several schools of nursing in Philadelphia.

A nationally recognized lecturer, Dr. Robinson has been honored by her professional associates to give memorial lectures—the Mary Swartz Rose lecture in New York City, the Lenna F. Cooper Memorial lecture in San Antonio, and the Anna dePlanter Bowes lecture in Philadelphia. Dr. Robinson's numerous publications in professional journals include research reports on blood proteins, acid-base balance, and metabolic studies on children. She has also published many papers on the practice of clinical nutrition. Her books include *Normal and Therapeutic Nutrition*, *Fundamentals of Normal Nutrition*, and *Case Studies in Clinical Nutrition*.

The Authors

EMMA S. WEIGLEY, B.S., M.S., Ph.D., R.D., F.A.D.A., is presently Adjunct Associate Professor at the School of Nursing, University of Pennsylvania. She received her master's from Drexel University and her doctorate from New York University, both in nutrition. Dr. Weigley has taught at Albright College, New York University, and Temple University, as well as at Penn. She began her career as a therapeutic dietitian at Reading Hospital, Reading, Pennsylvania. Subsequently she has been employed as a community nutritionist and as a renal dietitian in Philadelphia.

Dr. Weigley has published dozens of articles and reviewed many more, several of them on historical subjects connected with nutrition and cookery. She has served as a consultant on diet and nutrition for the government and for local institutions in Philadelphia. In addition to coauthoring the three previous editions of this text, Dr. Weigley has served as coeditor of *Essays on the History of Nutrition and Dietetics* and has authored *Sarah Tyson Rorer: The Nation's Instructress in Dietetics and Cookery.*

DONNA H. MUELLER, B.S., M.S., Ph.D., R.D., F.A.D.A., is Associate Professor of Nutrition and Food Sciences, Department of Bioscience and Biotechnology, Drexel University. She received her master's in nutrition education from Drexel University and her doctorate in health education and health administration from Temple University. She completed the dietetic internship at The New York Hospital–Cornell Medical Center. She was Chief Therapeutic Dietitian at Temple University Hospital before joining St. Christopher's Hospital for Children to practice pediatric nutrition.

She began her teaching career at Thomas Jefferson University and has taught at the Community College of Philadelphia, Immaculata College, Temple University, and University of Pennsylvania. Dr. Mueller has educated students entering the health professions, including dental hygienists, dentists, dietitians, nurses, occupational therapists, physical therapists, physicians, and social workers. The American and Pennsylvania Dietetic Associations selected her as an Outstanding Dietetics Educator in 1996.

Dr. Mueller's research area is human nutrition, especially pediatrics. Her publications focus on cystic fibrosis and dental nutrition. She has contributed chapters to other texts, including *Krause's Food, Nutrition and Diet Therapy; Medical Nutrition and Disease; Nutrition and Metabolism in Patient Care;* and *Nutrition in Oral Health and Disease.*

■ Part I lays the foundation for food and nutrition in health and disease. The eleven chapters focus on foods and the nutrients they contain; national dietary guidelines; contemporary issues related to foods and nutrients; and practical methods used by health care providers to determine the nutritional status of their clients and to counsel them about their diet.

Together, Chapters 1 and 2 provide the underpinnings for the text. These comprehensive chapters discuss the kinds and amounts of foods and nutrients necessary for people to maintain health. Various standards for foods and nutrients are described and illustrated. The concept of a model "Basic Diet" based on the Food Guide Pyramid and adjustable for any person's nutritional needs is presented, then used throughout the remainder of the text.

Health care providers nutritionally assess clients as part of routine health checkups and present nutrition information as part of health delivery. Chapter 3 introduces the basic techniques of evaluation and counseling. These skills are incorporated into each following chapter.

Body homeostasis (balance) is the theme of Chapter 4. Several crucial topics are included. First is an overview of the processes used by the body to maintain its nourishment. Since water serves numerous functions in the body, yet often is overlooked, the second section discusses fluid balance. The final sections present electrolyte balance and acid-base balance.

Chapters 5, 6, and 7 detail the energy-containing nutrients—carbohydrates, lipids, and proteins. A standard chapter format is used for ease of comparing one nutrient group with another. The outline is: characteristics; functions; metabolic processes; meeting body needs; body balance; and health and nutrition care issues.

Chapters 8 and 9 focus on the body's ability to achieve energy balance and physical fitness. As major health issues in the United States, the problems of underweight and overweight receive dominant coverage.

Chapters 10 and 11 detail vitamins and minerals. The presentation style is unique in these two chapters. An overview of the nutrient category is provided. In unified formats, highlights of the individual vitamins and minerals are discussed in separate sections. This arrangement allows general facts to be presented, with attention given to each vitamin or mineral for more intense study or reference as desired.

Food, Nutrition, and Health

Chapter Outline

In ways often interrelated with patterns of physical inactivity, dietary factors are associated with 5 of the 10 leading causes of death in the United States: coronary heart disease, some types of cancer, stroke, noninsulin-dependent diabetes mellitus, and atherosclerosis.*

Increase to at least 85 percent the proportion of people aged 18 and older who use food labels to make nutritious food selections.*

Achieve useful and informative nutrition labeling for virtually all processed foods and at least 40 percent of fresh meats, poultry, fish, fruits, vegetables, baked goods, and ready-to-eat carry-away foods.*

 PERSONAL STUDY PREVIEW

In this chapter you will begin a personal nutrition evaluation, assessing your food intake and eating behaviors, using various evaluation standards and tools.

*Healthy People 2000: National Health Promotion and Disease Prevention Objectives. Public Health Service, U.S. Department of Health and Human Services, Washington, DC, 1991, p. 56; Objectives 2.13 and 2.14, p. 94.

F rom birth to death food is a dominant factor in our lives. The tasks of acquiring, preparing, and consuming food are so innate that we rarely stop to think that foods are highly complex chemical substances, furnishing the 50 or more nutrients we need.

Early in the twentieth century scientists found that classic nutritional deficiency diseases such as scurvy, rickets, beriberi, pellagra, and nutritional anemias could be prevented by consuming adequate diets. Likewise, through immunization and improved sanitation, many infectious diseases could be prevented. Because of the greatly reduced prevalence of infectious and nutritional deficiency diseases, people now live longer and are more prone to suffer chronic diseases, which have their greatest incidence in the middle and later years. Obesity, hypertension, coronary heart disease, stroke, cancer, diabetes mellitus, osteoporosis, and diseases of the liver are major health problems in the United States and other developed countries and are beginning to appear in underdeveloped countries as some of the population become more prosperous. In this chapter you will begin evaluation of your own food intake and lifestyle in relation to health promotion and prevention of chronic disease.

FOOD IN OUR LIVES

Food has many meanings. We know that the food we eat is necessary for our existence. We know that it provides energy and that it builds, maintains, and regulates muscles, bones, nerves, the brain, eyes, hair, and all our physical being.

But food does much more than nourish the body, for most of us enjoy eating. Food makes us feel secure and happy; we use food as a link in our friendships, as an expression of pleasure during holidays, and as a symbol of our religious life.

Food is the world's biggest business. A large part of the world's work is concerned with the growing, processing, and preparation of food. We spend an important percentage of our income on food. In the United States and in other developed countries there is food in abundance and variety.

Most of the world's people spend the greater part of their days growing food and much of their income buying food. In some countries, three-fourths or more of the working population is directly concerned with growing food, yet seldom manages to produce enough. Millions of the world's people are chronically hungry. Is it any surprise if these people are diseased, discontented, and have a short life span?

NUTRITION: SOME BASIC CONSIDERATIONS

Nutrition as a Science

Throughout recorded history people have commented about food and its effects on the body. Ancient Egyptian writings record the use of food for the treatment of various diseases. Hippocrates, the famous Greek physician, wrote about the proper foods for treating disease. He observed, for example, that people who are very fat tend to die earlier than those who are slender.

Through the ages people learned that some foods would prevent diseases and that some plants were poisonous. Along with this experience a great deal of superstition about foods arose, some of which persists today.

The science of nutrition developed only after the groundwork had been laid for the sciences of chemistry and physiology, beginning in the late eighteenth century. Most of our understanding of the functions of nutrients in the body, the nutritive values of foods, the body's requirements for nutrients, and the role of nutrition in health and disease has resulted from research done since 1900. It must be emphasized that much still remains to be learned.

THE STUDY OF NUTRITION

Your Responsibility for Good Nutrition

First, you have a responsibility to yourself. You will benefit personally from good nutrition, for you will look and feel better and will be better able to meet the demands of your profession.

You have a responsibility to your family. This may lie in planning and preparing nutritious meals for them, in helping a child to develop good food habits, or in assisting an older person in making necessary dietary adjustments. This is a responsibility that you may assume more fully at some later time.

You have a responsibility as a nurse, nursing assistant, dietitian, dietetic technician, dental assistant, or other health care provider. Nutrition is an essential part of the total care of the patient. You will need an appreciation of what food means to your patients, how food can promote health, how illness changes their feelings about food, and how to help them with day-to-day food intake. See Chapter 16 for specific details on the nutritional care of patients.

You have a responsibility as a citizen. Because you are a health care provider, people will look to you for advice and as a role model. You will need to know how to answer simple questions. But you will also need to know when it is appropriate to refer people to other members of the health care team. You can support programs at the local, state, or national level that improve people's nutrition.

The Nutrition Team

The most important member of the nutrition team is the client. Indeed, the client is the only reason for the team's existence. Health professionals must remember that they must work with the client, permitting him or her to share in all decision making and respecting the client's decisions. Many professional people work with the client: physicians, nurses, dietitians, dietetic technicians, social workers, pharmacists, physical therapists, occupational therapists, speech language pathologists, laboratory technicians, and others. (See Figure 1–1.)

Some Definitions

Throughout your study of nutrition you will come across many terms that are important to your understanding. Here are a few basic terms.

Food is anything that nourishes the body. No two foods are identical in their ability to nourish for no two foods contain exactly the same amounts of nutrients.

Diet is the kind and amount of food and drink consumed each day. It includes the normal diet consumed by various age groups and the modified or therapeutic diets used in the treatment of disease conditions.

Nutrients are those 50 or more chemical substances in food that are needed by the body.

Nutrition refers to the processes in the body for utilizing food. It includes eating the appropriate kinds and amounts of foods for the body's needs; digestion

NUTRIENT CLASSES

Water
Carbohydrates
Lipids
Proteins
Vitamins
Minerals

FIGURE 1–1
A client meets with a nurse, a dietitian, a physician, and a social worker to review his progress and to make plans for continuing care. (*Source:* Dunwoody Village, Newtown Square, PA, and Peter Zinner, photographer.)

of foods so the body can use the nutrients; absorption of the nutrients into the bloodstream; use of the individual nutrients by the cells for production of energy and the growth and maintenance of cells, tissues, and organs; and elimination of wastes.

Health is defined by the World Health Organization (WHO) as the "state of complete physical, mental and social well-being and not merely the absence of disease and infirmity." Another term sometimes used to denote this state of health is **wellness.**

Nutritional status is the condition of health as it is related to the use of food by the body.

Malnutrition is an impairment of health resulting from a deficiency, excess, or imbalance of nutrient intake or body utilization. It includes **overnutrition,** such as an excess of calories or vitamin A toxicity. It also includes **undernutrition,** resulting from a deficiency of calories and/or some other nutrient or nutrients. Malnutrition may be *primary,* that is, caused by some fault in the diet, or *secondary,* that is, caused by some error in metabolism or some interaction between nutrients, medications, and so on. (See Figure 1-2.)

Epidemiology is the science of epidemic diseases, which are diseases that affect many people in a community. Through epidemiological studies the relationships between diseases and environmental factors including dietary components can be established. By this means the cause, treatment, cure, and ultimately the prevention of a disease can be achieved. For example, cancer of the colon and diverticulosis are common in the United States and other developed countries. Researchers found that these diseases were rare in rural Africa. Therefore, they tried to determine differences between the African and Western environments. They discovered that one variable was fiber content of the diet—Africans consumed a diet very high in fiber while diets in the Western world were low in fiber content. Based on these data, a **hypothesis** or theory

Epidemic: Greek *epi* = upon + *demos* = the people

Hypothesis: Greek *hypo* = under + *thesis* = a placing

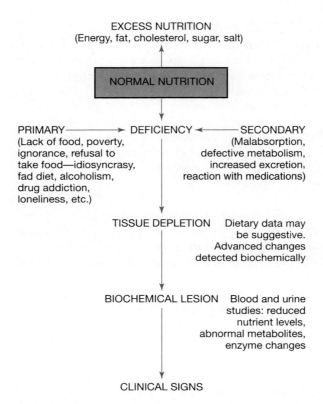

FIGURE 1–2
Normal nutrition is achieved when the intake of calories and nutrients is neither deficient nor excessive. Moderation and variety of food choices are key elements in achieving this balance.

was generated that a high fiber diet protects against a number of gastrointestinal diseases. With this working hypothesis, researchers can conduct further studies to determine whether the theory can be proved.

NUTRITION CHALLENGES FOR THE TWENTY-FIRST CENTURY

The Economic Crisis of Health Care

Research on the causes and treatment of diseases and the development of a technology for diagnosis and therapy provide opportunities for health care in developed nations that are unparalleled in the rest of the world. But sophisticated technology requires very large expenditures.

Attempts are being made to control these costs at levels that are sustainable in the national economy. Many Americans have no health insurance nor means to pay for health care. For some older citizens, health care costs may deplete their life savings.

The high costs of treating disease should be a powerful incentive to people of all ages to modify their behaviors to prevent diseases and disabilities.

The major dietary problem in the United States and other developed countries is overconsumption of certain dietary components. Excessive intake of fats and sugars has resulted in diets high in calories yet low in nutrient density.

Some dietary and environmental factors increase the possibility of developing chronic diseases, whereas others reduce it. Several risk factors are common to many chronic diseases. For example, obesity increases the chances of developing heart diseases, some cancers, and a type of diabetes mellitus. One in three Americans is obese. Selected risks for some prevalent diseases are outlined below and are discussed in more detail in the chapters pertaining to the specific nutrients and to each disease.

Dental Caries **Dental caries** or tooth decay is so common that people often do not think of it as a disease, yet it is the most common chronic condition. The formation of sound, healthy teeth begins during fetal life and continues into childhood. During this time calcium, phosphorus, fluoride, and vitamins A, C, and D are especially important. Frequent exposure of the teeth to sugars is a major factor in dental decay. Heredity, lack of professional dental care, and poor dental hygiene practices are nondietary factors promoting tooth decay.

Cardiovascular Diseases Each year about 1.5 million Americans have heart attacks. Over half of all deaths in the United States are from diseases of the blood vessels and heart.

Atherosclerosis or narrowing and hardening of the arteries is the chief cause of heart attacks and stroke. It can begin in childhood and advances throughout life. It may be a silent disease for decades, but then can result in a heart attack without warning.

Three risk factors for cardiovascular disease are beyond an individual's control: family history, advancing age, and male gender. On the other hand, smoking, obesity, high blood pressure, sedentary lifestyle, and high serum cholesterol levels are five major risk factors that generally can be modified. High serum cholesterol can be partially controlled when the diet is adjusted to attain and maintain healthy weight; reduce total fat, saturated fat, and dietary cholesterol content; and increase soluble fiber levels.

Hypertension or elevated blood pressure is the most common circulatory problem. It can frequently be controlled through weight management, moderation of sodium intake if the person is salt sensitive, and adequate intake of dietary calcium and potassium. If alcohol is consumed, moderation is the byword. Hypertension is a risk factor for stroke and heart disease.

Cancer Cancer ranks second to heart disease as the cause of death in the United States, claiming about half a million lives each year. Today many cancers are curable if they are treated before they have **metastasized** or spread to other parts of the body. Up to 80 percent of cancers have their origins in environmental factors and therefore theoretically can be prevented if the environment is modified, for example, by reducing unprotected exposure to sunlight to prevent skin cancers. Occupational exposure to such materials as asbestos must also be taken into account in determining how best to prevent cancers. Tobacco use is a potent risk factor. Heredity and advancing age are factors that currently cannot be modified.

DIETARY RISK FACTORS ASSOCIATED WITH CARDIOVASCULAR DISEASE

High saturated fat intake
High total fat intake
High cholesterol intake
Low soluble fiber intake
Excessive caloric intake
Excessive alcohol intake
High sodium intake in
  salt-sensitive persons
Low calcium intake
Low potassium intake

POSSIBLE DIETARY RISK FACTORS FOR CANCER

High fat intake
Low fiber intake
Excessive caloric intake/
  obesity
Excessive alcohol intake
High intake of smoked,
  salt-cured, or charbroiled
  meat or fish

Diabetes Mellitus Diabetes mellitus is a major public health problem; in the United States about half a million cases are diagnosed each year. About 2 percent of deaths result from diabetes. Heredity and obesity are major risk factors for non–insulin-dependent diabetes mellitus (NIDDM), which primarily begins in the middle years. The onset of NIDDM can be prevented or postponed by maintaining a healthy weight. Diabetes mellitus increases risks for diseases of the heart, blood vessels, and kidneys.

Osteoporosis **Osteoporosis** is a reduction in the quantity of bone, resulting in skeletal atrophy. Each year about a quarter of a million persons, mostly women, suffer hip fractures. One in five of these persons dies from complications and some of the others find that their quality of life is diminished. The primary cause of these fractures is osteoporosis.

National Initiatives in Chronic Disease Prevention

The Surgeon General's Report on Nutrition and Health reviews the role of nutrition in promoting health and preventing chronic diseases. This report is especially useful for policymakers and in program planning at the federal, state, and community level. *Diet and Health* is a comprehensive review of the scientific basis for the relationship of diet to chronic disease. These books, which represent the contributions of hundreds of nutrition professionals, provide the basis for setting health objectives for the American population.

Healthy People 2000 outlines health issues for various segments of the United States populace and provides more than 200 objectives for health promotion, health protection, preventive services, and surveillance and data systems to be achieved by the year 2000. "Nutrition" and "Physical Activity and Fitness" are among the eight priority areas for health promotion. Most of the objectives require individuals to make some type of behavioral change.

In this textbook, nutrition-related objectives relevant to chapter content are highlighted at the beginning of many chapters. These objectives, in part, apply to your own nutritional behavior. They also relate to the nutritional care you provide for your clients. Throughout this book you will find practical ways to select foods and diets that promote health and, at the same time, reduce the risks of chronic diseases.

STANDARDS OF NUTRIENT INTAKE

Countries throughout the world have established standards for desirable nutrient intake. These standards differ because they are intended for specific populations with different climatic conditions, varying levels of activity, differing food resources, and unique dietary practices. For example, Canada has developed values that are a "single, unified set of nutrition recommendations for the maintenance of health and the reduction of risk factors for chronic disease." (See inside back cover.) The World Health Organization (WHO) together with the Food and Agriculture Organization (FAO) have also set standards for a number of nutrients.

Recommended Dietary Allowances (RDAs)

Nutrition standards for the United States are developed by the Food and Nutrition Board of the National Research Council. Since 1943 the recommendations have been revised at approximately five-year intervals to reflect the latest nutrition research.

Philosophically, the RDA standards are intended to provide for the needs of almost every healthy person. With few exceptions they are not averages, but they

---

RISK FACTORS FOR OSTEOPOROSIS

Hormonal change—
  postmenopausal
Lack of weight-bearing
  exercise
Small stature
Long-term low calcium intake
Smoking
Genetics

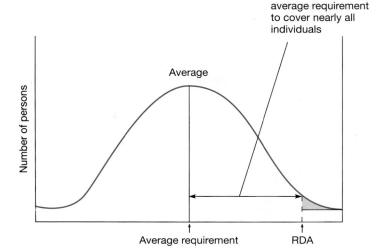

FIGURE 1-3

Individuals vary in their nutritional requirements. A few people will have very low requirements for a given nutrient and a few will have very high requirements. Most people have requirements that fall around the average. Note that the RDA is set high enough to provide the needs for practically all of the healthy population.

are set high enough to fulfill the nutritional needs of 97 to 98 percent of the healthy population. (See Figure 1-3.) They are not intended to cover the requirements of ill people whose conditions may necessitate more or less of a nutrient. Since RDAs are set high they provide more of most nutrients than many people need. Also, since individual needs vary, RDAs cannot be viewed as the exact requirements for any one person.

The major RDA table appears inside the front cover of this book. The reference population is divided into age and age-gender groups with additional categories for pregnant and lactating women. Recommendations for protein, 11 vitamins, and 7 minerals are given. Since a large body of information exists about these nutrients, allowances are stated as single figures, which include a margin of safety.

Another table, also inside the front cover, lists an additional two vitamins and five minerals. Less is known about these nutrients, so intake is presented as a range. Also inside the front cover are estimated minimum requirements for sodium, chloride, and potassium.

Energy or caloric requirements for the age-gender groups are stated as averages and do not include a margin of safety. (See inside back cover.) Because obesity is a major health problem and a risk factor for many chronic diseases, adding extra calories would be inappropriate.

The RDAs are used by health care professionals, nutrition researchers, the food industry, and public policy planners. They can be utilized to plan and/or evaluate the intake of groups of people, determine how well the food supply meets nutritional requirements, provide a standard for food labeling, and serve as a basis for food guidelines for educating the public. Many nutrient analysis computer software programs use the RDA as a standard for assessing dietary adequacy.

There is no RDA for carbohydrates, lipids, and some vitamins and minerals known to be essential to humans but about which information is limited. The RDAs have been criticized because they are planned to prevent nutrient deficiencies but do not address the promotion of health or the prevention of chronic diseases.

DIETARY GUIDELINES FOR WELLNESS AND CHRONIC DISEASE PREVENTION

Consumers need a practical guide that will help them make food choices to meet their nutritional and health requirements. During this century, and especially in recent years, several food group guides as well as numerous diet and health guides have been published by governmental agencies and the private sector.

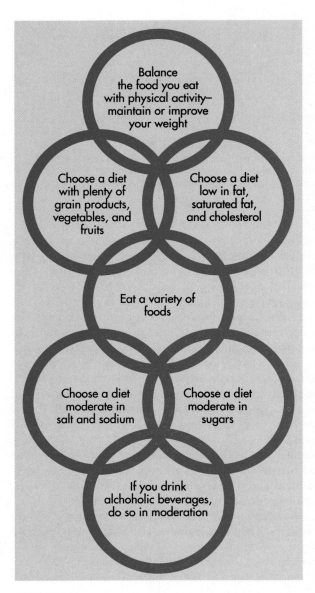

FIGURE 1–4
The 1995 Dietary Guidelines for Americans. (*Source:* U.S. Departments of Agriculture and Health and Human Services, *Nutrition and Your Health: Dietary Guidelines for Americans*, 4th ed., 1995.)

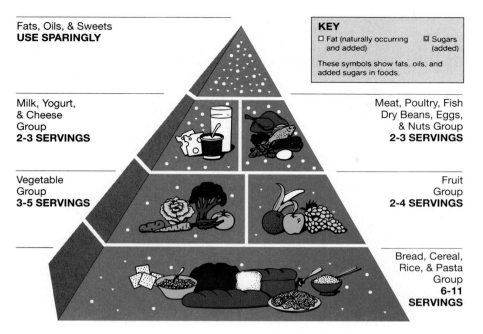

Fats, Oils, & Sweets
USE SPARINGLY

KEY
☐ Fat (naturally occurring and added)   ☒ Sugars (added)

These symbols show fats, oils, and added sugars in foods.

Milk, Yogurt,
& Cheese
Group
2-3 SERVINGS

Meat, Poultry, Fish
Dry Beans, Eggs,
& Nuts Group
2-3 SERVINGS

Vegetable
Group
3-5 SERVINGS

Fruit
Group
2-4 SERVINGS

Bread, Cereal,
Rice, & Pasta
Group
**6-11
SERVINGS**

FIGURE 1–5
This Food Guide Pyramid emphasizes the importance of eating a variety of foods in moderation. (*Source:* U.S. Department of Agriculture, Washington, DC, 1992.)

Dietary Guidelines for Americans

In 1995 the fourth edition of *Nutrition and Your Health: Dietary Guidelines for Americans* (see Figure 1-4) was issued jointly by the U.S. Department of Agriculture (USDA) and the U.S. Department of Health and Human Services (DHHS). This publication contains general health recommendations and dietary suggestions based on current information about nutrition. By following these guidelines people can both promote and maintain wellness and prevent nutrition-related chronic diseases or at least delay their onset.

The Food Guide Pyramid

The Food Guide Pyramid was developed to assist individuals in implementing the Dietary Guidelines for Americans. Consumer research was utilized to create a format that users would both understand and remember. The pyramid emphasizes variety both among and within its six groups, moderation in portions, and proportionality in eating more foods from the larger groups near its base and relatively fewer foods from other groups. (See Figures 1-5 and 1-6.) You should become thoroughly familiar with the Food Guide Pyramid, a widely used teaching aid.

Exchange Lists

This system of food lists was developed collaboratively by the American Dietetic Association and the American Diabetes Association for use by individuals with diabetes mellitus. (See Table A-2 in the Appendix.) These exchange lists are also

The amount of food that equals 1 serving is listed below. If you eat a larger portion, count it as more than 1 serving. For example, a dinner portion of spaghetti would count as 2 or 3 servings of pasta. No serving size is given for the fats, oils, and sweets group because the message is USE SPARINGLY.

FOOD GROUPS

Milk, Yogurt, and Cheese

| 1 cup of milk or yogurt | 1½ oz of natural cheese | 2 oz of process cheese |

Meat, Poultry, Fish, Dry Beans, Eggs, and Nuts

| 2–3 oz of cooked lean meat, poultry, or fish | ½ cup of cooked dry beans, 1 egg, or 2 T of peanut butter count as 1 oz of lean meat |

Vegetable

| 1 cup of raw leafy veg-etables | ½ cup of other veg-etables, cooked or chopped raw | ¾ cup of veg-etable juice |

Fruit

| 1 med. apple, banana, orange | ½ cup of chop-ped, cooked, or canned fruit | ¾ cup of fruit juice |

Bread, Cereal, Rice, and Pasta

| 1 slice of bread | 1 oz of ready-to-eat cereal | ½ cup of cooked cereal, rice, or pasta |

FIGURE 1–6
What counts as 1 serving?

useful in making food choices according to a meal pattern or for estimating an intake of calories, carbohydrates, fat, and/or protein. Each of the lists includes foods that, in stated portions, supply approximately the same amounts of these nutrients. The following lunch menu gives an example.

LUNCH BASED ON EXCHANGE SYSTEM

Turkey Sandwich	*Exchange List*
2 slices whole wheat bread	2 starch
2 oz white meat turkey	2 very lean meat
1 tsp regular mayonnaise	1 fat

Salad	
Lettuce	Free
⅛ avocado	1 monounsaturated fat
1 tomato	1 vegetable
1 tbsp French dressing	1 polyunsaturated fat

1 orange	1 fruit

1 cup 2% milk	1 low-fat milk

An exchange is defined as a serving of food with approximately the same amount of carbohydrates, protein, fat, and calories as other foods on the same list. For example, in the lunch menu, 1 teaspoon margarine could be "exchanged" for 1 teaspoon mayonnaise and an apple for the orange. However, the orange could not be replaced with margarine. (See Fruit List and Fat List, Appendix Table A-2.)

It is important to remember that the amount of calories and grams of carbohydrate, fat, and protein are approximations for ease of use. (See Chapter 26.) Exchanges assume that the individual consumes a varied diet.

FOOD COMPOSITION

Tables of Food Composition

The nutritive values of foods have been determined by chemical analyses in research laboratories of colleges, universities, the United States Department of Agriculture (USDA), and the food industry. Table A-1 in the Appendix of this book, tables in other professional and popular books, and tables in computerized data programs are derived for the most part from the USDA's National Nutrient Data Bank. (See Figure 1-7.)

To use tables of food composition it is important to understand the quantitative relationships that exist between metric units and to be able to interpret them with respect to commonly used household measurements. (See Table B-1 in the Appendix for conversions to and from metric measures.)

Note that food composition tables cannot take into account the way each nutrient is processed by the body in conditions of health and disease. Also, laboratory analyses can identify nutrients chemically present in food, but cannot determine the amount of the nutrient available to the individual consuming that food.

Bioavailability refers to the amount of nutrients absorbed from the intestinal tract and therefore available to the body. Bioavailability can vary from individual to individual and within individuals under different circumstances.

FIGURE 1–7
A student analyzes her nutrient intake. (*Source:* Drexel University, Philadelphia, PA, and Peter Groesbeck, photographer.)

Healthy people absorb more than 90 percent of carbohydrate, fat, and protein. However, a person with a lipid malabsorption problem, for example, might absorb considerably less fat.

Bioavailability of vitamins and minerals may be influenced by the source of the nutrient, the form of the nutrient, the nutritional status and needs of the individual, and other components in the diet. Vitamin C, for example, increases utilization of iron; phytate may decrease bioavailability of some minerals. Some disease processes and medications (see Chapter 18) affect the bioavailability of nutrients.

Nutrient Density

Nutrient density refers to the quantity of one or more nutrients supplied by a food in relation to its caloric content. Study the following example:

EXAMPLE OF NUTRIENT DENSITY

	Calories	*Calcium, mg*
Custard pie, ⅛	250	110
Cheddar cheese, ½ oz	60	102
Broccoli	60	103

You can see that the three foods are comparable in calcium content, but the custard pie has four times as many calories as the cheese and broccoli. Thus, broccoli and cheese have the higher nutrient density for calcium, and custard pie raises the question, "Can you afford the extra calories?"

TABLE 1–1

Nutrient References for Food Labels

REFERENCE DAILY INTAKES (RDIs)

Nutrient	Amount	Nutrient	Amount
Protein	50 g	Vitamin C	60 mg
Thiamin	1.5 mg	Vitamin A	5,000 IU
Riboflavin	1.7 mg	Vitamin D	400 IU
Niacin	20 mg NE	Vitamin E	30 IU
Biotin	300 μg	Calcium	1,000 mg
Pantothenic Acid	10 mg	Iron	18 mg
Vitamin B_6	2 mg	Zinc	15 mg
Folate	400 μg	Iodine	150 μg
Vitamin B_{12}	6 μg	Copper	2 mg

DAILY REFERENCE VALUES (DRVs)[a]

Food Component	DRV	Calculation
Fat	65 g	30% of kcalories
Saturated fat	20 g	10% of kcalories
Cholesterol	300 mg	Same regardless of kcal
Carbohydrate	300 g	60% of kcalories
Fiber	25 g	11.5 g per 1,000 kcal
Protein	50 g	10% of kcalories
Sodium	2,400 mg	Same regardless of kcal
Potassium	3,500 mg	Same regardless of kcal

[a]DRVs are appropriate for adults and children over 4 years of age. Values for energy bearing nutrients are based on 2,000 kcal a day.

FOOD LABELING

The Nutrition Labeling and Education Act (NLEA) of 1990 had three objectives: to clear up terminology confusion; to provide consumers with information to assist them in making healthy food choices; and to encourage food product innovation that would give the public healthier foods for a contemporary lifestyle. This legislation also provides for nutrition education to make available accurate and useful information to the public about foods they choose.

Ingredient Labeling

REQUIRED INGREDIENT IDENTIFICATION

FDA-certified color additives
Sources of protein hydrolysates
Statement of caseinate as a
 milk derivative
% of juice in beverages containing juice

Packaged foods that contain two or more ingredients must list them on the label, along with the name and address of the manufacturer. The ingredients have to be listed in decreasing order according to weight, that is, the most plentiful ingredient has to be listed first, the next most plentiful listed second, and so on. To assist consumers with food allergies or intolerances, sources of certain ingredients must be stated.

Descriptive Terms A dictionary of consistent and uniform terminology now permits better understanding of and confidence in the contents of the food. Terms such as "low," "high," "less," "fresh," "reduced fat," "cholesterol-free," and "low calorie" are clearly defined. For example, to be labeled "low sodium" a serving may contain no more than 140 mg sodium.

Serving Sizes To assist consumers in making informed nutritional comparisons between brands of the same food item, standardized, consistent serving sizes in amounts people actually eat have been developed. These sizes must be used in labeling.

Nutrient References Nutrient reference standards fall into two categories, **Reference Daily Intakes (RDIs)** and **Daily Reference Values (DRVs).** The quantities for vitamins, minerals, and protein are designated as RDIs. DRVs include quantities as well as percentages of calories from nutrients not currently included in the RDA, such as total fat, saturated fat, and carbohydrates. References for fiber, protein, cholesterol, sodium, and potassium are also designated DRVs. (See Table 1–1.) To make the label more "consumer friendly," only DRV standards must appear. Standards appropriate for infants and for children up to age four have been developed for use on labels of foods intended for this age group.

A list of mandatory and voluntary nutrients and the order in which they must be listed appears in Figure 1–8. Nutrients involved in prevention and causation of chronic diseases are emphasized. The label must utilize a consistent format titled "Nutrition Facts." For a sample label for whole grain cereal, see Figure 1–9.

Restaurant Meals and Fresh Foods

"Limited menu" restaurants (that is, fast-food restaurants) must provide nutritional information for their customers. Nutritional information about the most commonly purchased fresh foods in markets must be displayed at the point of purchase.

- total calories
- calories from fat
- *calories from saturated fat*
- total fat
- saturated fat
- *polyunsaturated fat*
- *monounsaturated fat*
- cholesterol
- sodium
- *potassium*
- total carbohydrate
- dietary fiber
- *soluble fiber*
- *insoluble fiber*
- sugars
- *sugar alcohols (e.g., xylitol, mannitol, and sorbitol)*
- *other carbohydrates*
- protein
- vitamin A
- vitamin C
- calcium
- iron
- *other essential vitamins and minerals*

FIGURE 1–8
Mandatory and *voluntary* (in italics) components in the order in which they must appear on a label.

Nutrition Facts

Serving Size 1 cup (35g)
Servings Per Container 10

Amount Per Serving	Cereal	Cereal with 1/2 cup Skim Milk
Calories	130	170
Calories from Fat	0	0

	% Daily Values**	
Total Fat 0g*	0%	0%
Saturated Fat 0g	0%	0%
Cholesterol 0mg	0%	0%
Sodium 200mg	8%	11%
Total Carbohydrate 30g	10%	12%
Dietary Fiber 4g	16%	16%
Sugars 18g		
Protein 3g		

Vitamin A	25%	25%
Vitamin C	25%	25%
Calcium	10%	25%
Iron	10%	10%
Thiamin	25%	30%
Riboflavin	25%	35%
Niacin	25%	25%
Vitamin B6	25%	25%

* Amount in Cereal. One half cup skim milk contributes an additional 40 calories, 65 mg sodium, 6g total carbohydrate (6 g sugars), and 4g protein.
** Percent Daily Values are based on a 2,000 calorie diet. Your daily values may be higher or lower depending on your calorie needs:

		Calories:	2,000	2,500
Total Fat	Less Than		65g	80g
Sat Fat	Less Than		20g	25g
Cholesterol	Less Than		300mg	300mg
Sodium	Less Than		2,400mg	2,400mg
Total Carbohydrate			300g	375g
Dietary Fiber			25g	30g

Calories per gram:
Fat 9 • Carbohydrate 4 • Protein 4

INGREDIENTS: WHOLE WHEAT, SUGAR, MALT EXTRACT, CORN SYRUP, TRISODIUM PHOSPHATE, VITAMIN C (SODIUM ASCORBATE), IRON, NIACINAMIDE, VITAMIN A (PALMITATE), CALCIUM CARBONATE, VITAMIN B6 (PYRIDOXINE HYDROCHLORIDE), RIBOFLAVIN, THIAMIN, BHT TO PRESERVE FRESHNESS.

FIGURE 1–9
Nutrition facts box panel for whole grain cereal. (*Source:* Kurtzweil, P: "The New Food Label: Better Information for Special Diets," *FDA Consumer*, 29:19–25, January/February 1995.)

The NLEA authorizes inclusion on the label of claims about the labeled food and specific health or disease conditions. Such claims are designated and only relationships based on clear scientific evidence supporting the claim are allowed. Examples include fat and cardiovascular disease, fat and cancer, sodium and hypertension, and calcium and osteoporosis. As scientific evidence accumulates and demonstrates a clear relationship between a nutrient and a specific health or disease condition, other categories may be added.

PUTTING IT ALL TOGETHER

Nutrition education has shifted from an emphasis on preventing deficiencies to an emphasis on preventing the excesses that have been linked with various chronic diseases. Many standards and guides emphasizing healthy eating to promote wellness and reduce risks of chronic disease have been developed to assist consumers, including the Dietary Guidelines for Americans, the Food Guide Pyramid, and "user friendly" food labeling.

? QUESTIONS CLIENTS ASK

1. When I went to school we learned about the Basic 4. Why did they change to the Food Guide Pyramid?

The Basic 4 was developed to prevent deficiencies, but did not attempt to control fat or increase fiber. With present-day emphasis on chronic disease prevention and promotion of healthy eating, the Pyramid is planned to increase complex carbohydrates and reduce intake of fats and sweets.

2. I couldn't possibly eat six servings from the bread group each day. I'd get fat.

This group contains not only bread, but cereal, rice, pasta, and crackers. There are only about 75 calories per serving. Many people eat portions of pasta, for example, that contain two or more servings. Emphasize variety and whole grain choices. This group provides a good source of complex carbohydrates.

3. I don't see how all fruits can be grouped together. We hear all the time that you can't compare apples and oranges.

Apples and oranges are different, but an apple is more similar in nutrient content to an orange than it is to meat or bread or fat. By emphasizing variety within each food group, you can help ensure an adequate diet.

4. I am interested in reducing fat in my diet. How can the food label help me?

Food labels are required to list the following information about fat: total fat, saturated fat, cholesterol by weight (grams or milligrams), and Percent Daily Values. Total calories from fat must also appear. If you are keeping a tally of any of these types of fat, labels will help you determine the total. Checking such information when food shopping will alert you to foods that are simply too high in fat for your diet plan. The terms *fat-free* (less than 0.5 g per serving), *low-fat* (3 g or less per serving) and *reduced-fat* (25 percent less fat than the reference food) have been defined to give these expressions a consistent meaning.

1. Keep a record of your food and beverage intake for three days. Be sure to include size of portions.

   a. Group the foods you consumed according to the Food Guide Pyramid. In which groups was your intake satisfactory? Were any groups deficient? How could you improve your diet? Be sure suggestions are practical—foods that you can afford, that are available, and that you are able and willing to eat.

   b. Group your intake according to the Exchange Lists. How many exchanges did you have from each list? Indicate any items you consumed that do not appear on an Exchange List.

2. Using Table A–1 in the Appendix, record the vitamin C value in your food intake according to the amount eaten. Total the amount for each day and compare it to your RDA for vitamin C (inside front cover). If indicated, make suggestions for dietary changes.

3. Evaluate your eating habits and lifestyle according to the Dietary Guidelines. What changes, if any, might you make to promote health?

4. List risk factors for chronic diseases that might pertain to you. Which can be modified and which cannot? What steps, over the course of a year, could you take to reduce your risks? Develop a plan for such a change.

5. How do you differ from the reference man/woman of the RDA in height, weight, activity, or in other ways? How would you expect these differences to affect your allowances that should be provided in the diet?

6. Using the intake for any one day recorded for question 1 above, determine the levels of cholesterol and sodium. Use the Food Facts label from food packages where available; for any other items, use Table A–1. How does your intake compare to the recommendations in the Food Facts? If your total of cholesterol or sodium is excessive, how could your diet be modified to reduce cholesterol and/or sodium?

American Journal of Clinical Nutrition (Am J Clin Nutr)

American Journal of Nursing (Am J Nurs)

FDA Consumer

Journal of the American Dietetic Association (J Am Diet Assoc)

Journal of the American Medical Association (JAMA)

Journal of Nutrition Education (J Nutr Educ)

New England Journal of Medicine (New Engl J Med)

Nursing (Nurs)

Nutrition Reviews (Nutr Rev)

Nutrition Today (Nutr Today)

RN

Topics in Clinical Nutrition (Top Clin Nutr)

SOME PERIODICALS FREQUENTLY CITED IN THIS BOOK

FOR ADDITIONAL READING

Achterberg, C, *et al:* "How to Put the Food Guide Pyramid into Practice," *J Am Diet Assoc,* **94:**1030–35, 1994.

Committee on Diet and Health, Food and Nutrition Board: *Diet and Health: Implications for Reducing Chronic Disease Risk.* National Academy Press, Washington, DC, 1989.

Dollahite, J, *et al:* "Problems Encountered in Meeting the Recommended Dietary Allowances for Menus Designed According to the Dietary Guidelines for Americans," *J Am Diet Assoc,* **95:**341–47, 1995.

"Final Food Labeling Regulations," *J Am Diet Assoc,* **93:**146–48, 1993.

Gourlie, KE: "Food Labeling: A Canadian and International Perspective," *Nutr Rev,* **53:**103–05, 1995.

Healthy People 2000: National Health Promotion and Disease Prevention Objectives. U.S. Department of Health and Human Services, Public Health Service, Washington, DC, 1991.

"How Should the Recommended Dietary Allowances be Revised? A Concept Paper from the Food and Nutrition Board," *Nutr Rev,* **52:**216–19, 1994.

Kennedy, E, Meyers, L, and Layden, W: "The 1995 Dietary Guidelines for Americans: An Overview," *J Am Diet Assoc,* **96:**234–37, 1996.

Kennedy, ET, *et al:* "The Healthy Eating Index: Design and Applications," *J Am Diet Assoc,* **95:**1103–08, 1995.

Kurtzweil, P: "The New Food Label: Better Information for Special Diets," *FDA Consumer,* **29:**19–25, January/February 1995.

Popkin, BM: "The Nutrition Transition in Low-Income Countries: An Emerging Crisis," *Nutr Rev,* **52:**285–98, 1994.

Porter, DV: "Health Claims on Food Products: NLEA," *Nutr Today,* **31:**35–38, 1996.

"Proposed Nutrient and Energy Intakes for the European Community: A Report of the Scientific Committee for Food of the European Community," *Nutr Rev,* **51:**209–12, 1993.

Sims, LS: "Uses of the Recommended Dietary Allowances: A Commentary," *J Am Diet Assoc,* **96:**659–62, 1996.

Stamler, J: "Assessing Diets to Improve World Health: Nutrition Research on Disease Causation in Populations," *Am J Clin Nutr,* **59**(suppl):146S–56S, 1994.

The Surgeon General's Report on Nutrition and Health. U.S. Department of Health and Human Services, Public Health Service, Washington, DC, 1988.

Food Choices

Chapter Outline

Reduce infections caused by key foodborne pathogens. . . .*

✔ CASE STUDY PREVIEW

Picnics are a common source of foodborne illness. Whenever food is purchased, then prepared, stored, and transported over a period of time until consumption, food safety issues are paramount. What precautions do you take to prevent personally succumbing to foodborne illness?

**Healthy People 2000: National Health Promotion and Disease Prevention Objectives.* Public Health
Service, U.S. Department of Health and Human Services, Washington, DC, Objective 12.1, p. 107.

The particular foods each person chooses to eat depend on numerous nutritional and nonnutritional factors, including the person's age and cultural food preferences, the cost and appearance of foods, and the nutrients foods contain. Foods are highly complex chemical substances that furnish the more than 50 nutrients needed for life and health. Their eye-appealing colors and taste-tempting flavors are specific chemical compounds, as are the fibrous components that give them texture and shape.

Lifestyles differ for each individual and family. Many people enjoy preparing foods "from scratch," while others save time by using packaged foods and microwave ovens. When eating out, some people spend their money at prestigious gourmet restaurants; others eat in fast-food restaurants every day for one or more meals. Cultural diversity enables people to have a boundless array of tempting nutritious foods from which to make their food choices. A fundamental pleasure of life is the enjoyment of palatable foods attractively served in an inviting setting with family and friends.

The food supply in North America is as wholesome as any in the world. However, whether eating in their own home, in a restaurant, from a convenience store, or in a health care institutional setting, people depend on others to supply the foods they eat and the water they drink. When people lived in nomadic societies, they ate foods that were available day by day, meal by meal. Even then, people processed some foods to have something to eat in times of famine. Today, ready-to-eat foods are available daily from all corners of the world. Most issues concerning the wholesomeness of foods deal with matters invisible to the eye and insensible to the nose. Although the safety record is excellent, a breakdown at any point between the farm and the consumer's table, or a client's tray, can lead to malnutrition, foodborne illness, or death.

Millions of Americans have a renewed interest in the kinds and amounts of foods they eat. They want to select foods that not only meet their nutritional requirements, but that also carry lesser risks of chronic disease. Many people find getting the best nutritional value from each food dollar spent a continuing challenge. The use of current nutrition education tools greatly enhances people's chances of achieving their goals.

INTERNAL AND EXTERNAL FACTORS INFLUENCING FOOD CHOICES

Hunger and Appetite

When we have gone without food for a few hours we say we are hungry. **Hunger** is the stimuli within our bodies that indicate to us that we need to consume food. Upon eating food we say we are "full" or satisfied. This is called **satiety.** Our bodies tell us when we need food, but obtaining that food depends on numerous external factors. **Appetite** consists of the pleasurable sensations provided by, and associated with, the enjoyment of food—its appearance; its palatability; the people involved with its preparation and service; the mealtime environment; and the meanings of food in our lives.

The Five Senses

Each of the five senses helps govern our food choices. Such qualities as appearance, taste, smell, texture, and temperature are some of the most familiar sensory factors. Generally, food needs to look clean and attractive to us before we want to eat it. The tongue and palate provide the four taste sensations of sweet, sour, salty, and bitter. Taste sensations vary widely from one individual to another, and within an individual throughout life. These differences are explained in part by the number of taste buds, the interplay with the sense of smell, the habits we develop over the years, and certain disease processes. Infants appear to have a keen sense of taste, which the aging process diminishes. Diseases and their treatment, such as colds or radiation therapy for cancer, can alter taste.

Aromas increase or decrease our acceptance of food. The smell of peaches ripening in the kitchen, meat roasting, or bread baking may be appealing. On the other hand, many nonfood odors, such as the disinfectants used in institutions, are unpleasant, thereby interfering with the acceptance of food. Some medications interfere with both the taste and the odor of food. Even as you eat, your sense of smell makes a difference; if you hold your nose while eating a fresh succulent pear you probably will think that the pear has very little flavor.

The way food feels in the mouth influences whether we accept or reject it. We expect ice cream to be smooth and creamy, but raw carrots and celery to be crisp and crunchy. Some people like rice moist and sticky; others enjoy it dry and fluffy. We probably all complain about lumpy gravy, greasy meat, or stringy green beans. The temperature of food also makes a difference. In general, children prefer lukewarm to hot foods, whereas adults expect their foods to be piping hot or ice cold.

The Body's Systems, Organs, Muscles, and Cells

One of the body's major activities is processing food to release nutrients, thereby providing nourishment to sustain life or heal disease or trauma. These processing activities are tightly interconnected. For example, after a person decides to eat, the lips, teeth, tongue, and salivary glands begin the process by enabling mastication and swallowing of food. Another organ, perhaps less well known, is the hypothalamus. It is located at the base of the brain and exercises control over feelings of hunger and satiety. This organ is sometimes referred to as the "appestat." It behaves like the thermostat in your home—turning on and shutting off the feelings that make you want to eat. Numerous cells of the body produce many substances, like hormones, that serve in the regulation of food intake. One of the better known hormones is insulin, which is produced by the pancreas. When insulin lowers the blood sugar level, the brain registers the body's need to eat. The lungs and heart carry the oxygen needed for the body's cells to release energy from food. People with heart or lung conditions become exhausted just by attempting to eat because the cells do not receive the oxygen they require and cannot dispose of the carbon dioxide they produce. The process of transforming food for the body's multiple needs is discussed in Chapter 4.

Societal Influences

Both individually and especially in groups, people exert powerful influences over food choices. (See Figure 2–1.) For example, a person's age and gender, a family's social and economic status, the availability of foods in a certain locale, and people's religious and health beliefs all may influence food choices.

Family Customs The family environment greatly influences food choices and habits. A positive attitude about food is more likely to develop when the entire family comes together for meals in a happy, relaxed atmosphere. When a variety of foods are prepared in different ways, each member's food experiences

FIGURE 2–1
In today's fast-paced society, people often purchase food to eat from street vendors or food trucks. (*Source:* Drexel University, Philadelphia, PA, and Peter Groesbeck, photographer.)

are enriched. But negative attitudes about food may develop, too. Children are quick to imitate parents who do not eat certain foods. Children rapidly note signs of worry, dislike, or anger in their parents and develop antagonisms toward particular foods. They dawdle when they learn it is a way to gain attention. Adults sometimes punish children by refusing to give them dessert if they haven't finished their meal, or bribe them with a favorite food such as candy.

Social Patterns Foods are often classed as being for babies, young children, or adults. Milk, cut-up food, and peanut butter and jelly sandwiches are looked upon as children's foods; hamburgers, pizza, and huge sandwiches are teenage fare; tea and coffee are for adults. Foods can carry gender connotations. Meat, potatoes, and pie may be typical of masculine meals, whereas souffles, salads, and sorbet are classed as feminine foods. Some foods have more prestige value than others and are used as "company foods" to honor or impress friends or business associates. Often, these foods are more costly, are difficult to obtain, take much time to prepare, or are unusual. Examples of such status foods are filet mignon, wild rice, baked Alaska, and fine imported wines. Other foods are given low status, such as ground meat, margarine, powdered skim milk, dry beans, and canned tuna fish. Yet any of these latter foods are just as nutritious, if not more so, and can be prepared in many delicious ways.

Economic Factors Poverty can adversely affect the formation of satisfactory food habits. Inadequate income usually limits both the quantity and variety of foods available. However, people with larger discretionary incomes can make unhealthy food choices too.

Psychological Influences

Did you ever go out with your classmates for pizza after taking final examinations? Have your friends rewarded you for an accomplishment with an expensive meal in a restaurant? Do you recall with pleasure special meals prepared just for you at your birthday parties or on important holidays?

Food is often used to express feelings of happiness, love, and security, or to cover up emotions of worry, grief, and loneliness. The baby who is held while being fed associates food with warmth and security. But a child who is scolded for being messy at the table may associate mealtimes with unhappiness. Some teenagers overeat to compensate for a poor report card or unpopularity with their classmates. Elderly persons living alone often eat far too little, because they are lonely and unhappy. People who are bored, grieving, or who cannot face the problems that beset them may gain excessive weight because they find relief in eating. Others stop eating and lose weight.

Religious Influences

Foods have symbolic meanings in all religions. Some religions use food in celebrations, or place restrictions on the use or preparation of certain foods. Some foods may be permitted or prohibited for people of a certain gender or age, for those experiencing a certain period of life, or for those with an illness. For example:

- Roman Catholics may abstain from meat on fast days as a symbol of denial and penitence, although the regulations for fasting have been liberalized.
- Buddhists are vegetarians who do not eat the flesh of any animal.
- Seventh-Day Adventists are vegetarians who do not eat meat, but who use eggs, milk, nuts, and legumes as sources of protein. They do not consume coffee, tea, or alcoholic beverages.
- Muslims abstain from eating pork, although they do eat other animals. They do not drink alcoholic beverages. They fast for one month (Ramadan) each year. During this month no food is eaten from dawn until after dark.
- Orthodox Jews adhere to dietary laws based on the Bible and tradition. Animals and poultry are slaughtered according to ritual, and the meat is soaked in water, salted to remove the blood, and washed. This is known as koshering. Pork and shellfish are prohibited. Milk, sour cream, cottage cheese, and cream cheese are widely used, but no dairy foods are served at a meal with meat. Conservative Jews are less stringent in the observance of biblical dietary laws, and Reform Jews give minimum emphasis to the dietary practices.

Geographical Influences

Food habits result largely from foods available in various parts of the world. People from different cultures throughout the world have contributed to the richness and variety of the American diet. People everywhere tend to like the foods with which they are familiar. Even before tasting a food, they will look with suspicion and dislike at something that is unfamiliar. But the influences of advertising and the ease with which people travel from one part of the world to another are doing much to widen our food experiences and to make us more appreciative of other cultures.

Immigration Ethnic groups often quickly adapt their general food choices and eating patterns to the availability of foods in American supermarkets and to American lifestyles. Yet ethnic markets and restaurants also stock or serve foods characteristic of the country or region of origin of the people residing in their particular neighborhoods. Moreover, even though the actual daily diet of the residents may no longer closely conform to that of their former locale, traditional foods likely are served on holidays and at ceremonial events, and requested when the residents are ill. To accommodate such food preferences, local hospitals and institutional facilities plan their menus accordingly.

Regional Patterns Some regional differences still exist in the United States, even though, for the most part today, people living anywhere in the nation have foods from other regions available locally. This is due to the current popularity of sampling foods from different cuisines, as well as to the availability of fast-food and other franchised restaurants, national convenience stores, and advertisements in magazines. Nevertheless, New England is known for clam chowder, Boston baked beans, and lobster; the Pennsylvania Dutch regions for their many sweet and sour foods, scrapple, shoofly pie, and sticky cinnamon buns; the Midwest for dairy products, eggs, and meat; the South for cornbread, grits, fried chicken, hot biscuits and honey, turnip greens, and sweet potato pie; Louisiana for French and Creole cooking; the Southwest for Mexican dishes and mesquite-grilled fish; the West Coast for its luscious fruits and vegetables; and the Northwest for fresh salmon.

Contemporary Trends

Food availability, people's food selections, and eating habits depend on world, national, and local events. Droughts, floods, famine, war, and political unrest continue to affect food choices—by necessity.

Today people of all ages eat many meals outside the home—in schools, from street vendors, at food courts, in fast-food restaurants. Fast-food restaurant chains now are found throughout the world. Instead of cafeterias, some schools and hospitals now have these franchised restaurants and food court purveyors. The reasons given for their popularity are convenience, moderate cost, palatability, and the availability of a common meeting place. These restaurants also have their critics, who describe the foods as too high in calories, fat (especially saturated fat), and salt, and too low in fiber and some vitamins such as A and C.

With more and more families depending on all adults working to provide sufficient income, additional convenience foods are found in supermarkets and even in department stores and hospital cafeterias. Many such locations now have full-course meals that can be ordered by fax to be ready for pickup on the way home!

With the increasing interest in foods for a healthier lifestyle, each year hundreds of new foods are brought into the marketplace. Such foods are often lower in fat and cholesterol, lower in calories, and higher in fiber. Many of these foods enjoy only a short life before another "breakthrough food" takes their place on the shelf. This has led to confusion among the public and has created a challenge for nutrition educators who attempt to teach sound dietary principles.

To obtain optimal nutrition throughout life, everyone depends on choices from a wholesome (nutritiously processed, safe, and sanitary) food and water supply at each meal. Food and water are processed by a variety of methods to accomplish these goals. For example, nutrients are added to food and water to enhance their nutritional values; food is preserved to permit nonseasonal foods to be eaten year-round or to be transported all over the world; and milk is pasteurized and municipal water supplies are chlorinated to prevent illness from foodborne diseases.

Food and Water Protection Through Legislation

In the United States, food and water supplies are protected through legislation. At the federal level, four major agencies are mandated to focus on safeguarding interstate and imported supplies, with other agencies at local levels to control intrastate supplies.

Food and Drug Administration (FDA) The FDA, as an agency of the Department of Health and Human Services (DHHS), is responsible for the safety of all processed foods, other than meat, poultry, and eggs, that enter into interstate commerce. The FDA develops and enforces regulations that pertain to a safe food supply, conducts laboratory research, orders the seizure of products that do not comply with the regulations, and conducts educational programs for processors and the public. The FDA is responsible for food and dietary supplements labeling, setting standards for bottled water, and setting federal standards for food enrichment and fortification.

United States Department of Agriculture (USDA) The Food Safety and Inspection Service (FSIS) of the USDA develops regulations for meat and poultry products and inspects processing plants, equipment, and procedures. Animals and fowl are inspected before and after slaughter. This agency tests for chemical and drug residues, including exogenous hormones. The Agricultural Marketing Service (AMS) inspects eggs and egg products.

National Marine Fisheries (NMF) This agency, located in the United States Department of Commerce, conducts voluntary inspection programs for fish products. Services include boat and processing plant sanitation inspections, product laboratory analysis, and review of fish product labels. Fish processors, brokers, and retail and food service operators are eligible to participate.

Environmental Protection Agency (EPA) This agency develops regulations pertaining to air and public water quality and solid waste disposal. The EPA regulates the manufacture, use, and labeling of pesticides and monitors their presence in the environment, while the FDA and USDA enforce the tolerances in their respective food oversight responsibilities.

State and Municipal Food Safety Agencies States, counties, cities, and townships, primarily through their departments of health and agriculture, develop and enforce regulations pertaining to foods that do not enter into interstate commerce. Usually the standards are developed in cooperation with the respective federal agencies. Public eating establishments, from restaurants to hospitals, are inspected and can be closed for sanitary violations.

Food Preservation

People expect a wide array of food choices, yet a host of agents can cause food to spoil. Biological microorganisms, such as bacteria, parasites, yeasts, and molds, are major causes of food spoilage. Chemical changes that cause food spoilage are minimized by avoiding exposure to air or light. For example, enzymes normally present in food cause chemical changes that lead to softening of the food, development of "off" flavors, loss of some nutrients such as vitamin C, darkening of peeled fruit, or rancidity of fats. Physical factors produce undesirable changes in food, as when ice cream held too long in a freezer becomes grainy or gummy. Milk is packaged in opaque cartons or plastic containers to prevent the loss of riboflavin, a vitamin. Animals may contaminate food with hairs, feces and urine, or insect fragments.

Because food spoilage can occur by many means, methods to preserve food are multipronged. In fact, several methods might be used for one food, for example, milk. Milk is pasteurized to destroy disease-causing microorganisms, homogenized to intersperse the fat with the fluid portion, fortified with vitamin D to enhance its nutritional value, packaged in opaque containers to protect its riboflavin content, and refrigerated to preserve its shelf life from farm to consumption.

Traditional Food Preservation Methods One of the oldest methods is dehydration so microorganisms cannot grow. This is accomplished by exposing food to air or by adding salt or sugar. Fermentation is used to produce cheese and sour cream from milk, sauerkraut from cabbage, wine from grapes, and soy sauce from soybeans.

Adjusting temperature is another traditional method. Heating, whether by cooking, pasteurizing, canning, or sterilizing, is a universal method of killing microorganisms. Chilling, by placing foods in ice or cool streams of water and by refrigerating and freezing, slows microorganism growth. It is important to recognize that the microorganisms remain. In fact, many molds and some bacteria are known as psychrophilic because they are able to thrive at a relatively low temperature. These microorganisms can grow at refrigerator temperatures and thus can cause foods to spoil even while refrigerated. (See Figure 2-2.) A modification is freeze-drying. This process consists of rapidly freezing the food and then removing the moisture in a vacuum. Examples of foods preserved in this way are coffee, dried soup mixes, and foods prepared for campers, hikers, bikers, and boaters.

Adding chemical compounds is a method of preserving food that has been used for centuries. Sugar and salt are two of the oldest and most widely used preservatives. For example, sugar has some preservative effect when used in high concentrations for jams, jellies, and preserves, but molds will grow on the surface unless the foods are protected from air. Salt (sodium chloride) is used to pickle vegetables, fish, and meat. Sodium benzoate is used in many common food products, including margarine. Sulfur dioxide prevents the darkening of apples and apricots during dehydration. Calcium propionate in bread and sorbic acid in cheese wrapping retard mold growth.

Newer Food Preservation Methods The newest methods use high-technology chemistry, physics, and engineering. **Biotechnology** is a term that refers to altering living organisms, including plants and animals.

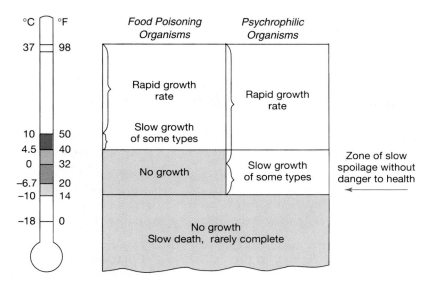

FIGURE 2–2
Food spoilage and poisoning organisms grow rapidly above 50° F. Food poisoning organisms do not grow below 40° F. (*Source:* Dr. Horace K. Burr and *Journal of the American Medical Association.*)

Although research on the topic of irradiation began after World War II, **irradiation** of food as a preservation method has had growing use in the United States. Starting in the 1960s, the FDA has permitted its use to control insects in foods, microorganisms in spices, parasites in pork, bacteria in raw poultry, and spoilage in fresh fruits and vegetables. Irradiation does not make the food radioactive. However, because of fears about nuclear energy, consumers have remained uncertain about this form of food preservation. In 1966, the FDA required all foods so preserved to be labeled, and since 1986, that the international logo be displayed.

Altering foods by genetic means, usually known by the term **genetic engineering,** is actually an older method performed in a high-technology manner. For centuries, farmers applied selective breeding to plants and animals. However, in 1973, scientists learned how to directly change DNA (deoxyribonucleic acid), a cell's genetic material. This capability has resulted in special bacteria that can be sprayed on fruits and vegetables to enable them to resist frost damage, grains that can withstand arid climates, tomatoes that destroy insects that prey on them, and livestock with more meat and less fat.

INTERNATIONAL FOOD IRRADIATION LOGO

Food Additives

Among the reasons that substances are added to food are to enhance its nutritional value, prevent disease, keep it from spoiling, add flavor and retain color, and prolong shelf life. The landmark Pure Food Law of 1906 was the first federal law related to additives. Since then, a number of other important laws have been enacted. In 1958, Congress passed the Food Additives Amendment to the Federal Food, Drug, and Cosmetic Act. The Delaney clause in this amendment specified that no substance may be added to food if it produces cancer. The Color Additives Amendment (1960) provided for the establishment of safe tolerance levels for all colors used in foods. Thus, **food additives** fall under the purview of the FDA. The United States food additive industry has annual sales of over $10 billion.

Reading the labels on food packages is crucial. However, fresh foods are not "labeled." Yet some additives are applied to them, such as waxes to the peels of cucumbers and apples to retain freshness. Today, many supermarkets display informative signs near fresh food counters.

Some important terms are **intentional additives, incidental additives,** and **indirect additives.**

Intentional Additives Intentional additives are those substances of known composition that are added to food to serve some useful purpose. (See Table 2-1.) Stabilizers, colors, and flavor enhancers fall into this category, as do nutrients added to foods. Important terms related to additives that affect the nutritive value of foods are:

■ **Restoration,** a general term for the replacement of nutrients lost during food processing to levels similar to the original levels. An example is wheat flour and certain vitamins and minerals. When "whole" wheat flour is processed to "white" wheat flour, nutrients unintentionally are removed; thus important nutrients would be lost for human consumption, unless they were returned.

TABLE 2–1
Some Uses for Intentional Additives

Purposes for Use	Typical Additives	Examples of Use in Foods
Improve Nutrition	Minerals, vitamins	Iodized salt
		Margarine with vitamin A
		Milk with vitamin D
		Enriched grain foods
Improve Keeping Quality		
Prevent mold	Sorbic acid	Cheese wrappers
	Calcium propionate	Breads, rolls
Prevent rancidity	Butylated hydroxyanisole (BHA)	Cereals, crackers, potato chips
	Butylated hydroxytoluene (BHT)	
	Tocopherol (vitamin E)	Candy, oils
Prevent darkening	Ascorbic acid (vitamin C)	Fruits to be frozen
Provide Leavening	Baking powder, baking soda	Cakes, cookies, quick breads
Enhance Flavor	Salt, spices, flavor extracts	Cheese, catsup, salad dressings
	Acetic acid, citric acid	
Artificial sweeteners	Saccharin (Sweet 'n Low)	"Diet" foods: soft drinks, desserts, candy
	Aspartame (NutraSweet, Equal)	
	Acesulfame-K (Sunette, Sweet-One)	
Maintain Texture		
Stabilizers	Carrageenan; disodium phosphate	Evaporated milk
Thickener	Xanthan gum	Gravy mix
Creaminess	Polysorbate 60	Salad dressing
Provide Coloring	Certified food colors	Baked foods, soft drinks, gelatin desserts
	Annatto; carotene	Butter, margarine

- **Enrichment,** the addition of nutrients to achieve concentrations specified by the standards of identity. For example, the vitamins riboflavin, thiamin, and niacin and the mineral iron are returned to some processed grain products (some white bread, pasta, crackers, rice) and thus the products are labeled as "enriched." Whether or not grain foods are enriched is mandated by each state, not by federal law.

- **Fortification,** the addition of nutrients at levels higher than those found, or never found, in the original food. More and more nutrients are being added to more and more foods. Examples include vitamin D added to milk; vitamins, minerals, and fiber added to ready-to-eat cereals, breakfast and sports bars; vitamin C added to dehydrated beverage powders; and fluoride added to municipal water supplies.

Incidental Additives Any substance that comes into contact with food during its growth or processing is called an incidental additive. Examples range from rodent hairs and insect fragments in grains to detergent residues remaining on dishes after washing.

Indirect Additives Something that is present in a food package, and thus potentially in the food when eaten, is called an indirect additive. For instance, foods may pick up substances from their wrappers or containers. Today, interiors of cans often are plastic coated to prevent unwanted minerals from leaching out into the food contents.

Using Additives in Foods Before any new additive can be included in a food product, the following conditions must be met:

- The additive must serve some useful purpose; it cannot hide an imperfect product.

- The manufacturer must test the safety of the additive by feeding it to animals and observing any physical, biochemical, or pathologic changes. If tests show that the additive produces tumors in animals at any level of intake, the additive cannot be used in any food.

- The manufacturer must describe how much of the additive will be used and to which products it will be added.

For legal purposes, additives fall into four categories:

- *Food additives.* This category also covers substances not previously used in foods, or which do not have a proven track record of safety. Examples of additives that have been reviewed and approved are the artificial sweetener aspartame and the emulsifying agent polysorbate 60.

- *Generally recognized as safe (GRAS) substances.* In 1958, when the Food Additives Amendment was enacted, about 600 substances were excluded from testing. The reasoning behind this exception was that these substances had been used over long periods of time without causing any known harm to people. The list included sugar, salt, baking powder, baking soda, spices, and minerals and vitamins added to food for enrichment, as well as other substances. In 1968 the public as well as some scientists became concerned when the artificial sweetener cyclamate, which was on the GRAS list, was found to produce bladder cancer in animals. The amount of cyclamate required to produce the cancer was so large that a human being would find it impossible to consume an equivalent amount. Nevertheless, cyclamate had to be removed from the GRAS list. As a result of the cyclamate research, the entire GRAS list has been under review to determine the safety of the substances originally excluded from scientific testing.

■ *Prior-sanctioned substances.* This category includes substances that, prior to 1958, had been sanctioned for use in a specific food. An example is nitrites, which could (and still can) be used in meat, but not in vegetables.

■ *Color additives.* This category includes dyes used not only in foods, but also in medications, cosmetics, and medical devices. These dyes are subject to premarketing testing. In 1960, nearly 200 colors were on the provisional approval list. Today, because of scientific testing since then, some colors have been removed.

Organically Grown Foods

Some people choose to eat organic foods. Although all food is "organic" because it contains the element carbon, organically grown foods are those grown in soil that has been treated only with animal manure and plant composts. Also implied is that no pesticides have been used on the crops. Organic meats are those obtained from animals that have been fed no antibiotics or hormones. During processing, no "artificial" additives are used. Foods so produced and processed usually are sold in specialty stores, farmers' markets, roadside stands, food co-ops, and through mail order catalogues. These foods generally are more expensive. Some people have home gardens or grow vegetables in neighborhood plots. Scientific research has not shown the nutritive or taste qualities of organically grown foods to be any better than that of foods grown in chemically fertilized soils. With consumers becoming more health conscious and environmentally minded, the organic foods industry has experienced growth. However, across the nation, there are no uniform standards regarding the production or certification of organically grown foods.

MEAL PATTERNS AND MENU STRUCTURES: BASIC DIET

Day after day, everyone eats. The people purchasing and preparing the food make decisions about what foods to purchase, from where, and how they are prepared and served. Table 2-2 summarizes some of the factors to consider in planning daily meal patterns, menu structures, and food selections, whether in a family or institutional setting.

Meal Patterns

For most people, eating includes meals and snacks. This may mean breakfast, lunch, dinner, and a snack before going to bed. It may mean eating three meals, with snacks (or coffee breaks) in the middle of the morning and afternoon. Some people eat this way and then snack all evening long while watching television, listening to music, or studying. Some people have stringent ideas about what constitutes breakfast and will not eat lunch without eating breakfast first. Many people eat their biggest meal at noon; others eat their biggest meal in the evening

TABLE 2–2
Meal Patterns, Menu Structures, Food Selections

Factors	Considerations
Nutritional needs	Include recommended amount of foods from each of the food groups of the Food Guide Pyramid. (Chapter 1) Complete the menu for adequate calories from foods that also supply essential nutrients. Consider nutritive value of meals away from home so that meals at home make up for any deficiencies.
Dietary guidelines	Increase intake of complex carbohydrates and fiber-rich foods: whole-grain cereal foods, fruits, and vegetables. Reduce intake of fat: use low-fat milk, smaller servings of meat, more frequent use of legumes, low-fat fish, poultry. Conclude meals with a light dessert such as fresh fruit, rather than a fat-laden pastry. Reduce intake of salt.
Family composition	Make menu adjustments for children of various ages (see Chapter 13), the pregnant woman (see Chapter 12), the older adult (see Chapter 15), variations in weight status, or specific health conditions.
Food habits	Consider psychological and cultural meanings of food.
Food costs	Plan menus that make use of seasonal foods, advertised specials.
Preparation facilities	Working people must budget time carefully, as well as money; many need to use convenience foods. Plan menus to make efficient use of energy, for example, oven meals. Plan for proper storage of foods.
Variety	Within each of the food groups select a wide variety of foods. *Color:* use foods that complement each other in color; make judicious use of garnishes; avoid meals of all one color. *Texture:* combine soft with crisp or chewy foods. *Flavor:* combine bland and highly flavored foods; use herbs and spices. *Preparation methods:* boiling, stewing, baking, broiling, roasting, stir frying; microwaving. *Climate/Season:* foods such as hot stews or thick soups may be more acceptable in cold climates; lighter foods such as salads in warmer climates.
Satiety	Protein and fat in each meal increase satiety, and reduce hunger between meals.

after completing work. Some people never seem to sit down for any meal—they are always "eating on the run" and consume their foods as "mini-meals."

Meal patterns are as numerous as there are people. Everyone has a different pattern. For example, people may work at night, travel, reside in a group setting, have tremendous family or work responsibilities, or experience physical limitations. People can have highly variable meal patterns, yet be adequately nourished and healthy.

Menu Structures

At each meal, the menu structure also can vary. For example, some people prefer the dinner menu structure to be appetizer, main course, dessert, and beverage, with a meat, a starch, and a vegetable as the food selections for the main course. Some people expect pasta at every dinner meal; others expect legumes and grain. Some restaurants serve "soup and sandwich" or "soup and salad" as a lunch menu special. Other restaurants are "limited menu" restaurants, including fast-food and other franchised chains as well as various ethnic restaurants. For most people, eating alone results in a different menu structure than when eating with family and friends. Holidays and other special occasions usually produce changes in the normal routine. Menus for people with disability or disease may contain suitable modifications while retaining individual choices.

Basic Diet

A "Basic Diet" at three calorie levels has been developed. (See Tables 2–3 and 2–4.) It meets the standards of the RDAs, Dietary Guidelines for Americans, and Food Guide Pyramid. Between 50 and 60 percent of calories are derived from carbohydrates. High fiber foods are emphasized. Less than 30 percent of calories come from total fat; unsaturated fats are emphasized. Cholesterol is less than 300 mg. Appropriate wholesome food selections will provide optimal levels of protein, vitamins, and minerals.

The Basic Diet can accommodate individual preferences for meal patterns, menu structures, and food selections. Components of the Basic Diet will be explained in subsequent chapters. It can be used as a foundation for evaluating and planning diets for healthy people and for people with certain diseases.

HEALTH AND NUTRITION CARE ISSUES

Illness from Food

The number of people who succumb to foodborne diseases is unknown since only the more severe outbreaks involving many people are likely to be reported to the Centers for Disease Control and Prevention (CDC) in Atlanta. Microorganisms are, by far, the most common source of illness from contaminated foods, with more than 33 million cases estimated to occur in the United States each year. Nearly half of these incidents result from people mishandling food in their own homes! Most persons ascribe resulting symptoms to the "flu" and suffer only short-term discomfort and absence from school or work. These acute illnesses, however, result in great economic losses, including an estimated $5 billion annually in medical expenses alone. The problem is that only when such illnesses strike infants, children, or infirm elderly people is there noticeable concern by the public, for these special people can die from foodborne illness that has less tragic results in otherwise healthy persons.

Sources of Contamination All food and water supplies, whether natural or processed, are susceptible to contamination. However, the biggest sources of contamination are raw or improperly processed foods and mishandled foods.

Contamination usually is categorized into three major classes: natural toxicants, chemical poisoning, and microorganisms.

TABLE 2-3
Basic Diet Nutrient Profile

MACRONUTRIENTS AND ENERGY

Food	Amount	Food Energy (kcal)	Carbohydrate (grams)	Dietary Fiber (grams)	Fat (grams)	Saturated Fat (grams)	Monounsaturated Fat (grams)	Polyunsaturated Fat (grams)	Cholesterol (mg)	Protein (grams)
Fruits	1 serving	69	17.3	1.6	0.3	0.1	0	0.1	0	0.9
Vegetables										
Dark-green	1 serving	18	3.3	1.9	0.2	0	0	0.1	0	1.9
Deep-yellow	1 serving	40	9.3	2.4	0.2	0	0	0.1	0	0.9
Dry beans and peas	1 serving	115	20.7	5.3	0.4	0.1	0.1	0.2	0	7.8
Other starchy	1 serving	94	21.3	2.4	0.3	0	0	0.1	0	2.5
Other	1 serving	15	3.2	1.0	0.2	0	0	0.1	0	0.8
Meat, Fish, Poultry	1 ounce	57	0.1	0	2.7	1.0	1.1	0.2	27	7.8
Egg	1 large	77	0.6	0	5.3	1.6	2.0	0.7	213	6.3
Bread, Cereal, Rice, Pasta										
Whole-grain products	1 serving	72	14.2	1.8	0.9	0.2	0.3	0.4	0	2.3
Enriched grain products	1 serving	83	15.9	0.6	1.1	0.3	0.4	0.3	4	2.1
Milk, skim	1 cup	86	11.9	0	0.4	0.3	0.1	0	4	8.4
Fats, Oils, Sweets										
Fats	1 teaspoon	37	0	0	4.2	1.3	1.4	1.3	3	0
Sugar (sucrose)	1 teaspoon	15	4.0	0	0	0	0	0	0	0

TABLE 2–3, *continued*

VITAMINS

Food	Amount	Vitamin A Value (mg)	Vitamin E (mg)	Vitamin C (mg)	Thiamin (mg)	Ribo-flavin (mg)	Pre-formed Niacin (mg)	Vitamin B-6 (mg)	Vitamin B-12 (µg)	Folate (µg)
Fruits	1 serving	32	0.4	33	0.07	0.05	0.4	0.15	0	26
Vegetables										
Dark-green	1 serving	363	0.8	24	0.06	0.13	0.4	0.15	0	76
Deep-yellow	1 serving	1693	0.7	6	0.05	0.05	0.5	0.15	0	13
Dry beans and peas	1 serving	0	0.2	0	0.11	0.05	0.3	0.09	0	78
Other starchy	1 serving	9	0.1	12	0.13	0.04	1.5	0.27	0	21
Other	1 serving	39	0.3	10	0.04	0.03	0.3	0.05	0	24
Meat, Fish, Poultry	1 ounce	62	0.1	0	0.05	0.09	1.6	0.10	1.09	3
Egg	1 large	84	0.5	0	0.03	0.26	0	0.06	0.55	22
Bread, Cereal, Rice, Pasta										
Whole-grain products	1 serving	0	0.2	0	0.08	0.06	0.8	0.04	0.01	8
Enriched grain products	1 serving	3	0.1	0	0.10	0.07	0.9	0.02	0.01	7
Milk, skim	1 cup	149	0	2	0.09	0.34	0.2	0.10	0.93	13
Fats, Oils, Sweets										
Fats	1 teaspoon	16	0.6	0	0	0	0	0	0	0
Sugar (sucrose)	1 teaspoon	0	0	0	0	0	0	0	0	0

MINERALS

Food	Amount	Calcium (mg)	Phos-phorous (mg)	Mag-nesium (mg)	Sodium (mg)	Potas-sium (mg)	Iron (mg)	Zinc (mg)	Copper (mg)
Fruits	1 serving	15	19	15	4	256	0.3	0.12	0.08
Vegetables									
Dark-green	1 serving	76	34	42	44	213	1.4	0.31	0.11
Deep-yellow	1 serving	21	27	11	46	227	0.4	0.21	0.09
Dry beans and peas	1 serving	55	106	47	224	417	2.6	1.04	0.22
Other starchy	1 serving	8	62	26	28	361	0.6	0.43	0.19
Other	1 serving	15	19	10	49	151	0.4	0.16	0.05
Meat, Fish, Poultry	1 ounce	5	60	7	39	87	0.6	1.29	0.04
Egg	1 large	25	86	5	63	63	0.6	0.52	0.01
Bread, Cereal, Rice, Pasta									
Wholegrain products	1 serving	18	57	21	81	59	0.7	0.47	0.08
Enriched grain products	1 serving	23	32	7	106	31	0.8	0.21	0.04
Milk, skim	1 cup	302	247	28	126	406	0.1	0.98	0.03
Fats, Oils, Sweets									
Fats	1 teaspoon	0	0	0	16	1	0	0	0
Sugar (sucrose)	1 teaspoon	0	0	0	0	0	0	0	0

Source: Adapted from "USDA's Food Guide: Background and Development," U.S. Department of Agriculture, Human Nutrition Information Service, Miscellaneous Publication No. 1514, September, 1993; February 1994.

TABLE 2–4

Basic Diet Nutrient Profiles at Three Calorie Levels

Food Groups and Nutrients	Pattern A	Pattern B	Pattern C
Bread Group (servings)	6	9	11
Vegetable Group (servings)	3	4	5
Fruit Group (servings)	2	3	4
Milk Group (servings)	2–3[a]	2–3[a]	2–3[a]
Meat Group (total ounces)	5	6	7
Total fat (grams)	53	73	93
Total added sugars (teaspoons)	6	12	18
Energy (kilocalories)	1,600	2,200	2,800
Macronutrients (percent of calories)			
Protein	20	17	16
Fat	30	30	30
Saturated fat	9	9	9
Monounsaturated fat	10	10	10
Polyunsaturated fat	8	8	8
Carbohydrate	52	55	55
Vitamins			
Vitamin A (RE)	1,973	2,513	3,059
Vitamin E (mg)	7.6	10.6	13.7
Thiamin (mg)	1.3	1.8	2.2
Riboflavin (mg)	1.8	2.2	2.5
Preformed niacin (mg)	15.8	21.3	25.8
Vitamin B-6 (mg)	1.5	2.0	2.4
Vitamin B-12 (μg)	7.2	8.3	9.4
Vitamin C (mg)	104	147	191
Folate (μg)	256	342	423
Minerals (mg)			
Calcium[b]	880	998	1,095
Iron	11.5	15.7	19.2
Magnesium	273	339	399
Phosphorus	1,244	1,464	1,654
Zinc	11.4	13.9	16.1
Potassium	2,780	3,470	4,130
Sodium	1,350	1,830	2,210
Copper	1.1	1.5	1.9
Other components			
Cholesterol (mg)	256	303	348
Fiber (g)	17	22.5	27.5

Source: USDA, 1994.

[a]Women who are pregnant or breastfeeding, teenagers, and young adults to age 24 need 3 servings.
[b]An additional serving of milk would add more nutrients.

■ *Natural Toxicants.* Many natural constituents of foods produce poisons when those foods are eaten. Numerous fungi, including molds, also produce toxins. The most recognized probably is the poisonous property of mushrooms and toadstools. Other examples are:

Oxalic acid	Rhubarb leaves
Solanine	Green portion of sprouting potatoes
Mycotoxins	Grains and nuts, produced by molds
Aflatoxins	Peanuts and Brazil nuts, produced by molds

■ *Chemical Poisoning.* Food may be contaminated accidentally by minerals and pesticides. Some examples are:

Lead	Food exposed to air or dust containing lead
	Food stored in containers made with lead solder, alloys, enamel, or glaze
Zinc	Acid foods (lemonade, lemon juice) stored in galvanized cans
Mercury	Fish and aquatic plants from rivers, bays, and oceans contaminated with industrial and agricultural mercury waste-product pollutants
Pesticides	Fruits and vegetables
	Foods with mistakenly mislabeled pesticides

■ *Microorganisms.* Microorganisms that cause foodborne illness include bacteria, viruses, protozoa, and other parasites such as worms. Some of these biological organisms produce toxins. Table 2-5 identifies some common microorganisms.

Typical Symptoms The usual symptoms from any source of food contamination relate to the digestive tract (nausea, vomiting, abdominal cramps, diarrhea), although other organs can be affected (liver, heart, brain, and muscles). Resultant medical complications include severe dehydration, blood acid-base imbalances, confused mental state, and death.

Prevention The first line of defense is selecting wholesome food and water (also ice), and the next is carefully handling foods and leftovers. (See Figure 2-3.) Cross-contamination can occur by transferring contaminants from hands, utensils, cutting boards, and other equipment to food and water.

In 1994, the FDA incorporated the process of Hazard Analysis Critical Control Point (HAACP) into its Food Code for commercial establishments. Some state and local regulators also mandate it. The procedure focuses on problem prevention by analyzing typical hazards, identifying critical control points, and establishing actions to monitor and correct any difficulties.

Food-Related Illness in Immunocompromised Persons Although everyone is at some risk from eating or drinking tainted foods and beverages, individuals with weakened immune systems are especially vulnerable. Such people include those with cancer, diabetes mellitus, or human immunodeficiency virus/acquired immunodeficiency syndrome (HIV/AIDS), as well as organ transplant recipients. (See specific chapters in Part III.) Not only are immunocompromised people more likely to become infected with foodborne microorganisms, the infection itself is more likely to recur, and the person's underlying condition makes the infection more difficult to treat. These individuals must be especially diligent in purchasing foods from stores, vendors, and restaurants with the best

TABLE 2-5
Examples of Foodborne Illness Caused by Microorganisms

Organisms	Symptoms	Foods	Prevention
		BACTERIA	
Campylobacter jejuni	**Onset:** 3–5 days or longer Digestive; fever; headache, muscle aches	Meat, poultry, eggs Dairy foods Untreated water	Thoroughly cook animal foods Use pasteurized milk and cheese Avoid cross contamination
Clostridium botulinum	**Onset:** 12–48 hours Digestive; double vision, unable to swallow, speak, breathe Potentially fatal	Canned, low-acid foods, e.g., green beans, mushrooms, olives Honey, corn syrup	Discard cans with dents or broken seal Avoid "raw" honey
Clostridium paerfringens	**Onset:** 8–24 hours Digestive; fever	Meat and poultry left at room temperatures	Keep hot foods >140°F during serving Promptly refrigerate gravies and sauces
Escherichia coli	**Onset:** 2–4 days Digestive (bloody diarrhea) Potentially fatal (e.g., 0157:H7)	Meat, esp. hamburger Dairy foods Fruits; Vegetables	Thoroughly cook all meats Use pasteurized milk, cheese, yogurt Wash fruits and vegetables
Listeria monocytogenes	**Onset:** 7–50 days Digestive; fever, chills, headache Potentially fatal	Raw and processed meats Raw seafood Dairy foods Coleslaw	Thoroughly cook animal foods Carefully prepare pâté Use pasteurized milk and cheese Use foods before expiration date
Salmonella	**Onset:** 6–72 hours Digestive; fever, chills, headache	Raw and undercooked meat, poultry, eggs, fish Dairy foods	Thoroughly cook all foods Avoid cross contamination, e.g., cutting boards, utensils, packaging

Organism	Onset/Symptoms	Source	Prevention
Staphylococcus aureus	**Onset:** 1/2–8 hours; Digestive; fever, chills, headache, weakness, dizziness	Foods contaminated with organisms from nose and skin	Wash hands; wear gloves. Avoid food preparation when sick. Cover food; use "sneeze guards"
Shigella	**Onset:** 1–6 days; Digestive; fever	Cold mixed salads and sandwiches, e.g., potato, chicken, tuna, shrimp. Contaminated water	Wash hands after bathroom use
Vibrio vulnificus	**Onset:** 1–7 days; Digestive; fever, chills, skin blisters	Warm-water shellfish	Avoid raw and undercooked shellfish

PARASITES

Organism	Onset/Symptoms	Source	Prevention
Crytosporidium	**Onset:** Days to weeks; Digestive; fever	Water	Use water and ice free of bird and animal fecal contamination
Trichinella spiralis	**Onset:** Weeks to months; Muscle weakness, fever	Pork, e.g., fresh pork, sausage, hot dogs. Wild game	Thoroughly cook pork and wild game
Tapeworms	**Onset:** Weeks to months; Digestive	Raw fish, beef, pork	Thoroughly cook fish, beef, pork. Avoid raw fish, e.g., sushi

VIRUSES

Organism	Onset/Symptoms	Source	Prevention
Hepatitis A	**Onset:** 1/2–2 months; Digestive; fatigue; Liver disease, e.g., jaundice; Potentially fatal	Raw fish and shellfish	Use fish from sewage-free waters. Use clean water

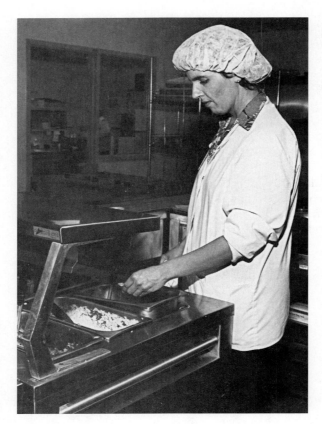

FIGURE 2–3
A dietary manager routinely checks the temperature of foods on the hot and cold tray lines during meal service. (*Source:* Anthony Magnacca/Merrill.)

reputations for cleanliness and sanitation, and must be highly observant of food safety practices at home and when traveling. In the institutional setting, health professionals need to monitor closely the foods served to patients.

Designer Foods, Functional Foods, Nutraceuticals, Pharmafoods

People worldwide desire to prevent disease, especially chronic impairments. Health care costs need to be contained. To fill these consumer and society demands, food and pharmaceutical manufacturers have developed foods with reported beneficial health effects. Most research supporting these hypotheses has been epidemiological. (See Chapter 1.) Some countries, especially those in North America, Asia, and Europe, have surpassed others in developing a strong market for these foods. Although the precise terminology has yet to be clarified, the underlying premise is that food can be modified to yield stronger health benefits or disease-allying properties. Examples include increasing the fiber content of processed foods; decreasing the total fat content, yet retaining the sensory characteristics, of processed foods; adding certain vitamins, especially the **antioxidant** vitamins A, C, and E, either to processed foods or as supplements to the diet; and extracting **phytochemicals** (plant substances) from vegetables and herbs to be supplied in tablet or powder form. Future scientific studies should lead to greater resolution of the debate.

Over the ages, most societies have developed traditional beliefs or traditions about food, and use foods as symbols. This so-called knowledge, passed on from one generation to the next, is known as **food lore.** Most food lore is based on observation and perceived cause and effect, not on scientific principles. Some people openly practice their beliefs and superstitions about foods; others do so covertly or when they have fears about health, aging, illness, or death. Even without realizing it, most people are susceptible to food lore. Prescientific beliefs about foods and their relationships to health and disease abound throughout the world.

For instance, some cultures ascribe certain properties to certain foods to ensure the successful outcome of pregnancy and delivery. Some people believe that certain categories of food must be kept in balance with others. Millions of people identify certain foods as "cold foods" that must be kept in balance with "hot foods" to protect health or to treat diseases.

Health professionals, educators, and nutrition and food science specialists must be aware of food lore—both their own beliefs and those of their clients, students, and customers. Food lore is steeped in early childhood experiences and reinforced throughout life. Although usually there is no evidence to support them, sometimes these beliefs can appear to work where modern medicine and health technology do not. Thus usually it is more prudent to counsel people about eating a variety of acceptable foods than to voice overt criticism of their beliefs.

Food and Nutrition Misinformation

Food Fads Just as fads in fashion come and go, so food fads come and go. A food fad is a food or nutrition style, practice, or craze that many people adopt for a short period of time. "Here today, gone tomorrow" is the adage.

An example is any one of the weight-loss scams that are recycled every year. Each time, a promise of immediate weight loss is the eye-catching appeal. Each time, there is a little variation on the theme. One season, it is special foods to buy; the next it's a protein powder or liquid beverage to purchase; for the third season, an expensive cooking utensil has just been discovered that can "burn off" the calories in food; and finally, after all else has failed, a specially formulated fat-dissolving capsule has been developed in a secret laboratory. Fortunately, most food fads are harmless, primarily because they are so short lived. Others create an economic hardship for people—especially for those on limited incomes—because the foods or supplements or gadgets may be quite expensive. Unfortunately, each plausible fad is lucrative for the sellers, who are encouraged to create another, slightly different, version.

Food Fallacies A food fallacy is a false or mistaken idea about a food or nutrient. In this case, usually special virtues are ascribed to a particular food or nutrient. Sometimes, a "half-truth" is the ploy. Examples abound because so many people possess incomplete knowledge about food, nutrition, health, and disease. As people learn more about these topics, one might expect the number of food fallacies to decrease. But, it seems, as each new scientific fact is discovered, additional potential areas for food and nutrition fallacies follow. Take "health foods" as an example. A "health food" usually is one for which a claim is made that it promotes health or is effective in treating disease. But, as you already know, no one single food possesses unique or miraculous qualities for achieving overall health. Rather, optimal nutrition is achieved by eating a combination of foods that provide the abundance of necessary nutrients at recommended levels. Although food fallacies may be harmless (except for the outlay of money), potentially they can be harmful. Following misguided advice could lead to serious nutritional deficiencies or even ill health, the very conditions the buyer is attempting to avoid!

Food Frauds Food and nutrition fraud is intentional deceit. This is the most serious kind of food and nutrition misinformation because the targeted person is sought after and knowingly tricked. In such cases, once the deception is uncovered, legal action usually is considered. Examples include selling expensive nutritional supplements advertised to promote or protect health, when all they contain is some inexpensive substance like skim milk powder or vegetable oil. Or people are tricked into buying special food preparation utensils, which do not live up to their miracle-producing characteristics. Or, hucksters sell a food as being fortified with "secret" nutrients, when absolutely nothing has been added, except perhaps a fancy wrapper. Food and nutrition fraud is the most harmful, not only because the hoax is intentional and usually exorbitantly expensive, but because most times the targeted person is extremely vulnerable, such as a teenager with a poor self-image from being overweight, or a pregnant woman who wants to deliver the healthiest and smartest baby possible, or an elderly person who is searching for eternal youth, or a dying patient who would do anything to find a cure for an incurable disease. Food and nutrition fraud usually encourages the person to substitute self-therapy for the advice and treatment of a competent health professional. Effective treatment may be delayed until it is too late.

Recognizing Food and Nutrition Misinformation It is not always easy to differentiate between legitimate vs. exaggerated claims made for a food, a dietary regimen, or a food-related product. Misleading claims primarily are emotional in their appeal. The item will grant you youth, beauty, glamour, health, physical or mental endurance and power, long life, and cure of disease. Specially grown or prepared foods, food supplements, vitamin and mineral pills, protein powders, diet books, weight-loss regimens, equipment or utensils for food preparation, or furnishings for dining pleasure are among the products offered to a gullible public. The claims are presented through an exciting and unusual approach: invigorating health lectures, magical demonstrations, fast-paced videos or television shows, and books, magazines, and audiences filled with compelling testimonials. The list is long, the practice has gone on for centuries, and it has preyed on the fears of all cultural groups throughout the world. No one, it seems, is immune to food and nutrition misinformation!

CHARACTERISTICS OF FOOD AND NUTRITION QUACKS

Emotional appeal
Exaggerated claims
Lack of scientific basis
Testimonials for support
Attacks on food industry
Criticism of health
  professionals and agencies

Recognizing the Food and Nutrition Quack A food and nutrition quack is a person who boasts or makes false claims for the health virtues or curative properties of certain foods, or the nutrients they contain; that is, the person is selling food and nutrition misinformation. Such persons appear highly reputable—all are charismatic and articulate, many are famous, most are highly educated, and more than a few are health professionals! Yet these purveyors of nutritional quackery seldom have the professional education to qualify as an expert in nutrition, food science, and health, although diplomas and certificates from nonaccredited schools often are displayed. The ones who do have authentic degrees from accredited colleges and universities may have majored in fields that have little or no relationship to food and nutrition. Nevertheless, not all of these "experts" are charlatans; many truly believe in their espoused principles.

Food and nutrition quacks usually are extremely critical of the foods from modern American farms that are available at supermarkets. They claim that these foods are not safe because they were exposed to pesticides or contain additives; or that the foods are low in nutritive value, either because they were grown in chemically fertilized soils or because they were robbed of their nutrients by overprocessing. Almost always, they claim that Americans are living at too hectic a pace to "eat right"; or that they are under so much stress that "extra" vitamins and minerals must be taken—and, of course, their product is just perfect!

Generally, nutrition quacks accuse traditional medicine, pharmaceutical companies, and the food industry of keeping the true facts about health and disease from the American public. Often, quacks complain of being persecuted by scientists and governmental agencies. At other times, quacks can sound very learned by quoting from the same reliable scientific books and journals that other quacks shun, but then lift ideas out of context from these publications and put their own incorrect interpretations on the "evidence." Yet, in nearly every case, quacks do not conduct their own valid, unbiased research that could justify the claims they make.

Defensive and Offensive Actions The best defense is being wary about any quick and simple cure or benefit or one that costs a lot of money. Food and nutrition appear to be such innocuous subjects that it is very easy to be duped. Remember that food fads, fallacies, and frauds are steeped in food folklore and so have been around a long time. Food and nutrition quacks perpetuate them by appealing to fear, tradition, and people's desire to achieve something better in life. We need to recognize and appreciate our own beliefs and customs as well as those of our clients. Sound nutrition knowledge is necessary, coupled with appropriate nutrition messages at all levels and in all forums. To combat nutrition misinformation, the best offensive position is relaying correct nutrition information. In this way, people will be as informed as possible and will have the opportunity to make the best possible choices about the foods they select for their nutritional well-being.

PUTTING IT ALL TOGETHER

Chapters 1 and 2 have discussed the kinds and amounts of abundant, wholesome foods available to meet people's individual food choices and nutrient needs, to promote wellness, and to prevent or delay the onset of chronic diseases. The food and nutrition theme is moderation, balance, variety, and enjoyment. The Recommended Dietary Allowances and the Dietary Guidelines for Americans are important as ways to promote health and reduce the risks of chronic diseases. The Food Guide Pyramid, the Exchange Lists, nutrition/ingredient labeling of foods, and the Basic Diet are sound nutrition education tools for applying food and nutrition principles to everyday meal planning, menu structure, and food selections.

As we enter the twenty-first century, research should provide additional answers to our many questions about health and disease. Strategies to improve health care should improve the day-to-day practices of our respective professions. As a health professional, you can convert the findings of the science and art of foods and nutrition into tangible, useful information. You can implement your knowledge on a daily basis to create a sound nutritional lifestyle for yourself, your family and your community, and your clients.

? QUESTIONS CLIENTS ASK

1. Is raw milk safe to drink?

There is no important nutritional difference between raw and pasteurized milk. Foodborne illness can result from drinking raw milk. Although certified raw milk is produced under highly sanitary conditions, milk not so inspected can contain harmful bacteria. Raw milk may spoil faster than pasteurized milk. Refrigeration only slows the growth of microorganisms; it doesn't kill them.

2. I enjoy a cup of herbal tea once in a while. Are these teas more healthful than regular teas and coffees?

Herbal teas do not possess any magical properties for improving health or treating disease. Some of them, such as rose hip, orange, and peppermint, are enjoyable alternatives to coffee or regular tea. Many herbs used in some tea concoctions, however, especially if used to excess, can have undesirable and even dangerous effects. A few examples are:

aloe	severe diarrhea
chamomile	asthma or anaphylactic shock in persons allergic to asters, chrysanthemums, ragweed
licorice	large amounts may cause sodium retention, loss of potassium, increased blood pressure, and heart failure
sassafras	contains safrole, a potent cancer-causing agent

Not enough is known about herbal teas to conclude that they are safe, nor at what levels of intake they will produce undesirable symptoms. If you enjoy herbal teas it is a good idea to purchase only those that are prepackaged and sold under the labels of nationally known food processors. Herbal teas that are sold loose may contain a mixture of herbs of unknown composition, along with unknown contaminants.

3. How should I thaw a frozen chicken or turkey?

Preferably the poultry should be thawed in the refrigerator. This takes 24–48 hours, depending on the size of the bird. When it is partially thawed, remove the giblets from the cavity to hasten thawing. If a shorter thawing time is desired, place the bird in a waterproof bag in cold water. Change the water frequently. NEVER thaw any food on a counter at room temperature since bacteria will grow rapidly on the surface before thawing is complete. Follow the "Safe Handling Instructions" label on the poultry package.

4. Could I stuff the bird a day ahead to save time?

This is risking the possibility of bacteria growth in the stuffing and bird cavity. It is better to mix your dry ingredients ahead of time, if you wish, and wait until just before the bird is to be put into the oven to do the actual stuffing. After dinner, remember to immediately remove any leftover stuffing from the cavity, and promptly refrigerate all leftovers.

✔ CASE STUDY: Food Poisoning Following a Family Picnic

On a hot day, Mr. and Mrs. H. and their children, Paul, Tony, and Nancy, went to the park for an all-day picnic. The menu was a favorite one—hamburgers on buns, potato salad, deviled eggs, lettuce and tomatoes, watermelon, and chocolate cake. Mrs. H. shopped earlier in the week and prepared the foods the previous day. She mixed the ground beef with onion and seasonings and shaped the patties so they would be ready to broil. She prepared the potato salad and deviled eggs. All these foods were kept in the refrigerator overnight. The family stopped for ice for their beverages on the way to the park.

At noon, Mr. H. broiled the hamburgers; Tony wanted his well done. Since the foods were favorite choices, everyone ate some of each. Later in the afternoon, Nancy stopped playing ball and complained that she had a stomachache. Mrs. H. thought Nancy might have become overtired playing on such a warm day. But soon Nancy began to vomit and have diarrhea. Then, one by one, the rest of the family started to suffer the same symptoms. Tony seemed to experience the mildest symptoms.

Questions

1. What foods might have contributed to the illness?

2. Which organisms most likely were responsible for the illness? Explain your answer.

3. What are some reasons why Tony experienced the mildest symptoms?

4. Although Mrs. H. had refrigerated the foods after she prepared them, some precautions apparently were overlooked. List several hazards that might explain the resulting illness.

5. From the viewpoint of your chosen career, identify your responsibilities in the prevention or treatment of this incident of foodborne illness. What precautions do you take to prevent personally succumbing to foodborne illness?

FOR ADDITIONAL READING

American Dietetic Association: "Position Paper: Biotechnology and the Future of Food," *J Am Diet Assoc,* **95:**1429–32, 1995.

American Dietetic Association: "Position Paper: Food Irradiation," *J Am Diet Assoc,* **96:**69–72, 1996.

American Dietetic Association: "Position Paper: Food and Nutrition Misinformation," *J Am Diet Assoc,* **95:**705–07, 1995.

Grivetti, LE: "Morning Meals: North American and Mediterranean Breakfast Patterns. Part 2: America at Breakfast," *Nutr Today,* **30:**128–34, 1995.

Knabel, SJ: "Scientific Status Summary. Foodborne Illness: Role of Home Food Handling Practices," *Food Technol.,* **49:**119–31, April 1995.

Kurtzweil, P: "HACCP. Patrolling for Food Hazards," *FDA Consumer,* **29:**5–10, January/February 1995.

McMahon, KE: "Consumer Nutrition and Food Safety Trends 1996: An Update," *Nutr Today,* **31:**19–23, 1996.

McNutt, K: "Medicinals in Food. Part I: Is Science Coming Full Circle?" *Nutr Today,* **30:**218–22, 1995.

McNutt, K.: "Medicinals in Food. Part II. What's New and What's Not?" *Nutr Today,* **30:**261–63, 1995.

Ollinger-Snyder, P, and Matthews, ME: "Food Safety: Review and Implications for Dietitians and Dietetic Technicians," *J Am Diet Assoc,* **96:**163–71, 1996.

CHAPTER 3

Nutritional Assessment and Dietary Counseling

Chapter Outline

Anthropometric Assessment
Clinical Assessment
Laboratory Assessment
Dietary Assessment
Dietary Counseling
The Counseling Process

Putting It All Together
Questions Clients Ask
Personal Study: Planning, Implementing,
   and Evaluating Indicated Changes in Food
   Intake and Behaviors

Increase to at least 75 percent the proportion of primary care providers who provide nutrition assessment and counseling and/or referral to qualified nutritionists and dietitians.*

 PERSONAL STUDY PREVIEW

In this chapter you will continue and expand upon the personal nutritional assessment begun in Chapter 1. You will also assess the nutritional status of a classmate or friend.

Healthy People 2000: National Health Promotion and Disease Prevention Objectives. Public Health Service, U.S. Department of Health and Human Services, Washington, DC, 1991, Objective 2.21, p. 95.

Nutritional assessment can be carried out in a school, senior citizens' center, weight management program, or fitness center as well as the outpatient clinic, hospital, or extended care facility. The scope of the assessment depends on its purposes and goals. It may be as brief and selective as height-weight measurements of schoolchildren, or comprehensive and detailed as for patients in critical care situations. Assessment in health care facilities is further covered in Chapter 16.

Nutritional assessment is discussed at this point so that you will learn to include assessment criteria in the study of normal and therapeutic aspects of nutrition. For example, in studying about iron, you will learn that physical signs such as pallor and easy fatigue, a diet poor in iron, and laboratory measurements such as low hemoglobin and hematocrit provide confirming diagnosis of deficiency.

Nutritional assessment encompasses anthropometric, clinical, laboratory, and dietary evaluations. Each component has its advantages and limitations, and any one is made more valuable by supporting information from one or more of the other facets of assessment.

ANTHROPOMETRIC ASSESSMENT

Anthropometry is the science of measuring the human body. The most common anthropometric measurements are height and weight. Triceps skinfold, subscapular skinfold, sometimes skinfolds from other parts of the body, elbow breadth, and midarm and wrist circumference values are frequently obtained. In infants and children, head and chest circumference are sometimes determined.

Height

Height is genetically determined, and each of us has a genetic potential, the maximum height a person might attain, assuming optimal nutritional and environmental conditions. Malnutrition and certain illnesses during the growing period can prevent an individual from reaching optimum height.

Height is best obtained with the back to a measuring device attached to a vertical surface, standing erect with the eyes looking forward. For infants and very young children, recumbent length is measured. With people who are unable to stand erect or for whom skeletal changes or other problems make height values questionable, leg or knee length is sometimes measured. Using equations, height can be projected from these values.

Weight

Weight should be measured using a beam-balance scale with nondetachable weights or a digital scale. For infants and children too young to stand, a pediatric model with a pan permits the child to be weighed while lying down. The scale should be calibrated periodically for accuracy. Spring-type bathroom scales are useful in the home, but not sufficiently precise for clinical purposes.

Height-Weight Standards

The Metropolitan Life Insurance Company height-weight tables have been the most commonly used standard for adults for some years. Based on mortality experience of Canadian and U.S. life insurance companies and involving over four million insurees, they were most recently revised in 1983. (See Table D–1 in the Appendix.)

Weights are presented by height, gender, and body frame—small, medium, and large. Frame size for these tables is based on wrist measurement. (See Figure D-5.) Height-weight guides have been developed by other individuals and groups using various databases. For an example, see Figure D-6.

A considerable range of weights for height is consistent with good health and well being. For this reason, "healthy" or "reasonable" weight has replaced "ideal" or "desirable" as the standard.

The National Center for Health Statistics (NCHS) has developed growth charts for boys and girls. (See Figures D-1 through D-4.) The individual child's height and weight are plotted on a graph and compared to those of other children representative of the same age and gender by percentile.

Body Mass Index

A value not directly measured but calculated from height and weight data is the **Body Mass Index (BMI),** an indicator of body fat content. It has the advantage of yielding a single figure, which simplifies comparisons among groups and individuals. This index is increasingly being recommended by national organizations as a measure of body fat or mass. Standards for the BMI appear in Figure B-1.

$$BMI = \frac{weight, kg}{height\ m^2}$$

Example: Individual 66 in. (168 cm or 1.68 m), 154 lb (70 kg)

$$BMI = \frac{70}{2.82} = 24.8$$

Skinfold Measurements

Since about half of the body's adipose tissue is subcutaneous, skinfold measurements are a means of assessing energy reserves in the form of fat. They also help distinguish between people such as professional athletes who are heavy because of large muscle mass and those who are overweight due to excess body fat.

Triceps (TSF) and subscapular (SSF) measurements are most frequently taken, but other areas such as the abdomen are sometimes used. A caliper is used (see Figure 3-1) and practice of the technique is necessary to develop accuracy. Customarily three measurements are taken and the readings averaged and compared to a set of normal values.

Midarm and Arm Muscle Circumference

Using a nonstretchable tape, the circumference at midpoint of the upper arm (MAC) is measured and compared to a standard. Midarm muscle circumference (MAMC) can be calculated if the midarm circumference (MAC) and triceps skinfold (TSF) are known. The midarm circumference gives some indication of protein and energy stores. The MAMC is an index of the body protein reserves and is decreased in protein malnutrition.

Body Composition

In addition to skinfold values, midarm circumference, midarm muscle circumference, and body mass index, other determinations of body composition are sometimes made.

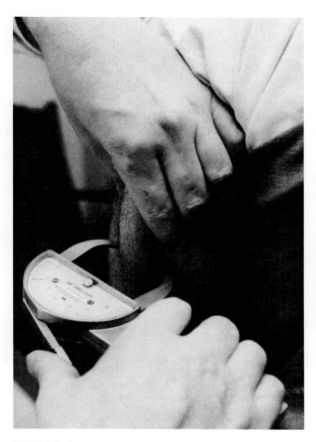

FIGURE 3–1
The degree of body fatness is determined by the thickness of a skinfold with a caliper.
(*Source:* Roche Medical Image, Hoffmann-La Roche, Inc.)

Dilution procedures to determine total body water and lean body mass involve injection of an isotope and measuring the concentration of the tracer in a sample of body fluids. Several techniques measure body fat. Underwater weighing or **densitometry** is based on the principle that a submerged body displaces its own weight. Bioelectrical impedance utilizes the principle that electrical conductivity of lean tissue is greater than that of adipose tissue. Ultrasound, computed tomography (CT), magnetic resonance images (MRI), and X-ray techniques have been used to evaluate body fat stores. These determinations are more precise than skinfold measurements but are more expensive and invasive.

Some individuals view all body fat as "bad" and believe that "the lower the better." Fat performs numerous essential functions, though, and a certain amount or percentage is necessary for health. (See Chapter 6.)

CLINICAL ASSESSMENT

During a physical assessment a physician, nurse, nurse practitioner, or physician's assistant may conduct a thorough clinical assessment or evaluation of physical signs. (See Figure 3–2). Dietitians, dentists, dental hygienists, and other members of the health care team sometimes observe physical signs that indicate possible

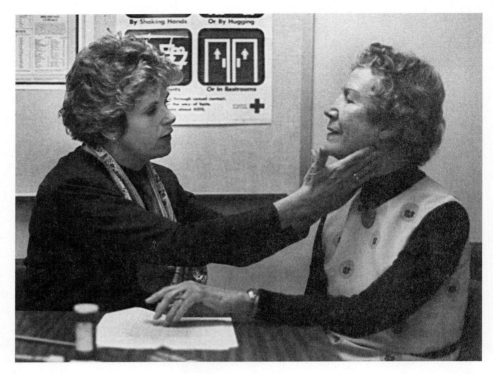

FIGURE 3–2
A nurse clinician completes a clinical assessment of a patient. (*Source:* Dunwoody Village, Newtown Square, PA, and Peter Zinner, photographer.)

nutritional deficiencies or excesses. Teachers may also be aware of clinical and behavioral factors that could be nutrition related. Such findings are not diagnoses, but should be viewed as clues useful for referrals for further testing.

The skin, hair, eyes, teeth, gums, and tongue are most frequently examined because the epithelial and mucosal cells have a rapid turnover rate and symptoms of nutritional deficiency may appear earlier. Some possible clinical signs of good and poor nutritional status appear in Table 3–1.

Clinical observation has the advantage of being nonthreatening and noninvasive. Frequently it can be conducted while talking to or working with the client about other matters. Disadvantages are subjectivity and nonspecificity. For example, skin that might be judged pale by one person would appear of normal color to another. Itching, burning eyes may result from allergy, smoke, extreme temperatures, infection, or lack of sleep as well as from a nutrient deficiency. Other factors such as cosmetics and personal hygiene can affect some of these findings.

LABORATORY ASSESSMENT

Laboratory analyses may be utilized in obtaining baseline data, establishing a diagnosis, or developing a care plan for a client. Whole blood, blood serum, and blood plasma are most commonly studied since nutrients, metabolites, and waste products are carried by the blood.

TABLE 3-1

Some Signs of Good and Poor Nutrition

	Good Nutrition	Poor Nutrition
Body Size	Healthy weight for height, body frame, and age; normal rate of growth	Excessive obesity or thinness, sudden unplanned loss or gain in weight, failure to grow in stature or to gain weight
Behavior	Alert, good attention span Regular attendance at school or work Cooperative, agreeable, cheerful, interested, has endurance	Apathetic, listless, short attention span Frequent absences Irritable, easily fatigued, inability to concentrate, poor work capacity
Skin	Firm, clear, slightly moist; healthy, pink mucous membranes Firm, pink nails	Dry, pale; scaly around ears and nose; bedsores Brittle, ridged nails
Hair	Soft, glossy; healthy scalp	Dry, brittle, thin, easily pulled out; change in pigmentation
Eyes	Clear, bright, not unduly sensitive to light	Red, swollen, or dry; itching or burning; poor vision in dim light; excessive sensitivity to bright light
Mouth	Moist, smooth Pink tongue with surface papillae Pink, firm gums Even teeth in well-formed jaw, clean, in good repair	Chapped, swollen; fissures at corners of lips Swollen tongue—scarlet or magenta; smooth in appearance Bleeding gums, receding from tooth line Decayed or missing teeth; inability to chew
Skeleton	Erect posture, arms and legs straight, abdomen in; chest up; chin in	Poor posture; deformities of long bones, chest, spine, pelvis; humpback; bowlegs or knock-knees
Neuromuscular System	Firm, strong muscles with moderate padding of fat; good muscular coordination	Flabby underdeveloped muscles; lack of fat padding or excessive fat; poor muscle coordination; reduced knee and ankle reflexes; burning and tingling of hands and feet
Gastrointestinal System	Good appetite and digestion, regular elimination	Poor appetite, indigestion, frequent or chronic diarrhea or constipation
Glands	No enlargement of thyroid	Enlarged thyroid
Immune Systems	Resistance to infections	Frequent colds and other infections, longer convalescence from illness, slower wound healing

Urine constituents are frequently tested. The level and excretion rate of a nutrient may be measured as well as excretory metabolites of the nutrient. Feces, hair, or biopsies of liver and bone are studied less commonly.

Types of Studies

The concentration of a nutrient in the blood may be low due to inadequate intake, impaired absorption, transport, or metabolism, increased excretion, or some combination of the above. Many medications affect laboratory values. A laboratory test will not differentiate among possible causes, and additional assessment may be necessary to determine the reason.

Blood tests used in assessing nutritional status measure nutrient levels such as glucose, lipids, or minerals, metabolic products such as hemoglobin, enzymes, or serum proteins, or response to a load or test.

Hemoglobin, hematocrit, erythrocytes and **leukocytes** are routinely measured. Hemoglobin, an iron-containing pigment in red blood cells, would be depressed in anemia, hemorrhage, and protein-energy malnutrition (PEM), while elevated in dehydration and **polycythemia** (excess red blood cells). Low hematocrit or packed red cell volume might be indicative of iron deficiency. White cell (leukocyte) counts would be increased in infections and leukemia. Red cell (erythrocyte) counts would decrease in anemia, chronic infectious diseases, and hemorrhage and would be elevated in dehydration and polycythemia.

Lymphocyte, serum albumin, and **serum transferrin** levels may be evaluated. Decreased lymphocyte (a white blood cell produced in the lymph) counts are associated with malnutrition, stress, trauma, and lowered immune response. The proteins albumin, which is involved in the maintenance of fluid and acid-base balance, and transferrin, which transports iron, would be depressed in protein malnutrition and overhydration. (See Chapter 16.)

Interpretation

Since methodology and normal values vary among laboratories, it is essential to evaluate test results by the reference standards of the laboratory that conducted the tests. Awareness of causes for abnormal values that are not nutrition-related is also important.

DIETARY ASSESSMENT

Evaluation of the diet provides a foundation for dietary counseling, and, together with anthropometric, clinical, and laboratory information, can provide a nutritional profile of the client.

Twenty-Four-Hour Recall

The most common method of obtaining dietary information is by 24-hour recall. Through interview or questionnaire, a client reports all food and beverages consumed in the past 24 hours. Drawbacks of this tool are that people may not remember fully or accurately what they ate, may have trouble describing portion sizes, may not for various reasons be completely truthful, or the previous day's food intake may not be representative of their customary diet. Sometimes a typical day's intake may be requested. This also is subject to memory lapses and untruthfulness, and is inappropriate for those with irregular eating habits.

Food Frequency List

Checklists indicating how frequently foods are eaten sometimes serve as a crosscheck on 24-hour recalls. Lists for double-checking purposes would probably include broad categories of foods. Depending on the purpose of the checklist, it might include detailed list of foods, or concentrate on foods of a given composition, such as those high in saturated fat.

Clients are asked to record everything eaten for a period of days or sometimes a week or more. Additionally, they may be asked to indicate circumstances surrounding the food intake such as time of day, location, mood, and companions.

An accurate record provides insight into the food habits and behaviors of the individual and is a more reliable indicator than the 24-hour recall. However, keeping the record is time consuming and compliance may be poor unless the client is highly motivated. Knowing that a record must be kept may prompt some people to modify their diets, either to simplify recording or to present a more adequate food intake.

Other Methods

Members of the health care team through observations of the client, especially at mealtime, can obtain information about food intake and factors affecting it, both in the hospital and at home. Occasionally in research situations foods may be weighed.

Diet History

A diet history, usually taken by a registered dietitian or dietetic technician, contains a record of food intake obtained by the methods previously described plus other information related to the individual's food consumption. It includes a medical history, anthropometric, clinical, and laboratory findings of nutritional relevance together with information about the socioeconomic status, family and employment environment, activity level, facilities for obtaining and preparing food, dental health, medications affecting nutrition, and any other factors that might affect the client's nutritional status. Much of this information is available from medical or health records. For a sample diet history form, see Figure C-2.

Dietary Evaluation

When the information has been gathered, the dietary intake is evaluated by comparing it to some standard. The Food Guide Pyramid (see Figure 1-5) is most commonly used. This guide is quick and easy to use, but is most useful when the individual eats a varied diet that does not include too many mixed dishes that are difficult to evaluate.

Scanning the food intake record can also alert the evaluator to possible deficiencies. For example, a diet devoid of fruits and vegetables will be low in ascorbic acid.

The Exchange Lists for Meal Planning (see Chapter 26 and Table A-2) are useful when a quick approximation of protein, fat, carbohydrate, and/or caloric content is desired. Information about foods not on the exchange lists can be obtained from food composition tables (see Table A-1) and food package labels. Some food packages indicate exchanges per serving and some fast-food companies publish exchange equivalents of their foods.

Occasionally, more detailed calculations of nutrients are desired. In the past, values had to be transcribed from food composition tables, a slow, time-consuming process. Today numerous pieces of computer software with nutrient databases are available to speed and simplify the process.

Such calculations are compared to RDA values (see inside front cover) for the reference person of the same gender and age category. Note limitations of the

Registered Dietitian (RD): has completed a baccalaureate degree in dietetics or a related area at a regionally accredited U.S. college or university, completed a supervised clinical experience, and passed a national examination. To retain RD status, continuing education activities are required. RDs are qualified to perform nutrition screening, assessment, and treatment.

Dietetic Technician, Registered (DTR): has completed a minimum of an associate degree in dietetics or a related area at a regionally accredited U.S. college or university, completed a supervised clinical experience, and passed a national examination. A DTR must fulfill continuing education activities. DTRs are qualified to perform nutrition screening and other nutrition services under the direction of an RD.

RDAs in Chapter 1. Some computer programs include comparisons to RDA values and recommendations for achieving dietary adequacy.

DIETARY COUNSELING

To many, dietary counseling means instructing patients about therapeutic diets. Although such counseling is essential, many healthy people will benefit from changes in their food intake and habits. With the current emphasis on preventive health care, risks of developing chronic diseases may be reduced through dietary counseling. The goal of counseling is to produce a desirable change in food behaviors.

Dietary counseling involves constant interaction between client and counselor. It is not a one-sided lecture by the counselor telling the client what to do.

Clients

Clients can be of any age and health status. They can be patients in a hospital, extended care facility, or outpatient clinic. They might be participants in federally funded nutrition programs such as food stamps, school lunch, or congregate feeding for the elderly. Some health maintenance organizations (HMOs) and worksite fitness programs include nutritional assessment and counseling.

The client takes an active role in dietary counseling by providing information about diet history and food intake, listening to the counselor's evaluation and reasons for recommending changes, setting realistic goals for change, studying materials, and asking questions. Ultimate responsibility for implementing any change is that of the client.

If the client does not have primary responsibility for acquiring and preparing food, the person or persons who perform this function should be included. Involving family members and significant others is helpful so that they understand any recommended diet and behavior modifications and can assist and support the client in reaching the desired goals.

Counselors

Registered dietitians are the professional group best qualified to provide dietary counseling. Nurses can reinforce counseling of the dietitians. Dietetic technicians are qualified to provide basic advice on normal and therapeutic nutrition.

Other members of the health care team may be involved in various ways. The physician who prescribes a diet should explain the need and reasons for it to the client and his or her family. A physical therapist or occupational therapist might assist a disabled person with food preparation, kitchen reorganization, or self-help devices to aid in preparing or consuming meals. A speech-language pathologist can assess and plan for a patient with dysphagia. Pharmacists might plan medication schedules to minimize undesirable drug-nutrient interactions. A clergyperson might assist in planning food intake to conform to religious dietary laws. Social workers can arrange for nutrition-related community services such as home-delivered meals or a home health aide to assist with food shopping and preparation. Dentists, dental hygienists, and others may contribute to the counseling process.

Effective Counseling

Counseling should take place in quiet, comfortable, private surroundings. Interruptions from telephone calls or other staff members not involved in the counseling should be avoided. Clients should be made to feel that they have the counselor's undivided attention and that they matter. In a hospital setting, the first counseling session should be scheduled well in advance of discharge to allow time for follow-up and avoid pressures and distractions that characterize last-minute sessions. Avoid times when your client is hungry, exhausted, or in pain. In a client's home, distractions such as television or radio should not intrude during counseling.

The counselor must be a good listener and must communicate at a level the client can comprehend without appearing to "talk down" to the listener. Medical jargon and acronyms should be avoided. Pictures and foods can facilitate communication, especially if there is a language barrier.

Questions to the client should be nondirective. Asking "Did you have a sandwich for lunch?" might well tempt a client to answer in the affirmative even if he or she had skipped lunch. The counselor must respond in a nonjudgmental way. Scolding a client for eating "junk" food will do nothing to encourage truthfulness or develop a good client-counselor relationship.

When the dietary assessment is completed, compliment the client on the positive aspects of the diet. These good attributes can become the basis for a diet plan and any changes can be built on the strengths of the present food intake. The client will feel less threatened and chances of compliance will increase.

Help the client set reasonable goals. Phasing in the goals gradually may be more realistic than attempting to make several changes at once. Even one small step toward improving food intake can have considerable benefit and will prevent the person from feeling overwhelmed. Understanding the client's needs and values, respecting the individual's dignity, and maintaining confidentiality are essential qualities of the successful counselor.

THE COUNSELING PROCESS

The counseling process comprises assessment, planning, implementation, and evaluation. Assessment involves the gathering and evaluating of information about the client's nutritional status.

Some assessments will reveal that the clients have adequate food intake, normal anthropometric, clinical, and laboratory values, along with satisfactory food behaviors. They will be assured that their present practices are appropriate and no further counseling will be necessary. For others the assessment will provide baseline information around which planning will be done.

Planning based on this assessment includes realistic objectives developed jointly by client and counselor with suggestions by the counselor for realizing these objectives. Customizing the diet and including the client in decision making are important factors in client satisfaction. There may be both short-term and long-range goals. Each client should receive a written list of these objectives and the diet changes necessary to achieve them. Other individuals involved in obtaining and preparing food for the client should participate in the planning. A professional should develop or obtain teaching materials suitable for the educational level of the client and others involved.

To implement this plan, the client must be willing to work toward these objectives. Understanding the dietary changes, applying them to planning meals, choosing foods in restaurants, and shopping for, preparing, and consuming food are essential. Support of family and significant others is necessary for success.

Progress toward these goals should be evaluated periodically by client and counselor. The evaluation will reveal the degree of success in implementing the plan. In some cases, evaluation will lead to revision of the plan. On occasion it will reveal that the client is not motivated to effect any changes.

PUTTING IT ALL TOGETHER

Nutrition screening and assessment are essential to identify individuals at risk and to obtain baseline information about the four components of nutrition assessment—anthropometric, clinical, laboratory, and dietary data. This information will indicate what interventions are appropriate. Nutrition counseling that involves the client in goal setting and planning and that is respectful of the client's values and preferences is essential to accomplish the objective of optimal nutrition.

? QUESTIONS CLIENTS ASK

1. I want to locate a reliable nutrition counselor. Should I look under "nutritionist" in the Yellow Pages?

Some "nutritionists" listed in the Yellow Pages have degrees from unaccredited institutions. Many states do not regulate nutrition counseling. The best way of ensuring you have a reliable counselor is to find a Registered Dietitian (RD). A local dietetic association, hospital, credentials listed in the Yellow Pages, or the American Dietetic Association's National Consumer Hot Line (800/366-1655) are possible sources. The RD must have at least a baccalaureate degree in nutrition or a related field from an accredited institution. The RD's education has included classroom and clinical work plus passing a national exam.

2. Can a hair analysis assess my nutritional status?

Hair analysis has very limited use—primarily in detecting heavy metals such as lead or mercury or in testing for drugs. It cannot provide you with a complete nutrition profile.

3. My family and I are very health conscious. We read and hear so much in the media about contradictory nutrition research results. What are we to do about applying this to our daily lives? We are confused.

Try to learn about clinical research. A single study rarely is reason to make major diet changes, but is a piece of the larger picture. Many studies viewed together can give direction for dietary modification. Remember, moderation and variety is the best philosophy for a healthy diet.

✔ PERSONAL STUDY: Planning, Implementing, and Evaluating Indicated Changes in Food Intake and Behaviors

1. Using the Sample Form for Diet History and Nutritional Assessment, Figure C-2, assess your own nutritional status.* Complete the personal data, as much of the anthropometric data as possible (at least height, weight, and BMI—others if possible), clinical findings, dietary history, and any laboratory data you may have. Anthropometric measurements may require the assistance of a classmate or instructor.

2. If you have access to a computer and nutrient analysis software, evaluate your food intake by this means and compare to the RDAs. (See inside front cover.) Is this analysis consistent with analysis of the Food Guide Pyramid? (from Chapter 1).

3. What do you consider the greatest strength of your food intake and behaviors? The greatest weakness?

*The form will also be used in question 6 of this exercise and in Chapter 16. You may want to photocopy the form or use different color pencils or pens to distinguish the answers.

4. Based on your assessment, form a written plan for modifying your food intake and behaviors. Set realistic goals and means of implementing them that are consistent with your lifestyle. If your food intake is adequate, are there any behaviors you would like to modify—eating more slowly, increasing activity, and so on?

5. Evaluate your plan and its implementation at the end of a week and then monthly. Are any modifications indicated?

6. Assess the nutritional status of a classmate, friend, or relative, using the same nutrition history form. Review the attributes of effective counseling before beginning. Work with your client, making sure he or she is involved in the planning.

7. Have five classmates independently assess your clinical signs, using items on the nutrition history form. After the assessments are complete, compare them. To what extent do they agree or differ? How might these differences be explained?

FOR ADDITIONAL READING

Chumlea, WC, and Guo, SS: "Bioelectrical Impedance and Body Composition: Present Status and Future Directions," *Nutr Rev,* **52:**123–31, 1994.

Eliades, DS, and Suitor, CW: *Celebrating Diversity: Approaching Families Through Their Food.* Arlington, VA: National Center for Education in Maternal and Child Health, 1994.

"Fine-Tuning Your Patient Teaching," *Nurs 95,* **25:**32J, January 1995.

Guigoz, Y, Vellas, B, and Garry, PJ: "Assessing the Nutritional Status of the Elderly: The Mini Nutritional Assessment as Part of the Geriatric Evaluation," *Nutr Rev,* **54:**S59–S65, 1996.

"Identifying Patients at Risk: ADA's Definitions for Nutrition Screening and Nutrition Assessment," *J Am Diet Assoc,* **94:**838–39, 1994.

Keenan, DP: "In the Face of Diversity: Modifying Nutrition Education Delivery to Meet the Needs of an Increasingly Multicultural Consumer Base," *J Nutr Educ,* **28:**86–91, 1996.

Neumark-Sztainer, D, and Story, M: "The Use of Health Behavior Theory in Nutrition Counseling," *Top Clin Nutr,* **11:**60–73, March 1996.

Thompson, FE *et al:* "Dietary Assessment Resource Manual," *J Nutr* **124** (suppl):2245S–2317S, 1994.

Trudeau, E, and Dubé, L: "Moderators and Determinants of Satisfaction with Diet Counseling for Patients Consuming a Therapeutic Diet," *J Am Diet Assoc,* **95:**34–39, 1995.

Warpeha, A, and Harris, J: "Combining Traditional and Nontraditional Approaches in Nutrition Counseling," *J Am Diet Assoc,* **93:**797–800, 1993.

Weigley, ES: "The Body Mass Index in Clinical Practice," *Top Clin Nutr,* **9:**70–75, June 1994.

Wichowski, HC, and Kubsch, S: "Improving Your Patient's Compliance," *Nurs 95,* **25:**66–68, January 1995.

Young, LR, and Nestle, M: "Portion Sizes in Dietary Assessment: Issues and Policy Implications," *Nutr Rev,* **53:**149–58, 1995.

Transformation of Food and Fluid for Body Needs

Chapter Outline

*Water is the most abundant constituent of the body.**

✔ PERSONAL STUDY PREVIEW

Each day you enjoy eating and drinking many different kinds of foods and beverages. You may or may not think about all the nutrients they contain or how your body processes them to keep you well nourished and healthy. How might you put this chapter to personal use and then apply your knowledge to someone else?

---

*National Academy of Sciences, *Recommended Dietary Allowances,* 10th ed., National Academy Press, Washington, DC, 1989, p. 247.

"You are what you eat" is an often quoted cliché. Yet it is reasonable to suppose that a relationship exists between the kinds of nutrients present in food and the substances present in the body. It is also reasonable to expect that the body can be properly built and maintained in health and treated when diseased only if the necessary foods and their nutrients are available to it.

The human body is an organized structure of chemicals. These chemicals combine to form cells. Cells are grouped into tissues, tissues into organs, and organs into related functional systems. Thus, the human organism is a highly interconnected structure. As in any smoothly functioning team, all parts depend on each other for excellence and success.

Four chemical elements—oxygen, carbon, hydrogen, and nitrogen—account for 96 percent of body weight. Water is by far the most abundant compound in the body and accounts for about two-thirds of body weight. Proteins and fats each represent roughly one-fifth of the total body weight. Carbohydrates are present in limited amounts. Slightly less than a pound of carbohydrate occurs in the adult body, as glycogen in the liver and muscles and as glucose in the blood. A wide variety of minerals account for only 4 percent of body weight. As for vitamins, the entire body store adds up to only a few grams, not even one ounce.

However, the proportions of major body constituents vary widely from one individual to another. Babies have a higher proportion of body water than do adults. Athletes have a lower body fat content than do sedentary individuals of the same body weight. Women have higher proportions of body fat than do men.

HOMEOSTASIS

Even though our bodies are complex, superbly designed structures of cells, tissues, organs, and systems, precisely integrated and controlled, we are unaware of the thousands of physical and chemical changes continually taking place to keep our bodies in balance. For example, suppose you are late for class, and you decide to run. Your increased activity means you breathe more rapidly to take in more oxygen, your heart beats faster, your blood pressure rises to speed the oxygen and glucose to your cells, your skin feels warm, and you may even perspire. But once you sit down in the classroom your respiration and heart rates fall, your blood pressure returns to normal, and you begin to feel cooler. You made the intellectual decision to run, but your body's response involved many physical and chemical changes that were made automatically for you. In healthy individuals, the body's internal environmental processes are balanced so that a state of equilibrium exists. This stable condition is called dynamic equilibrium or homeostasis.

Controls for Metabolic Processes

Body homeostasis is maintained by the precise regulation of:

- Concentration of gases, water, nutrients, electrolytes, acids, bases
- Temperature
- Pressure

Metabolism is an inclusive action-oriented term describing all the changes taking place in the body. Nutritionally, metabolism may be described for specific nutrients, for example, protein metabolism, vitamin E metabolism, or sodium metabolism. These processes take place in a precisely integrated and controlled fashion. The body uses nutrients from recently eaten food and reuses many nutrients, such as iron or amino acids, that are released during metabolic processes. At any given moment, an ubiquitous mix of nutrients from the diet and from body cells is available. This is referred to as the metabolic pool of available nutrients.

Metabolism can be divided into anabolic and catabolic activities. Anabolism refers to the building up of complex substances from simpler substances. An example is building proteins from amino acids, which occurs in growing children, during pregnancy, or during recovery from illness. By contrast, catabolism refers to the breakdown of complex substances into simpler substances. An example is the release of energy from body stores during periods of high fever, strenuous prolonged exercise, or food deprivation as in starvation.

Functions and Processing of Nutrients

The functions of nutrients fall into three major categories:

- Supply energy
- Build and maintain body tissues
- Regulate body functions

A series of steps are necessary to permit the nutrients in foods to perform their functions. They are summarized in Table 4-1 and are described more fully in later sections of this chapter and in the chapters pertaining to the various classes of nutrients.

The pathway of each single nutrient can be traced from ingestion through excretion. However, this is an oversimplification. The processes involving individual nutrients never occur in isolation but always interlinked and interdependent, affected by other processes preceding them or occurring at the same time. As you learned in Chapter 3, numerous physical and chemical methods have been developed for measuring the changes that occur. Part of your study of nutrition in health and disease is concerned with a knowledge of such changes and an interpretation of their importance in evaluating people's nutritional metabolic processing and balance.

TABLE 4-1

Steps in Metabolic Processing of Foods and Their Nutrients

Step	Definition
Ingestion	Taking food into body (e.g., eating, tube feeding into stomach)
Digestion	Breakdown of foods into their constituent nutrients
Absorption	Transfer of nutrients from the digestive (gastrointestinal) tract into circulation
Transportation	Movement of nutrients through the lymphatic system and circulatory system to the cells for their use and the movement of wastes to sites for excretion
Respiration	Provision of oxygen to the cells for the oxidation of some nutrients and removal of some metabolic waste products
Utilization	Use and function of nutrients in the cells
Storage	Deposits of nutrients for later use
Excretion	Elimination of wastes and excess nutrients

Source: Modified from Robinson, CH, Lawler, MR, Chenoweth, WL, Garwick, AE: *Normal and Therapeutic Nutrition,* 17th ed., Macmillan Publishing, New York, 1986.

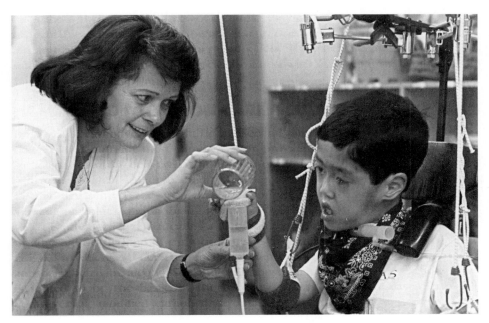

FIGURE 4–1
A caregiver assists a child in feeding himself a formula that flows into his stomach through a tube. (*Source:* Barbara Schwartz/Merrill.)

INGESTION

The one process under the direct control of most people is the act of taking food into the body, termed ingestion. Usually, this is accomplished by eating food and drinking beverages. However, there are numerous other processes that enable this seemingly simple task to occur. For example, food must be available, so a person must have sufficient money to grow or purchase food. Also, a person must be mentally and emotionally able to discern hunger, thirst, and satiety. A person must be physically able to shop for food, cook food, and move food from the plate into the mouth. Proper eye-hand-body coordination is necessary. Sometimes, disease processes prevent oral ingestion from occurring, thus requiring the person to be fed by alternative means. (See Figure 4–1 and Chapter 17.)

DIGESTION

Foods are highly complex packages and thus cannot be used by the body without undergoing change. Digestion includes the physical and chemical processes whereby the nutrients contained in food are made suitable in size and composition for transfer from the digestive tract, through absorptive cells, into the body's circulation. While food and its contained nutrients remain in the gastrointestinal tract, they are still external to the body for its direct use.

For digestion to occur, numerous processes and substances must function in an orderly manner. All along the way, muscles and nerves, enzymes and hormones, acids and bases must act in concert to transform food into its constituent parts. Figure 4–2 illustrates the digestive organs and their location in

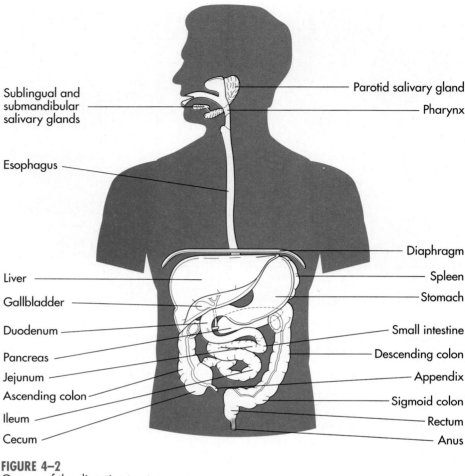

Sublingual and submandibular salivary glands

Parotid salivary gland

Pharynx

Esophagus

Diaphragm

Liver

Spleen

Gallbladder

Stomach

Duodenum

Small intestine

Pancreas

Descending colon

Jejunum

Appendix

Ascending colon

Sigmoid colon

Ileum

Rectum

Cecum

Anus

FIGURE 4–2
Organs of the digestive tract.

the body. Table 4–2 summarizes some of the major digestive enzymes; Table 4–3 summarizes some of the hormones that facilitate digestion; and Table 4–4 summarizes some acid and base functions.

Role of Some Digestive Organs

The initiation of digestive activity in the mouth is under conscious control; that is, you may chew your food well or you may eat rapidly, gulping it down. Once food is swallowed, digestive activity is under the internal controls of the body without your thinking about it.

Digestion in the Mouth and Esophagus The digestion of food begins in the mouth where food is masticated (chewed) and mixed with saliva. Chewing is important because it increases the surface area of the food particles for later digestive action. The teeth, lips, tongue, and other mouth and throat muscles must function properly. Saliva contains some enzymes, but food remains in the mouth for such a short time before it is swallowed that only a very small amount of digestion can occur. Once swallowed, food is moved through the esophagus by muscular, rhythmic, wavelike contractions, termed **peristalsis.**

TABLE 4-2

Some Enzyme Reactions That Occur During Digestion

Site of Activity	Enzyme	Substrate	Products of Enzyme Activity
Mouth	Salivary amylase	Cooked starch	Dextrins, maltose
Stomach	Protease (pepsin)	Proteins	Proteoses, peptones, polypeptides
	Rennin	Milk casein	Calcium caseinate
	Lipase	Emulsified fats	Fatty acids, glycerol
Small intestine	Pancreatic juice		
	Protease (trypsin)	Proteins	Proteoses, peptones, polypeptides, some amino acids
	Lipase	Fats	Di- and monoglycerides, fatty acids, glycerol
	Amylase	Starch	Maltose
	Intestinal juice		
	Peptidases	Peptones, polypeptides	Amino acids
	Sucrase	Sucrose	Glucose, fructose
	Maltase	Maltose	Glucose (2 molecules)
	Lactase	Lactose	Glucose, galactose

TABLE 4-3

Some Hormones That Regulate Digestive Activity

Hormone	Where Produced	Stimulus to Secretion	Activity
Gastrin	Pylorus; duodenum	Proteins, caffeine, spices, alcohol	Stimulates flow of gastric juice
Secretin	Duodenum	Polypeptides; acid chyme	Pancreas secretes thin, enzyme-poor juice into duodenum
Cholecystokinin	Duodenum; jejunum	Fats	Contracts gallbladder; bile flows into duodenum
		Polypeptides; acid chyme	Pancreas secretes thick, enzyme-rich juice into duodenum
		Chyme	Intestinal juices secreted by intestinal glands

Digestion in the Stomach Food passes from the esophagus into the stomach when the lower esophageal (gastroesophageal) sphincter relaxes. Following the entrance of food into the stomach the sphincter closes, thus preventing the regurgitation of food. The stomach serves as a temporary storehouse for food, begins partial digestion, and prepares food for further digestion in the small intestine. The food is continually churned and mixed with the enzymes and acid of the gastric juice until it reaches a liquid consistency known as chyme. Peristalsis moves the chyme toward the pylorus, where small portions are released gradually through the pyloric sphincter into the duodenum.

Digestion in the Small Intestine Most digestive activity takes place in the small intestine, which includes the duodenum, the jejunum, and the ileum. At this point, other so-called "accessory organs" (liver, gallbladder, and pancreas) play a major role. For example, bile is made by the liver and stored in the gallbladder. As soon as fats enter the duodenum the secretion of a hormone, cholecystokinin, is stimulated. Cholecystokinin causes the gallbladder to contract and release bile into the duodenum. The function of bile is to **emulsify** fats, that

TABLE 4–4

An Example of an Acid and Base Involved in Digestion

Name	Functions
Acid: Hydrochloric acid (Stomach)	Swells proteins to make them more easily acted upon by enzymes
	Provides the acid medium necessary for the action of stomach enzymes
	Increases the solubility of calcium and iron for greater absorption
	Reduces the activity of harmful bacteria that may be in food
Base: Bicarbonate (Small intestine) Secreted by pancreas into duodenum	Neutralizes acid chyme emptied from the stomach into the duodenum, thereby enabling digestive reactions to occur that require an alkaline environment

is, break them apart into tiny globules and keep them in suspension so that fat-splitting enzymes have greater contact with fat molecules. Also, bile is highly alkaline. Bile helps neutralize the acidic chyme to provide the alkaline medium necessary for the action of the digestive enzymes produced by the pancreas and certain cells of the small intestine.

Function of the Large Intestine The large intestine includes the cecum, colon, rectum, and anal canal. By the time the food mass reaches the large intestine, the digestion of food and absorption of nutrients essentially are completed. Digestive bacteria in the large intestine function to ferment remaining food; some synthesize vitamins, such as vitamin K. As most of the remaining water and digestive juices are reabsorbed, the intestinal contents gradually take on a solid consistency. The resulting feces contain the indigestible fibers of plants, small amounts of other undigested food, bile compounds, cholesterol, mucus, bacteria, and gastrointestinal tract cellular wastes.

Digestibility of Food

Digestibility refers to both the completeness of digestion and the ease or speed of digestion.

Coefficient of Digestion In general, the efficiency of digestion is remarkably high. However, plant fibers and seeds are not digested. Therefore, a diet made up of many fruits, vegetables, legumes, and whole-grain products could have a digestibility of carbohydrate of about 85 percent. The completeness of digestion is reduced greatly in some disorders of the gastrointestinal tract. For instance, some hereditary diseases such as celiac disease, cystic fibrosis, and lactose intolerance cause one or more enzymes for the digestion of carbohydrates, fats, and proteins to be missing, resulting in much starch and sugar, fat, or protein to be eliminated in the feces.

Motility Through the Tract Transit time (speed at which food moves through the digestive tract) varies widely according to the size and composition of the meal and is sometimes affected by emotional or psychological factors. As little as 9 hours or as many as 48 hours may elapse from the time food is eaten until feces are eliminated. Large meals remain in the stomach longer than do small meals, as do poorly chewed foods. High-fiber foods speed up the transit time.

You may have heard another phrase, that some foods "stick to your ribs." This is translated to mean these foods stay in the stomach longer and delay hunger contractions. Because such foods are more satisfying, they are said to have a high satiety value. But sometimes digestion is slowed so much that discomfort results. An example might be eating an excessive amount of fat, especially in the form of fried foods.

Other foods increase or decrease the flow of digestive juices. Secretion of digestive juices may increase with the pleasant sight, smell, and taste of food. Conversely, secretion may decrease when foods are unattractively served or when mealtime surroundings are unpleasant. An individual who is excessively tired or who is under emotional stresses such as fear, grief, or anger often experiences digestive upsets.

ABSORPTION

Absorption is the process whereby the nutrients released from food by digestion pass from the intestinal tract, through cells and their membranes, into the blood or lymph circulation. Most nutrient absorption occurs from the duodenum and jejunum. By the time the digested mass reaches the end of the ileum, practically all nutrients—more than 90 percent—have been absorbed.

Absorption is a selective process. Some nutrients, for example, glucose, are absorbed totally; others, for example, calcium and iron, are absorbed according to body need. In this way, the body attempts to protect itself from nutrient deficiencies or toxicities.

Intestinal Villi

The small intestine is lined with 4 to 5 million tiny fingerlike projections called villi. (See Figure 4-3.) Each villus is a complex structure with a layer of epithelium over a layer of connective tissue (lamina propria) that is supplied with capillaries (for blood circulation) and lacteals (for lymph circulation).

On the surface of each villus are 500 to 600 microvilli also known as the brush border. Thus, the villi and microvilli give an immense surface through which nutrients can be absorbed—an area comparable to the size of a third of a football field!

Mechanisms for Absorption

Nutrients are absorbed by complex mechanisms classified as passive processes or active processes. Most nutrients are absorbed by active processes. Table 4-5 summarizes the major ways nutrients are absorbed.

The passive processes move nutrients into the body from an area of greater concentration (inside the small intestine) to an area of lesser concentration (circulation). Sometimes protein carriers are necessary, but no energy is required. Active processes "pump" nutrients from an area of lower concentration to one of higher concentration. Not only do most of these active processes require carriers to help the nutrients be absorbed, they all require body energy to fuel the process. (See Chapter 8.)

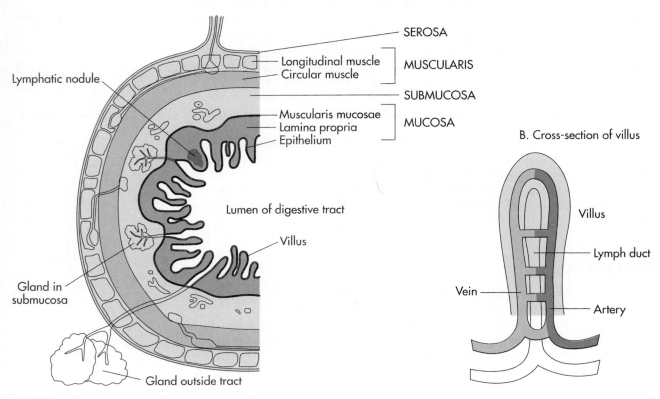

A. Cross-section of small intestine

SEROSA

Longitudinal muscle — MUSCULARIS
Circular muscle

Lymphatic nodule

SUBMUCOSA

Muscularis mucosae
Lamina propria — MUCOSA
Epithelium

B. Cross-section of villus

Lumen of digestive tract

Villus

Villus

Lymph duct

Vein

Gland in
submucosa

Artery

Gland outside tract

FIGURE 4–3
Absorption of most nutrients occurs from the small intestine into the blood or lymph
circulations.

TRANSPORTATION (CIRCULATION)

From the absorptive cells in the digestive tract, nutrients are carried to the liver
and then to all cells of the body by two forms of transportation (circulation).

Water-soluble substances like the amino acids, glucose, minerals, and water-
soluble vitamins can enter the blood circulation (bloodstream) directly since it is
water-based. (See Figure 4–3.)

Fat-soluble substances like the fatty acids, cholesterol, and fat-soluble
vitamins first enter the lymphatic circulation (lacteals) and then the blood
circulation near the heart. They travel to the liver, then to the body's cells. (See
Figure 4–3.)

RESPIRATION

To utilize some nutrients, cells require oxygen and produce carbon dioxide,
which is toxic to the body. Respiration is the process of gas exchange between
the cells and the blood, and between the blood and the atmosphere. The proper
coordination of the respiratory and circulatory systems is necessary to supply
sufficient oxygen for the oxidation of food and for the elimination of carbon
dioxide and water, both of which are cellular metabolic by-products.

TABLE 4–5

Examples of Nutrient Absorption Mechanisms from Inside Small Intestine into Villi

PASSIVE PROCESSES	
(Higher to lower concentration; require no energy)	
Osmosis	Water molecules pass easily through cell membrane "pores" (protein channels) from a higher to a lower concentration. *Example:* Water
Simple Passive Diffusion	Nutrients pass easily through cell membrane "pores" from a higher to a lower concentration. *Example:* Short-chain fatty acids, vitamin C, B-complex vitamins, fat-soluble vitamins
Carrier-Facilitated Passive Diffusion	Nutrients cross membrane attached to a "carrier" (protein) from a higher to a lower concentration. *Example:* Fructose "ferried" by protein molecule
ACTIVE PROCESSES	
(Lower to higher concentration; require energy)	
Active Transport	Nutrients cross membrane attached to a carrier and are "pumped" from a lower to a higher concentration. Requires energy (ATP). Most nutrients. *Example:* Glucose, galactose, amino acids, dipeptides, tripeptides, calcium, chloride, iodide, iron, magnesium, potassium, sodium
Pinocytosis	Nutrients are engulfed by membrane. Rare occurrence. *Example:* Intact proteins

UTILIZATION (CELLULAR METABOLISM; INTERMEDIARY METABOLISM)

Within every cell nutrients function in terms of supplying energy, building or maintaining body tissues, and regulating body functions. This dynamic process of finally using nutrients at the cellular level for the body's activities may be identified as utilization (cellular metabolism or intermediary metabolism). Specific enzymes speed each step of every reaction. Many of the exact mechanisms are known, but others still are under intense investigation. Each cell must also process the nutrients it receives, that is, nutrients are absorbed, transported from place to place inside the cell, used, stored, and excreted. Thus, metabolism goes on inside the cell as well as inside the body as a whole.

For example, in a series of complex catabolic steps, glucose combines with oxygen to release the energy for the body's work and the waste products of water and carbon dioxide. Anabolic processes incorporate fatty acids into cell membranes; cholesterol is converted into sex hormones. Amino cells are used to build new cells, to form hormones or enzymes, or as a backup source of energy. Minerals and vitamins become part of the cell's structure or enter into any one of its many regulatory activities.

Interconversion of some nutrients occurs during cellular utilization. This process is important to maintain the body's homeostasis. For instance, some amino acids can be converted into other amino acids. An excess of glucose or some amino acids can be converted into fat.

STORAGE

Although the cells of the body need nourishment every second of the day, most people eat only a few times each day. Occasionally, people fast and eat or drink nothing for a day or more, yet life is maintained. To function, cells depend on the storage deposits of energy and other nutrients.

The body has many reservoirs. Energy is stored for the short term as ATP (adenosine triphosphate) in cells, for the intermediate term as glycogen in the liver and muscles, and for the long term as fat in adipose tissue. The liver stores many different kinds of nutrients including protein, iron, and vitamin A. The bones store many minerals such as calcium and phosphorus.

The body stores large quantities of some nutrients, such as energy in the form of fat. On the other hand, it has the capacity to store only small quantities of other nutrients, such as energy in the form of glycogen. The body conserves and reuses nutrients, such as water and iron.

A very important fact to keep in mind is that "stored" nutrients are released and replenished. Once stored, they do not stay indefinitely. Rather they are constantly being released and replaced. All storage reservoirs are very active places!

Thus, the body is extremely adaptable. For optimal health, as well as recovery from disease and injury, the body ultimately depends on the food choices made by the individual person, day by day, over the course of a lifetime.

EXCRETION (ELIMINATION)

The metabolic processes result in products that ultimately must be discarded from the body. The removal of excess, toxic, or other end-products that the body no longer needs is termed excretion (elimination). Body wastes are removed mainly by four organs:

- *Bowel:* indigestible fiber and undigested food, bile pigments, cholesterol and other by-products of metabolism, intestinal bacteria
- *Kidneys:* most nitrogenous wastes, water, minerals, excesses of water-soluble vitamins, detoxified substances
- *Skin:* water by visible and invisible perspiration, small amounts of some minerals and nitrogenous wastes
- *Lungs:* carbon dioxide, water

FLUID (WATER) BALANCE

Next to oxygen, water is most important to sustain life. Of course, we can live only moments without oxygen. Although we can survive for weeks without food, at best, we can survive only a few days without water. Persons lost in the desert can perish within 24 hours. Fluid balance (amount and proportion of water distributed in the body) is critical for life. (See Figure 4–4.)

Characteristics

Water is classified as a **macronutrient,** that is, a nutrient found in the body in large quantities. Water is the largest single constituent of the body, about 45–75 percent of total body weight. Every cell of the body contains water. Blood plasma is about 90 percent water, muscle tissue about 75–80 percent, fat tissue 20

FIGURE 4–4
People of all ages need to drink enough water each day to maintain optimal fluid balance. (*Source:* Anthony Magnacca/Merrill.)

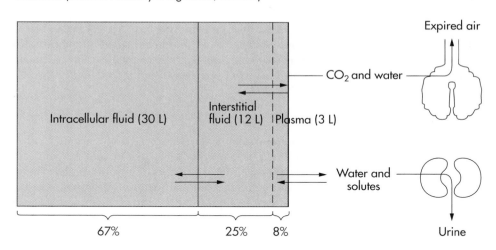

FIGURE 4–5
Distribution of body fluids. Note that fluid balance is maintained between compartments and between the blood and interstitial fluid. The kidneys are the final regulators of fluid balance.

percent, and bone about 20 percent. Water is a compound of hydrogen and oxygen. It is measured in liquid volumes, such as liters.

Body fluids are separated into sections or compartments, divided by membranes. **Intracellular** fluid (within the cells) accounts for about two-thirds of the body's water. The amount of water in this compartment remains fairly constant. **Extracellular** fluid (outside the cells) accounts for the remaining one-third. Extracellular fluid is further divided into the supply in the circulation (blood and lymph) and the fluid situated between cells or tissues (**interstitial** fluid). (See Figure 4–5.)

Functions

Water is the solvent for materials within the body. Almost nothing takes place in the body without water. To summarize the functions of water:

- The foods we eat are moistened by saliva and digested in an abundance of gastric and intestinal juices.
- The nutrients are carried in solution during absorption across the intestinal membrane.
- The lymph and blood transport nutrients and other dissolved substances (e.g., oxygen) to all cells.
- Materials dissolved in water are transported across membranes to reach the interior of all cells.
- Some chemical reactions require or produce water.
- Body wastes are dissolved and excreted in water.
- Water has a lubricating effect to help avoid friction between body parts (e.g., bone joints).
- Water maintains body moisture (e.g., saliva, tears, mucus).
- Body temperature is regulated through evaporation of water from the skin.

Metabolic Processes

In healthy people, the majority of water is ingested through the foods and liquids they consume. Water is undigested and is absorbed from the small intestines by osmosis into the blood circulation. Cellular function both uses and creates water. The water produced when carbohydrates, lipids, and proteins are metabolized is termed metabolic water. Water is stored mainly in the extracellular spaces and is excreted by all excretory organs, but mainly by the kidneys. (See Table 4–6.)

Meeting Body Needs

Requirements and Recommendations There is no RDA for water. Studies done to measure people's fluid intake and output, known as "I and O," indicate the normal daily requirement is about 1 ml/kcal of food consumed. (See Figure 4–6.) This calculation leads to the general recommendation that adults need to consume about 2 liters (about 2 quarts) daily. However, infants require about 1.5 ml/kcal because they have proportionately more body water and a larger surface area of skin.

TABLE 4–6
Typical Water Balance for an Adult

Sources of Water		Losses of Water	
Ingested liquids	1,600 ml	Kidneys	1,500 ml
Ingested foods	700 ml	Skin	500 ml
Metabolic water	200 ml	Lungs	300 ml
		GI tract	200 ml
Total	2,500 ml	Total	2,500 ml

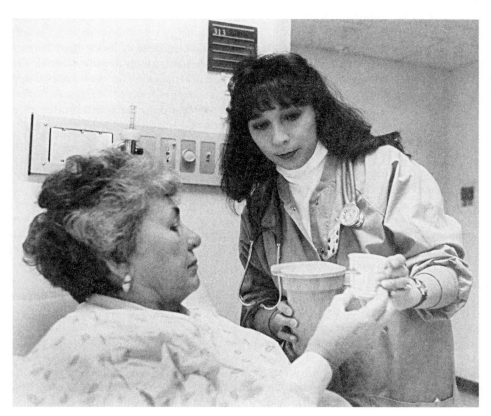

FIGURE 4–6
Accurate measurements of a patient's fluid intake and output ("I and O") is an
important part of the daily nutritional assessment in the clinical setting. (*Source:* Tom
Wilcox/Merrill.)

TABLE 4–7
Water Content of Foods

	Water (%)
Fruits and vegetables	70–95
Milk	87
Cooked cereals, pasta, rice	60–85
Eggs	74
Cooked meat, poultry, fish	
Well done	40–50
Medium to rare	50–70
Cheese, hard	35–40
Bread	35
Dry cereals, crackers	3–7
Nuts, fats, sweets	0–10

Sources The body replenishes its water content in three ways. The fluids we
drink account for the chief source of water. Water is contained in most foods.
Table 4–7 lists the water content of some beverages and foods. Metabolic water
provides the smallest source. (See Table 4–6.)

Body Balance

Optimum/Adequate Signs and Symptoms Ordinarily, body fluid balance is recognized as optimal when the body fluid compartments are maintained. The body's main systems for regulating water balance include the nervous (sending signals to the brain), endocrine (hormones), kidneys (conserve or excrete water), and cardiovascular (transport more or less water) systems. Signs and symptoms of adequacy include the integrity of the skin, lack of any quick shifts in body weight, and correct color and volume of the urine.

Insufficiency/Deficiency Mild water inadequacy is termed water depletion; more severe deprivation is termed **dehydration.** A diagnostic category is fluid volume deficit. Even when fluid intake is restricted or ceases, obligatory losses of water through the kidneys, skin, and lungs continue. The earliest symptom of water depletion is thirst. Unfortunately, this occurs when already about 2 percent of the body weight is lost. Loss of 10 percent of body water results in severe dehydration; loss of 20 percent is fatal.

Dehydration is a serious medical problem requiring prompt attention. Fluids are given by mouth when possible. Intravenous fluids are given when the patient is unable to take sufficient fluids by mouth.

Healthy people most at risk include babies because of their naturally high body water content, athletes, pregnant/lactating women, and older adults. People in hot climates always need to focus on their fluid balance. People who have illnesses require close monitoring. Such persons are those unable to recognize symptoms of thirst, such as those in coma or who have Alzheimer's disease; those with fever, increased respiratory rate, or skin burns; premature infants under radiant heat warmers; people who have foodborne poisoning resulting in vomiting and diarrhea; and clients receiving nourishment by enteral or parenteral feedings.

Excess/Toxicity **Edema** is the accumulation of excess interstitial fluid. It occurs when the body is unable to excrete water in sufficient amounts. Nutritionally, healthy people usually cannot consume too much water because it is easily excreted by the kidneys. Treatment for people with medical conditions focuses on correcting the underlying medical problem. Medications like diuretics (Chapter 18) and dietary restriction of sodium (Chapter 11) often are used. Individuals most at risk are those who suffer prolonged protein deficiency (Chapter 7) or who have various organ failures, such as failure of the liver, heart, or kidneys (Chapters 23, 24, 28).

Health and Nutrition Care Issues

Since water is the nutrient in the largest quantity in the body, it is usually the first nutrient to assess. But many people forget to think about their water intake and output. Thus, health care providers must be highly aware of clients' water status and provide close monitoring and education.

ELECTROLYTE BALANCE

A fuller discussion of fluid balance requires an explanation of electrolytes. The body fluids contain a variety of dissolved chemicals, some of which are termed **electrolytes.** These are compounds that, when dissolved in body fluids, can separate their bonding and exert a positive or negative electric charge. The human body conducts electrical currents through the chemicals dissolved in its body fluids. An example might be when you touch something and feel an "electric shock" or see an "electric spark." Specific electrolytes are discussed in Chapter 7 and especially Chapter 11.

Characteristics

Electrolyte particles are termed **ions** when dissolved in water. **Cations** are ions with a positive electrical charge (can give away electrons), such as sodium (Na^+). **Anions** are ions with a negative charge (can receive electrons), such as chloride (Cl^-). In an electrolyte solution, the total number of cations exactly equals the total number of anions.

Salts, acids, and bases are classified as electrolytes. Most electrolytes are inorganic compounds (mineral elements), but a few are organic, such as some proteins. The concentration of an ion in body fluids is commonly expressed as the number of electrical charges in each liter of solution, that is, **milliequivalents** per liter (mEq/L). Knowing the weight of an electrolyte does not tell anything about its electrical combining power in solutions. But knowing the calculated mEq/L number does express the ion's electrical combining capability. (See Table 4-8.)

TABLE 4-8
Concentrations of Solutions Calculated as Milliequivalents Per Liter

Principles

Any cation can pair off with any anion.
Concentrations of solutions are more easily compared in milliequivalents rather than weights.

Examples

Concentration calculated by weight of ion:

| 23 mg sodium (Na^+) | combines with | 35 mg chloride (Cl^-) |
| 39 mg potassium (K^+) | combines with | 35 mg chloride (Cl^-) |

Concentration calculated by chemical combining power:

| 1 mEq of sodium | combines with | 1 mEq of chloride |
| 1 mEq of potassium | combines with | 1 mEq of chloride |

Calculation

mEq/L =

$$\frac{\text{milligrams of ion per liter of solution}}{\text{atomic weight}} \times \text{number of charges on one ion}$$

(The atomic weight of an element can be found in the periodic table. The number of charges on one ion is its valence.)

Problem

Calculate the milliequivalents of calcium in 1 liter of plasma.

Given: One liter of plasma   = 100 milligrams of calcium
Atomic weight of calcium =   40
Valence of calcium   =   2

Substituting values in formula: $\frac{100}{40} \times 2 = 5$

Answer: There are 5 mEq of calcium in 1 liter of plasma.

Functions

Electrolytes have three main functions in the body:

- Serve as essential minerals
- Control the osmosis of water between body compartments
- Help maintain the acid-base balance throughout the body

Health and Nutrition Care Issues

Sodium, potassium, and chloride are the most abundant electrolytes in the body. Because individually and collectively electrolytes have important functions in health and disease, health care providers must perform careful assessment and monitoring of electrolyte status.

Table 4-9 lists the electrolyte concentrations of some body fluids of healthy people. The numbers listed may be considered examples of "normal" laboratory values. Compare the number of electrolytes in each category of fluid. Note that the total number of cations exactly balances the total number of anions. Compare the differences in kinds of electrolytes. Note that the major cation in blood plasma is Na^+, but the major cation in intracellular fluid is K^+; the major plasma anion is Cl^-; but the major intracellular fluid anion is HPO_4^{2-}.

When people are ill, electrolyte imbalance occurs. For example, changes in electrolyte metabolism characterize the onset of acute infections. Malnutrition causes severe electrolyte imbalances. Such patients experience a loss of the intracellular ions, potassium, magnesium, and phosphorus, balanced by a gain in

TABLE 4–9
Electrolyte Composition of Blood Plasma and Intracellular Fluid

	Blood Plasma	Intracellular Fluid
Cations	mEq/liter	mEq/liter
Sodium (Na^+)	142	10
Potassium (K^-)	5	150
Calcium (Ca^{2+})	5	
Magnesium (Mg^{2+})	3	40
Total Cations	155	200
Anions		
Chloride (Cl^-)	103	
Bicarbonate (HCO_3^-)	27	10
Phosphate (HPO_4^{2-})	2	150
Sulfate (SO_4^{2-})	1	
Organic acids	6	
Protein anions	16	40
Total Anions	155	200

Source: Adapted from Robinson, C. H., et al.: *Normal and Therapeutic Nutrition.* 17th ed. New York: Macmillan Publishing Co., Inc., 1986, p. 140. Used with permission.

sodium and water. Electrolyte levels must be followed closely when these patients are nutritionally rehabilitated, especially by refeeding with alternative means. (See Chapters 11 and 17.)

ACID-BASE BALANCE

Normal metabolic processes produce acids and bases. In solutions, acids dissociate into hydrogen ions (H^+) and bases into hydroxyl ions (OH^-). However, body fluids must maintain a fairly constant balance of acids and bases for reactions to occur. Thus, maintaining the acid and base balance of body fluids is another method by which the body attempts to maintain its homeostasis.

Characteristics

A solution's acidity or alkalinity is expressed on a pH scale that runs from 0 to 14. The acronym **pH** is used to denote the degree of acidity or alkalinity of fluids. The number 7 denotes neutrality. Any pH below 7.0 indicates acidity; the lower the pH, the greater the acidity. For instance, a solution with a pH of 3 is much more "acid" than one with a pH of 5. Conversely, the higher the pH number, the greater the solution's alkalinity.

Each body fluid has a different acidity or alkalinity (basicity). However, the normal limits for each fluid are specific and narrow. The body uses four major methods to maintain balance:

- *Dilution:* Since fluids represent about two-thirds of the body weight, excess acids and bases can be distributed easily to lower their concentrations.

- *Buffer systems:* A **buffer** is a substance that reacts chemically with either acids or bases to prevent swings in the pH of the solution. One example of a buffer in the blood to maintain its proper pH is known as the bicarbonate carbonic acid (HCO_3^- H_2CO_3) system. In this way, excess acids or bases produced by the cells are transported to the lungs or kidneys for excretion.

- *Lung respirations:* The respiratory rate regulates the losses of carbon dioxide (CO_2) (dissolved as HCO_3^-) and the intake of oxygen.

- *Kidney excretions:* The kidney makes the final adjustment that keeps the body fluids' pH within their normal limits by excreting either an acid or alkaline urine.

Health and Nutrition Care Issues

The various steps of metabolism are extremely sensitive to even small changes in the acidity or alkalinity of the environment in which they occur. For example, normal saliva is pH 6.35 to 6.85; normal gastric juice is pH 1.2 to 3.0; normal pancreatic juice is pH 7.1 to 8.2; and normal blood plasma is pH 7.35 to 7.45. Healthy individuals have the four efficient mechanisms discussed above for maintaining proper acid-base balances throughout the body. But healthy individuals who impose undue imbalances on themselves or patients with illnesses can experience acid-base upsets that may impair optimal health or cause death. For instance, the lower limit of blood pH at which a person can live more than a few hours is about 7.0, with the upper limit about 8.0.

Acidosis and Alkalosis Acidosis is a condition of excess acidity; alkalosis is one of excess alkalinity. The conditions are classified as resulting from either respiratory or metabolic causes. Respiratory acidosis results from retention of CO_2; respiratory alkalosis results from lowered blood CO_2. Metabolic acidosis is caused by either overproduction or inadequate metabolism of acids; metabolic alkalosis is caused by either an acid loss or a bicarbonate gain. Figure 4–7 illustrates some examples of these four conditions.

Reaction of Foods Foods exhibit two points that need to be considered when they are selected because of their acidity or alkalinity. First, the pH of a food can affect the upper gastrointestinal tract. For example, ingesting foods with a low pH (acid) may elicit an uncomfortable response for people who had oral surgery. Some medications need to be either administered or not given with acid foods, based on their pH. Foods with a low pH include grapefruit juice, vinegar, beer, wine (pH = 3.0); pineapple juice, orange juice (pH = 3.5); tomato juice (pH = 4.2); coffee (pH = 5.0). Milk has a nearly neutral pH (6.6–6.9); eggs are alkaline (pH = 7.6–8.0).

Second, after nutrients are metabolized in the cell, they may yield an excess of mineral elements that are potentially acid or base. These food metabolic residues may influence the pH of the urine. In healthy people, the body is unaffected. (See "Questions Clients Ask.") Most people who experience acid-base imbalances are ill from an abuse of drugs or from an underlying disease. In these situations, health care professionals administer intravenous solutions to correct the imbalance rapidly.

Death	Acidosis pH 6.8 to 7.3	Normal pH 7.35 to 7.45	Alkalosis pH 7.5 to 8.0	Death

Respiratory:	Decreased ventilation Pulmonary edema Pneumonia Asphyxia Obstruction in emphysema Injury to respiratory center Morphine		*Respiratory:* Increased ventilation Hysteria Salicylate poisoning Infections	
Metabolic:	Uncontrolled diabetes mellitus Starvation Severe diarrhea Chronic renal failure		*Metabolic:* Loss of HCl (severe persistent vomiting) Excessive intake of bicarbonate Loss of H ions in renal dysfunction	

FIGURE 4–7
The body maintains normal blood plasma pH within a narrow range of 7.35–7.45. Abnormalities of respiration or metabolism can lead to acidosis or alkalosis.

Food and its contained nutrients promote health, or assist in the body's healing when injured or diseased. Normal life processes depend on the body's ability to maintain a constant internal environment. Countless complex and marvelously coordinated steps occur to enable all this to happen.

? QUESTIONS CLIENTS ASK

1. During hot weather, I am concerned about my father who is elderly and lives alone. What can I do, nutritionally, to maintain his proper fluid balance and prevent dehydration?

Some actions you can take are:

- Ensure a 24-hour intake of at least 1,500 ml of oral fluids. If the weather is especially hot, add another 250–500 ml. The water from food intake and metabolism should be enough to supply the additional water needed for good hydration.
- Suggest he drink fluids, especially water, hourly during the day and evening, and when he gets up during the night. Fluids need to be taken throughout the 24 hours.
- Since tea, coffee, and alcoholic beverages are diuretics, remind him to drink less of these or make up fluid loss with other beverages.
- Suggest he keep a pitcher of water and other preferred beverages in the refrigerator, or in an insulated carafe by his bed and favorite chair.
- Advise he eat foods with high water content, such as berries and melons; cooked pasta and rice; gelatin salads and desserts; puddings and sherbets.

2. I read in a magazine that I should eat only certain food combinations at certain times to prevent my body from getting too acidic or basic. What should I do?

First, it is important to highlight that a healthy individual maintains acid-base balance regardless of the composition of the diet or the time when foods are eaten. But some points to consider are:

- Some foods are said to be **alkali-producing** since they contain more cations (calcium, sodium, potassium, and magnesium) than anions.

Other foods have a different mineral content and contain more anions (sulfur, chloride, and phosphorus) than cations, and so are potentially **acid-producing.** Other foods are so low in mineral elements or contain a similar balance of cations and anions that they are considered to be neutral. Examples are:

Acid producing:	meat, poultry, fish, eggs, cheese, legumes, cereal foods, corn, almonds, chestnuts, coconut, prunes, plums, cranberries
Alkali producing:	fruits, vegetables, milk, peanuts, walnuts, Brazil nuts
Neutral:	butter, margarine, oils, sugar, syrup, starch, tapioca

- Certain fruits such as lemons, grapefruit, oranges, and peaches contain organic acids that give a sour (acid) taste. These organic acids are "weak" and thus do not increase the strong acidity (HCl) of the stomach. In the cell, they are metabolized just as are carbohydrates to yield energy, carbon dioxide, and water. The cations remaining after this oxidation has taken place contribute to an alkaline reaction. Thus, in spite of their acid taste, these fruits actually are alkali-producing.
- The exceptions are three fruits: plums, cranberries, and prunes. They contain two organic acids (benzoic and quinic acids), which are converted in the liver to other acids (hippuric and tetrahydoxy hippuric acids). Thus, these acids increase the acidity of the urine.

Select one day from the three-day food record you maintained for your "Personal Study" in Chapter 1.

Questions

1. Choose one or more foods you ate. Diagram the metabolism of the food from ingestion to excretion.

2. Determine your daily fluid requirements. Explain each part of your calculation.

3. Maintain an "I and O" on yourself. What are your results? Were you in fluid balance that day?

If not, how might you have remedied the imbalance?

4. Choose a classmate, friend, or relative as your "client." Repeat questions 2 and 3. What suggestions might you recommend to your "client"?

5. From the viewpoint of your chosen career, identify your responsibilities in the prevention of electrolyte imbalances.

6. From the viewpoint of your chosen career, identify your responsibilities in the treatment of acid-base imbalances.

FOR ADDITIONAL READING Cirolia, B: "Understanding Edema," *Nurs 96,* **59:**66–70, February 1996.
Mays, D: "Turn ABGs into Child's Play," *RN,* **58:**36–40, January 1995.
Reedy, DF: "How Can You Prevent Dehydration?" *Geriatric Nursing,* 9:224–26, 1988.
Remer, T, and Manz, F: "Potential Renal Acid Load of Foods and Its Influence on Urine pH," *J Am Diet Assoc,* **95:**791–97, 1995.
Stringfield, YN: "Acidosis, Alkalosis, and ABGs," *Am J Nurs,* **93:**43–44, November, 1993.

Carbohydrates

Chapter Outline

*Increase complex carbohydrate and fiber-containing foods in the diet of adults to 5 or more daily servings for vegetables (including legumes) and fruits, and to 6 or more daily servings for grain products.**

 CASE STUDY PREVIEW

Dental cavities is the number one disease in the United States. Yet, because it is not considered life-threatening, the mass media give it little attention. What food and nutrition factors are implicated in the prevention and treatment of this disease?

Healthy People 2000: National Health Promotion and Disease Prevention Objectives. Public Health Service, U.S. Department of Health and Human Services, Washington, DC, 1991, Objective 2.6, p. 93.

o you choose the breads and breakfast cereals you eat because they are high in complex carbohydrates and fiber? Are you convinced that complex carbohydrates are good for you, and that dietary fiber can prevent a number of chronic diseases? Do you find that such choices provide interesting variety to your meals?

Do you eat many foods with sugar even though you have been told that it is bad for you? As a child were you told that eating candy would spoil your appetite for dinner or might rot your teeth? When you want to reward yourself, do you celebrate with carrot sticks or a piece of chocolate cake with chocolate icing? When you notice children misbehaving, do you think it is because they are filling up on sugary foods?

If you are trying to lose weight, are you substituting sodas, candy, and cereals made with "no calorie" sweeteners instead of ones made with sugar? If you are watching your fat intake, what ingredients are taking the place of fat in your favorite foods to retain their flavors?

Of course we are faced with such choices daily, and our selection of foods cannot be classified as entirely good or bad for us. Sugars have their place in the diet, as do starches and dietary fiber.

In the study of foods and nutrition, alcohol (ethanol) often is presented as a separate topic for discussion, usually along with lipids. Since this compound is derived from carbohydrate foods, it is covered in this chapter.

CHARACTERISTICS

Nomenclature

Green plants use energy from the sun, water from the soil, and carbon dioxide from the air to make carbohydrates. This process is known as **photosynthesis.**

Chemical Structure/Composition/Measurement

Figure 5–1 illustrates the basic structure of some common carbohydrates. Note that glucose is the predominant compound. All carbohydrates contain the chemical elements carbon, hydrogen, and oxygen. Hydrogen and oxygen are present in the same proportions as in water. Ethanol is made by fermenting the carbohydrate in foods and, chemically, is an alcohol.

Carbohydrate, as a macronutrient, is measured in grams. Each gram provides 4 kilocalories (kcal). (See Chapter 8.) Because of its chemical structure, alcohol is processed differently by the body. In the cell, ethanol is metabolized somewhat like a fat. By weight, 1 gram yields 7 kcal, and by volume, 1 milliliter yields 5.6 kcal. Appendix Table A–1 includes the energy values of hundreds of carbohydrate- and alcohol-containing foods.

Distribution

Foods vary in the amount and kind of carbohydrates present. In the adult body, the amount of carbohydrate is about 300 g (¾ lb) or less. Most is stored in the liver and muscles as glycogen; some is found in the blood, mainly as glucose.

Classification

Table 5–1 shows a typical classification of the major carbohydrates. Simple carbohydrates (sugars) include the monosaccharides (single sugars) and disaccha-

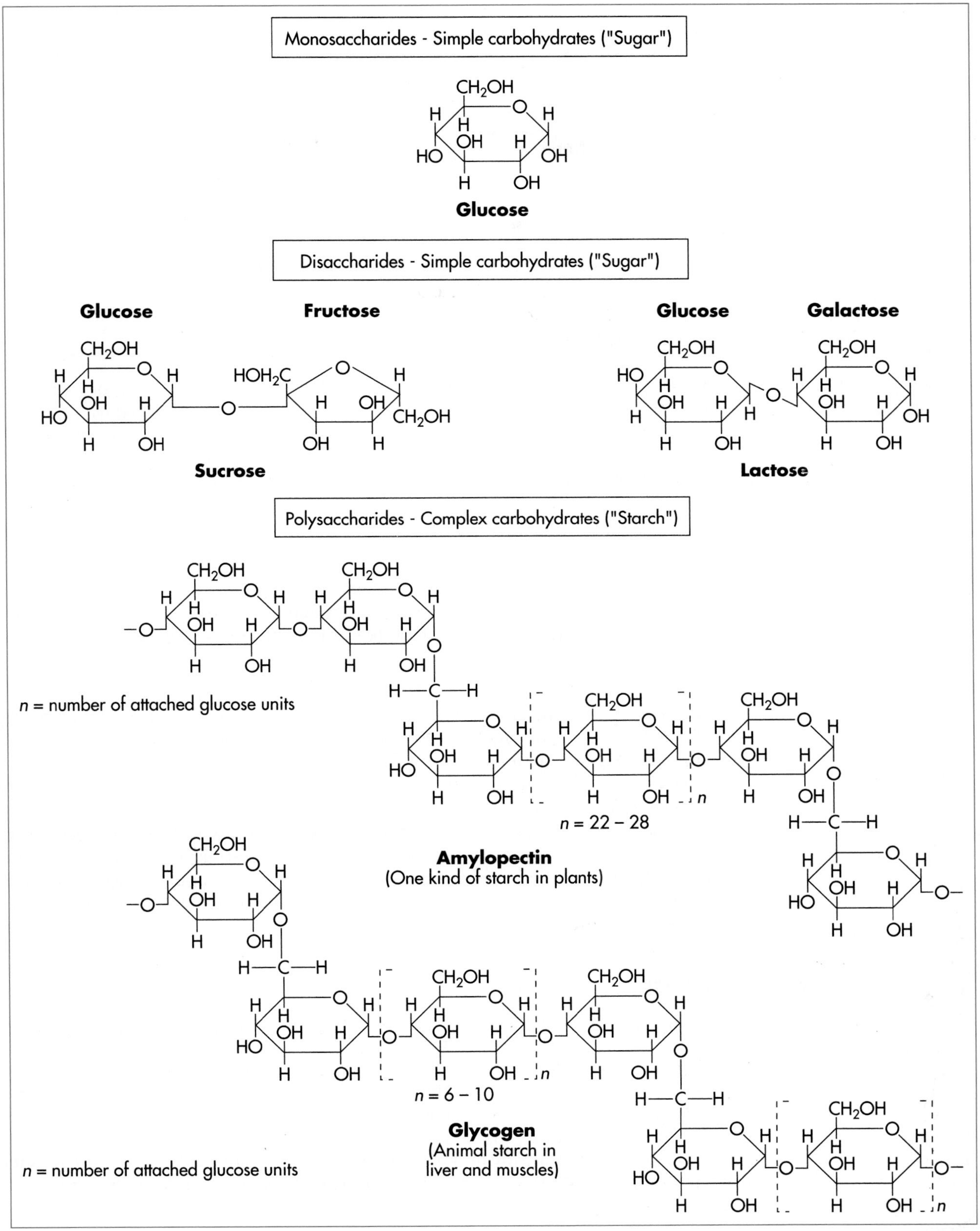

FIGURE 5–1
Structure of some common carbohydrates.

TABLE 5-1

Classification of Carbohydrates

Class	Examples	Some Food Sources
Monosaccharides (single sugars)	Glucose (dextrose, grape sugar, corn sugar, blood sugar)	Fruits, vegetables, corn syrup, honey
	Fructose (levulose, fruit sugar)	Fruits, vegetables, corn syrup, honey
	Galactose	Occurs only from the hydrolysis of lactose
Disaccharides (double sugars)	Sucrose	Cane, beet, maple sugar; small amounts in fruits, vegetables
	Maltose	Malting of cereal grains; acid hydrolysis of starch
	Lactose *only in milk*	Milk only
Polysaccharides (complex carbohydrates)	Starch	Grains and grain foods, legumes, potatoes and other root vegetables, unripe fruits
	Glycogen (animal starch)	Liver and muscle of freshly killed animals; freshly opened oysters
	Dextrin	Partial breakdown of starch by heat or in digestion
	Dietary fibers[a]	

[a] See Table 5-2.

rides (double sugars), whereas complex carbohydrates are the polysaccharides (starches and dietary fibers). Other compounds, like sorbitol and xylitol, are derived from carbohydrates and are found in foods and in the body.

Dietary Fiber **Dietary fibers** are the parts of plants that cannot be digested by the human intestinal tract. These fibers often are called "roughage" or "bulk." Dietary fiber includes (a) fibers that are not soluble in water and that provide the structural parts of plants and (b) fibers that are soluble in water and that are viscous rather than fibrous in nature. (See Table 5-2.) Except for **lignin,** which is the woody part of plants and not a carbohydrate, dietary fibers are complex carbohydrates that vary considerably in their physical nature, structure, and behavior in the gastrointestinal tract. Being indigestible, dietary fiber does not provide any nutrients. However, some fibers such as hemicelluloses and the water-soluble fibers are slightly fermented by bacteria in the intestinal tract and may yield a small amount of calories.

Properties

Carbohydrates may be ranked according to sweetness, from fructose, the sweetest, to starch, which doesn't taste sweet at all. This is important to food flavors. For example, starches are bland in flavor and not sweet. A green banana is high in starch content; as it ripens the starch changes to glucose and the sweetness increases. By contrast, when corn ripens, it becomes less sweet as the sugars are converted to starch.

Starches vary greatly in their solubility in water. Since sucrose is both sweeter and more soluble in water than glucose, for instance, it is used in beverages.

RELATIVE SWEETNESS COMPARED TO SUCROSE

Carbohydrate	Percent
Fructose	150–175
Sucrose	100
Glucose	70
Dextrin	30
Lactose	15
Starch	Not sweet

TABLE 5–2

Dietary Fiber: Food Sources and Action

Classification	Food Sources	Action in Digestive Tract
Water Insoluble: Structural Fibers Cellulose Hemicelluloses Lignin (not a carbohydrate)	Whole wheat and rye, wheat bran, seeds, nuts, vegetables	Requires more chewing Increased sense of fullness Holds water Reduces transit time May reduce pressure in lumen of colon Only hemicellulose is fermented by colon bacteria Binds some minerals such as calcium, magnesium, iron, zinc, copper
Water Soluble: Nonstructural Fibers Pectins Gums Mucilages	Apples, grapes, plums, citrus fruits, other fruits Oatmeal, oat bran, barley Legumes: beans, chickpeas	Holds water Delays gastric emptying May reduce pressure in lumen of colon Partially fermented by colon bacteria Binds bile acids and cholesterol May modify bacterial flora in colon

FUNCTIONS

Functions in Foods

Some functions of carbohydrates in foods are:

- Sugars and starches provide consistency; fiber provides texture.
- Sugars and alcohol give flavor and aroma.
- Carbohydrate foods provide vivid and pastel colors because of the pigments they contain, thus making them eye-appealing.

Functions in the Body

Some functions of carbohydrates in the body are:

- The chief function of carbohydrates is to provide energy to carry on the body's work and heat to maintain the body's temperature. Glucose is the only form of energy used by the central nervous system, but other tissues can use fats for energy. (See Chapters 6 and 8.)
- Carbohydrates spare proteins. Known as the **protein-sparing** effect, this means the body uses carbohydrates to meet its energy needs rather than protein from the diet or body tissues. (See Chapter 7.)
- Carbohydrates furnish chemical structures that combine with nitrogen to manufacture nonessential amino acids.
- Carbohydrates are required for the complete cellular metabolism of fats. When too little carbohydrate is available, some products of fatty acid metabolism, known as ketones, accumulate. (See Chapter 6.)
- Carbohydrates form part of body structural compounds, such as cartilage, bone, and nervous tissue.

■ Lactose, when present in the small intestine, increases the absorption of calcium and phosphorus. In the large intestine, lactose favors the growth of certain bacteria that synthesize some of the B-complex vitamins.

■ Dietary fiber absorbs and holds water. This increases the water content of the feces to produce a softer, bulkier stool.

METABOLIC PROCESSES

Ingestion

The healthy gastrointestinal tract is exceedingly efficient in processing ingested carbohydrate-containing foods. In the mouth, their flavors add taste appeal. Fibrous foods provide the sensory sensation of "crunch" and exercise for the mouth's muscles, bones, and teeth.

Digestion

About 98 percent of the carbohydrate in the typical diversified American diet is completely digested to simple sugars. When diets contain large amounts of fiber, for example, as in some strictly vegetarian (**vegan**) diets, the percentage of carbohydrate digested may be about 85 percent.

The goal of carbohydrate digestion is to reduce the disaccharides and polysaccharides to the single sugars—glucose, fructose, and galactose. This requires enzymes contained in the digestive juices and in the intestinal cell. (See Table 5–3.). Although digestion can occur in each portion of the gastrointestinal tract, starting in the mouth, nearly all occurs in the small intestine.

Alcohol is a molecule so small in size and uncomplicated in chemical structure that it requires no digestion.

Absorption

Likewise, absorption can occur in various sites, but the major site is the small intestine, mostly the jejunum. Both passive and active processes are required. Fructose enters the blood circulation by passive diffusion. Glucose and galactose enter by active transport.

Each sugar is absorbed at a different rate. This fact led researchers to study the **glycemic response** of different carbohydrates in various foods. Some interesting results were found in controlled laboratory settings. But for all practical purposes, little effect could be found in healthy people (and even for people with diabetes mellitus) who eat meals containing a wide array of food combinations. (See Chapter 26.)

TABLE 5–3
Summary of Carbohydrate Digestion

Site of Activity	Enzyme	Carbohydrate	Products of Enzyme Activity
Mouth	Salivary amylase	Cooked starch	Dextrins, maltose
Small intestine	Pancreatic amylase	Starch	Dextrins and maltose
	Intestinal juice		
	Sucrase	Sucrose	Glucose + fructose
	Maltase	Maltose	Glucose + glucose
	Lactase	Lactose	Glucose + galactose

Alcohol can be directly absorbed from the stomach. The remainder is absorbed from the small intestine.

Transportation

The blood circulation transports absorbed glucose, fructose, and galactose to the liver. The body does use galactose for a few direct functions, but nearly all is converted to glucose by the liver. Thus, all absorbed carbohydrates are converted to glucose. It does not matter whether the carbohydrates started as rice and beans, cereal and milk, berries and nuts, mashed potatoes and gravy, soda and candy, or as carbohydrate dietary supplements of beverages and sports bars.

The blood carries glucose to each cell of the body. Homeostasis of blood glucose is maintained by balancing the amount removed by the cells or discharged into it from the liver. The liver does store some glucose as the compound glycogen, which can be easily converted back to glucose. This catabolic process is termed **glycogenolysis.** The liver is very versatile and also can make glucose from protein's amino acids and fat's glycerol; this process is called glyconeogenesis. (See Chapters 6 and 7.) Several hormones help regulate the removal and entrance of glucose into the blood circulation. Insulin aids in lowering blood glucose levels, which is balanced by glucagon, epinephrine, and the steroid, thyroid, and adrenocorticotropic hormones. (See Figure 5-2.)

After fasting, an adult's blood glucose concentration ranges from about 60 to 100 mg/dl (3.3-5.6 mmol/L). Following a meal, the blood glucose rises to about 140 mg/dl (7.7 mmol/L).

Alcohol is transported to the liver.

Respiration

To emphasize, the chief function of glucose is to supply the body's cells with a continual source of energy. To do this, the lungs must supply the cells with oxygen and remove the carbon dioxide and some of the water that are formed when the body's cells produce energy from glucose.

Utilization

Glucose is the primary carbohydrate used by the body. A highly skilled team of many organs, enzymes, hormones, and other nutrients function to enable the body's cells to use glucose. For example, the major organs are the liver, the pancreas, and the adrenal, pituitary, and thyroid glands. Trace amounts of thiamin, niacin, riboflavin, iron, magnesium, phosphorus, chromium and other vitamins and minerals are necessary at all the numerous transformation points along the way.

A problem with any member of the skilled body team affects the body's utilization of glucose. For instance, the pancreas produces insulin, the hormone known to help glucose enter the cell. (See Chapter 26.) Thiamin, niacin, and riboflavin are vitamins required by enzymes that oxidize glucose to energy. (See Chapter 10.) Phosphorus, a mineral, is part of the ATP (adenosine triphosphate) compound. (See Chapter 11.)

Alcohol depends on the liver and two sets of specific enzymes to process it. Since it is so quickly and easily absorbed, alcohol can quickly overpower the numbers of enzymes available to catabolize it. The liver metabolizes alcohol at the rate of about ½ ounce per hour. This is equivalent to about the amount of alcohol contained in one drink per hour. Alcohol is toxic to the liver, thereby interfering

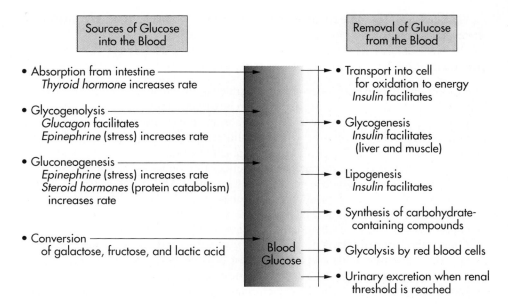

Sources of Glucose into the Blood		Removal of Glucose from the Blood

- Absorption from intestine
  Thyroid hormone increases rate

- Glycogenolysis
  Glucagon facilitates
  Epinephrine (stress) increases rate

- Gluconeogenesis
  Epinephrine (stress) increases rate
  Steroid hormones (protein catabolism) increases rate

- Conversion
  of galactose, fructose, and lactic acid

Blood Glucose

- Transport into cell
  for oxidation to energy
  Insulin facilitates

- Glycogenesis
  Insulin facilitates
  (liver and muscle)

- Lipogenesis
  Insulin facilitates

- Synthesis of carbohydrate-containing compounds

- Glycolysis by red blood cells

- Urinary excretion when renal threshold is reached

FIGURE 5–2
Normally after a meal the blood sugar rises to about 140 mg/dl. After a night's fast the blood sugar level is about 60 to 100 mg/dl. The blood sugar level increases by the release of glucose from glycogen, the synthesis of glucose from glycerol and from some amino acids, and of course through absorption of sugars from the intestinal tract. The blood glucose level fails as the liver removes some of it for the synthesis of glycogen, the conversion of glucose to fats, and the synthesis of other compounds.

with the liver's other crucial work. When alcohol reaches the body's other cells, like those of the brain and nervous system, it acts as a narcotic. (See Chapters 18 and 23.) The body can use up alcohol only in a very slow, methodical way. Thus, nothing anyone can do will overcome a hangover—even walking around, drinking other fluids like fruit juices, or ingesting stimulants like black coffee or caffeine tablets.

Storage

Once glucose enters cells, it is either immediately used for some specific function or it is stored. Since the body has a constant need for energy, glucose is stored only as a few compounds in a few compartments. As quick reserves, each cell stores energy as ATP right in the cell. As intermediate reserves, glucose is converted into glycogen and stored in the liver and muscles. The synthesis of glycogen is an anabolic metabolic process termed **glycogenesis.** For long-term remote storage, the liver synthesizes fats from glucose, termed **lipogenesis,** and sends it mainly to adipose tissues under the skin and around body organs. Adipose tissues can expand as much as necessary to accept stored energy and can shrink to release it.

Alcohol also can provide energy, but not very effectively or efficiently. It is, however, easily and quickly converted to fat, which is then stored in the liver and adipose tissues.

Excretion

When the body normally uses glucose for energy, the only products needing to be eliminated from the body are carbon dioxide and water. Carbon dioxide is exhaled through the lungs and water is removed through the lungs, skin, and kidneys.

The **renal threshold** is the point at which a substance in the blood is excreted in the urine. In the case of glucose, some might be eliminated when the level rises above 160 mg/dl (8.8 mmol/L). This occurs in certain diseases, such as untreated diabetes mellitus.

Excess alcohol is excreted by the kidneys and lungs. The alcohol levels in the urine and breath are tests for drunkenness.

Since dietary fiber is undigested and therefore unabsorbed, it is directly excreted as part of the feces.

MEETING BODY NEEDS

Requirements and Recommendations

The exact carbohydrate needs of the body have not been established; thus, there is no RDA for carbohydrates. Some people in the world (in Asia, for example) maintain good health with very high intakes of carbohydrates, whereas other (many Eskimos) are equally healthy on diets very low in carbohydrates.

Generally, the daily diet for adults should contain a minimum of 50–100 g of digestible carbohydrates. This amount is sufficient to supply the glucose needed by the central nervous system and for the effective oxidation of fats. In the United States, because of the diseases for which the population is at risk, nutrition researchers and clinicians suggest a preferred intake of about 60 percent of recommended calorie levels, with about ⅚ coming from foods containing complex carbohydrates and natural sugars. Sixty percent is 300 g of carbohydrate of a 2,000 kcal intake diet. The daily intake of dietary fiber is advised to be slightly more than 12 g per 1000 kcal, which is about 25 g for a 2,000 kcal intake.

The body has no need, and thus there is no requirement, for alcohol.

Sources

Of course, the body relies on foods to supply it with carbohydrates. All plants contain carbohydrates. Animal foods, except for milk, do not contain carbohydrates.

However, the body does not depend solely on recently ingested carbohydrates for glucose. As mentioned, the liver can convert some amino acids of proteins and the glycerol part of fats into glucose. These amino acids and glycerol can come from the foods just eaten or can already be present in the body.

Complex Carbohydrates: Starch All grain foods are excellent sources of complex carbohydrates and many contribute protein, vitamins, and minerals. Grain foods include breads and rolls of all kinds: whole grain, white, enriched, rye, pita, pumpernickel, bagels, English muffins; quick breads: muffins, biscuits, pancakes, waffles; breakfast cereals, cooked and ready-to-eat; grits; crackers: graham, rye crisp, saltines; snacks such as pretzels; pastas: macaroni, spaghetti, noodles; and rice. (See Food Guide Pyramid, Figure 1–5.)

Legumes are high in complex carbohydrates. Besides soybeans, other legumes are dried beans such as kidney, navy, and pinto beans, and dried peas such as black-eyed peas, split peas, chickpeas, and lentils.

Some vegetables such as white and sweet potatoes, corn, taro, plantain, cassava, and breadfruit are also important sources of starch.

Although liver and muscle of freshly butchered animals and just-opened oysters contain some glycogen, the glycogen rapidly catabolizes and does not contribute to dietary intake.

Complex Carbohydrates: Dietary Fiber Dietary fibers, both water insoluble and soluble, are present naturally in whole, unprocessed grains, vegetables, fruits, and legumes. For example, while all bread contains starch, only whole-grain breads like whole-wheat or whole-grain cereals such as shredded wheat contain water-insoluble dietary fiber. Unpeeled vegetables are important sources of cellulose. Whole fruits (not juices) supply pectins and gums. Legumes are good sources of soluble fibers. Food sources are preferred over fiber supplements because foods supply a variety of different dietary fibers, along with vitamins and minerals, whereas most supplements do not.

Naturally Occurring Sugars All fruits and vegetables contain simple sugars, such as glucose, fructose, and sucrose, in varying amounts. Milk is the only food of animal origin to provide carbohydrate (lactose) in the diet.

Added Sweeteners Added sweeteners are simple carbohydrates added to foods. These sugars include cane sugar, corn sugar, corn syrup, and high fructose corn syrup. Soft drinks, table sugar and sugary foods, and bakery products account for more than half of the added sugars in the diet. Americans consume, per person, more than 400 12-oz cans of sweetened soft drinks each year. Sugary foods include table syrup, jams, jellies, gelatin desserts, Popsicles, presweetened cereals, and table sugar. Examples of bakery foods are cakes, cookies, pastries, donuts, pies, and other sweet desserts that so many people like. In addition, sugars are added intentionally to numerous foods. (See Chapter 2.) The list is extensive. Examples of such foods range from breads and cereals, to fruit ades, drinks, and punches, to salad dressings, catsup, frozen meals, and fruit yogurt. Even many vitamin tablets and medications contain added sugars as binders and flavors!

Alternative Sweeteners Because so many people are interested in lowering their calorie level or in modifying their sugar intake for medical reasons, commercially produced substitutes for sugars have become increasingly popular. Such additions to the daily diet may be termed nutritive sweeteners if they add calories or other nutrients, or nonnutritive sweeteners if they do not. Food and pharmaceutical companies constantly research new compounds. Some widely used products are aspartame, saccharin, cyclamate, and acesulfame K.

- Aspartame, known commercially as Equal® or NutraSweet®, and by other names, is an example of a nutritive alternative sweetener. It is a compound containing a dipeptide of two amino acids (aspartic acid and phenylalanine) and a methyl group, methanol. Before the FDA approved aspartame for general use, the question of its safety was extensively examined. Because it contains phenylalanine, the one group of people who need to be cautious are those who have the inherited condition **phenylketonuria** (PKU). (See Chapter 27.) Thus, a statement on the label of foods is required. Since the dipeptide is unstable when heated, aspartame is not used in cooked or baked foods.

- Saccharin and cyclamate are examples of nonnutritive alternative sweeteners. Cyclamate is banned in the United States, but may be used in Canada.

Because both are heat stable, they are used in many foods. Saccharin became available about 1900, so it has been used extensively by people for almost a century. One objection to its use is that it leaves a bitter aftertaste. Although high intakes of saccharin produced bladder cancer in some animals, there is very little evidence of a risk for humans even at relatively high intakes.

■ Acesulfame K (trade name Sunette®) is another example of a nonnutritive artificial sweetener. It was approved for limited uses by the FDA in 1988. It has wide application in a variety of foods because it is stable in heat.

Alcohol Nearly all cultures consume fermented foods in one form or another. Sources of alcohol are fermented grains, fruits, and vegetables. Beer is made from fermented grain, usually barley; wine is made from fermented fruit; vodka is made mainly from potatoes or corn; and soy sauce is made by fermenting soybeans.

Basic Diet

The Basic Diet Nutrient Profile (Table 2-3) identifies both total carbohydrate and dietary fiber. Note the carbohydrate content of plant foods. You can see the importance of grains, vegetables, and fruits as the foundation groups of the Food Guide Pyramid (Figure 1-5).

BODY BALANCE

Optimum/Adequate Signs and Symptoms

In assessing nutritional status regarding carbohydrates, two evaluations should have priority. (See Chapter 3.) First, assess the effects on the gastrointestinal tract, namely, the condition of the mouth and the pattern of bowel movements. Nutritionally, a healthy oral cavity may reflect a limited intake of sugars. Adequate bowel movements may reflect a sufficient intake of dietary fiber. Because fibers absorb water, monitor water intake. Counsel clients that additional water should be taken when fiber is eaten. Second, assess urine and blood glucose levels. Because the healthy body maintains homeostasis, normal urine and blood glucose levels vary only slightly.

Insufficiency/Deficiency

Conditions Since carbohydrates are found in all plant foods and since most people eat these, carbohydrate deficiency is rare. However, the typical American diet contains only about 46 percent of calories from carbohydrates instead of about 60 percent. Moreover, typical intake of dietary fibers is only about 8-12 g, less than half of the suggested 25 grams. Thus, the carbohydrate intake of many healthy people is insufficient, but not deficient.

Signs and Symptoms Obvious signs and symptoms of an insufficient intake of total carbohydrate are limited. When the body is deficient in carbohydrate, blood glucose levels fall. This condition is termed **hypoglycemia,** or low blood sugar. When this occurs, excessive lipid catabolism occurs, resulting in an elevation of ketones in the blood and urine. Regarding fiber, people who consume insufficient quantity or quality can exhibit acute and chronic disorders. For example, alterations in bowel function such as constipation, diverticulosis, and hemorrhoids are common. Also in the intestinal tract, pectins and gums are involved in triglyceride and cholesterol excretion in the feces. Soluble fibers have been shown to lower blood levels of these compounds. (See Chapter 6.) Although a decreased fiber intake has been implicated in colon and rectal cancer risk, the more likely association is the high fat intake that usually accompanies low carbohydrate intake. Theories relating the benefits of fiber intake to health are summarized in Table 5–4.

The timing of eating carbohydrates is important. Severely restricting carbohydrate intake or eating them sporadically may impair proper utilization of fat and protein. For example, if not enough carbohydrate is available, the body will metabolize fat and protein to meet its crucial need for energy.

People Most at Risk At risk for an insufficient intake of total carbohydrates are people who follow restricted diets, such as people who are starving, on weight loss diets, or who follow erratic eating patterns. Most at risk for insufficient dietary fiber intake are people whose diet is highly refined. As you recall the examples of fiber-containing foods, you remember that even though all grains contain the complex carbohydrate starch, only whole grains contain fiber. Vegetables and fruits with the most fiber are ones that are eaten with their skin and seeds. Juices of fruits and vegetables, while popular with people of all ages, do not contain fiber.

Excess/Toxicity

Conditions Perhaps the greatest single problem associated with a high added sugar intake is that sugar contributes only simple carbohydrates and

TABLE 5–4
Possible Effects of Fiber in Reducing Disease Risks

Problem	Hypotheses Regarding Benefits
Constipation	Water held by fiber increases stool bulk and produces softer stools; improves regularity
Hemorrhoids	Larger, softer stools reduce straining
Diverticulosis	Larger, softer stools reduce pressure within colon; less out-pouching to form diverticuli
Obesity	More chewing and increased dietary bulk provide greater sense of fullness Slower rate of carbohydrate absorption reduces insulin response, thus reducing appetite
Diabetes mellitus	Slower gastric emptying slows the rate of carbohydrate absorption; improves regulation of blood sugar; lowers insulin requirement
Atherosclerosis	Pectins and gums bind cholesterol and bile acids in the intestinal tract to increase cholesterol excretion; lowers the blood cholesterol and triglyceride levels
Colon/rectal cancer	Large, soft stools may dilute potential carcinogens More rapid transit may reduce contact of carcinogens with colon mucosa May modify bacterial flora; possibly reduce breaking down bile acids and other substances that could become carcinogens

calories. Thus, the popular sodas, iced tea drinks, flavored juice beverages, candies, jellies, honey, and table sugar contribute little nourishment for the body. Also, many highly sugared foods such as pastries, donuts, pies, puddings, and ice cream contain large amounts of fat, a nutrient with twice as many calories as sugar. (See Chapter 6.) Several major nutritional concerns affecting healthy people are discussed in the next paragraphs.

Signs and Symptoms Dental caries is the most prevalent disease in the United States. Dentists, dental hygienists, and dental assistants usually warn their patients about the effects on the teeth of fermentable carbohydrates, especially the simple sugars. (See Figure 5-3.) Bacteria in the mouth digest sugars, leaving a transparent, sticky substance called dental plaque on the surface of the teeth. The bacteria thrive in the plaque and produce acids from the sugars, leading to tooth decay. Regarding sugars, three factors are especially important:

- Frequency of exposures to simple sugars. The most important factor is the number of times the teeth are exposed to sugars. Frequent snacking on sweets provides many opportunities for bacterial growth and increases the risk of tooth decay.

- Retentiveness of the sugars on the teeth. Chewy sweets like fudge cookies stick to the teeth for long periods so that bacteria have ample time to grow and produce acids that cause decay. On the other hand, a sweetened

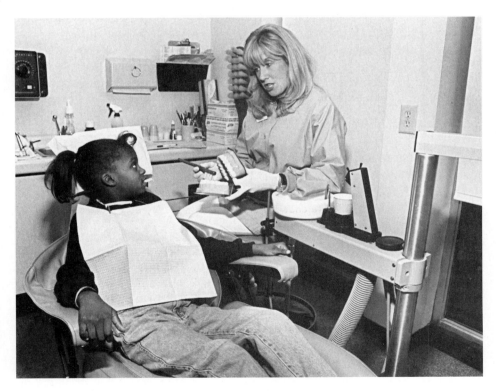

FIGURE 5–3
Demonstrating the connection between food intake and oral hygiene practices helps promote a lifelong healthy mouth and attractive appearance. (*Source:* Anthony Magnacca/Merrill)

beverage is quickly swallowed and so leaves little sugar in contact with the teeth. Also, foods with high water content are less damaging to the teeth because the sugars are diluted by the water. For example, fresh grapes are less damaging to the teeth than are raisins.

■ Quantity of sugars in the food or beverage. Of these three factors, the least important factor in the initiation of dental decay is the total amount of sugars contained in a particular food or beverage.

Overweight and obesity, from excessive food consumption, is a growing national health problem. Sugar in coffee or other drinks sweetened with sugar, honey on waffles, candy bars, jam on bread, a piece of cake—all of these contribute calories, but few other essential nutrients. It is very easy to eat excess amounts of these sugar-rich foods so that the caloric intake is above the requirement. Overweight and obesity can result. For example, if you were to drink one 12-oz can of soda each day beyond your caloric requirement, the daily caloric excess would be about 145 kcal, with a potential weight gain of almost 15 pounds in one year. (See Chapter 9.)

Lactose intolerance is common in some ethnic groups who produce insufficient lactase, the intestinal enzyme that splits lactose. When such individuals consume what for them are excessive amounts of lactose (usually in the form of milk), they experience flatulence, abdominal cramping, and diarrhea. Many modified foods and digestive aids are available to combat this problem. (See Chapter 22.)

Excessive fiber intake can be a problem, especially in people who attempt to correct past dietary deficiencies by going overboard now. Cramping, diarrhea, and excessive intestinal gas are common complaints when the fiber intake is increased. An excessively high intake of fiber may lead to intestinal obstruction. These undesirable effects are minimized or avoided if the fiber content of the diet is increased gradually over a period of six to eight weeks. For example, starting to eat more fresh, whole fruits and vegetables and adding whole-grain breads and cereals over the course of weeks might be logical advice. A liberal intake of water must be included if fiber is to provide maximum benefits. The absorption of several minerals is reduced somewhat when the fiber intake is high. However, studies of vegetarians, who customarily have a higher fiber intake than other Americans, show that signs of mineral deficiencies are rare. Only if diets fail to meet the recommended levels of nutrient intake is the lower absorption of minerals of any importance.

Alcohol intake, if moderate and occasional, poses little risk for most persons. But people who abuse alcohol derive most of their calories from it. Since alcohol does not supply other necessary nutrients and there is the possibility that insufficient food will be eaten to make up for the lack, this dietary lifestyle usually leads to nutritional deficiencies. (See Chapters 18 and 23.) Another example of alcohol's danger is fetal alcohol syndrome (FAS), which can occur in babies born to women who drink alcohol during pregnancy. The babies show abnormal facial development, stunted growth, and retarded mental development. (See Chapter 12.)

People Most at Risk Consumption of excessive quantities or kinds of carbohydrates generally is found among people who are paying either too little or too much attention to their diets. In both cases, people just do not have all the facts about their food choices. For example, children may not realize the harm of fermentable sugars eaten frequently on a daily basis, and people who are attempting to cut down on another component of the diet, like fat, may be fooled into believing that carbohydrates can be eaten with freedom from liability.

HEALTH AND NUTRITION CARE ISSUES

After water, carbohydrate might be the second nutrient to assess. Since carbohydrate is the nutrient most important for providing energy for body needs, a sufficient quantity delivered throughout each day is necessary. Health care providers can teach people about wholesome sources of carbohydrates. In the institutional setting, they can provide appropriate food choices on menus, in vending machines, and in dining rooms.

It is often claimed that eating a diet either very high or low in carbohydrates, or adjusted in the kind of carbohydrates, is the cause of disorders ranging from fatigue to allergies, diabetes mellitus, depression, abnormal child behavior, antisocial behavior, alcoholism, and drug abuse. In other instances, carbohydrate is touted to be the basis for the cure of cancer or heart disease. Although controversial, modified carbohydrate intake is not the cause or cure for any disorder. Rather, the underlying factors include hereditary, infectious, viral, immunological, and/or environmental issues. Nevertheless, when carbohydrate food choices are restricted unnecessarily, any condition can be made worse. On the other hand, people are healthier when their food contains carbohydrates and other crucial nutrients. Again, the body strives for homeostasis.

PUTTING IT ALL TOGETHER

Overall, because of the body's many different requirements for them, carbohydrates form a basis of a healthful diet. Selecting and eating a wide variety of carbohydrate-containing, minimally refined foods is the foundation of a sensible daily diet.

? QUESTIONS CLIENTS ASK

1. What is sorbitol? Can I use it in place of sugar?
Sorbitol is a sugar alcohol. It is absorbed from the intestinal tract more slowly than are sugars, and therefore it will cause less rise in the blood sugar. Each gram of sorbitol, like each gram of sugar, furnishes 4 kcal. Excessive amounts of sorbitol can cause bloating, gas, and diarrhea.

2. Is it a good idea to substitute honey, molasses, brown sugar, raw sugar, or maple sugar for cane sugar?
Good rules to remember: Choose your sugars for the flavors you enjoy, and consider their calories and costs, if that is important to you. Each gram of each of these sugars contributes 4 kcal—about 16 kcal for one teaspoon. Dark molasses is a relatively good source of iron, but most people do not use enough to make much difference. The other sugars contain such tiny traces of minerals and vitamins that the amounts are not important in the diet, but the calories these sugars contain are.

3. What is the high-fructose corn syrup that I see listed on food labels?
This is a syrup made from the chemical breakdown of cornstarch. It has a high concentration of fructose and supplies 4 kcal/g. It is widely used in soft drinks because it is cheaper than sucrose.

Robert, nine years old, was seen by the school dental hygienist. He has many carious teeth. The dental hygienist asked him what he had eaten the previous day. Robert reported:

Breakfast:	Orange drink, scrambled egg, whole wheat toast with butter and honey, and whole milk.
Before school:	Gum and jelly beans. (His mother gave him money to buy milk for lunch, but on his way to school he bought these foods.)
Lunch:	Bologna sandwich with catsup and lettuce on white bread, banana, cupcake, and water from drinking fountain.
After school:	Chocolate chip cookies and milk.
Dinner:	Hamburger, French fries, peas, cherry pie, and soda.
Watching TV:	Peanut butter and jelly sandwich, chips, and bottled iced tea drink.
Bedtime:	Brushed teeth.

Questions

1. List the foods in Robert's diet that supply carbohydrates. Subdivide the list of foods according to simple carbohydrates and complex carbohydrates.

2. What foods in Robert's diet are good sources of dietary fiber? Why are these foods important for Robert?

3. What foods in Robert's diet might especially contribute to dental decay? Why?

4. Which of these is likely to be more harmful to teeth:
   a. Eating six caramel candies at one time or eating one caramel at six different times during the day? Why?
   b. Drinking a soda or eating a handful of raisins? Why?

5. Evaluate the timing of Robert's eating schedule.

6. What other factors are important for healthy teeth?

7. What are five changes Robert might make to reduce tooth decay? (Remember that Robert is a nine-year-old boy.)

8. Prepare an improved meal pattern, menu structure, and food selections with Robert for optimal dental health.

9. Compare the ingredient list and nutrition label of three sugar-coated cereals with three unsweetened cereals. Identify the various kinds of carbohydrates each contains. What is the size of one serving? How many calories does one serving of each provide? What percentage of calories is carbohydrate?

FOR ADDITIONAL READING American Dietetic Association: "Position Paper: Oral Health and Nutrition," *J Am Diet Assoc,* **96:**184–98, 1996.

"Complex Carbohydrates: The Science and the Label," *Nutr Rev,* **53:**186–93, 1995.

Clydesdale, FM: "Workshop on the Evaluation of the Nutritional and Health Aspects of Sugars," *Am J Clin Nutr (Suppl 1),* **62:**161S–296S, July 1995.

Lipids

Chapter Outline

*Reduce dietary fat intake to an average of 30 percent of calories or less and average saturated fat intake to less than 10 percent of calories among people aged 2 and older.**

✔ CASE STUDY PREVIEW

"Fat of the land," "fat cat," "fat wallet," "delicious rich foods." The words *fat* and *rich* often carry positive cultural meanings, often as idioms for wealth and health. But what images do the words convey when used in different expressions, like "fat person" or "blood rich in fat"?

---

* *Healthy People 2000: National Health Promotion and Disease Prevention Objectives.* Public Health Service, U.S. Department of Health and Human Services, Washington, DC, 1991, Objective 2.5, p. 93.

Look at the great old paintings and read novels written long ago. What is so different from similar ones today? In the past, fat denoted health, well-being, education, prosperity, prestige. In some cultures, fat still is desirable in foods and in a person's appearance. But fat now often means just the opposite. Fat is associated with overweight and obesity, and recognized as a major risk factor for several chronic conditions like heart disease and cancer. In some cultures, it means being less than desirable, having lack of will power, having limited education, being poor. Annually, thousands of hours are spent by scientists and health professionals and billions of dollars are spent by food manufacturers and media experts in the attempt to do something about fat!

CHARACTERISTICS

Nomenclature

Lipids are a class of nutrients, commonly known as fats and oils. Actually, lipids include fats and oils, but many more compounds, too. Lipids are organic compounds.

Chemical Structure/Composition/Measurement

Figure 6-1 illustrates the basic structure of lipids. Like carbohydrates, lipids are composed of three chemical elements: carbon, hydrogen, and oxygen. Lipids, however, contain much smaller proportions of oxygen than do carbohydrates. Some lipids also contain other constituents, especially phosphorus, nitrogen, carbohydrates, and proteins.

The major constituent of most lipids is **fatty acids.** A fatty acid consists of a chain of carbon and hydrogen atoms with a carboxyl (COOH) group at one end

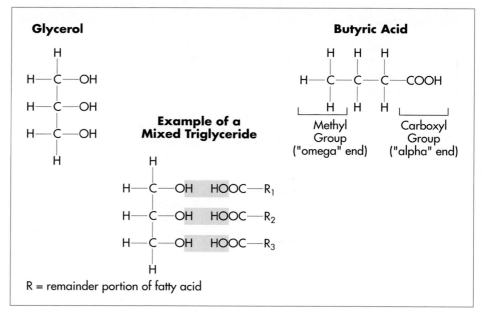

FIGURE 6–1
Basic structure of lipids. R_1, R_2, and R_3 denote different fatty acids.

(alpha) and a methyl (CH$_3$) group at the other end (omega). (See Figure 6–1.) About 20 fatty acids commonly are found in foods and body tissues.

Fatty acids usually contain an even number of carbons, but vary in the number of carbon atoms present: short chain (4–6 carbons), medium chain (8–12 carbons), and long chain (more than 12 carbons). Most of the fatty acids in foods and in the body are long chain.

Fatty acids also vary in the kind of carbon-to-carbon bonds, indicating if they are "saturated" or "unsaturated." A fatty acid is saturated if there is a single bond between each carbon atom. Like a wet sponge that cannot hold any more water, so the carbon cannot take up any more hydrogen. A fatty acid is unsaturated if there is a double bond between two carbon atoms. Such a fatty acid can take up more hydrogen. Those with one double bond are classified as monounsaturated (MUFA); those with two or more double bonds are called polyunsaturated (PUFA). (See Figure 6–2.)

Unsaturated fatty acids can exist as different **isomers** (shapes) at the location of the double bond(s). (See Figure 6–2.) This structure and composition configuration is important. In nature, most unsaturated bonds form a curved geometric pattern ("cis" isomers). During commercial food processing known as **hydrogenation,** hydrogen is added to make liquid oils into more solid fats. Unintentionally, as a by-product of the process, many double bonds become straightened, resulting in a linear pattern ("trans" isomers). Thus, although still unsaturated, these trans-fatty acids behave like saturated fatty acids. (See succeeding sections, also.)

A number of naming systems exist for the unsaturated fatty acids, based on the location of the double bond(s). One of them, the **omega system,** is based on the site of the first double bond, counting from the omega (methyl, CH$_3$) end. The unsaturated fatty acids thus are placed in three groups, designated as omega-3, omega-6, or omega-9. (See Figure 6–3.)

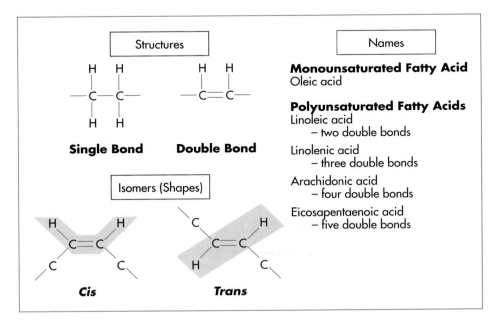

FIGURE 6–2
Structures, names, and shapes of unsaturated fatty acids.

$$CH_3(CH_2)_{16}COOH$$
Stearic acid (saturated) 18:0

$$CH_3(CH_2)_7 \quad CH = CH \quad (CH_2)_7COOH$$
Oleic acid (monounsaturated) 18:1 (omega-9)

$$CH_3(CH_2)_4 \quad CH = CH \quad CH_2 \quad CH = CH \quad (CH_2)_7COOH$$
Linoleic acid (2 double bonds: polyunsaturated) 18:2 (omega-6)

$$CH_3 \quad CH_2 \quad CH = CH \quad CH_2 \quad CH = CH \quad CH_2 \quad CH = CH \quad (CH_2)_7COOH$$
Linoleic acid (3 double bonds: polyunsaturated) 18:3 (omega-3)

FIGURE 6–3

All of these fatty acids contain 18-carbon atoms. They differ in their properties, depending on how unsaturated they are. Stearic acid with no double bonds is important in many animal fats. Oleic acid, with one double bond, occurs abundantly in all fats; it is especially high in olive oil. Fatty acids with two or more double bonds are found in most vegetable oils.

Lipids, as a macronutrient, are measured in grams. Each gram provides 9 kcal. (See Chapter 8.) Thus, lipids provide more than twice the number of calories as do carbohydrates and proteins. (See Chapter 7.) Appendix Table A–1 includes the energy values of numerous lipid-containing foods.

Distribution

In both foods and the body, lipids are widely distributed. The total amount and the kind vary with each food or part of the body. For example, in healthy nonobese men, body fat accounts for 15–20 percent of body weight, and in healthy nonobese women, 18–25 percent of body weight. The proportion of body fat is higher in sedentary and obese individuals, and it increases with age.

Classification

Although we eat lipids, all components of the lipids, with one exception, can be (and are) manufactured by the body. That one exception is three long-chain unsaturated fatty acids. They are linoleic acid, linolenic acid, and arachidonic acid. These three are considered dietary essential fatty acids, that is, the body is unable to synthesize them at all or in such inadequate amounts that they must be provided by the diet.

The essential classification of these fatty acids changed periodically, indicating fluctuations in scientists' state of knowledge. Linoleic acid (an omega-6 fatty acid) always has been considered to be essential. Traditionally, linolenic acid (an omega-3 fatty acid) was considered to be an essential fatty acid, too. Later, it was thought that linolenic acid could be synthesized in sufficient quantities as long as linoleic acid was supplied. However, as more roles of linolenic acid became better identified and deficiencies were recognized even when linoleic acid was part of the diet, linolenic acid again was reclassified as essential, that is, to be supplied by foods. Arachidonic acid (an omega-6 fatty acid) remains classified as essential only for infants, since it appears that sufficient amounts can be synthesized from linoleic acid by adults.

Lipids are divided into three main categories, simple lipids, compounds lipids, and fatlike compounds. All three are present in foods and in the body.

Simple Lipids Simple lipids are what we think of as fats and oils, or just "fat." Fats and oils have different compositions and properties, some of which overlap. Visually, fats are considered to be solid at room temperature and oils to be liquid. Thus, food examples are beef fat and corn oil. Simple lipids stored in the body's adipose tissue sometimes are called "body fat." Importantly, the simple lipids are the class of lipids that yield energy, whether from foods or the body stores.

Simple lipids ("fat") contain only two compounds: fatty acids attached to glycerol. Glycerol is a 3-carbon alcohol with a structure similar to glucose. Each of the hydroxyl (–OH) groups can combine with a fatty acid. (See Figure 6–1.)

Linked to the glycerol molecule are one, two, or three fatty acids (the maximum number). **Monoglycerides** consist of a single fatty acid linked to the glycerol molecule; **diglycerides** consist of two linked fatty acids; and **triglycerides** consist of three linked fatty acids. The fatty acids may be the same one or any combination of different ones. About 98 percent of food fats and 90 percent of body tissues are mixed triglycerides. This means that at least two of the fatty acids attached to the glycerol molecule are different.

Scientists use the more precise term *acyl,* such as triacylglycerols or diacyl-glycerols. But the older terms persist in everyday language, on food labels, and in clinical laboratory reports.

Compound Lipids Compound lipids are lipids in which sometimes another alcohol replaces glycerol, and another chemical group replaces at least one of the fatty acids. Chemical group substitutions include phosphate, nitrogen, carbohydrate, or protein.

One example of a **phospholipid** (a phosphate and nitrogen group is substituted for one fatty acid) is lecithin, which the liver rapidly synthesizes to help transport fats throughout the body. Another example is cephalin, which is required for the formation of brain and nervous tissue.

An example of a **glycolipid** (lipid and carbohydrate, like galactose) are the sphingolipids, which are found throughout the brain and nervous system. **Lipoproteins** (lipid and protein) are the form in which lipids are transported in the blood.

Fatlike Compounds (Derived Lipids) Fatlike compounds contain carbons in the shape of ring structures. Most of these compounds originate from the simple and compound lipids. Examples are the sterols, such as cholesterol, bile acids, and sex hormones; and the eicosanoids, like prostaglandins, leukotrienes, and thromboxanes. The fat-soluble vitamins A, D, E, and K also fall into this class.

Properties

As a class, one major unique property of lipids is that they are insoluble in water. But both foods and body fluids are mainly water. Special considerations are made for the lipids so they can be distributed into watery environments. For example, in foods one liquid can become distributed throughout another liquid by a process known as emulsification. Mayonnaise is made from oil and vinegar; egg yolk, which contains lecithin, acts as an emulsifier to keep the two liquids interspersed together. In the blood circulation, lipids are held in solution by being attached to proteins (lipoproteins).

Another important property difference between lipids and the other macro-nutrients relates to their unsaturated nature or their ability to combine with other substances. Hydrogenation, oxidation, and saponification are three such processes with application to foods and in the body.

Hydrogenation Hydrogenation is the addition of hydrogen atoms to double-bond carbons. This process converts unsaturated lipids, such as fatty acids, into saturated ones. Hydrogenation is used to manufacture "spreadable" foods such as soft and solid margarines, shortenings made from oils, and peanut butters in which oil does not rise to the top of the jar. However, as mentioned previously, this processing also changes the isomers from the *cis* shape to the *trans* shape.

Oxidation Oxidation is the process of adding oxygen. Unsaturated carbon bonds are unstable, and other substances can react at the site. One is oxygen. In foods, the outcome is known as **rancidity.** The result is an unpleasant flavor and odor. You may have noticed these unpleasant characteristics in many foods that have been exposed to air, such as nuts, vegetable oils, butter, whole grains, and coffee. Certain chemicals called antioxidants, such as vitamins C and E, as well as butylated hydroxyanisole (BHA) and butylated hydroxytoluene (BHT), are added to foods to protect against oxidation. Check the labels of processed foods for these ingredients. Many oils naturally have a high concentration of vitamin E and thus are protected against oxidation. In a related fashion, the high and prolonged heating of fats, such as during commercial frying of foods, leads to the breakdown of the glycerol molecule. The substance formed, **acrolein,** can be irritating to the stomach and small intestine and, when vaporized, to the eyes, nose, and lungs. In the body, oxidative processes can affect unsaturated lipids, including unsaturated fatty acids and cholesterol. The body has a number of metabolic safeguards against oxidation, including various enzyme systems and vitamins A, C, and E. (See Chapter 10.)

Saponification Saponification is the combination of a fatty acid with a cation, such as calcium, to form an insoluble compound. From its first four letters, you may think the word looks like the word "soap." You are right. When fats combine with minerals, they form soap. In the small intestine, free fatty acids can combine with calcium and other minerals. These biological "soaps" are insoluble, resulting in fewer minerals being absorbed. (See Chapter 11.)

FUNCTIONS

The one class of nutrients that seems to have taken on negative connotations is lipids. Yet this is far from the truth. Fat is not "bad." No food or nutrient is "bad" or "good"— they all have a place. Lipids have numerous functions in foods and in the body.

Functions in Foods

Lipids have many functions in foods:

- Fatty acids give flavor and aroma, thus palatability. The short-chain fatty acids are especially volatile and distinctive, as in butter. Heating foods releases the fatty acids into the air and nose.
- Smooth crystals formed by fats provide appearance, as in ice cream or chocolate candy bars.
- Fats provide satiety, a sense of satisfied fullness, to the diet. Of the macronutrients, fats take the longest time to be digested, thereby delaying the hunger sensation.

- Like carbohydrates, different kinds of fats provide different textures, from creamy sauces, to fine-textured cakes, to coarser textured breads.
- Fats provide the most concentrated source of energy in the smallest amount of food.
- Lipids in food supply essential nutrients for the body, especially the essential fatty acids and the fat-soluble vitamins. Lipids also enhance the absorption of the fat-soluble vitamins.

Functions in the Body

Simple Lipids Simple lipids ("fat"), like triglycerides, have several functions in the body:

- After glucose, fat is the body's preferred source of energy.
- Fat, like carbohydrates, is protein-sparing because its availability reduces the body's need to use protein as a source of energy.
- The body's deposits of fat (adipose tissue) are a built-in reserve of stored energy.
- Fat underneath the skin provides insulation to help maintain constant body temperature.
- Fat provides a cushion of support around vital organs, such as the kidneys.

Essential Fatty Acids (EFAs) These fatty acids are also necessary:

- Linoleic acid is required for normal growth in children.
- Linoleic acid is required for healthy skin in both children and adults. Dietary lack leads to eczema, a form of dermatitis that is characterized by an inflammation of the skin marked by dry, scaly lesions. It is predominantly seen on the cheeks of the face.
- Linoleic acid, arachidonic acid, and linolenic acid are precursors of the eicosanoids, a group of 20-carbon-chain-length, unsaturated, ring structure compounds. These derived lipids are a group of hormonelike substances with seemingly wide and varied functions, including contraction of smooth muscles, regulation of blood pressure, platelet aggregation (clumping), transmission of nerve impulses, and immune responses. Thus, these compounds are being closely studied for their role in health and disease. This research, in turn, has spurred tremendenous interest in foods that contain these essential fatty acids, such as some vegetable oils and fish.

Cholesterol The great emphasis on cholesterol as a culprit in cardiovascular diseases has led many people to believe that it is a substance the body is better off without. This is not so. Cholesterol in the body performs vital functions:

- Body cholesterol is a normal constituent of tissues and is especially important for the formation of brain and nervous tissues, such as myelin.
- Body cholesterol is a precursor of vitamin D. Through a series of steps, the compound 7-dehydrocholesterol, which is present in the skin, is converted into vitamin D. (See Chapter 10.)
- Body cholesterol is a precursor of bile acids and hormones, such as those of the gonad glands (sex hormones) and adrenal glands (aldosterone, cortisol).

METABOLIC PROCESSES

Ingestion

Lipids, especially the simple lipids, impart important sensory characteristics to foods, enticing a person to select them. In most taste tests, people's expectation of what a food is supposed to feel like in the mouth (mouth feel) as well as its taste and aroma are related to the fats present in the food.

Digestion

The digestion of lipids is more complicated than that of other nutrients because lipids are insoluble in water. Although fat-digesting enzymes (lipases) exist in the mouth and stomach, almost all fat digestion occurs in the small intestine. (See Chapter 4.) Efficient digestion depends on fats being emulsified, which occurs in the small intestine.

When fats enter the small intestine, a number of hormones and enzymes are released. For example, fats cause the duodenum to secrete a hormone, cholecystokinin, that circulates to the gallbladder. The gallbladder in turn contracts and releases bile (which had been previously synthesized in the liver and now is stored in the gallbladder) into the small intestine. Bile acts as an emulsifier, dispersing the large fat spheres into small drops and suspending them within the watery contents of the small intestine so the fat-digesting enzymes can more easily reach them.

Each kind of lipid has designated digestive enzymes; for example, pancreatic lipases digest triglycerides, and cholesterol esterase digests cholesterol. By the end of digestion, the end products are glycerol, diglycerides, monoglycerides, and free fatty acids. (See Figure 6–4.)

After enzymatic digestion, the fat products and bile products combine to form tiny water-soluble spheres called **micelles,** which carry the digested lipids to the intestinal cell for further processing and absorptive release.

Absorption

About 95 percent of dietary fat and 10 to 50 percent of dietary cholesterol are absorbed from the small intestine into the circulation. Lipids are absorbed in two ways. First, glycerol and the short- and medium-chain-length fatty acids, being water soluble, are absorbed directly into the bloodstream.

Second, within the intestinal cells, synthesis of new triglycerides and other lipids, like cholesterol, takes place. Afterwards, they combine with a protein covering to form **chylomicrons,** one of the lipoproteins. This complex enters the lymph circulation. (See Figures 4–3 and 6–4.)

Transportation

The fatty acids that were absorbed directly into the blood circulation attach to the blood's circulating albumin as their protein transporters.

Chylomicrons are transported through the lymph circulation that then empties into the bloodstream via the thoracic duct. It is these chylomicrons that give a milky appearance to lymph and blood after a person eats fat.

Besides chylomicrons, several other classes of lipoproteins are formed by the intestines and liver to transport lipids. These classes are subdivided and have different functions. For example, low-density lipoproteins (LDLs) transport cholesterol from the liver to the body's cells, and high-density lipoproteins (HDLs) transport cholesterol from the body's cells to the liver.

ENTEROHEPATIC CIRCULATION OF BILE

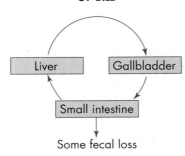

CLASSES OF LIPOPROTEINS

VLDL, very-low-density lipoproteins: consist primarily of triglycerides; rapidly removed by the tissues

IDL, intermediate-density lipoproteins: contain residues remaining after removal of triglycerides from VLDL

LDL, low density or beta-lipoproteins: synthesized by liver from IDL residues; chief carrier of cholesterol; levels increase with diets high in saturated fat

HDL, high-density or alpha-lipoproteins: high in protein; appear to reduce the risk of coronary heart disease

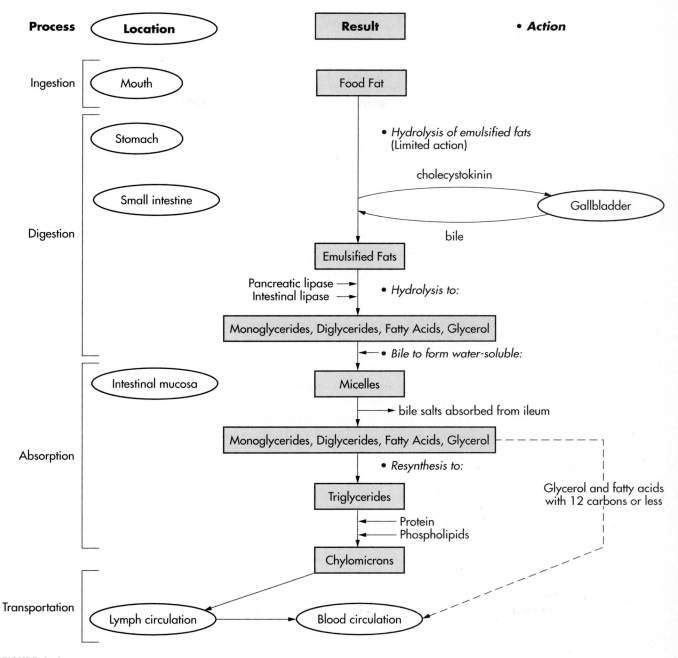

FIGURE 6–4
Summary of the multiple complex steps related to the ingestion, digestion, absorption, and transportation of fats and oils.

Respiration

The major role of fats is as the most concentrated source of energy. As with carbohydrates, for this to happen, the lungs must supply the cells with oxygen and remove the by-products, carbon dioxide and water. Of the three macronutrients, fats produce the least amount of carbon dioxide to be exhaled.

Utilization

Many different enzymes, hormones, and other regulatory substances are required for the body to utilize lipids. **Lipogenesis** (synthesis of new fats) and **lipolysis** (breakdown of fats) occur continuously in the body. All cells use lipids in some way, such as for energy or for making cell membrane structures. These processes take place in different parts of the cell. For example, all cells of the body except red blood cells and those of the central nervous system can use fatty acids to provide energy. When fatty acids arrive in the cell to be used for energy, they must go to the part of the cell known as the mitochondria. To traverse the membrane into this compartment, a special carrier is required, carnitine. By a step-by-step process, termed beta-oxidation, the fatty acids are broken down to two-carbon units that then enter the common energy-releasing cycle for oxidation into energy, carbon dioxide, and water. (See Figure 6–5.)

Some organs have special roles in the body's utilization of lipids. For example, triglycerides are synthesized by the cells of the intestines, liver, and adipose tissue. Importantly, all cells are able to synthesize cholesterol, but the liver is the chief regulator of the total body content of cholesterol and of the circulating serum cholesterol. The body's production of cholesterol goes on independently of dietary intake and can be formed from all of the macronutrients, not just fat.

Storage

Fat is stored in adipose tissues. Since triglycerides are the most concentrated source of energy in the smallest package, the body transforms extra energy into it for long-term reserves. Not only is excess fat stored in adipose tissue, but so is excess energy derived from dietary carbohydrates and proteins. These other two macronutrients are transformed into triglycerides and stored.

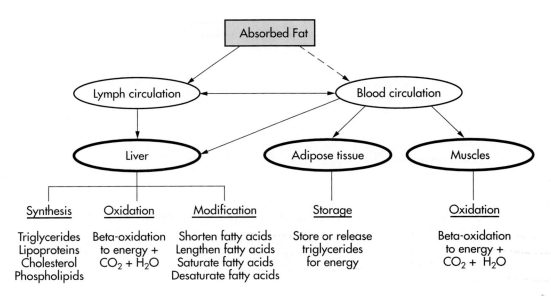

FIGURE 6–5
The liver, adipose tissue, and muscles perform many functions related to lipids to help the body maintain homeostasis.

There are two kinds of adipose tissue. The vast majority is white adipose tissue, commonly recognized as body fat found under the skin and around organs. Brown adipose tissue contains a greater blood supply (hence its darker color and name). It is found only in certain areas, especially the upper back in infants, and is important in heat generation. Some investigators are studying its effects on weight maintenance in children and adults.

Adipose cells contain enzymes that cause the synthesis or release of fats. The cells themselves have an amazing capacity to increase or decrease in size. In round numbers, one pound (0.45 kg) of adipose tissue contains about 3,500 kcal.

Excretion

The various components contained in dietary lipids as well as lipids synthesized by the body can be recycled. When the time comes for them to be eliminated from the body, several routes are used, again depending on their solubility. All four excretory organs are used.

As with carbohydrates, when the body normally uses glycerol and fatty acids for energy, the only products to be eliminated are carbon dioxide and water, eliminated by the kidneys, skin, and lungs. Ketones are water-soluble and volatile. Thus, they are excreted mainly in the urine, but also through the skin and lungs. Cholesterol is disposed of by the liver in bile, then via the gallbladder and intestine in the feces.

MEETING BODY NEEDS

Requirements and Recommendations

Good health can be maintained within wide ranges of fat intake. People of Asia obtain as little as 10 percent of their calories from fat, while other groups such as some Eskimos eat large quantities of fat. Americans typically consume about 34–42 percent of their calories from fat.

Some fat, of course, is needed to supply the essential fatty acids. For example, if 1–2 percent of dietary calories are supplied by linoleic acid, the signs of deficiency can be avoided. Infant formulas should provide 3 percent of calories as linoleic acid. Some cholesterol is essential for infants, so cholesterol-free diets for children under two years of age are not recommended. Thereafter, there is no need for dietary cholesterol, the body synthesizes enough for its purposes. Although there is no RDA for lipids, Table 6–1 summarizes the recommendations of several governmental and health organizations.

Sources

Total Fat Some fats are clearly visible, as butter, margarine, shortening, oil, in and around muscle fibers in meat and under the skin of poultry. Other fats are invisible, as in whole milk, cheese, egg yolk, and numerous food mixtures such as candy, pie, and cake.

Evaluated according to the Food Guide Pyramid (see Figure 1–5), about half of the total fat in American diets is furnished by meat and whole milk foods. The visible fats and oils group is the next largest source. Foods with negligible or very little fat are found in the vegetable and fruit groups (except olives and avocados), and the bread-cereal-grain group. (See Figure 6–6.)

TABLE 6–1

Current versus Recommended Average Lipid Intakes for Children, Adolescents, and Adults

	Current Intake (% of Calories)	Recommended Intake (% of Calories)
Total Fat	34–37	≤30
Saturated fatty acids	13–14	<10
Monounsaturated fatty acids	12–14	10–15
Polyunsaturated fatty acids	6–7	Up to 10
Cholesterol	193–435 mg	<300 mg

Source: Adapted from Report of the Expert Panel on Population Strategies for Blood Cholesterol Reduction, National Cholesterol Education Program, NIH 90-3046, U.S. Dept. of Health and Human Services, Washington, DC, 1990, pp. 17–20.

FIGURE 6–6
Chefs prepare gourmet low-fat meals with high taste and eye appeal. (*Source:* Courtesy of Francis McFadden and the Department of Hotel and Restaurant Management and Culinary Arts, Drexel University.)

Fatty Acids A single food contains fatty acids of different chain lengths and degrees of saturation/unsaturation. Whether or not a food is considered to be "high" or "low" in a particular kind of fatty acid is determined by the composite of all fatty acids contained in the food. It is important to keep in mind that most foods contain a variety of fatty acids, not just one kind. Examples of food sources of specific categories of fatty acids are summarized in Figure 6-7 and listed more fully in Appendix Table A-1.

> ■ *High in Linoleic Acid (Essential Fatty Acid)*
> Vegetable oils: safflower, corn, cottonseed, soybean, sesame, sunflower
> Salad dressings made from the above oils: mayonnaise, French
> Fatty fish: salmon, tuna, herring
> Special margarines: liquid oil listed first on label
> ■ *High in Omega-3 Fatty Acids*
> Fish, especially fatty fish: salmon, tuna, herring, mackerel, carp
> ■ *High in Monounsaturated Fatty Acid (Oleic Acid)*
> Olives, olive oil, avocados, cashew nuts, canola oil, peanut oil
> ■ *High in Saturated Fatty Acids*
> Whole milk, cream, ice cream, cheese made from whole milk, egg yolk
> Medium-fat or fatty meats: beef, lamb, pork, ham
> Bacon, beef tallow, butter, coconut oil, palm oil, lamb fat, lard, regular margarine, salt pork, hydrogenated shortenings
> Cakes, cookies, pies, puddings, chocolate candy

FIGURE 6–7
Examples of fatty acid content of some foods.

Although *trans*-fatty acids occur naturally in meat and dairy products, the vast majority are consumed in foods with partially hydrogenated fats, such as commercial margarines, fried foods, and baked products. Nutrition labels do not declare amounts of *trans*-fatty acids, but the typical phrase "partially hydrogenated vegetable oils" is stated in the ingredient listing. Health professionals and consumers can contact food manufacturers for more precise information. Restaurants, other retail outlets, and institutional food operations, however, do not usually have precise analyses for the foods they serve.

Cholesterol Cholesterol is found in animal foods, but not in plant foods. Cholesterol content has nothing to do with the fatty acid content of foods. Some foods have no cholesterol, but are high in saturated fatty acids. Examples include coconut oil and hydrogenated vegetable shortenings. Thus, some food labels can be very misleading to consumers. Sometimes, a label stating "no cholesterol" is interpreted erroneously to mean "no fat" or "no saturated fat" and even sometimes "no calories." Examples of food sources of cholesterol are summarized in Figure 6–8 and listed more fully in Appendix Table A–1.

Food selection and preparation suggestions to adjust the fat and cholesterol content of the diet are listed in Figure 6–9.

Fat Replacements Because many people want or need to reduce their fat intake even though they desire the taste and texture of fatty foods, there is tremendous opportunity for food manufacturers to develop foods that simulate the taste and texture of fatty foods.

A number of fat replacements have been developed in recent years. Some are formulated from proteins, and others from carbohydrates. Some are completely digestible and therefore contribute some nutrients and calories, however small. Others are indigestible and hence nutrient-free. The potential benefits of these replacements need to be assessed against their long-term safety and nutritional impact, such as the total nutrients they contribute to a person's daily intake versus the nutrients contained in the foods they replace. Moreover, many consumers mistakenly believe that a food product labeled "no fat" means "no calories." They do not realize that the other ingredients in the food contain

- *Absent*
  Egg white; all plant foods
- *Trace Amounts*
  Nonfat milk; cheese made from skim milk; low-fat and nonfat yogurt
- *Moderate Amounts*
  Whole milk, whole-milk cheese, cream, ice cream, butter, meat, poultry, fish, clams, crab, oysters, scallops
- *High Amounts*
  Egg yolk, liver, sweetbreads, brains, kidney, heart, fish roe, shrimp

FIGURE 6–8
Examples of cholesterol content of some foods.

- Use nonfat or low-fat milk—skim, 1%, or 2% fat. Select low-fat cheese.
- Substitute egg whites for whole eggs.
- Select lean cuts of meat, poultry, and fish. Trim off visible fat; remove skin from poultry. Broil or roast instead of frying; discard drippings. Refrigerate stews, soups, and nut butters; remove fat that rises to top.
- Use legumes and pastas frequently as entrees.
- Substitute herbs, spices, or lemon juice for butter or margarine to season vegetables.
- Use fruit or vegetable juices, like lemon, lime, and tomato, or low-fat dressings on salads.
- Substitute fruits for desserts that contain fat, like pies, pastries, cakes, cookies, or ice cream.
- Use vegetables and fruits and their juices for snacks. Limit high-fat cheese, dips, nuts, chips, and fatty crackers to rare occasions.
- Choose vegetable oils and margarines with poly-, mono-, or omega-3 unsaturated fatty acids. The first word on the ingredient list of the label should be "liquid," since liquid oils are less saturated than those that are partially hydrogenated. Use sparingly because all oils are 100% fat.

FIGURE 6–9
Food selection and preparation suggestions to adjust fat and cholesterol content of diet.

calories. Sometimes they also think they can add, rather than substitute, these foods to their diet; many times the total caloric value of the daily diet actually increases.

- Simplesse®* is processed from skim milk and egg white to simulate fat. It is completely digested and yields 1–2 kcal per gram. This product is used primarily by manufacturers to replace part of the fat in cold foods such as frozen desserts, mayonnaise, cream cheese, sour cream, and dips. It is not used in cooked products.

---

*The NutraSweet Company, San Diego, CA, a unit of Monsanto Company.

- Stellar®* is a product based on cornstarch that supplies about 1 kcal per gram. Up to 80 percent of the fat can be replaced by this product in foods. It can be used with hot or cold foods; for example, margarine, sour cream, salad dressings, ice cream, meat products, soups, dairy, and cheese products. Since it is a digestible modified starch product, it did not require additional Food and Drug Administration (FDA) approval.

- Olestra is a sucrose polyester, that is, fatty acids replace some of the hydroxyl groups of sucrose. The product is indigestible and therefore furnishes no calories. It can be used in hot or cold foods and can be added to oils and shortenings, thereby reducing the caloric content of these fats. Since it is indigestible, it required review by the FDA. In 1996, Procter and Gamble received FDA approval to market olestra under the trade name OLEAN®†.

Basic Diet

The Basic Diet Nutrient Profile (Table 2–3) has five columns related to lipids: total fat, the three categories of fatty acids—saturated fat, monounsaturated fat, and polyunsaturated fat—and cholesterol. The essential fatty acids are included in the polyunsaturated fat category. Make four special observations. First, note that all foods contain lipids, unless they are 100 percent carbohydrate, such as table sugar. However, some foods are especially low in total fat, such as fruits, vegetables, legumes, and skim milk. Second, see that each food usually has an assortment of fatty acids based on the kinds comprising the triglycerides that the food contains. Third, look at the column for "Food Energy." Some foods such as fats and oils are 100 percent fat and thus are concentrated sources of calories in a small quantity of food. This has implications for people who need to increase or decrease their total caloric intake. Fourth, recall that cholesterol is present in foods with animal fat, but not with plant fat. For a listing of the lipid content of specific foods, consult Appendix Table A–1.

BODY BALANCE

Optimum/Adequate Signs and Symptoms

In assessing nutritional status regarding lipids, prioritize three evaluations. (See Chapter 3.) First, evaluate the food intake. Screen the diet history and diet recall for intake of essential fatty acids. The diet must supply these. Calculate the total calorie intake and the percentage of total calories from fat. If warranted, compute cholesterol content. Second, perform serial anthropometric measurements, especially accurate height and weight measurements, and compare them to standards. Having baseline and follow-up measurements are valuable to determine trends over a person's lifetime. Using calipers to identify fat distribution in the subcutaneous regions of the body may provide additional information. Third, obtain laboratory blood profiles of lipids, as indicated by other elements of the overall history. For example, family history of cardiovascular disease might suggest the need to determine the levels of serum cholesterol, lipoproteins, and triglycerides.

*A. E. Staley Manufacturing Corporation, Decatur, IL.
† Procter and Gamble, Cincinnati, OH.

Insufficiency/Deficiency

Conditions Since lipids are found in nearly all foods, deficiency of essential fatty acids or total fat is rare.

Signs and Symptoms Deficiency of dietary intake of linoleic acid leads to eczema. Blood fatty acid profiles also are abnormal. In diets highly restricted in total fat, insufficient growth, weight gain, or weight maintenance might be the most obvious signs. Diseases that cause malabsorption may result in steatorrhea, an excessive amount of fats in the feces. This condition can lead to inadequate availability of total fat, calories, essential fatty acids, fat-soluble vitamins, and minerals.

People Most at Risk At most risk of deficiency of essential fatty acids are rapidly growing infants and children fed formulas and foods low in linoleic acid, but also in linolenic acid and arachidonic acid. Some families attempt to overly restrict lipids. Teenagers and adults following severely restricted fat diets may not realize the importance of lipids to dietary balance, overall health, and the body's homeostasis. Patients with maldigestion and malabsorption require careful assessment of their lipid status.

Excess/Toxicity

Conditions Lipids appear to be connected with several chronic diseases, which still have unclear etiologies and treatments.

- Overweight and obesity is associated with an excess ingestion of simple lipids (fats). Because fats are the most concentrated source of calories, it is very easy to consume calories in excess of one's energy balance needs. The mechanisms by which the body stores and releases fats, though, remain an enigma. (See Chapters 8 and 9.)

- Cardiovascular diseases, according to numerous studies, have as a major risk factor elevated *serum* (not dietary) cholesterol levels. Investigators have long explored the role of diet as one attempt to control the development of fatty materials from building up along blood vessel walls. Numerous theories have been studied, without resolution. The rank order of dietary lipid components associated with cardiovascular diseases indicates that some people with elevated *serum* cholesterol levels could reduce these levels by reducing their weight; reducing total fat intake; reducing saturated fat intake; modifying intake of unsaturated fatty acids, including polyunsaturated, omega-3, and monounsaturated fatty acids; and lastly, decreasing the cholesterol content of their diet. The role of other lipids, such as the antioxidant fat-soluble vitamins, are under intense investigation. (See Chapters 10 and 24.)

- Cancer and its relationship with lipids has been explored, primarily in animals and in epidemiological studies. Of all dietary factors, fat intake appears to be most directly associated with colorectal cancer, breast cancer, and prostate cancer. In animals, some studies showed that a high intake of fat, whether saturated or polyunsaturated, encouraged tumor growth. Certain people are advised to reduce their intake of fats to less than 30 percent of total calories. However, it must be emphasized that there are many risks for cancer, and diet alone may not necessarily reduce the risk for a given individual. (See Chapter 25.)

People Most at Risk Two groups of people are most at risk—those who consume high amount of fats and those who consume high amounts of total calories. All people living in countries with abundant supplies of food are at risk for consuming excess lipids. This is especially so because fats and oils are the most concentrated source of calories and just about everyone eats food without measuring it with scales and rulers!

HEALTH AND NUTRITION CARE ISSUES

After water and carbohydrates, lipids might be the third nutrient to assess. Probably, you will face a dilemma regarding lipids. Much of the public's perception about lipids is from the messages about fat and cholesterol delivered by health professionals, schoolteachers, food and pharmaceutical companies, media experts, and well-meaning family and friends. As with other nutrients about which there is imperfect knowledge, different messages are spoken, and people receiving them hear the messages differently. One message heard by many people is that lipids are "all bad for you" and should be sharply reduced or eliminated from the diet. Others believe they need to adjust their food (and dietary supplement) selections piecemeal, in the understanding that "some are bad," like saturated fat and cholesterol, but "others are good," like omega-3 fatty acids and lecithin. On the other hand, because of continuing controversy and inconclusion, other people disregard any information and think, "Forget it." On an individual client basis, to help resolve the dilemma, one approach might be to conduct a thorough nutritional assessment to guide outcome-based customized dietary counseling. (See Chapters 2 and 3.)

PUTTING IT ALL TOGETHER

Lipids are a class of nutrients that supply the most concentrated source of energy, both in foods and in the body. Plants and animals, including humans, synthesize triglycerides to be their long-term storage reservoir of energy. As a greater understanding of the many other functions of the individual lipids develops, professionals working in the health-related fields have the challenge of translating scientific theories into practical suggestions for the public and for their clients and patients.

? QUESTIONS CLIENTS ASK

1. Is taking a lecithin supplement a good idea?

Any lecithin eaten in foods or taken as a supplement is broken down during digestion into its chemical parts—glycerol, fatty acids, choline, and phosphoric acid. The body's need for lecithin is abundantly supplied through synthesis by the liver. Buying a supplement is a waste of money.

2. The ingredient label of many food products lists "vegetable oil" as a component. Can I assume that the oil used is high in polyunsaturated fatty acids?

No. Palm oil and coconut oil, both of which are highly saturated, are used in many food products; for example, they are ingredients in some coffee whiteners. Unless the kind of oil is stated, you cannot be sure. Also, if the oil is hydrogenated, it will contain a higher proportion of saturated fatty acids and *trans*-fatty acids.

Because of consumer concerns, many manufacturers are examining the kinds and amounts of fats used in food processing. Informed consumers must carefully read labels or write to the specific food company for detailed information.

3. I have read that olive oil can lower blood cholesterol. Is this true?

Olive oil is very high in the monounsaturated fatty acid, oleic acid. When foods high in monounsaturated fatty acids are substituted for saturated fatty acids, the serum cholesterol level may be lowered. However, monounsaturated fatty acids have not been found to change the blood level of high-density lipoproteins (HDLs).

4. Should I take fish-oil supplements?

The first question to ask is, "Does the oil in the capsule come from the liver or the flesh of the fish?" If it comes from the liver, it may contain high amounts of some fat-soluble vitamins and fat-soluble pollutants. Some contain high levels of saturated fat, cholesterol, and toxic metals, such as lead, that became stored in the fish's liver. Many of the fish-oil capsules on the market contain less omega-3 fatty acids than their companies claim. Also, consuming fish-oil capsules can cause vitamin E levels in the body to drop far below normal, thus necessitating a vitamin E supplement. (Fish oils are highly unsaturated, thus easily oxidized. Vitamin E is an antioxidant and is used up trying to "protect" the unsaturated fatty acids from oxidation.)

From a calorie standpoint, these capsules contain only oil. Thus each capsule is high in calories and so contributes to the total fat and calorie content of the daily diet. Moreover, there have been reports that people who self-prescribe these capsules may unknowingly develop bleeding problems, which become evident after suffering a bruise or during surgery, drawing blood for tests, or dental prophylaxis procedures. Before any such procedure, these supplements must be discontinued. Although these capsules are sold over the counter without the need for a prescription, people should not take these supplements on their own, but only upon the advice of a physician who monitors beneficial and negative outcomes.

✔ CASE STUDY: A Family Learns Facts About Fats in Foods

Ella and George have three teenage children, two boys and one girl. During a routine family health examination, George, age 45, was found to have a serum cholesterol level of 265 mg/dl (6.85 mmol/L). The laboratory tests were within normal limits for the rest of the family. Twenty-four hour dietary recalls revealed that total fat was about 40 percent of the family's caloric intake and everyone's Body Mass Index (BMI) was about 25. (Review Chapter 3 and Appendix Figure B–1.) Their physician recommended that everyone start getting more physical activity and decrease their caloric total fat consumption. Based on other test results, George was to decrease his intake of saturated fat and cholesterol, too.

Ella, George, and Jeff, the younger son who wants to try out for his high school baseball team, are attending nutrition and fitness classes offered at the community center. They were especially interested in the topic "Facts About Fats in Foods." It's certainly timely for their family!

Questions

1. At the beginning of the nutrition and fitness class, each person was asked to list reasons why fat is useful in the diet, if any. What are some responses they might give? Why would the statements be true or false?

2. Next, everyone was asked to list the functions of fat (lipids) in the body. The participants said they thought fat was "bad" for the body. Help them list and explain functions of different lipids in the body.

3. One participant asked, "Aren't 'heavy' foods like fried chicken, french fried potatoes, and donuts hard to digest?" What do you think was meant by "heavy"? Explain the digestion of such foods.

4. All the participants agreed they wanted to know the differences among the different kinds of fats. Each day some new research seemed to urge them to eat more or less saturated fat, or monounsaturated fat, or polyunsaturated fat, or omega-3 fatty acids, or *trans*-fatty acids, or cholesterol. What are these lipids and in what foods are they chiefly found?

5. George wanted to know what foods he could order from the lunch menu when he ate with his co-workers. Obtain a menu (perhaps from your favorite restaurant) and identify the foods

with and without saturated fat and with and without cholesterol.

6. Jeff wanted to know about lower fat and calorie snacks he and his friends could eat when they came back to his house after baseball games. What are examples of such snack foods that might appeal to teenagers?

7. Ella brought in labels from packages of foods she purchased for her family. "The labels are a mystery!" she declared. She also is concerned that her daughter now only wants to eat foods with "no fat" in them. What information should she and her daughter look for on the ingredient list? Help them decipher the nutrition label with respect to fats. How would they select foods without labels, like fresh foods?

8. Hot debate ensued when the participants evaluated examples of magazine ads about the health benefits of various food products that had been modified in fat content. Obtain some ads. In what ways are they truthful? Misleading? Fraudulent? (Review Chapter 2.) What can consumers do?

FOR ADDITIONAL READING

Expert Panel: "Summary of the Second Report of the National Cholesterol Education Program (NCEP) Expert Panel on Detection, Evaluation, and Treatment of High Blood Cholesterol in Adults (Adult Treatment Panel II)," *JAMA,* **269:**3015–23, 1993.

Gershoff, SN: "Nutrition Evaluation of Dietary Fat Substitutes," *Nutr Rev,* **53:**305–13, 1995.

Lichtenstein, AH: "*Trans* Fatty Acids and Hydrogenated Fat—What Do We Know?" *Nutr Today,* **30:**102–07, 1995.

Porth, CM: "Understanding the Cholesterol Transport System," *Nurs95,* **25:**31U–32U, April 1995.

"WHO and FAO Joint Consultation: Fats and Oils in Human Nutrition," *Nutr Rev,* **53:**202–05, 1995.

Proteins

Chapter Outline

Both animal and plant proteins are made up of about 20 common amino acids. *

 CASE STUDY PREVIEW

> Protein is a nutrient held in high regard by many people. People are told protein is the body's nutrient of top priority, and that the term means "to take the first place." What facts and fables about protein have you heard?

---

*National Academy of Sciences, *Recommended Dietary Allowances,* 10th ed., National Academy Press, Washington, DC, 1989, p. 52.

In technologically advanced countries, most people consume far more protein than they need. Protein-rich foods such as meat, poultry, fish, milk, cheese, and yogurt rank high in people's diets. Some regard high protein intakes as essential for physical activity whether at the workplace or for athletic performance. Often such persons resort to protein supplements, believing that they need these special products for optimal health. Other people wonder whether the high consumption of animal-protein foods is justified for good health and whether increased use of plant proteins such as legumes and grains might be more healthful as well as help to conserve the earth's resources.

The opposite is the case in many other countries of the world. The lack of protein ranks second to lack of calories as a major cause of severe malnutrition and death. Infants and young children are the main victims. Although protein deficiency rarely is seen in the United States, health professionals need to recognize the possibility of protein malnutrition in people who are elderly, who have inadequate money or food intakes, or who have chronic diseases.

CHARACTERISTICS

Nomenclature

Proteins are large, complex compounds. The translation of the term does mean "to take the first place" or "of first importance." Since proteins are a part of every cell, this name certainly is appropriate. Proteins are organic compounds.

Chemical Structure/Composition/Measurement

Like carbohydrates and lipids, proteins contain carbon, hydrogen, and oxygen. What sets proteins apart from the other macronutrients is that proteins all contain nitrogen. About ⅙ of protein is nitrogen. Usually part of the protein structure are other mineral elements such as sulfur, phosphorus, copper, iron, and iodine. (See Chapter 11.)

The basic structural of proteins is **amino acids.** Amino acids are organic acids that contain a carboxyl (COOH) group as well as an amino (NH_2) group attached to the same carbon atom. About 20 amino acids exist in foods and the body. Figure 7–1 illustrates the general formula for an amino acid, along with the structures of some common amino acids, the building blocks of proteins.

The structure of proteins is extremely complex and is constructed under the direction of DNA. The structures are formed in a step-by-step fashion and are called primary, secondary, tertiary, and quaternary structures. As an analogy, just as you would use the alphabet to develop a dictionary full of words, the words would form different kinds of sentences, sentences would form chapters, and chapters would form books (the final functional product).

To start, amino acids connect to each other by peptide linkages, in which the amino group of one amino acid is linked to the carboxyl group of another amino acid. With 20 amino acids that could be used to construct a protein, it becomes evident that a tremendous number of proteins could be created. The kinds of amino acids, the sequence in which they are arranged, and the frequency with which each is used in an amino acid chain is enormous.

Once the amino acids are in the correct lineup, other bonding and connecting mechanisms occur. Ultimately this causes the protein to assume a three-

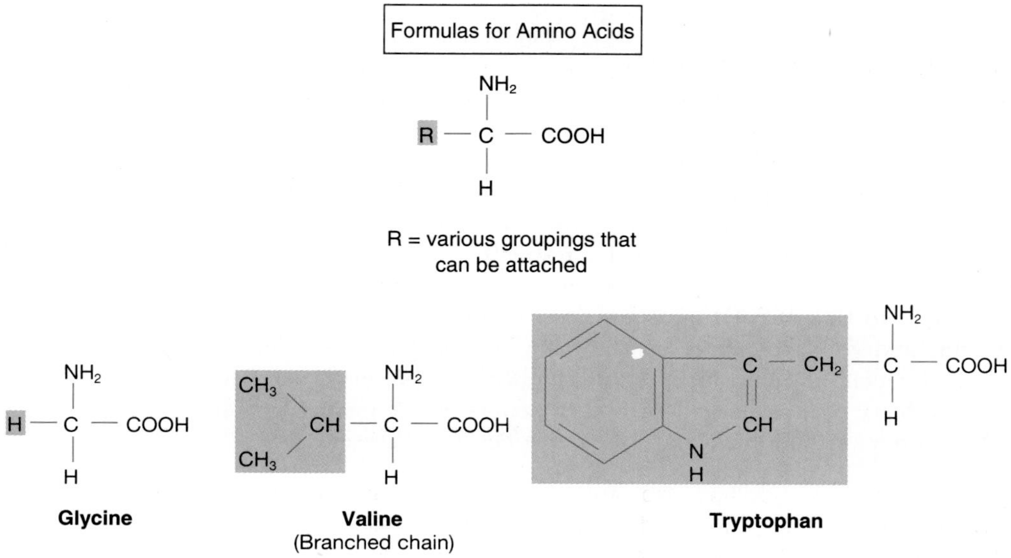

Formulas for Amino Acids

FIGURE 7-1
Each of the 20 amino acids has a different grouping attached to the carbon that holds the amino and carboxyl groups. Glycine, the simplest in structure, has only a hydrogen attached. Valine is one of three amino acids that has a branched chain. Tryptophan has a complex ring structure.

dimensional appearance, such as a helix or folded wad of paper. For example, bonds can connect between the mineral sulfur in one amino acid with the sulfur in another (disulfide bond) or several proteins can be combined together (hemoglobin).

Importantly, each animal and plant species makes its own kind of proteins. Moreover, within each species the proteins of each tissue are unique. For example, hair, bone, skin, liver, muscle, and blood have proteins that are distinctive for that tissue alone. Each body protein is constructed to perform a specific function and cannot be replaced by any other protein.

Proteins, as a macronutrient, are measured in grams. Each gram provides 4 kcal. (See Chapter 8.) Thus, proteins supply the same number of calories as does carbohydrate, and slightly less than half that of lipids. Appendix Table A-1 includes the energy values of many protein-containing foods. When laboratory determinations are made for the protein content of foods, or that is present in a tissue of the body, or the amount turned over by the body and eliminated in urine, nitrogen is measured since this is the unique element contained in proteins. The figure used for calculations is: 1 g nitrogen = 6.25 g protein. (Proteins are 16 percent nitrogen, thus $^{100}/_{16} = 6.25$.)

Distribution

Proteins are nearly everywhere. Unless a food is 100 percent water, carbohydrate, and/or lipid, proteins are present. Sometimes the total quantity is so small that an amount is not listed in food tables or on the nutrition labels of packaged foods. In the body, proteins are components of each cell and substance, such as hormones and enzymes.

The human body requires all 20 amino acids for the synthesis of its own protein. It can synthesize 11 of these, but is unable to make nine others. These nine must be supplied by the diet in adequate amounts and thus are termed dietary **essential amino acids.** (See Table 7-1.)

Proteins can be classified in several ways, including nutritional quality, composition, and shape.

Nutritional Quality of Proteins One classification of proteins is by the ability of the body to retain and use them for itself, known as a protein's **biological value.** As it turns out, not all foods are of equal quality in meeting body needs. Each cell must have a sufficient supply of all the amino acids it needs in order to build a new protein structure. All the necessary amino acids must be available at the right time and right place; if one is missing, the protein cannot be synthesized. This is sometimes referred to as the "all or none law."

Generally, foods are subdivided into two categories. If a food furnishes amino acids in the proportions and amounts needed by the body, it is said to furnish proteins of high biological value. Such proteins are referred to as complete proteins. Foods from animal sources, like eggs, milk, cheese, yogurt, meat, poultry, and fish, are examples of **complete protein** foods. Other protein foods do not provide sufficient amounts of one or more essential amino acids for optimal body protein synthesis. Such proteins are termed **incomplete proteins.** Foods from plant sources fall into this subdivision. Examples include legumes, nuts, grains, vegetables, and fruits. Protein quality can be ranked: Egg protein is best; then proteins of milk, flesh foods, legumes and nuts, grains, and vegetables; fruit protein is lowest.

Composition Like lipids, proteins can be divided according to composition. (See Chapter 6.) Simple proteins contain only amino acids, such as zein in corn or albumin in blood. Conjugated proteins are composed of simple proteins combined with another substance. Examples are casein (phosphoprotein) in milk,

TABLE 7–1
Classification of Amino Acids

Essential Amino Acids	Nonessential Amino Acids
Histidine	Alanine
Isoleucine[a]	Arginine
Leucine[a]	Aspartic acid
Lysine	Cysteine
Methionine	Cystine
Phenylalanine	Glutamic acid
Threonine	Glycine
Tryptophan	Hydroxylysine
Valine[a]	Hydroxyproline
	Proline
	Serine
	Tyrosine

[a] Isoleucine, leucine, and valine are often referred to as branched-chain amino acids.

lipoproteins (transporters of lipids) in blood, and metalloproteins (ceruloplasmin, containing copper) in blood plasma.

Shape Proteins, as three-dimensional shapes, are called globular proteins if the amino acids appear to be coiled, tightly packed balls. They are called fibrous proteins if the amino acids appear to be layered, parallel, linear shapes. Examples of globular proteins are hemoglobin, albumin, and insulin. Examples of fibrous proteins are collagen, myosin (muscle), fibrin (blood clot), and keratin (hair and nails).

Properties

Like carbohydrates and lipids, one of the important properties of proteins is solubility in water. Look again at the paragraph above. From the examples, you can notice that some travel in the blood—the globular proteins—thus, they are water soluble. The fibrous proteins are insoluble, which allows them to remain in their rightful location. Also, the large globular proteins cannot pass through membranes, like the kidney cell membranes, and so remain in the blood circulation rather than being eliminated when blood passes through the kidneys for filtration.

An important unique property relates to the folding characteristics of the amino acids. Denaturation is the process of splitting the cross-linkages with or without breaking the peptide bonds. When this process occurs, the proteins lose their biological activity. If the peptide bonds are not broken, the protein unfolding can be reversed and the biological activity of the protein is restored. Most times, however, this process destroys normal function. Heat, agitation, acids, and alcohol are destructive examples, with positive or negative side effects. For example, in the stomach, the acid pH destroys the function of foodborne pathogens and permits them to be digested into their smaller proteins and amino acids.

Another important unique property relates to the potential electric charge certain amino acids can exhibit as part of the protein. Some proteins are said to be amphoteric, that is, capable of functioning as either an acid or a base, depending on the pH of the medium in which the protein is located. (See Chapter 4.)

FUNCTIONS

Functions in Foods

Proteins have many functions in foods. Some of these functions are:

- Proteins provide stability in foods, such as in the formation of foams. A foam is created when a gas is dispersed in a medium such as a liquid. For example, foams commonly are formed in foods by incorporating air into egg whites, gelatin, and cream. If the foam is heated, the protein is denatured and the foam is stabilized.

- Proteins can be denatured and coagulated, such as during cooking. For example, raw eggs are turned into fried eggs and hard-cooked eggs, and harmful bacteria are killed to destroy foodborne pathogens. (See Chapter 2.)

- Because of their amphoteric properties, the texture of protein foods can be altered. For example, milk can be curdled to make cheese.

The body uses proteins for the structure of all body cells, for the regulation of many body processes, and as a source of energy.

Structural Role Proteins are essential components of the cells of all living things. Every tissue and fluid in the body except bile and urine contains some protein. Muscles account for about half of this protein. Collagen is a major structural protein in bones, tendons, ligaments, blood vessel walls, skin, and connective tissues. Collagen is the matrix into which minerals are deposited to provide the rigid structure of bones and teeth. Body cells are constantly being broken down, with new cells being built to take their place. The cells of some tissues such as those of the liver and intestine are replaced every few days. Red blood cells have a life span of about 120 days, while muscle and brain cells have a considerably longer life span. To maintain the delicate balance between breakdown and synthesis of cells, dietary protein is required throughout life. When new tissues are being built, as during pregnancy, infancy, and childhood, additional protein is needed.

Regulatory Functions Most of the regulatory materials of the body are protein in nature. Some regulatory examples include:

- *Nucleoproteins:* chromosomes
- *Hormones:* insulin, thyroxine, growth hormone
- *Enzymes:* lactase, cholesterol esterase, pepsin
- *Transport:* lipoproteins, albumin, transferrin, hemoglobin
- *Storage:* ferritin
- *Protective:* immunoglobulins, antibodies, interferon
- *Precursors:* tryptophan, the precursor of niacin (B-complex vitamin) and serotonin; phenylalanine and tyrosine, the precursors of catecholamines and melanin

Two other important regulatory functions exist. One relates to proteins' colloidal property. For example, serum albumin helps maintain the fluid balance of the body because of the osmotic pressure it exerts. The other regulatory function is proteins' ability to act as either acids or bases. In this way, they function as buffers to help maintain the body's acid-base (pH) balance. (See above and Chapter 4.)

Energy Once the amino group is removed for use by the body, the remaining carbon portion of amino acids can be used for glucose, fatty acid, or ketone production. Supplying energy for the body's activities takes priority over protein synthesis. If the diet does not supply sufficient calories, protein from the diet will be used to meet energy needs. For instance, in starvation the body uses its carbohydrate and fat stores, but also starts to attack its protein tissues. Muscle proteins are used before the proteins of the vital organs, like the heart, lungs, and brain. Conversely, when the diet supplies more protein than the body needs for building and repair of its tissues, the excess protein is available immediately for energy (as glucose) or is stored as body fat. The primary requirement of the body is for energy homeostasis. (See Figure 7-2 and Chapter 8.)

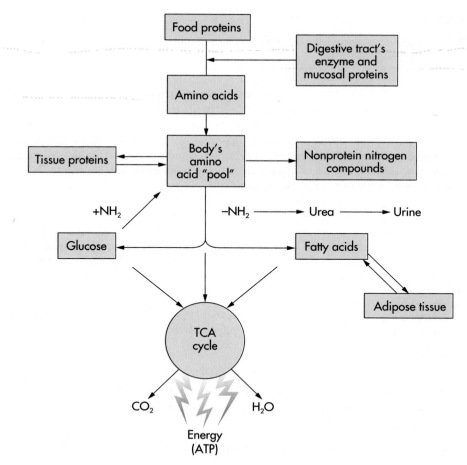

FIGURE 7–2
From the amino acid pool tissue proteins and special nonprotein nitrogen compounds can be synthesized. By removal of the amino acid group the rest of the amino acid molecule can be used to synthesize glucose or fatty acids. These, in turn, are potential sources of energy.

METABOLIC PROCESSES

Ingestion

Since nearly all foods contain proteins, ingestion of protein is usually commonplace. Many people overlook the sensory characteristics imparted by proteins to foods, like texture in fish, eggs, or yogurt; rather, they are more aware of the sensory attributes of carbohydrates and lipids.

Digestion

The purpose of digestion is to release the amino acids comprising the protein molecule. Each day protein from the diet (about 70–100 g) is digested. In addition, the worn-out digestive tract cells and enzymes account for about 70 g of protein that are reused. Thus, about 140–170 g of protein are subjected to digestion.

Although the mouth contains no enzymes to digest protein, food is masticated and mixed with saliva to assist later enzymatic action. In the stomach,

proteases such as pepsin split the protein molecule into polypeptides. In the duodenum and jejunum a series of proteases split the polypeptides into depeptides and amino acids. (See Table 4–2.) Each of the proteases acts at specific points of the amino acid chain.

Protein Digestibility The percentage of protein that is available for absorption is known as protein digestibility. About 97 percent of the protein in milk and eggs is digested; meat, fish, and poultry offer only slightly less. Plant proteins have a lower digestibility—about 75–85 percent. Decreased digestibility of proteins can occur from carbohydrate fiber and enzyme inhibitors. Enzyme inhibitors occur in such foods as navy beans and soybeans; heat inactivates the inhibitors. Typical American diets have a total protein digestibility of about 92 percent.

Absorption

Dipeptides and amino acids are taken up by the mucosal cells of the small intestine and transferred to the blood circulation. The digestive tract is unable to distinguish between the amino acids of food ingestion or body recycling. Almost all the amino acids are absorbed by active transport, thus requiring energy. Separate carriers exist for some amino acids and competition can occur for the same carrier. Thus, absorption can be impaired by an imbalance of amino acids present in the small intestine.

Transportation

At any given moment, the blood circulation and other body fluids contain an assortment of amino acids, as it were, a ready-made pool of available candidates just waiting to be called for work. This *concept* (not a location!) is referred to as the amino acid "pool." These amino acids include not only those just absorbed from the intestinal tract, but also those that were released during the breakdown of body proteins.

The amino acids are rapidly removed from the circulation by the cells of the liver and body tissues. At no time is the concentration of amino acids in the blood circulation high or low; the total content remains relatively stable.

Respiration

When amino acids are used by the body to perform their myriad functions, oxygen is required and carbon dioxide and water are produced. The amount of carbon dioxide produced falls between that of carbohydrates and lipids.

Utilization

To accomplish its functions, each cell utilizes amino acids available to it. As with carbohydrates and lipids, cells' abilities to synthesize new proteins or break down old ones depend on enzymes, hormones, vitamins, and minerals, along with other substances and a favorable internal environment, like temperature and pH.

The body's cells and organs function interdependently to make sure that amino acids are available for their functions. Several reactions are used, for example, deamination and transamination. **Deamination** is the removal of the amino (NH_2) group. The amino group can be used for synthesizing *nonessential* amino acids or other nonprotein nitrogen-containing compounds, or eliminated as urea in the urine. The remaining amino acid molecule can be catabolized to carbon dioxide, water, and energy, or converted to carbohydrates and lipids, or used to synthesize the non-nitrogen portion of new nonessential amino acids. **Transamination** is the transfer of the amino group from one amino acid to another amino acid. All these mechanisms help ensure the body's homeostasis regarding amino acids, hence proteins.

Storage

Other than being able to convert one amino acid into another and circulating them throughout the body, human beings cannot store amino acids. Thus, any excess amino acids are utilized for synthesizing carbohydrates and lipids, or excreted. However, the body's proteins may be considered a storage form of amino acids. In conditions of starvation, for example, muscles and organs, being protein, are catabolized for their amino acids and supply of substrates for energy. This, of course, is not a healthy or life-preserving situation.

Excretion

As with carbohydrates and lipids, amino acids and their waste products are excreted by the digestive tract, kidneys, skin, and lungs. Undigested dietary proteins and sloughed off digestive tract cells and enzymes are excreted in the feces. The kidneys play a pivotal role by excreting excess amino groups in the form of urea (produced in the liver) and water. Amino acids can be lost through perspiration, as well as through hair and nail clippings. Lungs excrete carbon dioxide and water. When amino acids and their metabolic by-products are excreted, it cannot be determined whether they originated in the diet or in body sources.

MEETING BODY NEEDS

Requirements and Recommendations

Nitrogen Balance Protein requirements and recommendations are based on the retention or loss of nitrogen by the body. Since the distinction between carbohydrates, lipids, and proteins is the nitrogen contained in amino acids, nitrogen is the element measured.

Theoretically, **nitrogen equilibrium** exists when the anabolic and catabolic activities are equal. Positive nitrogen balance exists when amino acids are used to build additional new body proteins. The opposite situation exists when body protein breakdown is in excess of tissue maintenance and repair and represents a state of negative nitrogen balance.

Nitrogen forms 16 percent of proteins, so the nitrogen contained in foods and the amount excreted can be measured and total protein balance calculated. Although this seems easy, it is not. Usually only the nitrogen excreted in the feces and urine is measured. There are small losses of nitrogen through the skin, hair, nails, and menses, and it is difficult to measure these. Most of the studies conducted were of short-term duration, so it is difficult to truly know longer-term outcomes regarding overall protein status. For example, losses from the skin may be considerable at high environmental temperatures accompanied by heavy perspiration, until the person acclimates. Apparently to compensate during this time period, the kidneys excrete less nitrogen.

Recommended Dietary Allowances (RDAs) Protein is the only macronutrient for which there are Recommended Dietary Allowances (RDAs). (See

Chapter 1 and inside front cover of text.) The concept of nitrogen balance forms the basis for the recommendations, which are based on the amino acids present in high biological value protein foods; protein digestibility; gender and age; and reproductive status of women. For example, healthy adults are assumed to be in nitrogen equilibrium, thus, the RDAs are established for amino acids replenishment plus a statistical factor as a margin of safety. Higher levels are established for periods of growth, such as during infancy, childhood, adolescence, pregnancy, and lactation, when people are expected to be in positive nitrogen balance.

Hard manual work and exercise do not increase the protein requirement. However, some laborers and athletes who are developing greater muscle mass require some additional protein during the conditioning period only. (Most people routinely ingest extra protein.)

Sources

In the United States, foods of high biologic value—eggs, milk, cheese, yogurt, meat, poultry, and fish—provide about ⅗ of the protein intake. Legumes and nuts are rich sources of protein. People of many cultures use these high-quality plant protein sources as dietary staples.

BIOLOGIC VALUE OF SOME FOODS	
Eggs	100
Milk	93
Rice	86
Casein, beef	75
Corn	72
Peanuts	56
Wheat gluten	44

Although the quality and amount of proteins present in grain foods is lower, the servings of breads, cereals, rice, and pasta that people eat provide an important proportion of the protein intake. Vegetables and fruit account for only a small part of the proteins in typical American diets.

A good rule to observe in planning diets for adults is to include at least ⅓ of the day's protein from complete protein foods. For children, ⅔ of the protein need should be met from complete protein foods. To help preserve body tissues, each meal might supply ¼ to ⅓ of the day's protein allowance so that a dietary supply of the essential amino acids is available for tissue synthesis.

Complementarity The protein (amino acid) quality of diets can be enhanced by combining a variety of foods into meals. This is especially important when diets contain limited quantities of animal foods, or the person is in need of extra protein, or the individual is a young child with large amino acid needs but a small stomach capacity.

When one food provides the amino acids that are missing or limited in another, it is said to complement (to make complete) the second food. For example, neither dry beans nor corn provide all the amino acids needed by the body. (Each is a plant protein, thus an incomplete protein source, and does not supply all the essential amino acids.) But if they are combined for a meal, such as black-eyed peas and corn bread, the two foods supply sufficient amounts of the essential amino acids. Similarly, smaller amounts of animal foods can be used to complement larger amounts of plant foods. For instance, cheese topping complements the protein in pizza crust (wheat flour); a small amount of milk makes up for essential amino acids missing in breakfast cereal.

Vegetarian Diets Complementarity forms the basis of vegetarian diets. Probably the most important reason worldwide for eating a vegetarian diet is the availability of certain foods. In many parts of the world plant foods are abundant, whereas animal foods are scarce or expensive. In the United States, vegetarianism has been adopted by about 1 percent of the population, primarily because of concern about the most efficient use of available land for food production; a belief that vegetarian diets are more healthful; or religious or ethical beliefs.

A vegetarian diet sometimes includes one or more of these animal foods: meat, poultry, fish, eggs (ovo), milk (lacto). As long as some animal foods are included, protein and other nutrient adequacy is highly probable. Lacto-ovo-vegetarian diets and lacto-vegetarian diets can be easily planned to be nutritionally adequate, since they include animal protein foods.

For people who are strict vegetarians (vegans), careful planning is required so that complementary protein sources are included. Sample menu structures are a legume food combined with grains, nuts, or seeds at the same meal:

- *Legumes:* lima beans, red beans, soybeans, white beans; black-eyed peas, chickpeas, peas; lentils
- *Grains, nuts, seeds:* barley, bulgur, cornmeal, millet, oatmeal, rice, rye, whole wheat; Brazil nuts, peanuts, almonds, cashews, filberts, pecans, walnuts; pumpkin, sesame, and sunflower seeds

However, there are many variations of strict vegetarian diets. Some can be nutritionally adequate, whereas others are seriously lacking in one or more nutrients besides proteins. Such an example is the Zen macrobiotic diet. This is a diet progressing through several stages, the last of which consists chiefly of brown rice. Because the final stages of this diet are lacking in essential amino acids, vitamins, minerals, and water, severe malnutrition and death can result from prolonged use.

Basic Diet

The Basic Diet Nutrient Profile (Table 2–3) lists the total amounts of protein in foods. Contrast the amounts in animal foods with that in plant foods. You see that foods such as meat, fish, poultry, eggs, and milk make the largest contribution of protein to the diet. These foods also are of the highest biological value since they supply the best assortment of the essential amino acids and they are highly digestible. Legumes provide a similar total amount of protein, but it is of slightly lower quality since one or more amino acids are limited and the foods are not as digestible.

Although the total amount of protein from bread, cereal, rice, and pasta appears small in comparison, the daily quantity eaten of these foods provides an important proportion of the total protein intake. For example, the suggested 6–11 servings of the group, according to the Food Guide Pyramid, provide a minimum of 12–22 g from this food group alone. (See Figure 1–5.)

BODY BALANCE
Optimum/Adequate Signs and Symptoms

In assessing nutritional status regarding proteins, prioritize four evaluations. (See Chapter 3.) First, evaluate food intake for total calories and determine that caloric requirements are being met. If they are not, protein will be sacrificed for energy instead of being utilized for tissue growth, maintenance, or repair and healing. Second, determine the total amount of protein, the sources of the protein (complete or incomplete protein food sources), and the combination of foods at meals and snacks. For best overall metabolism, a fairly even distribution throughout the day's meals is helpful. In conjunction with dietary protein intake, evaluate water balance. Excess amino acids are excreted as urea, so sufficient water intake is required for elimination. (See above sections and Chapter 4.) Third, perform serial anthropometric measurements. In children, protein is necessary for adequate linear growth (length or height) and weight gain; for adults, protein is necessary for weight. It may be important to monitor measurement of midarm muscle circumference. Retention of muscle mass in the arms and legs is an indicator of somatic (muscles) protein status. (See Chapters 3 and 16.) Fourth, evaluate laboratory data. Routine laboratory blood tests, such as the Sequential Multiple Analyzer (SMA) series, include values related to protein status, such as albumin, creatinine, creatine kinase, total protein, urea nitrogen, and uric acid. Albumin, prealbumin, and transferrin are indicators of visceral (organs) protein status. Since proteins cannot pass through membrane, no protein should be in urine. However, breakdown products can be excreted and may be useful indicators. However, interpret laboratory test results with caution, as these tests are indicators of multiple pathologies that secondarily may affect protein status. (See Chapters 3 and 16 and Appendix Table B–2 and Figure B–1.)

Insufficiency/Deficiency

Conditions Traditionally, two forms of protein deficiency are recognized—kwashiorkor and marasmus. These two conditions are known as protein-energy malnutrition (PEM). Both were (and are) prevalent in countries torn by war, famine, and political strife. Both deficiencies continue to be major health problems for infants and young children in many parts of the world. Both conditions are examples of negative nitrogen balance, that is, either protein intake is not keeping up with body demands, and/or the body is using up its own proteins to supply energy.

Kwashiorkor is characterized by the condition of insufficient protein, but sufficient calories. Classically, this condition appears after a baby is weaned, when a new infant is born. The food given to the older child is starchy and low in protein, leading to the findings summarized in Table 7–2. **Marasmus** is the condition of both insufficient protein and insufficient calories. This condition occurs in infants who are weaned early and not fed enough formula and food, for example, because the mother's milk supply is inadequate during lactation, or because there is no food available for the baby, or because the mother must return to work. Thus, the infant receives a diet lacking in both calories and protein. The plight of this kind of child is desperate and the death rate is high. (See Figure 7–3.)

Although not caused by the same reasons, these two forms of protein-energy malnutrition occur in other people as well. For example, either one or a combination of both can occur in children and adults admitted to hospitals from their homes where they ate inadequately; or in people who remain without nutrition therapy while in the hospital; or in people who live in residential settings where little attention is paid to their food and nutritional needs. (See Chapter 16.) Also, several diseases such as gastrointestinal diseases and diabetes mellitus, when untreated, can result in negative nitrogen balance. (See Chapters 22 and 26.)

Signs and Symptoms In kwashiorkor from any cause, when protein deficiency is prolonged, the plasma albumin level gradually falls to 3.5 g/dl (35 g/L) or less. The ability to regulate osmotic pressure is reduced so that fluids begin to shift from the correct body compartment and to accumulate in the body. (See Chapter 4.) Puffiness around the eyes and swelling of the abdomen and legs/ankles are common signs. This condition is known as nutritional edema. The

TABLE 7–2

Characteristics of Kwashiorkor and Marasmus in Children

	Kwashiorkor	Marasmus
Age of Onset	18–24 months	6–12 months
Dietary Lack	Protein Multiple minerals and vitamins	Calories; protein Multiple minerals and vitamins
Clinical Findings	Severe apathy Growth failure ↓Hair color; easily pulled out ↓Skin pigmentation; ulceration Round, full face Edema Diarrhea Frequent infections Enlarged fatty liver Anemia is common ↓Serum albumin ↓Immune response Some potassium deficiency	May seem alert Severe growth failure; emaciated No hair color change Pigmentation not usually changed Wrinkled face No edema Diarrhea Frequent infections No fatty liver; may be enlarged Anemia is common Serum albumin almost normal Normal to low immune response Severe potassium deficiency
Recovery with Therapy	Rapid; sudden death sometimes occurs	Slow; brain development sometimes inadequate

person appears bloated and in the early stages may appear slightly "chubby" to the casual observer. Protein deficiency also reduces the immune response so that the individual is less able to withstand infections. (See Chapters 16 and 19.)

Marasmus is characterized by adequate serum albumin levels, but loss of subcutaneous body fat and muscles. People are using their own proteins and adipose reserves for energy. Persons appear emaciated, the "skin and bones" look of people starving, which is exactly what they are doing. (See Chapter 16.)

People Most at Risk In technologically advanced countries, protein deficiency is rare. This is primarily because agriculture and food distribution systems are highly sophisticated, resulting in an abundance of available wholesome food, and people usually have adequate means to purchase some protein-containing foods.

Healthy people who select a vegetarian diet need special assessment and counseling, especially children and pregnant or lactating women. First, emphasize caloric adequacy and protein quantity and quality. The large amount of food that must be ingested to supply sufficient amino acids (and calories) may be too much for people with limited stomach capacities, like children. Second, highlight vitamin and mineral requirements. Vitamin B_{12} is present in animal foods. Children raised on strict vegetarian diets will not grow properly unless their diets are supplemented with vitamin B_{12}. In children's diets, soy milk fortified with vitamin B_{12}, vitamin D, and calcium is substituted for cow's milk. Although leafy green vegetables and whole-grain cereals are a good source of iron, some of this iron is poorly absorbed. The absorption of iron is improved if each meal supplies a generous amount of vitamin C. (See Chapters 10 and 11.)

Protein deficiency can occur when an individual does not eat enough protein-containing foods, or eats inferior quality protein foods, or obtains

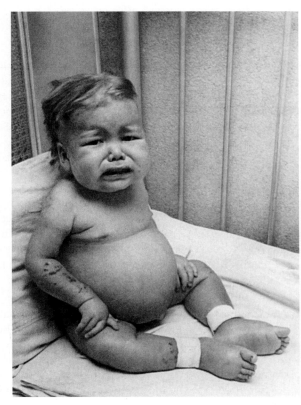 

FIGURE 7–3
The child on the left has the diagnosis of kwashiorkor; the child on the right has marasmus. Compare and contrast the clinical findings of the two conditions. As you examine the photographs of these two children, especially note the appearance of their faces, arms and legs, and bodies. *(Source:* Left, Nagata/UNICEF/United Nations; right, courtesy of Food and Agriculture Organization/United Nations.)

insufficient calories. Some persons use crash diets for weight reduction and thus have a low protein intake. (See Chapter 9.) Many elderly persons are unable to chew well, or do not drink milk, or believe they do not need protein foods such as meat or eggs. (See Chapter 15.) Fractures, burns, infections, and surgery increase the requirement for protein, but patients in these circumstances often have poor appetites. (See Chapters 20 and 21.) In fact, appetite is negatively affected by most diseases, just when the needs for nourishment are high. Health care providers must be particularly alert to the possibility of protein undernutrition in patients with poor appetites. Steps to improve food intake should be taken quickly before problems arise. (See Chapter 16.)

Excess/Toxicity

Conditions In healthy people, no specific term or diagnosis is applied to protein excess. The body uses several avenues to remove excess amounts before toxic situations appear. The amino group, if not converted into urea rapidly enough, can form ammonia, which is toxic to the brain. So, to preserve homeostasis, excess dietary intake of food proteins or commercial protein and amino acid supplements may not be absorbed and thus excreted directly in the feces, or may be deaminated by the liver and excreted by the kidneys, or may be converted into carbohydrates or lipids and stored as body fat.

Signs and Symptoms The healthy liver and kidneys have amazing capacities to perform their degradation tasks within wide ranges of protein intake, apparently without damage to either organ. However, when the liver or kidneys become diseased, the handling of excess protein becomes impaired or impossible; modification of the total amount and the quality of proteins is crucial. (See Chapters 23 and 28.)

If a high protein intake is based on increased amounts of meat, poultry, fish, whole milk, and whole-milk cheeses, the intake of total fat, saturated fat, cholesterol, and total calories may reach undesirable levels. (See Chapter 6.) Also, during short-term laboratory or epidemiological research studies, some investigators have seen increased excretion of calcium, leading to the speculation that a high protein diet may be a risk factor for osteoporosis. Others have theorized that long-term excess protein intake may lead to kidney damage over time. Both of these hypotheses remain in the research stage. However, there appears to be no benefit from excess protein intake.

People Most at Risk People most at risk for consuming unnecessary proteins are those who mistakenly believe proteins are superior to other nutrients. Examples are athletes (especially bodybuilders); people on weight-loss diets; and teenagers who want improved body contours and condition but who are especially duped by advertisements and personal testimonials. From an economic point of view, protein-rich foods are the most expensive. When income is limited, too much emphasis on protein may exclude other foods that are needed for calories, vitamins, and minerals.

HEALTH AND NUTRITION CARE ISSUES

After water, carbohydrates, and lipids, the fourth nutrient to assess might be protein. These four nutrients are considered macronutrients, that is, the body requires them in larger proportions than vitamins or minerals. Energy requirements of the body outdistance the body's other needs for protein. Thus, if the body must choose how it will use protein, it will metabolize it first for energy, even though this is an ineffective use of protein and an inefficient and expensive way to obtain energy. Therefore, as you screen clients' diets or assist in providing nutrition therapy, calculate the energy intake first to make certain it is sufficient for body needs, then focus on protein quantity and quality.

PUTTING IT ALL TOGETHER

Food proteins are unique because they contain amino acids with amino groups (NH_2) that the body uses to make its own species-specific proteins. All body cells contain proteins. Proteins' crucial functions relate to the growth, maintenance, repair, and healing of the body. In the event of protein intake beyond the body's needs, the proteins are dismantled into their constituents. The amino group is excreted as urea and the remainder is converted to carbohydrates or lipids. In times of energy demands, proteins and their amino acids are diverted to satisfy this primary requirement of the body.

1. I've heard that when I eat extra proteins I need to drink extra water. Why?

One of the ways the body gets rid of extra proteins is by changing them into a substance called urea. Urea is the main nitrogen-containing waste product (from the breakdown of amino acids) in urine. Several nonprotein nitrogenous compounds such as creatinine and uric acid are excreted in small amounts. To maintain the body in chemical balance, extra urine needs to be eliminated, so extra water needs to be taken in.

2. Are gelatin supplements of value?

The protein value of gelatin is overrated. Gelatin, although derived from animal sources, is deficient in nearly all essential amino acids. Dry, unsweetened powdered gelatin is about 90 percent protein, but the average gelatin dessert is mostly water and furnishes only about 2 grams of total protein.

✔ **CASE STUDY: Evaluation of a College Student's Protein Intake**

Alice, a 20-year-old college student, has decided she wants to follow a lacto-ovo-vegetarian diet and wants to make certain she is receiving enough protein. Her measurements are: height = 65 in (164 cm), without shoes; weight = 130 lb (59 kg), in lightweight indoor clothing; frame size = medium. Her serum albumin is 4.0 g/dl (45 g/L). The following is an example of a typical day's food intake, with the amounts listed in parentheses based on the servings from the Basic Diet (Table 2–3).

Breakfast:	Orange juice (1)
	Cooked oatmeal (1), with
	Skim milk (½)
	Coffee (–) with cream (1) and
	sugar (1)
Lunch:	Lentil soup (1)
	Egg salad sandwich
	Whole wheat bread (2)
	Egg (1)
	Mayonnaise (1)
	Carrot sticks (½)
	Skin milk (1)
After classes:	Peanut butter crackers
	Peanut butter (1)
	Enriched crackers (1)
	Banana (1)
Dinner:	Spaghetti (3)
	Tomato sauce (1)
	Spinach salad (1)
	Salad dressing (1)
	Herbal tea (–)
Study Snack:	Pretzels (2)
	Apple (1)

Questions

1. Evaluate Alice's anthropometric measurements and laboratory value. (See also Chapter 3 and Appendix Tables D–1 and D–2 and Figures B–1, D–5, and D–6.)

2. Based on 0.8 g protein per kilogram of recommended weight (RDAs), what protein intake is recommended for Alice?

3. Using the Basic Diet Nutrient Profile (Table 2–3), calculate Alice's approximate total energy (caloric) intake.

4. Using the Basic Diet Nutrient Profile (Table 2–3), calculate Alice's approximate total protein quantity. Determine the percentage of calories from protein. Interpret the findings. What changes, if any, would be desirable in her *total* caloric and protein intakes?

5. Classify each of the foods in Alice's diet that contain protein according to quality: complete or incomplete. What proportion of protein is from animal sources and from plant sources? What changes, if any, would be desirable in the *quality* of her protein intake?

6. What examples of complementary protein food combinations are evident in Alice's diet? How might the menus be adjusted?

7. Alice tells you she is interested in adding to her daily diet a powdered protein supplement of essential amino acids formulated for women. How might her body use the amino acids? How would you counsel her?

FOR ADDITIONAL READING

American Dietetic Association: "Position Paper: Vegetarian Diets," *J Am Diet Assoc,* **93:**1317–19, 1993.

Clifford, M: "Nutrition Counseling of the Vegetarian," *Top Clin Nutr,* **10:**44–47, March 1995.

Haddad, E: "Meeting the RDAs with a Vegetarian Diet," *Top Clin Nutr,* **10:**7–16, March 1995.

Johnston, P: "Vegetarians Among Us: Implications for Health Professionals," *Top Clin Nutr* **10:**1–6, March, 1995.

Energy Metabolism; Physical Fitness

Chapter Outline

Energy Transformation
Energy Release from Foods
Energy Needs of the Body
Fitness Factors
Health and Nutrition Care Issues

Putting It All Together
Questions Clients Ask
Case Study: A Man Wants to Be
  Nutritionally and Physically Fit

Increase to at least 40 percent the proportion of people aged 6 and older who regularly perform physical activities that enhance and maintain muscular strength, muscular endurance, and flexibility. *

✔ CASE STUDY PREVIEW

Energy resources are in the news daily—oil, gas, electricity, solar power, and the energy of the human body. People realize a delicate fit exists in nature between availability, use, and abuse. How do human beings achieve fitness?

Healthy People 2000: National Health Promotion and Disease Prevention Objectives. Public Health Service, U.S. Department of Health and Human Services, Washington, DC, 1991, Objective 1.6, p. 92.

The body requires a constant source of energy to keep its complex machinery functioning. Energy is defined as the capacity to do work. It is the power that keeps the body performing and at the best temperature. Energy is perhaps the most frequently discussed topic in nutrition. The original resource for the body's energy and heat is food.

In many parts of the world, far too many people do not get enough food to meet their energy needs. To survive, the body makes up the difference by cannibalizing its own parts for energy. The person reduces physical activity. Gradually the body adjusts to a lower level of metabolism. When the energy deficit becomes severe the time comes when the body runs out of its own tissues for survival. Hunger, starvation, and famine are the specters that often come to our attention on television and in magazines and newspapers. For these millions of faces, especially children, the human machine eventually stops.

In affluent countries there is an abundance of food and most people have options about their food choices to give them energy. Food is so plentiful that millions of people are concerned about their body's excess stores of energy. Many schemes help them to "count calories." They buy diet books, attend classes to learn about weight reduction, or purchase special products that are supposed to help them lose their energy deposits. (See Chapter 9.)

One definition of physical fitness is the efficient and effective use of energy by the body. The concept combines food and its contained energy, nutrients provided by food to release and store energy, and physical activity to help the cells, organs, and body systems operate in top-notch condition. The popular connotation is that thin or young people are physically fit and overweight or old people are physically unfit. This is not true. Outward appearances can be very deceiving about the inner workings of the body.

ENERGY TRANSFORMATION

Energy—The Body's Constant Need

The human body requires fuel every second of life to do its work. Since there can be no turning on and off of its energy demands, its energy supplies must be available as long as life continues. Food is the supply and adenosine triphosphate (ATP) is the fuel.

Metabolic Processes to Yield Energy

The energy needs of the body take priority over everything else, such as building and maintaining tissues or regulating body functions. Cells are the location of energy production, with specific compartments handling specific tasks. Many steps are involved. Some require oxygen and some do not; all require enzymes; most require vitamins and minerals. Many steps are prompted by signals from hormones and nerves. Some steps even require some of the body's own energy.

Glucose, glycerol, and fatty acids provide most of the body's energy sources, but if the supply of these materials is inadequate, amino acids will be used. Alcohol is a source of energy, too, but the production is inefficient. The yield of energy from food is sometimes summarized as:

$$\text{Glucose, glycerol, fatty acids, or amino acids} + \text{oxygen} \longrightarrow$$
$$\text{energy (ATP)} + \text{carbon dioxide} + \text{water}$$

Figure 8-1 diagrams energy release from food. Note that macronutrients start out along separate pathways, are transformed, and converge into the chemical known as acetyl coenzyme A (CoA). (There are a few exceptions.) Then this "common denominator" enters a pathway known as the tricarboxylic acid cycle (TCA cycle). This cycle also is known as the citric acid cycle, or Krebs cycle for the scientist who first described it. From here, the products enter their third and final pathway of the journey—the electron transport system—and undergo oxidative phosphorylation. All in all, the entire energy production system captures about 40 percent of food energy as body energy. The remainder is dissipated as heat.

ENERGY RELEASE FROM FOOD

Units of Energy

Strictly speaking we do not "eat" calories. Just as we can measure length in centimeters or inches and weight in kilograms or pounds, so we can measure the energy value of foods or the energy needs of the body in units called calories. In

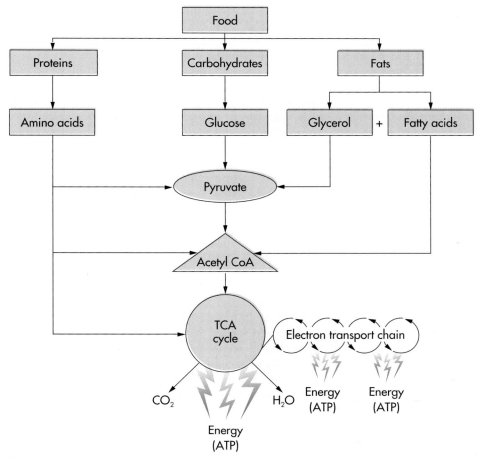

FIGURE 8–1

This simplified diagram shows that glucose, glycerol, fatty acids, and amino acids enter the tricarboxylic acid cycle by somewhat different routes to yield energy.

PHYSIOLOGICAL FUEL
FACTORS

Nutrient	kcal/g
Carbohydrate	4
Fat	9
Protein	4

CALORIC VALUE OF
ONE CUP WHOLE MILK

Nutrient	Amt., g	kcal Tot.
Carbohydrate	12 × 4 =	48
Fat	8 × 9 =	72
Protein	8 × 4 =	32
		152

nutrition, the "large" calorie, more correctly called a kilocalorie (kcal), is used, whether it is written with a small c (calorie) or a capital C (Calorie). This unit is 1000 times as great as the calorie unit used in chemistry and physics; that is, it is the amount of heat required to raise the temperature of 1000 g of water by 1 degree Celsius.

The energy value of food is measured in the laboratory with an instrument called a bomb calorimeter. A food is placed inside this double-chamber instrument, and water is placed between the inner and outer chamber walls. Oxygen is added to the chamber containing the food, ignited, and the increase in water temperature is noted. This method is known as direct calorimetry, because the energy contained in the food is measured. The caloric values for foods obtained with this instrument are higher than those actually realized by the body. A correction must be made for the imperfect digestibility of carbohydrates, fats, and proteins (see Chapter 4) and also for the losses of nitrogen in the urine. The corrected values, known as the physiological fuel factors, are used in calculations.

Energy Value of Foods

Compare the energy values in tables of food composition (for example, Tables 2–3 and 2–4, Appendix Table A–1, or a computerized nutrient database) with the numbers you see listed in modified food tables (for example, Appendix Table A–2), with the numbers you see printed on the nutrition label of packaged foods, and with the numbers you calculate from the grams of carbohydrate, fat, and protein contained in a food. You see there are discrepancies! The contrast reflects differences among the potential energy available in food, the potential energy yielded when food is transformed by the body, and the arithmetic method of rounding up or down to make calculations of meal patterns and menu structures as simple as possible. It also is important to remember that a range exists around each number (an average). Thus, when we nutritionally assess and treat, we do so in an estimated, approximate way.

Nevertheless, by examining the various tables you can draw some general conclusions about the caloric value of foods. First, foods containing the greatest proportion of water and fiber are lowest in calories. Vegetables and fruits, as food groups, fall into this category. The variation in calories between a tomato, for example, and a sweet potato lies in the much greater water content of the tomato compared to the sweet potato. Fresh fruits are much lower in calories than canned or frozen fruits packed in syrup. Many foods contain little water but appreciable amounts of digestible carbohydrates; among them are flour, cereals, bread, pastas, rice, sugar, and candy. Weight for weight, these foods rank higher in calories than vegetables and fruits. Legumes are examples of the storage reserves of plants and thus are a concentrated source of the energy-containing macronutrients, especially carbohydrates and proteins.

Second, the highest concentration of calories occurs in foods that contain the greatest proportion of fat. It doesn't matter if the fat comes from a vegetable or animal source; fats yield the same number of calories. Their almost 100 percent fat content makes oils, margarine, butter, and cream high calorie foods. Skim milk has about half the calories of whole milk; only the fat is removed. In an egg, the fat is located chiefly in the yolk, rather than in the white. Lean meat, poultry, and fish are moderate in caloric content, but fatty meats and fish are higher. Many cooked foods become higher in calories because of the ingredients used or the style of preparation. Cakes, cookies, pies, and pastries contain shortening, flour, sugar, eggs, milk, plus other ingredients like nuts and coconut. Deep-fat-fried foods are high in calories. A piece of lean meat may be only moderately high in calories if it is broiled, but it if is dipped in egg and bread crumbs and then fried, the calorie value could be twice as high.

Third, ranked after fat as a calorie source is alcohol. One can of regular beer (12 oz) contains 150 kcal, one glass of dessert wine (3½ oz) about 140 kcal, and one martini (1½ oz gin and ½ oz vermouth) about 135 kcal. Often coupled with alcohol consumption are high calorie foods like peanuts, olives, and chips.

ENERGY NEEDS OF THE BODY

The body's total energy requirement is the sum of the energy expended for basal metabolism, voluntary activity, and the influence of metabolizing food.

Basal Metabolism

Basal metabolism is the term given to the basic life-sustaining, vital-functions energy needs of the body. It is known as the basal metabolic rate (BMR) or basal energy expenditure (BEE). It is the largest proportion of the body's energy requirements and accounts for ½ to ⅔ of the total energy expenditure of most people. Included are the involuntary activities of the body, that is, those bodily activities over which we have no control: breathing, beating of the heart, circulation of the blood, metabolic activities taking place within all cells, maintaining muscles, and regulating normal body temperature. Of the basal energy expenditure, about ⅗ is due to the activities of the heart, lungs, liver, and kidneys, with about ⅕ due to the activities of the nervous system.

Basal metabolism is affected by many factors. Some are a function of others.

- *Body composition.* The greatest influence, about 80 percent, is correlated with the amount of lean body mass—the muscles and organs. This might make logical sense since these body parts are the most metabolically active, thus spending most of the body's energy. An example is the higher rate for an athlete with firm muscles and little body fat compared to a nonathlete with less muscle mass and more body fat.
- *Body surface area and size.* The external shape of the body affects the rate because more or less heat can be dissipated through the skin. With the same muscle mass and body fat, a tall person has a greater skin surface through which heat is lost than does a shorter, smaller person. Compared to adults, an infant has a larger surface area in proportion to total body weight and thus loses much heat through the skin, resulting in a higher basal metabolic rate.
- *Gender.* The basal metabolic rate is about 10 percent higher in men than in women. The higher percentage of lean body mass in men and the higher proportion of adipose tissue in women, as well as hormonal differences, are believed to account for the variation in basal metabolism.
- *Growth.* The building of new tissue accounts for a high rate in infants, during the rapid growth period of adolescence, and for the development of the fetus during the second and third trimesters of pregnancy.
- *Age.* The metabolic rate is highest in infants because of their large surface area and the cellular activity in building additional new tissues. The rate declines somewhat during childhood, then increases somewhat during the adolescent years, when the growth rate again accelerates. After about age

21 there is a steady decline of about 2 percent for each decade. Thus, a person 81 years old might have a metabolic rate 12 percent lower than that of a 21-year-old of the same gender and body surface area and size.

- *Body temperature.* For each degree of increase in body temperature above normal, the rate increases 7 percent for each degree Fahrenheit (13 percent for each degree Celsius).

Applications in Practice Originally, basal metabolic energy expenditures were (and still are) determined under strict research laboratory conditions. One method used is called indirect calorimetry because measurements are made of two gases (oxygen and carbon dioxide) that represent the body's energy at the cellular level. The inhalation of oxygen and exhalation of carbon dioxide is measured for about one hour while the person is at complete rest, but just awakened in the morning after a 12-hour fast.

For estimation purposes for initial screening of clients, we now apply the derived mathematical equations to calculate approximate basal metabolic rate. Computers and programmable hand-held calculators assist clinicians with the arithmetic. Two of the more commonly used equations for adults are:

Harris-Benedict equations (named for the scientists)

Men: $66.47 + (13.75 \times wt,kg) + (5.00 \times ht,cm) - (6.76 \times age,yr)$

Women: $655.10 + (9.56 \times wt,kg) + (1.85 \times ht,cm) - (4.68 \times age,yr)$

Weight basis equations

Men: $(wt,kg) \times (1.0 \ kcal) \times (24 \ hrs)$

Women: $(wt,kg) \times (0.9 \ kcal) \times (24 \ hrs)$

These equations may be modified for people whose illness or trauma increases basal metabolic rate. (See inside back cover.) Basal metabolism decreases during inadequate food intake, for example, during weight loss or starvation.

A variation of the research basal metabolic rate test can be performed. While a person is resting quietly, air exchange can be measured for about 10–20 minutes with a portable instrument commonly called a metabolic cart. This kind of measurement is known as the resting energy expenditure (REE). Since strict research conditions are not used, the resulting energy expenditure is slightly higher (about 3–10 percent higher) than the BMR. Other terms for the REE are Resting Metabolic Rate (RMR) and Resting Metabolic Expenditure (RME). Equations are listed in Table 8–1.

Voluntary Activity

The second largest proportion of energy expended by the body comes from the kind of physical activity in which we engage and the amounts of time spent in each activity. Our daily work may vary from sitting in front of a computer, to doing active housework, to doing hard manual labor. In our leisure time we might choose to watch television, take a leisurely or brisk walk, or go swimming or dancing. Mental effort, as in studying, is considered a very light activity. Of course, if you are tense and restless while you are studying, you will increase your caloric need somewhat. But if you are studying and nibbling food all the while, you can easily consume far more calories than your mental effort and physical activity expend.

Assigning exact values to any activity is difficult because individuals vary widely in the efficiency with which they use their bodies. Table 8–2 shows calculation estimates for adults. Table 8–3 provides examples.

The third and least proportion of energy expenditure is the energy required for food and nutrients' ingestion, digestion, absorption, transportation, utilization, storage, and excretion. This energy is known as the calorigenic effect of food. It is highly variable and amounts to about 5–10 percent of the energy ingested. Other names for this term are specific dynamic action of food, thermic effect of food, and diet-induced thermogenesis. Sometimes you will see this energy use added to the expenditure side of the equation and sometimes you will not. Part of the problem is that you need to know the calorie content of the foods ingested. Other times, you will see calculations using 5–10 percent of the BMR as the number to be added. Some investigators think the number is so small as not to be relevant over the long term.

Energy Allowances

The Recommended Dietary Allowances (RDAs) for kilocalories are shown in the table printed on the inside back cover of the text. These recommendations are only approximate because of individual variations. Screening and monitoring of weight, body composition, and physical endurance and conditioning are keys to identification of someone's energy requirements over the course of a lifetime.

Basic Diet Review Table 2–4 to see Basic Diet Nutrient Profile comparisons of three diets at different calorie levels, based on the Food Guide Pyramid (Figure 1–5).

FITNESS FACTORS

A general definition of fitness is having a balance of physical, mental, emotional, and spiritual factors in one's life. Nutrition plays a role in each of these factors. In the United States, from the 1960s to the late 1980s, fitness assumed the meaning of intense physical exercise requiring high-tech accoutrements of the "right" equipment, location, and wardrobe. Some people went after physical fitness with a vengeance.

The problem, though, slowly became apparent—an exceedingly small group of people (about ⅕ of the adult population) were estimated to be active enough to be considered physically fit. Along with no improvement in physical fitness for most Americans there continued to be an increasing percentage who were at unhealthy weights (too low or too high), plus there were increasing numbers of "sports injuries." For instance, people began an exercise program without consulting anyone, other than perhaps a friend; some read one of the popular exercise books or watched a videotape; some followed along with exercises demonstrated on early morning TV programs; others joined a newly constructed gym or spa to which they had to drive. Except for a minority of people, enthusiasm remained high for a few weeks, but soon, as the individual saw no clear benefits or found other uses for his or her time, the program was dropped. Strong motivation was required to maintain an intense exercise program over a period of time, outside of the routine of the typical day. Even more unsettling was the fact that the majority of people never even attempted to increase their physical activity.

By the early 1990s, health researchers and practitioners realized that exercise, as "physical activity," needed to be as consistent and integral a part of each individual's day as eating nourishing, enjoyable foods. Since everyone needed to

TABLE 8–1

Equations for Predicting Resting Energy
Expenditure from Body Weight

Gender and Age Range (years)	Equation to Derive REE in kcal/day
Males	
0–3	$(60.9 \times wt^a) - 54$
3–10	$(22.7 \times wt) + 495$
10–18	$(17.5 \times wt) + 651$
18–30	$(15.3 \times wt) + 679$
30–60	$(11.6 \times wt) + 879$
>60	$(13.5 \times wt) + 487$
Females	
0–3	$(61.0 \times wt) - 51$
3–10	$(22.5 \times wt) + 499$
10–18	$(12.2 \times wt) + 746$
18–30	$(14.7 \times wt) + 496$
30–60	$(8.7 \times wt) + 829$
>60	$(10.5 \times wt) + 596$

Source: Reprinted with permission from *Recommended Dietary Allowances: 10th Edition.* Copyright 1989 by the National Academy of Sciences. Courtesy of the National Academy Press, Washington, D.C.

[a] Weight of person in kilograms.

TABLE 8–2

Approximate Energy Expenditure for Various Activities in Relation to Resting Needs for Males and Females of Average Size

Activity Category	Representative Value for Activity Factor per Unit Time of Activity
Resting Sleeping, reclining	REE × 1.0
Very light Seated and standing activities, painting trades, driving, laboratory work, typing, sewing, ironing, cooking, playing cards, playing a musical instrument	REE × 1.5
Light Walking on a level surface at 2.5 to 3 mph, garage work, electrical trades, carpentry, restaurant trades, house cleaning, child care, golf, sailing, table tennis	REE × 2.5
Moderate Walking 3.5 to 4 mph, weeding and hoeing, carrying a load, cycling, skiing, tennis, dancing	REE × 5.0
Heavy Walking with load uphill, tree felling, heavy manual digging, basketball, climbing, football, soccer	REE × 7.0

Source: Reprinted with permission from *Recommended Dietary Allowances: 10th Edition.* Copyright 1989 by the National Academy of Sciences. Courtesy of the National Academy Press, Washington, D.C.

TABLE 8–3

Example of Calculation of Estimated Daily Energy Allowances for Exceptionally Active
and Inactive 23-Year-Old Adults

| | Step 1: Derivation of Activity Factor[a] | | | |
| | Very Sedentary Day | | Very Active Day | |
Activity as Multiples of REE	Duration (hr)	Weighted REE Factor	Duration (hr)	Weighted REE Factor
Resting 1.0	10	10.0	8	8.0
Very light 1.5	12	18.0	8	12.0
Light 2.5	2	5.0	4	10.0
Moderate 5.0	0	0	2	10.0
Heavy 7.0	0	0	2	14.0
TOTAL	24	33.0	24	54.0
MEAN		1.375		2.25

Step 2: Calculation of Energy Requirement, kcal per day

Gender	Resting Energy Expenditure[b]	Very Sedentary Day (REE × 1.375)	Very Active Day (REE × 2.25)
Male, 70 kg	1,750	2,406	3,938
Female, 58 kg	1,350	1,856	3,038

Source: Reprinted with permission from *Recommended Dietary Allowances: 10th Edition.*
Copyright 1989 by the National Academy of Sciences. Courtesy of the National Academy Press,
Washington, D.C.

[a] Activity patterns are hypothetical. As an example of use of the ranges within a class of activity
(Table 8–2), very light activity is divided between sitting and standing activities.
[b] From equations given in Table 8–1.

participate, revised media messages and focused programs were initiated. In
schools, physical education turned to emphasizing activities that were appropriate for lifetime participation. (See Figure 8–2.) Socially oriented physical activities
were incorporated into day programs and residential programs for older adults.
Physical activities were modified for children and adults with special health
needs, such as those who were mentally and physically challenged and those
with arthritis, heart conditions, or lung disorders.

Activity Plus Nutrition: Basis of Physical Fitness

Activity On the output side of the energy balance equation, any activity
increases energy expenditure and improves overall physical fitness. Increasing
energy expenditure by 100 kcal has the same immediate effect on body weight as
reducing the diet by 100 kcal.

Aerobic exercises are those exercises that require oxygen. They involve the
large muscle groups. The everyday activities of daily living like brisk walking,
climbing stairs, or heavy gardening are excellent. Other kinds of exercise include
swimming, biking, rollerblading, running, dancing, and calisthenics.

The amount of energy used in physical activity depends on the intensity of
the exercise, the duration of the exercise, and the weight of the individual. It is
also influenced by the level of skill, the amount of rest taken during the exercise
time period, and the environmental temperature. Generally, a total of 30 minutes
daily, either at one time or in divided segments during the day, is recommended.

FIGURE 8–2
Physical education classes focus on body conditioning exercises and activities that can be enjoyed both in the school gym and at home. (*Source:* Anthony Magnacca/Merrill.)

During exercise, glucose and fatty acids provide the main sources of energy. Glucose can be metabolized aerobically and anaerobically; fatty acid metabolism is aerobic. During aerobic exercise, mainly fats are burned. With an increase in aerobic activity the respiratory rate increases, so that more oxygen can be provided for the metabolism of fats and carbohydrates. With exercise there is a change in body composition. Fat stores are utilized for energy and lean body mass is increased. There is also a slight increase in the basal metabolism because of increased amount of muscle mass and improved tone.

Nutrition On the input side of the energy equation, fortunately, most people interested in physical fitness realize nutrition is a foundation. The goal for caloric intake is that level that will maintain healthy weight for the adult and permit normal growth for the child and teenager.

Water intake should increase to cover losses in sweat and from increased respiration, depending on activity and climate. Protein, vitamin, and mineral requirements for people who adopt a regular program of physical activity do not differ from the RDAs for persons of a given age and gender. Purchases of special supplements probably are unnecessary, although highly touted. (See Chapter 2.) Using the Food Guide Pyramid to choose delicious wholesome foods helps ensure optimal nutrient intakes. (See Figure 1–5 and Tables 2–3 and 2–4.)

HEALTH AND NUTRITION CARE ISSUES

With only 20 percent of adult Americans active, and this percentage remaining fairly constant over time, it seems out of reach to be able to meet the Healthy People 2000 goal of "increase to at least 40 percent the proportion of people aged 6 and older who regularly perform physical activities that enhance and maintain muscular strength, muscular endurance, and flexibility." Thus, this is a goal all people in the health fields can promote for others . . . and practice themselves.

PUTTING IT ALL TOGETHER

The body's demand for energy is unfaltering. But determining the body's exact day-to-day energy balance currently is imprecise. Tremendous interplay exists between physical activity, food choices, and optimal nutritional status. Daily enjoyment of nourishing foods and physical activities provides genuine benefits. Not only is energy balance improved, but so is body composition and overall health, leading to the attainment of "fitness" and "wellness."

❓ QUESTIONS CLIENTS ASK

Many high school and college students as well as adults engage in competitive sports. Athletes should train under the watchful eye of competent coaches and medical experts. Today, many athletes have the benefit of a personal registered dietitian and a personal exercise physiologist on the training team. Of great interest to the athlete are dietary factors to improve performance and endurance. Since nutrition and food misinformation abounds within the athletic world and is passed on to others through magazine articles, advertisements, and testimonials, health care providers often are asked athlete-nutrition-related questions. The questions and answers below relate to adults, not children.

1. What sports drink should I take?

WATER! Water is the greatest need of the athlete and dehydration the greatest potential problem. During training, water is lost mainly through sweat and also through respiration. Loss of 2 percent of body weight can impair performance, and a 5 percent loss creates the danger of dehydration and heatstroke.

Thirst is a poor indicator of the physiological need for water since exercise can interfere with the thirst sensation. Drinking only to satisfy thirst is not likely to replace actual fluid losses. The athlete should drink about 16 oz of cool water 2 hours before the competition, and another 16 oz about 15 minutes before the event. During the competition about 4–6 oz water every 15 minutes is recommended, depending on the kind of event and the climate. The athlete should continue to drink fluids after the competition until the pre-event weight is fully restored.

Of special concern are the practices sometimes engaged in by athletes such as wrestlers who compete in weight categories and have pre-event weigh-ins to determine their class. They often attempt to achieve weights considerably below their natural weight to give them a supposed competitive edge. Fasting, withholding liquids by mouth, using diuretics and laxatives, working out in plastic or rubber suits, and sometimes taking steam baths or saunas are common practices. These individuals are at risk for dehydration and electrolyte imbalance. Preferably, athletes should compete at weights that can be maintained without resorting to drastic measures.

2. I want to replace the electrolytes I lose through sweat because I don't want my muscles to cramp up. Should I take salt tablets and electrolyte replacement drinks?

Sodium and chloride are the main electrolytes lost through perspiration. To counteract this loss, salt tablets were once routinely used. However, the salt draws fluid into the stomach, causing cramping, dehydration, nausea, and vomiting. Nowadays, sports drinks containing electrolyte mixtures are heavily advertised. But most diets supply ample amounts of salt and other minerals. Thus, under most circumstances, athletes consume sufficient minerals by eating regular wholesome foods and, if necessary, eating more salty foods or using extra salt at meals. Although a number of commercial fluid electrolyte replacers are widely used, they probably needn't be.

3. How much extra protein do I need? I take amino acid supplements to help build my muscles.

The RDA of 0.8 g protein per kilogram of body weight may not be enough during periods of intense training and exercise. An allowance of 1.0 to 1.5 g protein per kg of body weight generally is recommended. Athletes eating according to the Food Guide Pyramid consume these levels or more. There is no need for protein or amino acid supplements because they have no effect on further increasing the body musculature. Only training will accomplish that.

4. What kind of "carbos" should I eat to "load up"?

The athlete depends on a store of glycogen in activities of short duration and high intensity. Throughout training about 50–55 percent of total calories should come from complex carbohydrates and an additional 10 percent from simple sugars. This requires emphasis on breads, pastas, rice, cereals, and starchy vegetables.

An endurance event is one in which there is continuous, uninterrupted activity for 1½ hours or more. Athletes competing in endurance events such as marathons, cross-country skiing, and long-distance running, cycling, walking, or swimming can deplete the glycogen stores during a competition and experience the fatigue and exhaustion known as "hitting the wall." To prevent this from occurring and to give a competitive edge, many endurance athletes engage in carbohydrate loading to increase the muscle glycogen stores. Carbohydrate loading is a disadvantage for performance in short, intense activities since a feeling of heaviness occurs from water retention. Athletes should use carbohydrate loading no more than three to four times a year. It is not recommended for children and boys and girls in their early teens.

5. My weight seems to fluctuate between seasons. What can I do?

The daily in-season caloric requirement for athletes ranges from 3,000 to 6,000 kcal for most sports. Unwanted weight gain is a frequent problem for many athletes during the off-season. Dietary counseling can help the athlete to avoid such gains and to maintain physical fitness through other forms of exercise.

6. As a female athlete, I want to know about "sports anemia."

Sports anemia is defined as transient low hemoglobin levels associated with strenuous activity. It has been reported in the early stages of physical training. Its causes are unknown, although increased destruction of red blood cells and dilution of hemoglobin by increased plasma volume have been suggested. Taking iron supplements to prevent sports anemia is not justified. True iron deficiency decreases the capacity of the blood to carry oxygen with resulting fatigue and weakness. Long-distance runners may be at risk, possibly due to abnormal low iron absorption and increased loss. The iron status of all women and girls should be monitored because even nonathlete females are especially vulnerable to developing iron deficiency. (See Chapter 13.)

7. I definitely find ergogenic aids helpful. How do they work?

Athletes often place great confidence in **ergogenic** aids (work-improving aids), believing them capable of improving performance. For example, caffeine appears to foster the mobilization of free fatty acids, thereby sparing muscle glycogen. Some athletes believe that caffeine taken before an endurance event improves performance, while others find it no help. Some experience side effects such as polyuria and dehydration. In some competitions the use of caffeine is illegal.

Other substances promoted as ergogenic aids include amino acid supplements, bee pollen, ginseng, lecithin, kelp, wheat germ, gelatin, and vitamin supplements. Although none of these has any proven physiological effect, athletes often are convinced that one or another of these substances improves their performance, and the psychological benefit may justify the use as long as there is no interference with a nutritionally adequate and overall healthy diet.

8. What should I eat at the precompetition meal?

The stomach should be empty at the time of competition to prevent distention and cramping, yet glycogen stores should be optimal. A meal high in complex carbohydrates, low in fat, and moderately low in protein should be eaten 3–4 hours before the event. Fruits, vegetables, pastas, breads, rice and other grain foods should make up most of the meal, and should contribute 300–800 kcal.

Some athletes have favorite meals or foods that they associate with some successful experience in the past and feel will improve performance. Within reason, such preferences should be satisfied. The precompetition meal is not the time to impose dietary changes.

Mr. Vic W., 39, decided it was time for him to pay better attention to his nutritional intake, nutritional status, and physical activity patterns. Anthropometric measurements revealed he is 72 in (183 cm), without shoes, has a large frame, and weighs 185 lbs (84 kg).

Physically, Mr. W. already made one change this past week—he started taking a walk after lunch with co-workers. Nutritionally, he wants to focus first on improving breakfast. Up to now, he always skips breakfast, except for a cup of black coffee that he drinks in the car on the way to work. Some of his co-workers take ergogenic aids and he wants to know if he should start on such a regimen.

Questions

1. Compare Mr. W.'s anthropometric measurements to standards. What is your interpretation? (See also Chapter 3 and Appendix Tables D–1 and D–2 and Figures B–1, D–5 and D–6.)

2. Using Table 8–1, calculate the predicted resting energy expenditure for Mr. W. Then, using the Recommended Energy Intake Table on the inside back cover of the text, what would be Mr. W.'s approximate daily energy allowance?

3. You request that Mr. W. keep a Daily Activity Log before his next appointment. What suggestions would you provide so that the log is practical to keep and as accurate as possible? How would you use the information? (See also Chapter 3 and Tables 8–2 and 8–3.)

4. Discuss breakfast options with Mr. W. What additional information do you need to obtain for your assessment? (See also Chapters 2 and 3.)

5. Depending on the answers to question 4, plan breakfasts for one work week with Mr. W. Evaluate nutritional balance according to the Food Guide Pyramid (Figure 1–5) or according to Tables 2–3 and 2–4 or Appendix Table A–1.

6. You request that Mr. W. keep a Food Diary before his next appointment. What suggestions would you provide so that the diary is practical to keep and as accurate as possible?

7. What other components of nutritional assessment might be appropriate to provide baseline information so that Mr. W. achieves his goal?

8. What are ergogenic aids? During the counseling session with Mr. W., what points would you prioritize?

FOR ADDITIONAL READING

Blair, SN: "Diet and Activity: The Synergistic Merger," *Nutr Today,* **30**:108–12, 1995.

Carpenter, RA: "Just Do Something," *Nutr Today,* **30**:112–13, 1995.

Pate, RR, *et al:* "Physical Activity and Public Health: A Recommendation from the Centers for Disease Control and Prevention and the American College of Sports Medicine," *JAMA,* **273**:402–07, 1995.

Saltzman, E, and Roberts, SB: "The Role of Energy Expenditure in Energy Regulation: Findings from a Decade of Research," *Nutr Rev,* **53**:209–20, 1995.

Nutrition and Weight Management

Reduce overweight to a prevalence of no more than 20 percent among people aged 20 and older and no more than 15 percent among adolescents aged 12 through 19.*

Increase to at least 50 percent the proportion of overweight people aged 12 and older who have adopted sound dietary practices combined with regular physical activity to attain an appropriate body weight.*

✔ CASE STUDY PREVIEW

Since 40 percent of women and 25 percent of men are on a weight-loss diet at any given time, many of your family members, friends, and possibly you yourself, have at some time tried to lose weight. You will help a college student plan a weight-reduction program.

**Healthy People 2000: National Health Promotion and Disease Prevention Objectives.* Public Health Service, U.S. Department of Health and Human Services, Washington, DC, 1991, Objectives 1.2, 1.7, pp. 91, 92.

Overweight, or body weight in excess of some set standard, and **obesity,** or excessive body fat, which affect more than one-third of adults and 21 percent of adolescents in the United States, have been singled out as major nutrition and public health problems. Underweight also, although it receives less media attention, can have severe health consequences.

ASSESSING BODY WEIGHT AND COMPOSITION

Several methods of assessing body weight and composition have been developed. (See Chapter 3.) Relative weight (RW) is sometimes used as a standard:

$$RW = \frac{\text{measured body weight}}{\text{midpoint of medium-frame weight on Metropolitan tables (Table D – 1)}}$$

Other Formulas

The Hamwi method is sometimes used. The result is at best an approximation.

Hamwi Formula

Men: 106 lb/5 ft + 6 lb/each additional inch
e.g., 5 ft 9 in. = 160 lb (106 + 54)

Women: 100 lb/5ft + 5 lb/each additional inch
e.g., 5 ft 5 in. = 125 lb (100 + 25)

For small frame, subtract 10%.
For large frame, add 10%

Somatotypes

Body types seems to be related to weight. Persons who have an **ectomorphic** or thin, angular body type seldom become obese, while those who are **endomorphic** or rounded are more likely to gain excessive weight. The muscular, athletic **mesomorph** takes in intermediary position. (See Figure 9 – 1.)

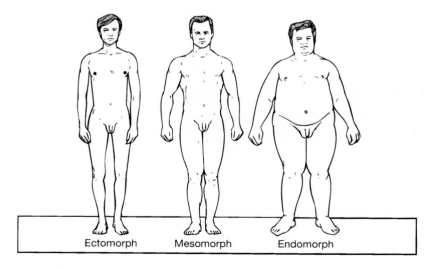

Ectomorph   Mesomorph   Endomorph

FIGURE 9–1
Types of body build.

Android Shape

Gynoid Shape

Overweight: 10–20% above accepted standard
Obese: 20% or more above standard
Grossly or Morbidly Obese: 100 lb (45 kg), or 100% or more above accepted standard

---

Waist-to-hip ratio: e.g., male, waist 55″, hips 45″
waist ÷ hips
     55 ÷ 45 = 1.22
Ratio = 1.22
Person would be at
  increased risk.

OBESITY AND OVERWEIGHT

Assessment

Determining who is overweight and obese is complex since both body weight and fatness are involved. Individuals obese by weight standards usually also have excess amounts of body fat, but there are exceptions such as the well-muscled athlete. Some commonly accepted categories are subjective. Body mass index (BMI) values indicative of overweight have been developed. (See Figure B–1 in the Appendix.)

There is evidence that not only excess adipose tissue, but its location is important in determining health risk factors. The **android** or apple configuration with abdominal or upper body fat deposits apparently carries greater risk than the **gynoid** or pear-shaped deposits on the thighs, hips, and buttocks. Although the reason for the greater risk is not clearly understood, it may be that abdominal fat cells release free fatty acids directly into the portal circulation, where they interfere with various metabolic processes.

A waist-to-hip (abdominal to gluteal) ratio is sometimes calculated; risk of chronic disease and mortality rises with the larger waist circumference in relation to the hip (see Figure 9–2). A ratio above 0.8 for women and 0.95 for men indicates risk.

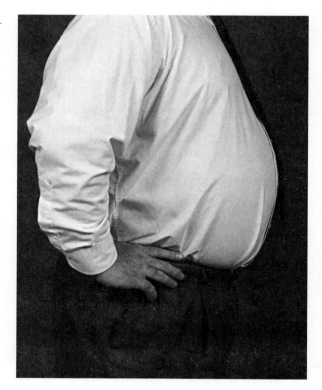

FIGURE 9–2
Measurement of this man's waist and hips would reveal a high abdominal-to-gluteal ratio with increased health risks. (*Source:* Barbara Schwartz.)

Psychological and physiological hazards accompany excess body weight and body fat. The psychological effects may be most painful in the short run. Being teased and picked last for teams, having difficulty in finding clothing, and experiencing a limited social life bring misery to many. Overweight individuals, especially women, may be discriminated against in college admissions and employment opportunities.

National Health and Nutrition Examination Survey (NHANES) data have shown increased incidence of hypertension, diabetes, and elevated blood cholesterol levels in overweight individuals. Other data link obesity to cancers of the colon, rectum, and prostate in males and cancers of the gallbladder, bile ducts, breast, and uterus in females. Complications of pregnancy and surgery occur more frequently. Gout and osteoarthritis are both associated with and aggravated by overweight. Fat deposits in the chest area make breathing more difficult. Insulating layers of fat cause discomfort in hot weather. Obesity also has a negative effect on longevity.

Weight reduction is frequently recommended for individuals with relative weights 20 percent or more above average. It may also be advisable for people with lesser degrees of overweight who have non-insulin-dependent diabetes mellitus or a family history of this condition, hypertension, or elevated blood cholesterol or triglycerides.

Since the relationship of obesity and increased mortality is most pronounced in people under 50 years of age, this group should be especially targeted for treatment. The increasing incidence of obesity in children and adolescents makes prevention and intervention programs aimed at them a priority item.

Causes

Simplistically stated, overweight and obesity result when energy intake exceeds output for an individual. This statement ignores the multiple causation and complex nature of excess body weight.

During the adult years the basal metabolic rate (BMR) decreases so that even individuals whose food intake and activity level remain constant will gain weight due to lessened basal energy needs. Excessive weight gain during pregnancy may not be completely lost, initiating a problem with excess weight.

Social and cultural influences can have a profound effect on weight. Parties without refreshments would be considered a failure, yet many guests have eaten an ample meal shortly before coming. The calories consumed by fans while watching the Super Bowl or World Series on television stagger the imagination. In some cultures, overweight is a standard of beauty or the mark of a good provider or successful person.

Food habits may foster excess weight. A family dinner table filled with platters and bowls of food is inviting, but can lead to consumption of seconds, thirds, and even more, and finishing up the leftovers just because the food is there. A "waste not, want not" philosophy may lead to consuming unneeded calories.

Eating during coffee breaks or study breaks may become habitual, unrelated to hunger or need. Unplanned snacking can cause a caloric surplus as can sheer availability of food. (See Figure 9–3.) See Chapter 2 for a discussion of hunger and satiety.

The set point theory of weight control holds that the body has a biologically programmed weight that it will attain and maintain regardless of external factors,

FIGURE 9–3
The energy expenditure of this man as he watches TV is about 100 kcal per hour. Ten potato chips will supply about 50 kcal and a 12-ounce bottle of beer about 150 kcal. (*Source:* Dunwoody Village, Newtwon Square, PA, and Peter Zinner, photographer.)

much as it maintains body temperature, pH, and so on, in spite of varying environmental conditions. The set point may be controlled by a fat-signaling hormone secreted by fat cells into the blood.

Although obese individuals are frequently less active than normal weight people, it is unknown whether inactivity is a cause or result of the obesity. Studies show that some obese persons eat no more than their normal weight counterparts; lack of activity is the critical factor. On the other hand, some obese people studiously avoid exercise, embarrassed to be observed at physical activities and possibly having humiliating recollections from the past associated with sports.

Most people at some time have found solace in food in response to joy, sorrow, stress, frustration, loneliness, boredom, or other emotional factors. There is no evidence that overweight individuals have a higher incidence of psychological problems. Rather, they may have different coping mechanisms.

Since most children grow up in the home of a parent or parents, separating genetic from environmental factors is difficult. What is known is that a child with obese parents is more likely to be obese than offspring of normal weight parents. If both parents are obese, there is an 80 percent chance that their children will be obese. If one parent is obese the percentage is 20 and if neither is obese, 7 percent. Preliminary research indicates that there may be a gene for obesity. Since obese children are at greater risk than normal weight children of becoming obese adults, isolating these etiological factors is important for screening those at high risk.

Some people exhibit little knowledge or considerable misinformation about food values, which can lead to excess consumption of high-calorie items. Table 9-1 lists some factors that may be involved in failure to balance caloric intake with body needs.

Types of Obesity

Obesity is sometimes classified according to fat cell morphology as **hyperplastic** with increased number of fat cells, or **hypertrophic** with increased size of fat cells. Obesity originating in adulthood is thought to be hypertrophic, caused by the expanding size of the **adipocytes** or fat cells as they fill with fat.

Prevention

Obviously, prevention of obesity is vastly preferable to treatment. Screening to pinpoint individuals at high risk will enable resources to be better allocated and should prove cost effective. Education programs directed at children, adolescents, and their parents and caregivers, especially those at high risk, should be provided.

Children should not be bribed or rewarded with food. Family recreation should include physical activity that all can enjoy and not be limited to sedentary activities such as television watching and automobile rides. Physical education programs with emphasis on lifetime activities and exercise that can be carried over into adulthood should be encouraged.

TABLE 9-1
Factors Leading to Imbalance in Caloric Intake

Some Possible Reasons for Excess Calorie Intake	Some Possible Reasons for Inadequate Calorie Intake
1. Family patterns of rich, high-calorie foods	1. Family pattern places emphasis upon low-calorie foods
2. Hearty appetite; likes to eat; likes many rich foods; may dislike fruits and vegetables	2. Small appetite; has little interest in eating; has many dislikes; unpalatable therapeutic diets
3. Ignorance or misinformation about calorie value of foods	3. Ignorance of essentials of an adequate diet
4. Skips breakfast; is a frequent nibbler; snacks on high-calorie foods	4. Skips meals; seldom makes up for skipped meals
5. Pattern of living a. Sedentary occupation; idleness b. Rides to work or school c. Little exercise during leisure d. Often sleeps more with increasing age	5. Pattern of living a. Often tense b. Overactive c. Not enough sleep and rest d. Smokes heavily
6. Emotional outlet: eats to overcome worry, boredom, loneliness, grief, stress	6. Emotional outlet: unhappy, worried, grieving, and so refuses to eat
7. Many social events with rich foods; frequently eats in restaurants and makes high-calorie food choices	7. Often lives alone; misses sociability; doesn't like to eat alone; insufficient resources to obtain adequate diet
8. Lower metabolism with increasing age, but failure to reduce intake or increase activity	8. Illness and infection; fever, diarrhea; hyperthyroidism; eating disorder; hypermetabolic condition
9. Influenced by pressures of advertising for many high-calorie foods	9. Affected by claims for fad diets; may get inadequate nutrient intake

Therapy

The safest program with the greatest potential for success emphasizes a healthy lifestyle incorporating a reduced-calorie, nutritionally adequate diet, tailored to the individual; increased activity, especially the sort that can become part of the daily routine; behavior modification; and possibly, depending on individual preferences, a self-help support program. Much failure can be traced to unrealistic expectations from fad diets, pills, or devices that promise to make fat vanish magically. It has been estimated that $33 billion is spent annually on such "quick fix" weight loss attempts!

Assessment and Planning

Screening clients can pinpoint those who have the motivation and support systems to make success likely. Others may opt to postpone a weight loss program. There is increasing emphasis on encouraging individuals to accept and feel positive about their body shape and size and concentrate on health promotion through a healthy lifestyle. The individual must realize the necessity of lifetime change rather than just "going on a diet" for some short-term goal such as looking slimmer for a reunion or swimming party. A weekly loss of ½ to 1 kg (1 to 2 lb) should be attainable without compromising nutritional adequacy, health, or well-being.

RECOMMENDED APPROACHES TO WEIGHT MANAGEMENT

Reduced-Calorie Diet

Dietary change is the main approach to weight management. Theoretically, a pound of adipose tissue equals 3500 kcal, so reducing intake by 500 kcal a day should result in a loss of 1 lb per week. Women usually lose weight on diets of 1200 to 1500 kcal while men will lose on up to 1800 kcal intakes. Very sedentary individuals may require less. Increased activity is an essential part of any weight management plan.

The exchange lists are often used in planning reduced-calorie diets. (See Table A–2 and Chapter 26.) Daily food allowances for 1000-, 1200-, and 1500-kcal diets are presented in Table 9–2. These are suggested diets only and can be adjusted to the preference of the individual. For example, the person who desires more skim milk could substitute a milk exchange for a meat or starch exchange since all are similar in caloric content. At lower calorie levels, it may be impossible to include the number of servings from each group suggested in the Food Guide Pyramid, Table 1–5. On intakes of 1000 kcal or less, multivitamin, iron, and possibly calcium supplementation may be necessary. The Basic Diet (see Table 2–4) can be used at the 1600 kcal level or adjusted downward for use as a reduced-calorie diet.

Usually the food allowance is divided into three approximately equal meals. If a snack is preferred, exchanges from the meals may be saved for it. Sometimes an occasional alcoholic beverage which would replace two fat exchanges may be permitted.

REDUCED-CALORIE DIETS SHOULD BE:

Adequate nutritionally
Adapted to individual lifestyles
Varied in choice of regular foods
Consistent with modifying eating habits
Appropriate as basis for weight maintenance

	NUMBER OF EXCHANGES		
Exchange List	1000 kcal	1200 kcal	1500 kcal
Starch/bread	4	5	6
Meat, medium-fat	4	5	6
Vegetable	3	3	3
Fruit	2	2	4
Milk, skim	2	2	2
Fat	2	3	4
Protein, g	62	72	82
Fat, g	30	40	50
Carbohydrate, g	129	144	189
Energy, kcal	1034	1218	1534

[a] Choices from the Exchange Lists are given in Table A–2.

Meals on a reduced-calorie diet should be attractive and palatable. Herbs and spices lend variety without calories. Labels on packaged foods must be read to determine caloric equivalents. A sample menu for a 1200-kcal diet is shown in Table 9–3.

Physical Activity

Increased activity is a necessary component of any weight management program. A cardiac stress test or physical examination before embarking on any program is essential and a stepwise schedule for previously sedentary individuals is wise. By increasing energy needs, exercise promotes weight loss on food intake that is more likely to be nutritionally adequate and aesthetically acceptable. In addition, exercise improves fitness. Physical activity is further discussed in Chapter 8.

Behavior Modification

Behavior modification is based on the belief that excess weight is caused by acquired maladaptive eating behaviors that can be modified. Changing food and activity behaviors that are inconsistent with weight maintenance is emphasized, rather than diets and calorie values.

Physicians, dietitians, nurses, or psychologists trained in the techniques of behavior modification provide guidance. To begin such a program, the individual keeps a food diary. (See Chapter 3.)

The completed diary is analyzed by the counselor, along with the client, to determine which behaviors require change. Modifications should be implemented gradually so the individual will not feel overwhelmed. Changes cannot be imposed on the individual. Rather, the person must see and accept the possibilities and act upon them. (See Figure 9–4.)

Reinforcement is important and occurs through appointments with the counselor. Sometimes a contract can be set up, earning points for behavioral changes. The points are then exchanged for some nonfood reward of the person's choosing. Support of family, friends, and social contacts is essential.

TABLE 9–3
Sample Menus for Reduced- and High-Calorie Diets

Reduced-Calorie Diet (1200-kcal)	High-Calorie Diet (3000–3200 kcal)

Breakfast

Grapefruit—½	Grapefruit—½
Dry cereal—¾ cup	Dry cereal—1 cup
Skim milk—1 cup	Low-fat milk—1 cup
Rye toast—1 slice	Poached egg—1
Margarine—1 teaspoon	Rye toast—2 slices
Coffee—no cream or sugar	Jelly—1 tablespoon
	Soft margarine—2 teaspoons
	Sugar for cereal and coffee—2 teaspoons
	Coffee with milk

Lunch

Salad plate:	Cream of mushroom soup—1 cup
Whole tomato stuffed with	Whole grain crackers—4
Salmon—1 oz	Salad plate:
Diced celery—2 tablespoons	Whole tomato stuffed with
French dressing—1 tablespoon	Salmon—3 oz
Lettuce	Diced celery
Roll, plain—1	French dressing—2 tablespoons
Skim milk—1 cup	Lettuce
Plums, unsweetened—2	Roll—1
	Soft margarine—2 teaspoons
	Jelly—1 tablespoon
	Low-fat milk—1 cup
	Plums in syrup—2
	Sugar cookie—1

Dinner

Sautéed veal cutlet with thyme and lemon (use 1 fat exchange for sautéeing)—3 oz	Sautéed veal cutlet with thyme and lemon
Corn—⅓ cup	Mashed potato—½ cup
Asparagus spears	Asparagus spears
Tossed green salad	Dinner roll—1
Low-calorie dressing	Soft margarine—3 teaspoons for
Tea with lemon	vegetables and roll
	Low-fat milk—1 cup
	Cherry pie—1 slice

Evening Snack

Saltines—6	Chicken sandwich:
Low fat cheese—1 oz	Bread—2 slices
	Mayonnaise—2 teaspoons
	Chicken—1 ½ to 2 oz
	Low-fat milk—1 cup

Self-Help Groups

Some people find groups such as Weight Watchers, Overeaters Anonymous, or TOPS (Take Off Pounds Sensibly) valuable aids in weight loss through support of others with similar problems. It is possible to form your own group by banding together with a friend or friends who wish to lose weight.

SHOPPING

Do not shop when hungry
Shop with a list and only enough money for list items
Avoid high-calorie snack foods
Store food out of sight

EATING

Prepare only as much food as will be eaten at one meal
Eat at regular times
Eat all food in one designated place
Do not watch TV or read while eating
Do not use serving dishes on the table
Do not be the food server
Leave some food on your plate
Leave the table when finished eating
Dispose of leftovers

EATING SPEED

Put down implements between mouthfuls
Chew food thoroughly before swallowing

PARTIES

Reduce alcohol intake
"Bank" calories by eating less earlier in the day
Do not stand by the food table or bar
Learn to say no to offered food

ACTIVITY

Walk instead of ride
Use stairs instead of elevators or escalators

FIGURE 9–4
Possible behavior modifications.

OTHER APPROACHES TO WEIGHT MANAGEMENT

Other Food Manipulations

Frequently new "diet" books appear, usually with some gimmick offering quick, painless weight loss, often touted by their authors on television talk shows. Many of these diets do nothing to change food behaviors permanently or create a weight maintenance program. Moreover, they are inappropriate for lifetime eating patterns, may be nutritionally inadequate, and possibly can be dangerous.

The "grandfather" of such regimens is the low-carbohydrate diet that surfaces periodically under some new name or promotion, promising unlimited intake as long as carbohydrates are restricted drastically or eliminated altogether. An initial diuresis leads to a dramatically lower reading on the scale and to promotional claims of "lose 8 pounds in 2 days." This is fluid loss, not true weight loss. Energy from protein and fat fosters satiety but produces ketosis, and fat can be a risk factor for atherosclerosis.

A fast may occasionally be appropriate for a very obese person under close medical supervision. Inadequately supervised, however, fasting can lead to loss of lean body tissue, ketosis, liver and kidney impairment, and electrolyte imbalance.

Controversy surrounds the use and value of very low calorie diets (VLCDs). The inadequate formulas of the past have been replaced with products containing less than 800 kcal per day with 50 to 70 g of high-quality protein, plus adequate amounts of vitamins, minerals and electrolytes. They have the advantage of producing rapid weight loss, but protein and potassium losses and the possibility of gallstone development are matters of concern.

VLCDs are not recommended for anyone in a growth period or with serious illness or for anyone less than 30 percent overweight (for whom a reduced calorie diet of regular foods is suggested). The VLCD should not be followed for more than 12–16 weeks. Individuals should be closely monitored and follow a maintenance plan after weight is lost.

Medications and Weight Loss

Many medications and drugs have been tried as aids to weight loss. Some are addictive, some useless, and others potentially dangerous.

Treatment of overweight and obesity with medications and drugs is controversial. Some researchers and clinicians consider medications useless and potentially dangerous. Others who view obesity as a chronic disease point out that medications are part of the therapy for most chronic diseases. They believe that used in conjunction with decreased food intake, increased activity, and behavior modification, medications have a role in weight management programs. Further research is needed. Currently available appetite suppressant medications are *dexfenfluramine, diethylpropion, fenfluramine, mazindol, phenylpropanolamine,* and *phentermine.*

Other Weight Loss Procedures

Surgery in the form of gastric bypass or gastric partitioning is sometimes used to treat extreme obesity. (See Chapter 21). **Liposuction** or suctioning of fat removes unwanted fat accumulations, but is more of a cosmetic than a weight loss procedure.

Weight Loss Maintenance

The record for long-term maintenance is poor. Many people are constantly going on and off diets and losing weight that is promptly regained. By this yo-yo effect, called weight cycling, an individual may lose and regain hundreds of pounds in a lifetime. Some studies suggest that weight cycling may increase metabolic efficiency, requiring even greater energy deficits to achieve weight loss in the future. Concern abut the dangers of such gain and loss lead some clinicians to recommend that if weight loss cannot be maintained, it is preferable not to attempt it. Others believe that the health risks of obesity are such that obese individuals should attempt moderate weight reduction despite concerns about weight cycling.

To maintain weight loss, individuals must view weight management not as "going on a diet" for a short period, but rather as making lifelong changes in food behaviors and activity patterns. Following the food intake pattern that produced the weight loss with slightly larger portions or a few additional items should produce an adequate, satisfactory maintenance diet. Behavior modification techniques and increased activity must be continued.

Societal obsession with slenderness contributes to the development of eating disorders, including anorexia nervosa and bulimia nervosa. The recently reported increased incidence may be partly an actual increase and partly due to improved diagnostic criteria, increased awareness, and reporting of cases.

Anorexia Nervosa

Anorexia nervosa, self-imposed starvation, is seen primarily in females 12–18 years of age from white upper and middle class families. It affects about 1 percent of this population. The term **anorexia nervosa** should be used for this disorder since **anorexia,** lack of appetite, is a symptom seen in many conditions. Girls with anorexia nervosa tend to be perfectionists whose families are highly controlling. Self-starvation may be a means of gaining control over their lives and also of denying sexual maturity. There may also be intense exercise, self-induced vomiting, and abuse of diuretics and laxatives.

Clinical Findings In addition to extreme weight loss and muscle wasting, symptoms include **amenorrhea** or cessation of the menstrual cycle, **bradycardia** or lowered heart rate, **hypotension** or decreased blood pressure, **hypothermia** or lowered body temperature, and growth of fine hair (**lanugo**) on the body. Cardiac arrhythmias, electrolyte imbalance, osteoporosis, constipation, and delayed gastric emptying also occur. See Figure 9–5 for diagnostic criteria.

Therapy No single treatment has proven best, so a multidisciplinary program of nutritional rehabilitation, behavior modification, family therapy, psychotherapy, and medications is used. (See Figure 9–6.) Treatment may occur in a hospital, residential center, or on an outpatient basis. Hospitalization may be necessary if outpatient therapy has failed or if life-threatening medical or psychological problems exist.

Nutritional rehabilitation is the first priority. Initially increasing calories in a stepwise progression is the goal. Sometimes contracts are made and the individual is rewarded for weight gain. If the patient is in acute danger, tube feeding or total parenteral nutrition may be necessary. (See Chapter 17.) Antidepressant medications are sometimes prescribed.

After weight gain has occurred, a plan for maintenance of healthy weight is devised. Follow-up and continued support are essential.

OBJECTIVES OF THERAPY

Achieve and maintain a
   healthy weight
Regain health
Resolve psychological
   disturbances and family
   problems

Bulimia Nervosa

This disorder of binging and inappropriate compensatory behaviors is most common in white, college-educated females in their late teens to early thirties, although some 10 percent of those affected are male. More prevalent than anorexia nervosa, it may affect 5 percent of this age group. Frequently it has its origins in late adolescence following attempted weight loss.

In response to stress, boredom, or loneliness, the bulimic will gorge on high-calorie, soft, easy-to-eat foods such as cakes, cookies, or ice cream. The binge is followed by feelings of guilt and disgust. To relieve abdominal discomfort, prevent weight gain, and achieve a sense of control, the individual induces vomiting. Abuse of diuretics, laxatives, and **emetics** such as *ipecac* to induce

■ ANOREXIA NERVOSA

A. Refusal to maintain body weight at or above a minimally normal weight for age and height (e.g., weight loss or failure to make expected weight gain during period of growth, leading to body weight less than 85% of that expected).
B. Intense fear of gaining weight or becoming fat, even though underweight.
C. Disturbance in the way in which one's body weight or shape is experienced.
D. In postmenarcheal females, amenorrhea, i.e., the absence of at least three consecutive menstrual cycles.

Specify type:

Restricting Type: during the current episode of Anorexia Nervosa, the person has not regularly engaged in binge-eating or purging behavior (i.e., self-induced vomiting or the misuse of laxatives, diuretics, or enemas)

Binge-Eating/Purging Type: during the current episode of Anorexia Nervosa, the person has regularly engaged in binge-eating or purging behavior (i.e., self-induced vomiting or the misuse of laxatives, diuretics, or enemas)

■ BULIMIA NERVOSA

A. Recurrent episodes of binge eating. An episode of binge eating is characterized by both the following:
   1. eating, in a discrete period of time, an amount of food that is definitely larger than most people would eat during a similar period of time and under similar circumstances.
   2. a sense of lack of control over eating during the episode.
B. Recurrent inappropriate compensatory behavior in order to prevent weight gain, such as self-induced vomiting; misuse of laxatives, diuretics, enemas, or other medications; fasting; or excessive exercise.
C. The binge eating and inappropriate compensatory behaviors both occur at least twice a week for 3 months.
D. Self-evaluation is unduly influenced by body shape and weight.
E. The disturbance does not occur exclusively during episodes of Anorexia Nervosa.

Specify type:

Purging Type: during the current episode of Bulimia Nervosa, the person has regularly engaged in self-induced vomiting or the misuse of laxatives, diuretics, or enemas

Nonpurging Type: during the current episode of Bulimia Nervosa, the person has used other inappropriate compensatory behaviors, such as fasting or excessive exercise, but has not regularly engaged in self-induced vomiting or the misuse of laxatives, diuretics, or enemas

FIGURE 9–5
Diagnostic criteria for anorexia nervosa and bulimia nervosa.

Source: Adapted from American Psychiatric Association, *Diagnostic and Statistical Manual of Mental Disorders,* 4th ed., rev. Washington, DC: American Psychiatric Association, 1994, pp. 544–45, 549–50.

FIGURE 9–6
A registered dietitian and family therapist meet with a young woman with anorexia nervosa and her mother. (*Source:* Anthony Magnacca/Merrill.)

vomiting also sometimes occur. Fasting and/or excessive exercise may also be practiced. Since binging and compensatory behaviors are done in private and the person is usually of normal weight, family and friends may be unaware of the problem.

Clinical Findings There are not always overt clinical signs of bulimia, so diagnosis may be based on self-reported symptoms. However, serious medical complications can result from extended binging and purging. Inflammation of the esophagus and erosion of tooth enamel can result from regurgitation of the very acid gastric contents, as can electrolyte imbalance. Hands may be scarred. Laxative abuse may affect gastrointestinal motility and damage the colon. *Ipecac* poisoning is a possibility.

Bulimics and their families have a high incidence of substance abuse, problems with impulse control, disorganized family environments, and depression. Eating habits of bulimics are often chaotic and they may not know how to eat "normally," fearing such a pattern will cause obesity. Society's emphasis on slimness, along with changing gender roles and expectations for women may be factors. Since bulimics are aware of the disorder and usually do not deny it, some will seek treatment and will cooperate with recommended therapies. See Figure 9–5 for diagnostic criteria.

Therapy The complexity of bulimia dictates a multidisciplinary approach. Most bulimics can be treated on an outpatient basis, but sometimes hospitalization is recommended to break the binge-purge cycle.

A weight maintenance diet, usually utilizing food exchanges (see Chapter 26) to create a structured eating pattern, is planned. Many of the behavior modification techniques used for weight control help the individual substitute positive behaviors for binging and purging. Counseling focuses on psychological problems. Group therapy and self-help groups assist the person in feeling less alone with her problem. Some bulimics are clinically depressed, so monoamine oxidase inhibitors and tricyclic antidepressants may be prescribed. (See Chapter 18.)

Strategies for preventing eating disorders are unknown. Hopefully educational programs to teach children and young people safe and effective ways to maintain healthy weight will decrease the incidence of eating disorders.

UNDERWEIGHT

Widespread attention is paid to overconsumption of food and its effects on weight and health, while underconsumption and underweight receive little thought. Yet they can be major problems, especially to the poor, elderly, and chronically ill. Underweight is usually defined as 15–20 percent below accepted standards and seriously underweight as 20 or more percent less than accepted standards.

Causes

Table 9–1 outlines some of the factors contributing to inadequate calorie intake. An energy deficit is often accompanied by inadequate intake of protein, vitamins, and minerals. Smokers weigh less than their nonsmoking peers.

Long illness frequently leads to weight loss because of nausea, lack of appetite, and inability to eat. In some individuals vomiting and diarrhea lead to failure to absorb nutrients, so that weight loss and undernutrition become severe. A high fever contributes to weight loss because each degree Fahrenheit rise in body temperature increases the rate of metabolism by about 7 percent.

Body Composition

In mild undernutrition, changes appear primarily in the protein stores, while body fat is not affected. Severe malnutrition results in losses of protein stores, muscle mass, and body fat. (See Chapter 7.)

Therapy

About 500 kcal daily above normal requirements are needed to gain 1 lb per week. Ordinarily, a 3000- to 3500-kcal diet is considered to be high in calories for the adult, but in cases of marked weight loss and greatly increased metabolism, 4000–4500 kcal is indicated. A sample menu for a high-calorie diet is shown in Table 9–3. Note that the menu items used in the low- and high-calorie diets are, for the most part, the same. The increase in calories was brought about by substituting a high-calorie dessert and adding margarine, sugar, jelly, bread, soup, and so on. Since fat is a concentrated source of energy, it may be necessary to include fat in excess of 30 percent to achieve sufficiently high calorie intake.

TO INCREASE CALORIES

Light or coffee cream
Sour cream
Whipping cream
Half and half
Ice cream and ice milk
Butter or margarine
Mayonnaise and other salad dressings
Jam, jelly, marmalade
Honey, sugar, candy
Cakes, cookies, puddings, pie, pastry
Concentrated formulas (see Chapter 17)

Additions can be made to Pattern C of the Basic Diet (see Table 2-4) to increase protein and calories.

Since weight loss is often accompanied by loss of protein tissue, it is necessary to provide a liberal protein allowance—usually 100 g/day. Supplements may be necessary to correct vitamin and mineral deficiencies.

Usually the person who requires 3000 kcal or more will be found to be consuming much less. To suddenly present a tray loaded with food will probably result in further loss of appetite and reluctance to eat. The high-calorie diet must begin with the patient's present intake with gradual increases.

Some people find an excess of fats or sugars to be nauseating, so these foods should be used with discretion. On the other hand, the individual should avoid filling up on bulky, low-calorie foods, such as vegetables and fruits.

Three meals a day plus an evening snack are, as a rule, preferable to three meals plus midmorning and midafternoon feedings, as the latter often blunt the appetite for the next meal. However, quickly digested and absorbed snacks such as fruit juice with cookies or crackers will increase calorie intake without interfering with the appetite.

CALORIE-RICH SNACKS

Chicken salad sandwich, milk
Chocolate milk shake,
  oatmeal raisin cookies
Strawberry ice cream or
  frozen yogurt, pound cake

PUTTING IT ALL TOGETHER

Extremes of weight, both over and under, are associated with increased morbidity and mortality. In overweight individuals, even modest weight losses can reduce health risk factors. Some eating disorders have their origins in attempts at weight control.

QUESTIONS CLIENTS ASK

1. My friend is on a very low carbohydrate diet and has lost six pounds in five days. It sounds too good to be true.

Low carbohydrate diets lead to depletion of glycogen and, with it, fluid loss. What your friend sees on the scale reading is mostly fluid loss, not loss of adipose tissue. A very low carbohydrate diet is not nutritionally balanced, usually is high in fat, cannot and should not be followed on an ongoing basis, and can cause ketosis. Remember, when something sounds too good to be true, it often is!

2. My high school reunion is coming up in two weeks and I want to weigh what I did in high school to impress my friends. That was 20 lb less than I weigh now. What can I do to achieve this?

Losing 20 lb in two weeks is unrealistic and any means of attempting it would be unsafe. A balanced low-calorie intake along with increased activity and behavior modification should enable you to lose 2-3 lb by then. If you wish to reduce your weight for the long term, lose it gradually and maintain the loss for your next reunion and beyond.

3. Snacking while watching television is a problem for me. I get so engrossed in the programs that I'm not even aware of how much I am eating.

You might try doing all your eating and snacking at one designated place—the kitchen table or dining room table, for instance. Do not watch television while eating and you will be more aware of your intake. Choose low-calorie snacks.

4. I think my friend may have an eating disorder. I am concerned about her and think she needs help, but I'm afraid if I suggest this I'll turn her off.

At some opportune time, mention your concern, emphasizing that you care about her and that she has your support. Others who share your concern can join you in encouraging your friend to seek assistance.

A Student Plans to Lose Weight

Arlene A., 35, is (66 in.) 165 cm tall and weighs (165 lb) 75 kg. She and her husband have three children, 12, 9, and 5 years of age. Before her first pregnancy, Arlene weighed 130. She gained some 50 lb during each pregnancy and never returned to her prepregnant weight. Boredom and availability of food around her home led to a lot of unstructured snacking. Her youngest child will soon enter school and Arlene has begun taking courses at the local community college to become a registered nurse. She decides she would like to lose some weight and makes an appointment with the registered dietitian at the student health service at her college.

Questions

1. Her wrist measurement is 6 inches. What is her body frame? (See Figure D–5.)

2. What is her Body Mass Index? What does the BMI measure?

3. What would be a healthy weight range for Arlene? (See Table D–1 and Figure D–6.)

4. Her daily energy needs are 1700 kcal. If she reduces her calorie intake to 1200 kcal, how long will it take her to reach the median point of the weight range mentioned in question 3?

5. By generally accepted standards, is she overweight or obese? Explain.

6. Is her overweight probably hyperplastic or hypertrophic? Explain.

7. Using the 1200 kcal plan in Table 9–2, plan menus for 3 days for Arlene. She usually takes a packed lunch to school. Include lunches that can be packed.

8. Sometimes Arlene and her classmates go to lunch at a fast-food restaurant. Suggest foods she might choose there.

9. On weekend evenings Arlene and her husband sometimes have an alcoholic beverage before dinner. How might this be worked into her meal plan?

10. Arlene and her husband will soon be attending a wedding reception. Give some tips for eating at this affair where food will be plentiful.

11. Arlene and her husband like to sit at the table after dinner and talk about the day. This is very important to them. While sitting there, Arlene often finishes any leftover food from the serving dishes and the children's plates. Suggest some behavior modification techniques to change her practices.

12. Arlene would like to increase her activity. Suggest some practical ways to do this, keeping her busy schedule and limited budget in mind.

FOR ADDITIONAL READING

Atkinson, RL, and Hubbard, VS: "Report on the NIH Workshop on Pharmacologic Treatment of Obesity," *Am J Clin Nutr,* **60:**153–56, 1994.

Foreyt, JP, and Goodrick, GK: "Weight Management without Dieting," *Nutr Today,* **28:**4–9, March/April 1993.

Heaton, AW, and Levy, AS: "Information Sources of U.S. Adults Trying to Lose Weight," *J Nutr Educ,* **27:**182–90, 1995.

Himes, JH, and Dietz, WH: "Guidelines for Overweight in Adolescent Preventive Services: Recommendations from an Expert Committee," *Am J Clin Nutr,* **59:**307–16, 1994.

Kucamarski, RJ, *et al:* "Increasing Prevalence of Overweight Among US Adults: The National Health and Nutrition Examination Surveys, 1960 to 1991," *JAMA,* **272:** 205–11, 1994.

Kushner, RF: "Body Weight and Mortality," *Nutr Rev,* **51:**127–36, 1993.

Larkin, M: "Losing Weight Safely," *FDA Consumer,* **30:**16–21, January/February 1996.

Lavery, MA, and Loewy, JW: "Identifying Predictive Variables for Long-Term Weight Change After Participation in a Weight Loss Program" *J Am Diet Assoc,* **93:** 1017–24, 1993.

National Task Force on the Prevention and Treatment of Obesity: "Very Low-Calorie Diets," *JAMA,* **270:**967–74, 1993.

National Task Force on the Prevention and Treatment of Obesity: "Weight Cycling," *JAMA,* **272:**1196–1202, 1994.

Perri, MG, *et al:* "Strategies for Improving Maintenance of Weight Loss: Toward a Continuous Care Model of Obesity Management," *Diabetes Care,* **16:**200–09, 1993.

Position of the American Dietetic Association: "Nutrition Intervention in the Treatment of Anorexia Nervosa, Bulimia Nervosa, and Binge Eating," *J Am Diet Assoc,* **94:**902–07, 1994.

Roberts, SB, and Greenberg, AS: "The New Obesity Genes," *Nutr Rev,* **54:**41–49, 1996.

Robison, JI, *et al:* "Obesity, Weight Loss, and Health," *J Am Diet Assoc,* **93:**445–49, 1993.

Wing, RR: "Behavioral Treatment of Obesity: Its Application to Type II Diabetes," *Diabetes Care,* **16:**193–99, 1993.

CHAPTER 10

Vitamins

Chapter Outline

Characteristics
Functions
VITAMIN HIGHLIGHTS: FAT-SOLUBLE VITAMINS
Vitamin A
Vitamin D
Vitamin E
Vitamin K
VITAMIN HIGHLIGHTS: WATER-SOLUBLE VITAMINS
Vitamin C
Thiamin

Riboflavin
Niacin
Vitamin B_6
Vitamin B_{12}
Folate
Biotin
Pantothenic Acid
Health and Nutrition Care Issues
Putting It All Together
Questions Clients Ask
Case Study: A Young Woman with Signs of Vitamin Toxicity

*Reduce the prevalence of gingivitis among people aged 35 through 44 to no more than 30 percent.**

✔ CASE STUDY PREVIEW

People are curious about the possible role of vitamins in preventing or treating medical/dental conditions and in enhancing overall health. The supplement industry estimates that consumers pay about $4 billion each year to purchase vitamin and/or mineral supplements. How would you decide if controversial nutrition statements were true or false?

Healthy People 2000: National Health Promotion and Disease Prevention Objectives. Public Health Service, U.S. Department of Health and Human Services, Washington, DC, 1991, Objective 13.5, p. 108.

Although they are found in all body cells, vitamins and minerals account for only a small percentage of body weight. Thus, compared to water, carbohydrates, lipids, and proteins, vitamins and minerals are classified as **micronutrients.** Only recently have scientists been able to develop instruments precise enough to measure the minuscule amounts of these micronutrients in foods and the body. People are often fascinated by these "magical" nutrients that occur in such tiny amounts. Diseases caused and cured by vitamins have intrigued people for centuries:

- Scurvy . . . this scourge took the lives of nearly half the sailors during the fifteenth to seventeenth centuries, until a British physician discovered that foods such as limes could prevent it. But the reason why limes worked wasn't known until the early twentieth century.

- Beriberi . . . this disease paralyzed people in Asia, until in the late nineteenth century physicians noted that people (and animals) who did not succumb were the ones who ate the outer layers of the rice kernel, and not just the "polished" white rice.

- Pellegra . . . this fatal disease was seen in many parts of the world and was diagnosed by the "4 D's" (dermatitis, diarrhea, dementia, and death). At the turn of the twentieth century in the United States, it was a leading cause of mental illness and death, especially among poorer groups of people. Through the concerted efforts of many scientists and physicians, the cure for and prevention of this disease was identified.

Perhaps the greatest lesson learned from the victory over vitamin deficiency diseases is that a varied diet of ordinary wholesome foods can cure, and better yet, prevent them. This eradication of so many debilitating and deadly diseases has been hailed as a major factor in promoting wellness and prolonging life. In the parts of the world where people have a sufficient food supply, these diseases rarely are seen. But they still run rampant anywhere that food is unavailable. As a health care provider, you may see them secondary to other medical conditions or treatments.

Because of vitamins' powerful history, in technologically advanced countries people now are asking new questions about them. Can vitamin A cure cancer? Will vitamin E protect against heart disease? Does vitamin C prevent colds, help the immune system, and aid in longevity? Can any of the vitamins stop the complications associated with mental retardation in children? Should everyone take megadoses of the B-complex vitamins to prevent harm to the body from modern-day stress? Will a public health policy requiring flour to be fortified with folate protect pregnant women from having babies with neural tube defects? Answers to these questions have proved to be more elusive.

CHARACTERISTICS

Nomenclature

Vitamins are a group of organic substances that are essential in trace amounts for the body's metabolic processes to occur. Almost always, vitamins must be supplied by the diet. When scientists first identified vitamins, it was thought they were vital proteins because of their chemical structures, food sources, and functions, hence the previously used terms "vital amines" and "vitamines."

8ni

Preformed vitamins are the active form of the vitamins. **Precursor vitamins** are compounds that can be changed into the active form of the vitamins.

Chemical Structure/Composition/Measurement

Figure 10-1 illustrates the structure of some of the vitamins. Note that each vitamin has a very specific chemical structure. Some structures are simplistic, but most are complex. All vitamins contain carbon, thus classifying them as organic compounds. Many contain nitrogen. Some contain minerals like sulfur or cobalt. Several exist in different forms, each with a different biological activity.

Until the 1980s, because of imprecise methods of measuring the quantity and bioavailability of vitamins, some vitamins were measured as "International Units" (IU). You still see this earlier designation in some food composition tables and on some food/supplement labels. The current measurement units are milligrams (mg) or micrograms (µg). (See Figure 10-2.) Appendix Table B-3 demonstrates conversions for these vitamin measurement units.

Distribution

Except for highly refined foods, vitamins are found in all foods. Throughout the body, vitamins are found in varying amounts, depending on their specific functions.

Classification

The group classifications and names of individual vitamins can be confusing. Part of the problem is because the quantities found in foods and the body are so tiny that they are difficult to discern and study—both in the past and even today! Sometimes this fact seems to make learning about the vitamins difficult. But weave together their definition, discovery, sources, and common and distinct functions in health and disease, and you will see their overall relevancy as nutrients. For example, if you have responsibilities for clients, think about problems they might encounter if they eat only certain kinds of foods or have a condition that interferes with the way the body can metabolize a nutrient. Or if you have responsibilities for food production, think about the properties of vitamins in food and how various processing, storage, and cooking methods affect their integrity.

Group Classifications Vitamins are classified in several ways. The traditional method is by a chemical property—their solubility, that is, whether they are soluble in fat or in water. Since you already have studied other nutrients classified this way, keep them in mind as you learn about the vitamins. Table 10-1 compares the two subgroups of vitamins classified according to their solubility. The four vitamins that are fat-soluble are vitamins A, D, E, K. The water-soluble vitamins are vitamin C and the eight B-complex vitamins.

Another major classification is by their general or most predominant functions. As a group, the B-complex vitamins are known as the "energy-releasing vitamins" because they are involved in the production of ATP from the macronutrients. A subgroup of the B-complex vitamins, especially vitamins B_6, B_{12}, and folate, are involved in blood formation and so are termed the "hematopoietic vitamins." Several vitamins have antioxidant functions in both foods and in the body. An **antioxidant** is a natural or synthetic compound that is itself oxidized

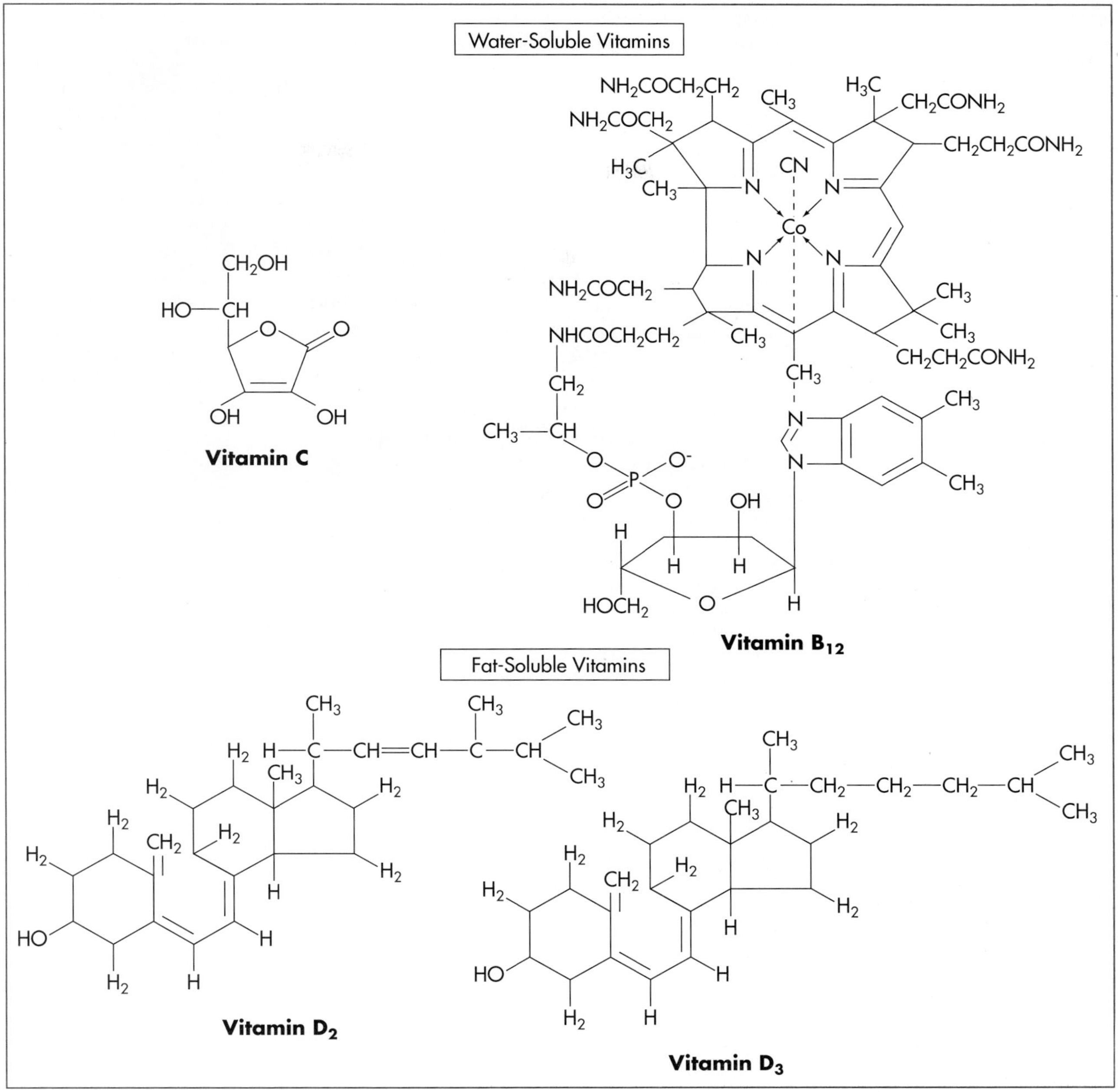

FIGURE 10–1
Structure of some vitamins. Each vitamin has a unique structure; most are complex. Since vitamins contain carbon, they are classified as organic compounds. Some vitamins exist in multiple forms.

readily, thus sparing another compound from being oxidized. The body uses several antioxidant mechanisms to maintain homeostasis, including the "antioxidant vitamins"—vitamins A, C, and E.

A third classification is by their food source. A particular vitamin usually is present in either plant or animal foods, or in both. This classification is especially

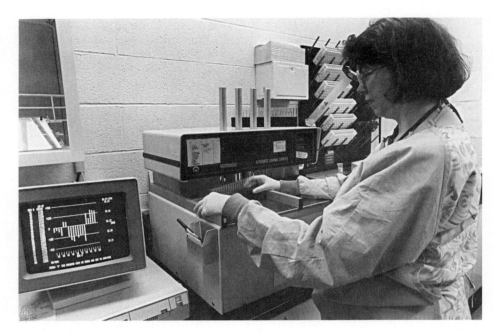

FIGURE 10–2
Blood analysis for vitamin content requires sophisticated instruments and trained staff. Concentration reflects the body's processing of the vitamin—one moment in time, when the blood was drawn. Levels can reflect the amount absorbed from food, released from body storage, or moving from one body location to another. Since test results indicate amount circulating, not direct use by cells, interpret blood level values cautiously. Laboratory tests are one component of the nutritional assessment process. (*Source:* Tom Wilcox/Merrill.)

TABLE 10–1
Some Comparisons of Fat-Soluble and Water-Soluble Vitamins

Fat-Soluble Vitamins	Water-Soluble Vitamins
Soluble in fat solvents	Variable solubility in water; vitamin C highly soluble; riboflavin and niacin less readily dissolved
Stable in usual cooking procedures	Variable losses in cooking waters; losses by exposure to alkali, heat, enzyme activity
Bile and pancreatic lipase required for absorption	Most are readily absorbed with other water-soluble nutrients; vitamin B_{12} requires special mechanism
Absorption reduced by mineral oils, laxatives, bile-binding agents, and other medications	
Transported in circulation by lipoproteins	Transported in circulation with other water-soluble nutrients
Excessive intakes stored in liver and other organs	Excessive intakes excreted in the urine; body reserves are limited
Vitamins A and D are toxic in megadoses	Megadoses of niacin and vitamins B_6 and C can lead to dependency and other side effects

helpful for teaching children and adults about selecting a nourishing diet, using such tools as the Food Guide Pyramid. For example, vitamin C is present mainly in fruits and vegetables, but vitamin B_{12} is present only in foods of animal origin. Vitamin A is present only in animal foods, but a precursor form of vitamin A, namely carotene, is present only in plant foods. Many vitamins are present in both. Some vitamins are present in unprocessed foods, but are removed when foods are processed. For example, vitamin E is present in whole wheat flour, but is removed when the germ portion is removed in making white flour.

Names of Individual Vitamins The naming of the individual vitamins seems like a language all its own. Over time, the names of individual vitamins have changed. Again, this demonstrates the confusion surrounding the discovery of vitamins.

The most common designation is by letters of the alphabet, for example, A, B, C, D, E, K. Terminology became more complicated when researchers discovered some vitamins they thought were the same were really different ones with different structures and functions, or had different sources and formations but with similar functions. Vitamin B is a typical example. As different structures with different functions were discovered, the practice began of chronologically adding numbers to the letters, for instance, vitamin B_1, vitamin B_2, vitamin B_3, etc. Vitamin D has several forms, but with similar functions. Sometimes you will see the names vitamin D_2 and vitamin D_3. Adding to the complexity is that sometimes scientists originally thought they had discovered a vitamin, only to find out later that the substance did not fulfill all the necessary requirements to be designated as a vitamin and the letter and number had to be dropped from the list. Other times the letter denoted a curative property. For example, the letter K stands for the Danish word *koagulation* because a Danish biochemist discovered vitamin K prevented hemorrhage in chickens by being a necessary factor in blood coagulation.

A final example is using the chemical name as the preferred name. Vitamin B_5 now commonly is known as pantothenic acid. Of course, once people became accustomed to a certain name, it often persisted. Such is the case with vitamin B_6 (pyridoxine, pyridoxal, pyridoxamine), vitamin B_{12} (cobalamin, cyanocobalamin), and vitamin C (ascorbic acid).

Properties

The properties of vitamins generally relate to their solubility, curative capabilities, biochemical reactions, sources, or retention in foods. For example, fat-soluble vitamins usually are located in the lipid parts of plants and animals. In the laboratory, fat-soluble vitamins can be altered to be more water-soluble. Such vitamins are then termed **water-miscible** and are the kinds prescribed for clients with malabsorption.

The macronutrients and vitamins are similar because all of them are organic. They share another property, inasmuch as some macronutrients and some vitamins have minerals as part of their structures. Sulfur is present in the amino acids methionine, cystine, and cysteine and in the two B-complex vitamins thiamin and biotin. As classes of nutrients, their solubility properties are similar— some vitamins are soluble in fat or water, like the macronutrients.

However, as a class of nutrients, vitamins have different properties from the macronutrient groups. One important difference is that one vitamin cannot substitute for another, nor can different vitamins be converted one into another.

For instance, the body can convert water-soluble glucose to fat when there is excess glucose to store, but the body cannot convert water-soluble vitamin C to any fat-soluble vitamin when excess vitamin C is present. Nor can vitamin C be converted to any of the other water-soluble vitamins either. Thus, the body does not store some vitamins to any great extent. Another contrast between the macronutrients and vitamins exists. Vitamins do not supply energy themselves, but rather are involved in the points along the metabolic pathways where energy is released from carbohydrates, lipids, and proteins, and, in some cases, for the conversion of one macronutrient to another.

FUNCTIONS

Throughout the world researchers have spent their entire professional lives in the pursuit of all the many functions of each vitamin; yet many answers remain unknown. The story about the vitamins as a class of nutrients and as individual nutrients continues to be written. This makes it difficult for us as health care providers because questions clients ask usually do not have straightforward, simple answers.

Overall, as a class of nutrients, vitamins are involved some way or another in metabolic reactions. These functions are found both in foods and in the body. Some vitamins function independently, while others behave like members of an assembly line. Again, one vitamin cannot substitute for another.

VITAMIN HIGHLIGHTS

Fat-Soluble Vitamins

VITAMIN A

Characteristics
Vitamin A is a group of compounds, generally organized into two major classes: retinoids (preformed vitamin A) and carotenoids (vitamin A precursors). Retinoids exist as retinol, retinyl esters, retinal, and retinoic acid. Carotenoids include α-, β-, and γ-carotene (*alpha-, beta-,* and *gamma-carotene*), and cryptoxanthin. Of these precursors, β-carotene is the most important. All forms are polyunsaturated, which makes them susceptible to destruction by oxidation.

Measurement of vitamin A activity in foods is expressed as retinol equivalents (RE), which accounts for the bioavailability and biological activity of the many forms of vitamin A.

Functions
Vitamin A has a variety of functions, some of which remain to be fully explained. Some forms of vitamin A appear to have no biological function. The most commonly recognized function of vitamin A is its role in vision. Vitamin A is essential for the visual cycle to occur. In this way, a person is able to see in bright or dim light.

(See Figure 10–3.) Another essential function is its role in the ability of cells to differentiate. All the mechanisms by which vitamin A does this still are being studied, although some involve gene expression. This function leads vitamin A to be involved in reproduction, growth, bone and tooth development, synthesis and maintenance of healthy epithelia, and integrity of the immune system.

Metabolic Processes
Vitamin A is digested and absorbed like the lipids. In the intestine, the presence of fat, protein, and bile facilitate its absorption and vitamin E prevents the oxidation of polyunsaturated vitamin A that might otherwise occur. The amount absorbed differs between preformed vitamin A and the carotenoids. First, preformed vitamin A is more bioavailable. Second, when the amount ingested increases, the amount of preformed vitamin A absorbed increases, too. But the opposite occurs with the carotenoids—the amount absorbed decreases. Of the β-carotene that is absorbed, most is converted to vitamin A in the intestinal cells. Nearly all vitamin A is transported by the lymphatic circulation.

About 90 percent of body stores of vitamin A are in the liver. Carotenoids are deposited in body fat, espe-

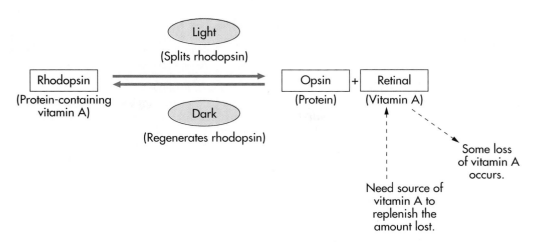

FIGURE 10-3

Role of vitamin A in vision. Rhodopsin is the eye pigment sensitive to light. When light strikes rhodopsin, it splits into its protein part (opsin) and its vitamin A part (retinal). Some vitamin A is lost in the process. In the dark, opsin and retinal rapidly recombine, provided there is an adequate supply of new vitamin A to replace the lost amount. If there is not enough vitamin A, regeneration is slow and the eyes cannot adapt to changes in light—as when a person goes from a lighted room into a dark one. The condition sometimes is called "night blindness."

cially noticeable under the skin, which takes an orange hue.

The majority of preformed vitamin A and the carotenoids are excreted into the bile and thus are removed from the body via the feces. Some vitamin A is excreted by the kidney.

Meeting Body Needs

Requirements/Recommendations
The RDAs for vitamin A for different groups of people are listed in the table inside the front cover.

Sources
There are many food sources of vitamin A. Preformed vitamin A is found in animal foods, whereas the carotenoids are found in plant foods.

Whole milk, cream, butter, whole-milk cheeses, and egg yolk are good sources of vitamin A. With consumers' interest in eating lower fat- and cholesterol-containing foods, commercial low-fat and skim milks and margarines fortified with vitamin A compare with the levels found in whole milk and butter, respectively, and thus are good choices. These and other fortified food products make important contributions to Americans' intake. Liver, of course, is a rich source. Although not ordinary components of the diet, fish-liver oils are excellent sources. These oils have been offered as dietary supplements. Advise caution in their use because of the potential for toxic vitamin A levels and contamination with environmental pollutants, which the fish stored in their liver.

About one-third of the vitamin A in the United States food supply comes from the carotenoids in vegetables and fruits. Emphasize those with deep yellow/orange/red and dark green colors. But consult tables of nutrient composition for actual vitamin content. (See Appendix Table A–1.) For example, both sweet potatoes and tropical yams are somewhat similar in color, but only sweet potatoes are a source of biologically active carotenoids. Examples of other foods with active carotenoids include:

- *Yellow vegetables*—carrot, pumpkin, sweet potato, winter squash
- *Yellow fruits*—apricot, cantaloupe, mango, papaya
- *Green leafy vegetables*—green tops of beets, turnips, mustard; chard, collards, kale, spinach, dark salad greens
- *Green stem vegetables*—asparagus, broccoli
- *Red fruits and vegetables*—tomato, red sweet pepper

Retention of vitamin A is stable in ordinary processing and cooking temperatures. Frozen and canned foods retain vitamin A for nine months or longer. Wilted and dried vegetables and fruits and rancid fats have negligible levels due to oxidation.

Basic Diet
The Basic Diet Nutrient Profile (Table 2–3) identifies the importance of the vegetables and fruits as the main sources of the carotenoids and whole milk or

vitamin A-fortified skim milk as the main sources of preformed vitamin A. In egg, the vitamin is located in the yellow yolk, along with the other lipids.

Body Balance

Optimum/Adequate Signs and Symptoms
Methods to determine an adequate state depend primarily on determination of blood level, either of the vitamin directly or of carrier proteins. When vitamin A is released from the liver to meet body needs, it is linked to specific proteins, retinol-binding protein (RBP) and transthyretin (TTR). In this way the fat-soluble vitamin A can be carried in blood, which is water-soluble. Serum carotene levels reflect recent dietary intake, not body stores.

Insufficiency/Deficiency
Dryness of the skin, impaired visual dark adaptation, eye discomfort, and low vitamin A levels are the earliest clues. **Keratinization** (drying, scaling and roughening) of the skin occurs. Soft, moist epithelia normally offer protection against bacteria, but as drying occurs, infections of the eyes, mouth, respiratory, and genitourinary tracts result. **Xerophthalmia** (dryness of the eye) develops in stages, the earliest of which is night blindness. This is followed by **xerosis** (drying and opaqueness) of the cornea. If treated at this point the condition can be reversed. However, untreated xerosis progresses rapidly to the deep layers of the cornea with scarring and loss of sight.

In healthy individuals who consume diets adequate in vitamin A, reserves are gradually built up so that adults have sufficient stores to meet body needs for several months to more than a year. Infants and children, of course, have not yet built up such stores. They are therefore more vulnerable to the effects of dietary lack.

Vitamin A deficiency rarely is seen in countries where food is available or where foods fortified with vitamin A are eaten. However, the possibility of vitamin A deficiency should be considered in persons with limited incomes or access to food, such as those who are considered "homeless"; in clients who eat a limited variety of foods or restrict fat intake; and in patients with any condition of the intestinal tract that interferes with absorption, such as intestinal surgery or conditions resulting in malabsorption.

In technology-limited countries or in those that experience famine, vitamin A deficiency ranks second in prevalence only to protein-energy malnutrition. When the two deficiencies are present together, the prognosis is grave. From 20 to 40 million children throughout the world are estimated to have mild vitamin A deficiency. About half a million persons, especially children, become blind each year. Under the sponsorship of UNICEF and other organizations, massive oral doses of vitamin A in quantities of 100,000 µg (300,000 IU) are given intermittently to young children. Other important public health strategies have been fortification of commonly consumed foods and nutrition education to promote production and consumption of locally grown vitamin A–rich foods.

Excess/Toxicity
Long before researchers described the toxic effects of vitamin A, Eskimos learned not to eat polar bear liver. It turned out that a single gram of polar bear liver contains about 6,000 RE of vitamin A.

Toxic effects of vitamin A, **hypervitaminosis A,** have been observed in children and adults when supplements of 6,000–15,000 RE have been taken daily for several months. Symptoms include nausea, vomiting, abdominal pain, failure to grow in children or weight loss in adults, drying and scaling of the skin, thinning of the hair, swelling and tenderness of the long bones, joint pains, headaches, and enlargement of the liver. The condition is corrected when the individual stops taking the supplement, although full recovery may take weeks. Pregnant women, however, are a special concern. They can experience spontaneous abortions or birth defects in their infants.

Animal studies have shown that tumor growth by certain experimentally produced cancers could be slowed or prevented by giving vitamin A. Some human studies have shown that large doses of carotenoids over long periods of time have some therapeutic value. Claims have been made that eating foods rich in carotenoids may provide some protection against certain forms of cancer, primarily because of their antioxidant properties. To date, public health organizations continue to recommend the consumption of fruits and vegetables, rather than pill supplements, since it remains unknown what factors associated with foods actually may be producing the effect.

VITAMIN D

Characteristics
Vitamin D is a group of distinct sterol compounds. It occurs in several forms, the two most important being vitamin D_2 (ergocalciferol), produced under ultraviolet light from its plant precursor ergosterol, and, in humans, vitamin D_3 (cholecalciferol), produced when 7-dehydrocholesterol in the skin is exposed to the ultraviolet rays of the sun. (The precursor of 7-dehydrocholesterol is cholesterol.) (See Figure 10–1.) Thus, vitamin D became known as the "sunshine vitamin." Both of these forms of vitamin D are known by the general name calciferol. These forms are transformed by the body to the active form of vitamin D.

Vitamin D differs from the other fat-soluble vitamins since it can be synthesized completely by the body, so that it may not be required as part of the diet for some people.

Vitamin D activity in foods is measured in micrograms (µg). Previously, the IU designation was used. You still may see this older term on food and supplement labels. (See Appendix Table B–3 for conversions.)

Functions

Vitamin D's best known function is its role in relationship with the minerals calcium and phosphorus. Vitamin D aids the absorption of these minerals from the intestine and then prompts the cells of bones and teeth to incorporate the minerals as the structures are being formed. Along with a hormone produced by the parathyroid gland, vitamin D aids in the regulation of the blood levels of calcium and phosphorus. Vitamin D influences the kidney's reabsorption of calcium and/or phosphorus. Vitamin D plays an indirect role in the proper functioning of the muscles, including the heart, and the nerves by its relationship with these minerals. (See Chapter 11.) Vitamin D also may be involved in cellular activity of several organs, including the skin, pancreas, and ovaries. Because of this role, vitamin D functions as a hormone since it is synthesized by one organ but has its effects on other body organs.

Metabolic Processes

To be digested and absorbed, dietary vitamin D requires the same substances and gastrointestinal environment as other lipids. After absorption, it also is transported in the lymph circulation, then into the blood circulation, then to the liver. Vitamin D produced from its precursors in the skin also is transported to the liver.

Several important steps occur next. In the liver, vitamin D undergoes transformation to another form, known as 25-hydroxycholecalciferol (calcidiol), which then can be transported by the blood circulation to the kidneys where the active form of vitamin D is synthesized, known as 1,25-dihydroxycholecalciferol (calcitriol, vitamin D_3 hormone). Thus, many organs are involved in the formation of the active form of vitamin D, whether it is obtained from food or from a precursor in the body.

Although vitamin D is found in several organs, it appears that the blood, liver, and adipose tissue are storage locations. Vitamin D is excreted mainly through bile into the feces, with a small amount found in the urine.

Meeting Body Needs

Requirements/Recommendations

Because of its role in bone and tooth formation, higher requirements are expected during periods of rapid growth. The RDAs for vitamin D for different groups of people are listed in the table inside the front cover.

Sources

Most people obtain enough vitamin D through exposure of the skin to sunlight. However, clothing, soot, fog, window glass, and sunscreens cut off ultraviolet light, thereby preventing the body from synthesizing vitamin D. Compared to light-skinned people, dark-skinned people produce less vitamin D. People who work at night and sleep during the day, people who live in cold climates and stay indoors, invalids who do not get out in the sun, and those whose clothing covers all the skin surface (e.g., "veiled" women, swaddled infants and children) may depend more on dietary sources.

Only small amounts of vitamin D are found naturally in foods. Thus, tables of nutrient composition rarely list food sources. Fatty fish, such as herring, salmon, and sardines, and fish liver oils are the main food sources. However, several kinds of foods are fortified. Almost all fresh, powdered, evaporated milk, and infant formulas are fortified with 10 µg (400 IU) vitamin D per quart or equivalent. Human milk contains only about one-tenth of this amount. Some breakfast cereals are fortified.

Vitamin D is stable in foods during storage, processing, and cooking.

Basic Diet

The Basic Diet assumes exposure of people to sunshine and the consumption of vitamin D-fortified milk. When you evaluate a person's diet, note the substitution of other dairy foods for fortified milk, since most commercially manufactured cheeses, yogurts, and ice creams are not made from vitamin D–fortified milk. Ingredient listings on food packages should provide information about vitamin D–fortified milk use. Nutrition Facts food labels need not list vitamin D. (See Figures 1–8 and 1–9.)

Body Balance

Optimum/Adequate Signs and Symptoms

An easy and inexpensive method to determine an adequate state has yet to be developed. Currently used methods measure blood concentrations of vitamin D. Since vitamin D is stored in the body, test results can be misleading.

Insufficiency/Deficiency

Vitamin D deficiency usually occurs in people who have insufficient exposure to sun, or maintain rigidly restricted diets without vitamin D-fortified foods, or have a condition that prevents the absorption of vitamin D from foods or conversion in the body. The major outcome is that deficiency leads to lowered absorption of calcium, thus resulting in serious complications. (See Chapter 11.)

Rickets is the deficiency disease seen in infants and children who fail to get enough vitamin D. Premature infants and dark-skinned children are more susceptible. Rickets is characterized by soft, pliable bones that yield to pressure, enlargement of the joints, and delayed closing of the skull bones, The child may have an enlarged skull, chest deformities, poor muscle development and pot belly, spinal curvature, and bowed legs. (See Figure 10–4.)

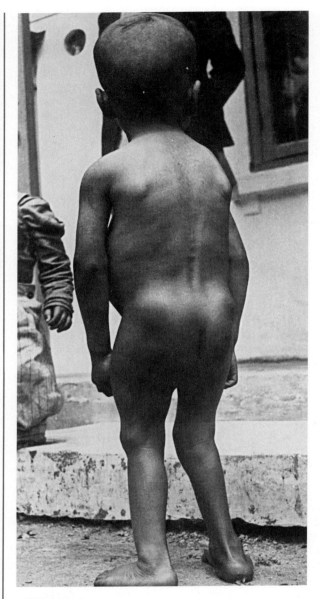

FIGURE 10–4
Early skeletal deformities of rickets often persist throughout life. Bowed legs that curve as shown here indicate that the weakened bones have bent as a result of standing. (*Source:* Courtesy of WHO/FAO/The United Nations.)

In adults, **osteomalacia** (adult rickets) signs and symptoms include pain in the legs and lower back, and deformed bones of the spine, pelvis, and limbs, leading to spontaneous fractures. It may be a primary deficiency, such as in women after multiple pregnancies, coupled with an extremely inadequate intake of vitamin D, calcium, and phosphorus. Osteomalacia may be caused secondarily by other conditions. For example, it may be induced by medications having an adverse effect on the absorption and synthesis of vitamin D. Among them are bile-binding agents used for their cholesterol-lowering effects, anticonvulsants, muscle relaxants, sedatives, and mineral oil. The severity depends on the dosage and duration of the medication. (See Chapter 18.) Patients with severe malabsorption disorders such as sprue, colitis, and cystic fibrosis often develop signs of osteomalacia. (See Chapters 22 and 23.) Patients with chronic renal disease can develop bone pain and bone disease (**osteodystrophy**) because the damaged kidney is unable to convert vitamin D to its active form, and osteodystrophy is complicated by medication side effects. (See Chapter 28.)

Excess/Toxicity

Excess consumption of vitamin D for extended periods of time can be toxic. Hypervitaminosis D has been seen in infants, children, and adults. It is characterized by loss of appetite, increased thirst, vomiting, diarrhea, fatigue, drowsiness, and growth failure in children. The serum calcium level increases, and calcium salts are deposited in the soft tissues including the heart and blood vessels, lungs, and kidneys.

As little as 45 µg taken daily has been reported to be toxic. Thus, it is important to determine all the possible sources of vitamin D in the diet, such as vitamin drops and pills, fortified milk or commercial formulas, and fortified cereals.

VITAMIN E

Characteristics

Vitamin E is a group of eight related compounds having an alcohol chemical structure. They are divided into two groups—tocopherols and tocotrienols. Alpha-tocopherol is the most biologically active form and is the most widely distributed in foods.

Measurement of vitamin E activity in foods is expressed as tocopherol equivalents (TE), based on the varying activity of the different forms of vitamin E present.

Functions

The primary function of vitamin E is its role as an antioxidant, both in foods and in the body. As you already read, vitamin E protects vitamin A from oxidation. It protects lipids in cell membranes from oxidation and also prevents the hemolysis of red blood cells.

Related to this is vitamin E's ability to act as a scavenger to inactivate certain chemical substances known as **free radicals,** which are atoms or molecules that have one or more unpaired electrons in their outer orbital. This imbalance makes them highly reactive. Free radicals normally are produced all the time in the body as by-products of normal metabolism. They also are environmental pollutants, like smog and cigarette smoke. Besides vitamin E, the body has a variety of systems it uses to inactivate these free radicals, including specialized enzymes and hormones.

Metabolic Processes

The digestion, absorption, and transportation of vitamin E is similar to the previously discussed fat-soluble vitamins. Absorption of α-tocopherol is relatively inefficient, ranging from 20 to 80 percent. In tests of absorption with supplements in capsule form, absorption efficiency declines as the dosage increases.

Vitamin E is stored in most organs, with the largest concentration found in adipose tissue. It is excreted in the bile and possibly the urine and skin.

Meeting Body Needs

Requirements/Recommendations

The RDAs for vitamin E are listed in the table inside the front cover. The requirement for vitamin E increases when intake of polyunsaturated fatty acids increases because of vitamin E's antioxidant function.

Sources

Vitamin E generally is present in plant foods with polyunsaturated fatty acids. Examples of plant foods are some vegetable oils (cottonseed, safflower, corn), margarine, whole grains and wheat germ, dark green leafy vegetables, nuts, and legumes. Most fruits and vegetables and foods of animal origin are relatively low in vitamin E. Egg yolk is an exception—it is a good source.

Retention in food can vary dramatically. Although vitamin E is stable at normal cooking temperatures and in acids, there is loss whenever air (oxygen) or certain minerals (iron) come into contact with the food. This can occur in fried foods, frozen foods, and foods remaining in storage for periods of time. Vitamin E is removed when whole grains are processed.

Basic Diet

Compare the foods that are sources of vitamin E with those that contain polyunsaturated fat. Note that they are similar.

Body Balance

Optimum/Adequate Signs and Symptoms

Neither deficiency nor toxicity related to vitamin E is common. Methods to evaluate vitamin E status are fraught with problems in procedure or interpretation. Two commonly used clinical tests either measure serum vitamin E concentration compared to serum lipids, or measure red blood cell resistance to rupture.

Insufficiency/Deficiency

Deficiency may be evident in two groups of people: infants, especially premature ones, and people who have malabsorption of fat. Small premature infants (1,500 g or less) who are given oxygen are especially susceptible to **retrolental fibroplasia,** which is an abnormal growth of fibrous tissue behind the crystalline lens. This can lead to poor vision or blindness. By reducing the undesirable oxidation causing this condition, vitamin E therapy has been found to be beneficial. At birth infants have about half of the adult blood levels of vitamin E. As a consequence, infant formulas high in polyunsaturated fatty acids and iron may cause red cell hemolysis unless adequate vitamin E is supplied. Any person with malabsorption, such as persons with impaired pancreatic digestive function, diminished bile flow, or intestinal surgery, should be considered at risk for deficiency. Such patients usually are treated with a water-soluble vitamin E preparation.

Excess/Toxicity

Even though many healthy people take oral doses many times higher than the RDAs, little toxicity has been reported. This may be accounted for by inefficient absorption of high doses. Some researchers have reported that megadoses of vitamin E, especially when administered intravenously, can lead to elevation of blood lipids, interference with the vitamin K clotting mechanism, and impairment of white blood cell activity.

VITAMIN K

Characteristics

Vitamin K, generally known as quinone, exists in at least three related forms: phylloquinone (vitamin K_1), commonly found in plants; menaquinone (vitamin K_2), synthesized by bacteria in the intestinal tract; and menadione (vitamin K_3), a synthetic water-soluble form.

Vitamin K is measured in foods in micrograms (μg).

Functions

The principal function of vitamin K is its involvement in several steps of the blood clotting mechanism, including factors II (prothrombin), VII, IX, and X, and proteins C and S. Vitamin K also appears to be important for bone mineralization because it is necessary for certain calcium-binding proteins to function.

Metabolic Processes

The digestion and absorption of vitamin K, whether from dietary or intestinal bacteria sources, requires the same favorable conditions as for the other fat-soluble vitamins already described. Transportation is first by the lymph system. Vitamin K is stored in the liver and excreted mainly into the bile and the urine.

Meeting Body Needs

Requirements/Recommendations

The requirements for vitamin K appear to depend on sufficient synthesis by intestinal bacteria, supplemented with dietary sources. The RDAs are based on a dietary intake of about 1 μg/kg body weight. (See inside front cover of text.)

TABLE 10–2

Summary of Fat-Soluble Vitamins

Vitamins	Metabolism and Function	Deficiency or Excess	Meeting Body Needs
Vitamin A Retinol Retinal Retinoic acid Provitamin A: carotenes	Bile needed for absorption Mineral oil prevents absorption Stored in liver Bone and tooth structure Healthy skin and mucous membranes Vision in dim light	Night blindness Lowered resistance to infection *Severe:* drying and scaling of skin; eye infections; blindness Overdoses are toxic: skin, hair, and bone changes	**RDA:** Men, 1,000 RE (5000 IU); women, 800 RE (4,000 IU) Liver, kidney Egg yolk, butter, fortified margarine Milk, cream, cheese Dark-green leafy and deep-yellow vegetables Deep-yellow fruits

Vitamin D
 	
Precursors:
■ Ergosterol in plants
■ 7-dehydrocholesterol in skin
	Some storage in liver
	Liver synthesizes *calcidiol*
	Kidney converts calcidiol to *calcitriol*
	Functions as hormone in absorption of calcium and phosphorus; mobilization and mineralization of bone
		Rickets in babies
		  Soft bones
		  Enlarged joints
		  Enlarged skull
		  Deformed chest
		  Spinal curvature
		  Bowed legs
		Osteomalacia in adults
		Osteodystrophy in renal disease
		Overdoses are toxic
			RDA: Infants, children, adolescents, and pregnant women: 10 µg (400 IU)
			Fortified milk
			Concentrates: calciferol, viosterol
			Fish-liver oils
			Exposure to ultraviolet rays of sun

Vitamin E Tocopherols	Prevents oxidation of vitamin A in intestine Protects cell membranes against oxidation Protects red blood cells Limited stores in body Polyunsaturated fats increase need	Deficiency not common Red cell hemolysis in malnourished infants Low toxicity	**RDA:** Men, 10 mg α-TE; women, 8 mg α-TE Salad oils, shortenings, margarines Whole grains, legumes, nuts, dark green leafy vegetables
Vitamin K	Forms prothrombin for normal blood clotting Synthesized in intestines	Prolonged clotting time Hemorrhage, especially in newborn infants, and biliary tract disease Large amounts toxic	**RDA:** Men, 80 µg; women, 65 µg Synthesis by intestinal bacteria Dark-green leafy vegetables

Sources

Because of imprecision of laboratory techniques, vitamin K content of foods has yet to be accurately analyzed. Therefore, this nutrient does not appear in standard food composition tables. Available data do indicate the primary dietary sources of vitamin K are green leafy vegetables, especially broccoli, spinach, brussels sprouts, turnip greens, lettuce, cabbage, and dried green tea leaves.

In food, vitamin K appears stable in heat and to oxidation, but retention seems to decline when exposed to light, acid, and alkali. Vitamins A and E are antagonists of vitamin K in the body, but the exact mechanism of their interference is unknown.

Basic Diet

The Basic Diet Nutrient Profile (Table 2–3) does not list vitamin K, since its actual levels in food are unknown. Although green leafy vegetables are the emphasized dietary source, the tiny amounts present in other foods may provide an adequate total amount of vitamin K.

Body Balance

Optimum/Adequate Signs and Symptoms
Like vitamin E, neither vitamin K deficiency nor toxicity is common. In clinical practice, vitamin K status continues to be measured indirectly by tests indicating the clotting abilities of blood.

Insufficiency/Deficiency
Several groups of people are at risk. Newborn infants have a sterile intestinal tract, so sufficient synthesis of vitamin K may not occur until they are one week old. Breast milk is a poor source of vitamin K, leading to the possibility of hemorrhagic disease during the first days of life. To prevent such hemorrhage, a single dose of vitamin K, usually as an injection of phylloquinone (vitamin K_1), is given shortly after birth.

Severe liver disease may inhibit the synthesis of the clotting factors even though dietary vitamin K is adequate. Diseases of malabsorption, biliary obstruction, and oral use of sulfa or antibiotic drugs can lead to deficiency. Hemorrhage could be a serious problem if such an at-risk individual experiences an injury or has surgery, including oral surgery or dental prophylaxis.

Various prescription and over-the-counter medications have vitamin K side effects. Anticoagulating agents, such as *dicumarol*, are used in treating several forms of heart disease. Clinicians usually attempt to balance doses of medication with vitamin K intake to reduce hemorrhagic risks. Mineral oil and bile-binding medications interfere with the absorption of vitamin K. Aspirin and salicylates are widely known for their ability to induce hypoprothrombinemia and hemorrhage. (See Chapter 18.)

Table 10–2 summarizes the fat-soluble vitamins.

VITAMIN HIGHLIGHTS

Water-Soluble Vitamins

VITAMIN C

Characteristics
Vitamin C has a chemical structure similar to glucose, thus making it soluble in water. (See Figure 10–1.) Plants and many animals use glucose to synthesize vitamin C, but humans lack the liver enzyme that does this. In foods, two forms with biological activity exist, L-ascorbic acid and dehydroascorbic acid.

Measurement of vitamin C activity in foods is of L-ascorbic acid and is measured in milligrams (mg). Some tables include analysis of both forms.

Functions
Many functions of vitamin C have been identified, although the exact mechanisms still are under investigation. Of unique importance is that vitamin C, compared to the other water-soluble vitamins, does not appear to function directly in the release of energy from the macronutrients nor does it act as a coenzyme. Rather, vitamin C has a wide range of functions to keep the body in homeostasis.

Vitamin C maintains healthy cell function; holds cells together; heals wounds; builds strong bones and teeth; forms and maintains the cardiovascular, neurological, and digestive systems; keeps the immune system intact to fight infection and disease; and detoxifies medications and environmental pollutants. One major way vitamin C accomplishes all this is by being involved in **hydroxylation reactions,** in which the –OH chemical group is introduced during the conversion of one compound to another. For example, this process is necessary for the synthesis of collagen, serotonin, norepinephrine, and bile. A second way is by being an antioxidant. For example, it protects vitamins A and E and polyunsaturated fatty acids (such as in cell membranes) from excessive oxidation. Vitamin C enhances calcium and nonheme iron absorption.

Metabolic Processes
Most absorption of vitamin C is active, but some occurs by simple diffusion. Since it is water-soluble, it passes directly into the blood circulation. Because of its numerous functions, vitamin C is present throughout the body. Storage location is considered to be each cell, but concentration is especially high in the adrenal glands and eye retina. On an experimental diet devoid of vitamin C, healthy adults have storage supplies for about one month, when deficiency signs may become evident. The kidneys excrete water-soluble vitamin C and its breakdown products.

Meeting Body Needs

Requirements/Recommendations
Experimental evidence indicates that healthy adults maintain a body supply of about 1,500 mg vitamin C. Agreement seems to be that 10 mg/day prevents scurvy and 100–200 mg/day results in direct urinary excretion. Recommendations for vitamin C are set at different levels in different countries. The RDAs are listed inside the front cover.

Sources

Fruits and vegetables are the main source of vitamin C, with citrus fruits leading the list. In general, the active parts of plants contain higher amounts; mature or resting seeds are lacking in vitamin C. Thus, sprouting peas and beans have higher amounts than dormant ones. Human milk from healthy mothers supplies sufficient amounts for infants, but pasteurized milk contains only traces. In the American diet, the majority of vitamin C is supplied from citrus and other fruits, potatoes, dark green and deep yellow vegetables, and other vegetables including tomatoes.

Retention of vitamin C is very unstable. Of all the vitamins, vitamin C is the most easily destroyed—by light, heat, air (oxygen), alkali, and the presence of minerals such as iron and copper. Thus, enormous care must be taken when harvesting, storing, processing, cooking, and serving fruits and vegetables. For example, vitamin C leaches out into any water in which fruits or vegetables soak or cook, so these foods are best prepared by microwave rather than by boiling in large amounts of water. Keeping vegetables in steam tables, a common practice in restaurants, cafeterias, and health care facilities, causes loss of vitamin C in those vegetables. Baking soda, an alkali, sometimes is added to green vegetables to retain their vibrant color. This practice destroys the vitamin C. Instead, a short cooking time in a small amount of water or by microwave retains their attractive color. Dried fruits contain negligible vitamin C. Of course, selecting freshly harvested fruits and vegetables and eating them raw overcomes the retention problem. Nowadays many processed foods are fortified with vitamin C, including fruit juices and drinks, breakfast cereals, and infant formulas.

Basic Diet

The Basic Diet Nutrient Profile (Table 2–3) illustrates the importance of fruits and vegetables as the sources of vitamin C. Selecting suggested servings from these two groups of the Food Guide Pyramid and preparing such foods with vitamin C retention in mind provides levels greater than the RDAs.

Body Balance

Optimum/Adequate Signs and Symptoms

The traditional method to assess vitamin C status was by a balance technique, that is, the measurement of the amount of vitamin C ingested by a person compared to the amount excreted in the urine. Newer techniques are used now, namely measuring the amount of vitamin C in blood plasma and within white blood cells. No matter what test is used, caution is necessary in interpreting laboratory results.

Insufficiency/Deficiency

The vitamin C classic deficiency disease is **scurvy.** This disease is characterized by easy bruising and hemor-rhaging of the skin, loosening of the teeth, bleeding of the gums, delayed wound healing, joint pains and disruption of the cartilages that support the skeleton, and increased susceptibility to infections. In the United States, scurvy occurs rarely, although some of the signs may be seen in people with grossly inadequate diets. It has been reported in infants who have had a cow's milk formula for several months without vitamin C supplementation. One of the symptoms seen in these infants is the extreme tenderness of the skin to touch.

The need for vitamin C increases under certain conditions. Cigarette smoking increases the need by as much as 50 percent. Patients with infections such as pneumonia, rheumatic fever, and tuberculosis require increased amounts. Persons having higher demands for collagen synthesis, such as those with severe burns or fractures, or who are undergoing surgical procedures, have higher needs.

Inconclusive results from many studies make it difficult to determine if people who are at risk for developing or who have the common cold, cancer, or cardiovascular disease have increased metabolic needs for vitamin C. Conducting scientific studies of the sophistication required to answer these questions is technically difficult, painstakingly time consuming, and extremely expensive. Thus far, investigators have been unable to isolate the effects of vitamin C from the numerous other factors that influence the incidence and course of these diseases.

Medications can interfere with vitamin C absorption, or increase its urinary excretion, or increase the rate of its metabolism. (See Chapter 18.)

Excess/Toxicity

In many countries, healthy people regularly ingest greater than the RDAs for vitamin C, but usually from vitamin supplements, not food. Some take 1–5 grams daily (20–80 times the RDAs). Reports of risks periodically appear in the professional literature. For example, reported negative side effects include nausea, vomiting, and diarrhea; excessive absorption of iron, leading to iron overload; interference with the activity of anticoagulants such as *heparin* and *coumarin;* increased excretion of uric acid, a problem for persons susceptible to gout; false-negative test results for urinary glucose in patients with diabetes mellitus; formation of oxalate kidney stones in persons prone to stone formation; and scurvylike withdrawal symptoms when the person abruptly stops taking the high dosages.

THIAMIN

Characteristics

Thiamin has the distinction of being the first of the eight B-complex vitamins to be discovered. It is unique inasmuch as it is one of the few that contains a mineral, namely sulfur. There is one form in nature and one commercially produced form, known as thiamin hydrochloride. Thiamin is measured in food in milligrams (mg).

Functions

Like most of the B-complex vitamins, thiamin mainly functions as a coenzyme. This coenzyme, thiamin pyrophosphate (TPP), is necessary for several reactions that transform macronutrients, especially glucose, into energy. TPP also is required as a coenzyme for the enzyme transketolase. This enzyme functions in the pathway that produces ribose, a sugar essential for the formation of DNA and RNA, the carriers of the genetic code. Apart from its coenzyme functions, thiamin is necessary for the proper functioning of nerves; the mechanism is unclear.

Metabolic Processes

Thiamin in food is found in a bound compound, which is separated during digestion. Thiamin is absorbed by active transport into the blood circulation. When high levels are ingested, some is absorbed by passive diffusion. There appears to be no specific storage location. The majority of thiamin is found in muscles. Thiamin and its degradation products are excreted in the urine.

Meeting Body Needs

Requirements/Recommendations

Requirements are based on the diet's total energy content, especially the energy yielded from carbohydrate. The RDAs are listed inside the front cover.

Sources

Thiamin is widely distributed in small amounts in most plant and animal foods. The greatest contributions are from whole and enriched grains. Thiamin is one of the three vitamins added to flour, breads, cereals, pasta, and rice labeled as enriched. (See Chapter 2.) Thiamin is found in dry beans and peas and vegetables, such as potatoes. Of the meats, lean pork is especially high in thiamin.

Thiamin is water-soluble and destroyed by heat, so losses of thiamin occur during soaking or cooking of food. Since the parts of grains containing thiamin are removed during refining, enriched products need to be selected whenever whole grain ones are not.

Basic Diet

The Basic Diet Nutrient Profile (Table 2–3) illustrates the wide distribution of thiamin in foods. Choosing the number of servings identified in the Food Guide Pyramid easily provides the RDAs for thiamin.

Body Balance

Optimum/Adequate Signs and Symptoms

Thiamin status is assessed by measuring the red blood cell levels of the enzyme transketolase. Such a laboratory test is an indirect measurement of thiamin levels in the body.

Insufficiency/Deficiency

The classic thiamin deficiency disease is **beriberi.** The disease is progressive in adults, but in infants, the onset can seem to be sudden, with the outcome fatal. Beriberi affects the gastrointestinal, cardiovascular, and nervous systems. Like many medical conditions, symptoms are vague and general. Early symptoms include fatigue, irritability, depression, and loss of appetite, weight, and strength. As the deficiency continues, the adult may describe symptoms related to the affected systems, such as indigestion, constipation, headaches, unusually rapid heart rate after mild exercise, numbing of the feet, and cramping or weakness of the legs.

Beriberi is most common in countries with inadequate food supplies or in countries where nearly all energy is derived form carbohydrate, especially from grain foods that have been highly milled ("refined" or "polished"). Otherwise, thiamin insufficiency or obvious deficiency occurs rarely. The exceptions are in people having a long-term inadequate intake of nourishing foods, or experiencing decreased absorption or increased cellular utilization. For example, health professionals in the United States are recognizing beriberi among malnourished homeless people and among refugees from war-torn countries. Some elderly people with narrow food variety may be at risk for symptoms of thiamin insufficiency. Wernicke-Korsakoff syndrome is the classic thiamin deficiency seen secondary to alcohol abuse. Some healthy people, believing they have insufficient thiamin levels for their body needs, self-medicate with supplements of thiamin to improve energy levels or combat emotional stress. Scientific evidence of increased thiamin absorption and body retention or improved cellular utilization is lacking.

Excess/Toxicity

Excessive oral ingestion of thiamin appears to have no adverse side effects. Patients receiving excessive doses intravenously have been reported to experience problems such as headache, irregular heartbeat, and convulsions.

RIBOFLAVIN

Characteristics

Riboflavin is unique inasmuch as it is yellow-green fluorescent in color. These kinds of pigments are designated as "flavins." Contrasted to several other vitamins, riboflavin is stable in heat. In fact, this is how it was identified as being a vitamin separate from thiamin, since thiamin is destroyed by heat. Riboflavin was the first vitamin identified to function as a coenzyme.

Riboflavin in food is measured in milligrams (mg).

Functions

Now riboflavin is known to be a constituent of several enzymes, especially riboflavin monophosphate (FMN) and flavin adenine dinucleotide (FAD). For example, not only is riboflavin involved in releasing energy from the macronutrients, it is a coenzyme in a number of their other transformation pathways, even often assisting other vitamins in their functions.

Metabolic Processes

In foods and during its metabolism in the body, riboflavin exists either in the free state or in combination with other compounds, especially protein and phosphate. Part of the digestive process liberates riboflavin. It is absorbed by active transport and carried by the blood circulation. No site serves as a special storage location. Riboflavin and its metabolites are excreted in the urine, which takes on a bright orange-yellow color.

Meeting Body Needs

Requirements/Recommendations

Information on riboflavin requirements come from feeding studies in which intake and urinary excretion were measured. The RDAs are based on energy intakes and needs during positive protein balance. The figures are listed inside the front cover.

Sources

Although riboflavin is found in plants, animal foods, especially milk, are the best sources. The faint green tinge you can see when you look through a glass of skim milk is riboflavin. Riboflavin is added to enriched grains.

Retention in foods is good. Different from some of the other vitamins, it dissolves sparingly in water and is stable during cooking and in the presence of air and acids. However, it is destroyed by alkali (like baking soda) and is quickly decomposed by ultraviolet rays and light. This knowledge led milk processors to package milk in paper cartons instead of glass bottles. This is an example of a small change in food processing having enormous nutritional benefits for people.

Basic Diet

The Basic Diet Nutrient Profile (Table 2–3) demonstrates the wide distribution of riboflavin in foods. Observe the special contribution of milk and other animal foods, dark green vegetables, and whole and enriched grains.

Body Balance

Optimum/Adequate Signs and Symptoms

The activity of one of the red blood cells' enzymes, glutathione reductase, of which riboflavin is a constituent, is the laboratory test used to measure riboflavin status.

Insufficiency/Deficiency

No deficiency disease is associated with riboflavin. Some signs and symptoms are associated with decreased intake, such as cracking of the skin at the corners of the lips (**cheilosis**), redness and swelling of the tongue (**glossitis**), and scaliness of the skin around the nose and ears. The eyes also are affected: increased capillary blood vessels in the cornea give a "bloodshot" appearance; eye fatigue, itching, burning, and watering, and extreme sensitivity to bright light are other symptoms. Growth failure is a theorized sign. Some patients on medications, such as *chloramphenicol,* have been reported to exhibit signs of riboflavin deficiency. Usually, rather than a separate condition, inadequate riboflavin status is associated with other nutrient deficiencies.

Excess/Toxicity

Riboflavin toxicity has not been reported, probably since riboflavin absorption appears limited.

NIACIN

Characteristics

Niacin is the general term applied to the two active forms of the vitamin: nicotinamide and nicotinic acid. Niacin is the constituent of the coenzymes nicotinamide adenine dinucleotide (NAD) and nicotinamide adenine dinucleotide phosphate (NADP).

Niacin in foods is measured in milligrams of preformed niacin. However, in the body, something unique about niacin can happen. The body is able to transform tryptophan, an essential amino acid, into niacin. To do this conversion, riboflavin and another B-complex vitamin, vitamin B_6, are necessary. The combined amount of niacin is termed **niacin equivalents (NE).**

Functions

Like thiamin and riboflavin, niacin is involved in the release of energy from the macronutrients. Niacin is essential for the synthesis of the macronutrients, too.

Metabolic Processes

Niacin is freed by digestion, absorbed passively, and circulated in the blood. Niacin is in every cell and does not appear to be stored anywhere particular. Niacin's metabolic by-products, rather than niacin itself, are excreted in the urine.

Meeting Body Needs

Requirements/Recommendations

Determination of the requirements for niacin is complicated by the fact that the body can use tryptophan as a precursor of niacin. The RDAs take this conversion into consideration and thus are expressed as NE. Generally, 1 mg of niacin can be produced from 60 mg of tryptophan. The RDAs for niacin are based on energy intake. (See table inside front cover of text.)

Sources

Meat, fish, and poultry are the chief sources of preformed niacin. Some niacin is present in vegetables. The niacin in whole grains is bound, making much of it (up to 70 percent) biologically unavailable. This chemically bound niacin, though, can be released in an alkaline environment. Niacin is added to refined grains labeled enriched; such grains provide biologically available preformed niacin.

Retention of preformed niacin in foods is very good. Unlike some other vitamins, niacin is stable in heat, acids, light, and oxygen (air).

Based on the above information, the following story illustrates the fascination of finding the answer to one of nutrition's great puzzles. Early research on the niacin content of foods showed that milk was not a good source of preformed niacin. Yet milk was found to be an effective food for people whose diets were deficient in niacin. Why this seeming discrepancy? Answer—milk is an excellent source of tryptophan. But some people whose diets were based on corn developed niacin deficiency and some did not. Why? First, corn is not a good source of tryptophan, so the body did not have the niacin precursor. Second, preformed niacin can be liberated by alkali. So people whose staple food was unrefined corn tortillas (prepared with an ingredient with an alkaline pH, thus enabling the niacin to become bioavailable) did not develop niacin deficiency, but people whose staple food was grits (boiled refined cornmeal devoid of niacin from milling) did.

Basic Diet

The Basic Diet Nutrient Profile (Table 2–3) shows the amount of preformed niacin in foods. Foods in typical American diets furnish about 1 percent tryptophan from good quality proteins. Thus, a 60 g protein diet could supply an additional 10 NE.

Body Balance

Optimum/Adequate Signs and Symptoms

Niacin status is assessed by measuring the amount of niacin metabolic by-products excreted in the urine.

Insufficiency/Deficiency

The classic niacin deficiency disease is **pellagra.** (See Figure 10–5.) Pellagra often is called the disease of the progressive "4 Ds"—diarrhea, dermatitis, dementia, and death. Because the cells of the gastrointestinal tract turn over so frequently, the first signs are evident there. An extremely sore and swollen tongue makes eating difficult, and diarrhea is common. Next, the skin changes appear. Dermatitis is symmetrical, that is, on both hands, both forearms, both legs. Also, the exposed parts of the body are more affected, leading to sensitivity to the sun. As the disease advances, mental changes become evident, such as depression, disorientation, delirium, and dementia. Without treatment, the ultimate outcome is death.

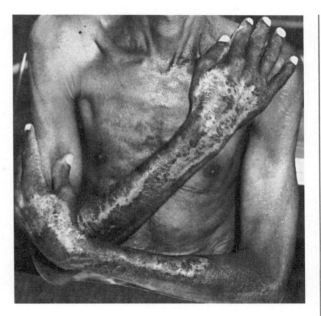

FIGURE 10–5
Dermatitis, a sign of pellagra, appears on both hands and forearms. Exposed parts of the body are more affected than are parts covered by clothing. (*Source:* Courtesy of FAO/The United Nations.)

In the early 1900s, when treatment for pellagra was discovered, some people in institutions for the mentally ill could be "cured." Nutritional attempts are tried periodically to treat children and adults with various disorders affecting the brain. Usually this includes huge doses of vitamins, including niacin. Up to now, unless the person actually had been specifically vitamin deficient, no improvement has been detected.

One medical condition that has shown benefits from treatment with nicotinic acid (but not nicotinamide) is elevated blood cholesterol. The treatment involves daily large doses (greater than 3 grams). Total blood cholesterol and LDLs decrease and HDLs increase.

Excess/Toxicity

Obvious side effects experienced from excess niacin intake are nil, except from nicotinic acid. Toxic effects are observed during treatment for heart disease. The most immediate and obvious symptoms are uncomfortable sensations from flushing of the skin, accompanied by skin redness. More serious side effects include heartburn, gastric ulcers, and elevated blood glucose levels.

VITAMIN B$_6$

Characteristics

Vitamin B$_6$ consists of a group of related compounds, all with some vitamin activity: pyridoxine, pyridoxal, and pyridoxamine. Pyridoxine is the most stable. Vitamin B_6 is a constituent of the coenzyme pyridoxal phosphate.

Measurement of vitamin B_6 in foods is in milligrams (mg).

Functions

Vitamin B_6 is unique because most of its enzyme systems are involved in the catabolism or anabolism of proteins, rather than in the release of energy. For example, as previously noted, vitamin B_6 is necessary for the conversion of tryptophan to niacin. Vitamin B_6 also is required for the formation of heme, the protein portion of red blood cells. Thus, vitamin B_6 can be classified as a hematopoietic vitamin. Riboflavin is required to activate its major coenzyme.

Metabolic Processes

Vitamin B_6 usually is present in a bound form in food. It is liberated during digestion, passively absorbed, and transported in blood circulation. Although the liver does store some vitamin B_6, most of it is located throughout the body, especially in the muscles. The vitamin itself, along with metabolic by-products, is excreted in the urine.

Meeting Body Needs

Requirements/Recommendations
The RDAs are based on protein intake. When protein intake increases or when the body's needs for protein increase, as during pregnancy, vitamin B_6 requirements increase. (See table inside front cover.)

Sources
Vitamin B_6 is found in plant and animal foods, but appears to be more bioavailable from animal sources. Vitamin B_6 binds with fiber and with other substances in plants.

Food retention varies. It is destroyed by heating and freezing. Vitamin B_6 is lost when whole grains are milled; it is not one of the three vitamins added to enriched grains.

Basic Diet
The Basic Diet Nutrient Profile (Table 2–3) lists the existence of some vitamin B_6 in all foods, except fats and sugar. However, its bioavailability and retention are best in minimally processed animal foods.

Body Balance

Optimum/Adequate Signs and Symptoms
To evaluate vitamin B_6 status, a number of urine tests are performed. At this time, any one specific test alone does not provide sufficient information to make a correct interpretation.

Insufficiency/Deficiency
The crucial role of vitamin B_6 was brought to light in the 1950s when vitamin B_6 was inadvertently destroyed during sterilization of infant formula. Since infancy is a time of rapid formation of body proteins, deficiency soon became apparent. The infants fed this formula had reduced growth rates, nervous irritability, hypochromic anemia, and convulsions. They recovered promptly when adequate vitamin B_6 was provided. All commercial formulas for infants now supply sufficient vitamin B_6 to meet infants' needs.

Several medications interfere with the use of vitamin B_6. *Isoniazid,* used for the treatment of tuberculosis, binds vitamin B_6 so that it cannot be utilized by the body. If the intake of vitamin B_6 is not increased, neurological problems may occur. *Penicillamine,* a medication used in the treatment of Wilson's disease, increases the loss of the vitamin in the urine. Renal dialysis treatments remove the vitamin. In these conditions, supplements of vitamin B_6 are recommended.

Excess/Toxicity
Overt symptoms with continued use of daily doses over 100 mg have been reported. Most symptoms relate to the nervous system, including hands and feet numbness and difficulty walking. People taking these doses of vitamin B_6 were attempting to prevent or treat various diseases, including premenstrual syndrome, depression, and heart disease.

VITAMIN B_{12}

Characteristics

Vitamin B_{12} is unique in several ways, starting with its structure. (See Figure 10–1.) It has, by far, the most complex chemical structure of any of the vitamins and it is the only one containing the mineral cobalt (Co). It appears to be formed only by microorganisms and to be present only in animal foods. Plant foods may be a source only if they inadvertently contain animal particles or microorganisms.

The amount of vitamin B_{12} in foods is so small that it is measured in micrograms (µg).

Functions

Vitamin B_{12}, as a constituent of two major coenzymes, methylcobalamin and 5-deoxyadenosylcobalamin, performs five vital reactions. Major functions include DNA formation, conversion of one amino acid to another, and nervous system development. The division and maturity of red blood cells is related to this vitamin's role in DNA formation, which provides its classification as a hematopoietic vitamin.

Metabolic Processes

Another unique characteristic is the complex series of steps taken by the body to transform vitamin B_{12}. In food, the vitamin is held with other substances. Part of the digestive step requires its release and then recombination with a substance secreted by cells in the stomach. This substance is termed **intrinsic factor;** vitamin B_{12} is termed **extrinsic factor.** In this combined state, the vitamin is absorbed into the blood circulation. There it is bound to proteins. This actually prevents vitamin B_{12} from being readily excreted by the kidneys. Instead, some is excreted into bile, thus eliminated from the body in the feces. Also unlike other water-soluble vitamins, vitamin B_{12} is stored, chiefly in the liver. Even if adults completely stop eating animal foods, they have supplies estimated to last 20–30 years. The same does not hold true for children—they have not had enough time to develop stores, so are more at risk for deficiency.

Meeting Body Needs

Requirements/Recommendations
Requirements are based on worldwide studies of people who eat different levels of animal or plant foods. The RDAs reflect needs related to growth and development. (See inside front cover.)

Sources
Only animal foods are sources. As previously mentioned, plant foods are lacking in vitamin B_{12}. Vitamin B_{12} associated with vegan diets (Chapter 7) is from a microorganism source connected with the plants, such as on legumes.

Retention can be preserved by limited processing and heating of animal foods.

Basic Diet
The Basic Diet Nutrient Profile (Table 2–3) clearly shows the division of sources of vitamin B_{12} as being from foods of animal origin vs. plant foods.

Body Balance

Optimum/Adequate Signs and Symptoms
Blood levels of vitamin B_{12} or its metabolites are examples of laboratory tests used to measure status.

Insufficiency/Deficiency
A number of different situations can create an inadequacy of vitamin B_{12}. **Pernicious anemia** is the classic deficiency secondary to a defect in which the intrinsic factor is missing. Dietary vitamin B_{12} cannot be absorbed, so the bone marrow is unable to produce mature red blood cells. This results in the development of large and immature cells, known as **macrocytic (megaloblastic) anemia.** Symptoms include pallor, soreness of the mouth, anorexia, dyspnea, weight loss, and prolonged bleeding time. Gradually there is degeneration of the peripheral nerves, unsteadiness of gait, and mental depression. To bypass the block in absorption, injections of vitamin B_{12} must be given.

Pernicious anemia can occur if the stomach's structure or function is damaged, such as following gastrectomy. (See Chapter 21.) The anemia may not become

evident for years, until liver stores have been depleted. Also, people over 60 years of age produce less stomach secretions and are at risk for not being able to digest and absorb vitamin B_{12}.

Vitamin B_{12} is absorbed in a specific portion of the small intestine, the ileum. Interference can occur in diseases of malabsorption or following surgical removal of the specific section of the ileum. Again, injections are required.

For people who are strict vegetarians (vegans), oral supplements or fortified foods, such as fortified soy milk, are possible modes of therapy. Since vegans do not have intrinsic factor deficiency to impair digestion and absorption, they only need to be supplemented with the extrinsic factor—vitamin B_{12}—in an acceptable oral source.

Excess/Toxicity
No toxicity has been reported.

FOLATE

Characteristics
Folate is the general term applied to the several forms of the vitamin. Like many other vitamins, its structure is highly complex. Because of its role in blood formation, like vitamins B_6 and B_{12}, it is classified as a hematopoietic vitamin.

Like vitamin B_{12}, the amount in foods and the body is so small that it is measured in micrograms (μg).

Functions
Folate functions as coenzymes that are especially involved in amino acid conversions. As such, folate activities are essential for DNA synthesis and cell division. Thus, folate functions in rapidly dividing cells, such as red blood cells and the fetus, and body systems where cells turn over rapidly, such as the gastrointestinal tract. In a related fashion, folate participates in the regeneration of a number of enzymes.

Metabolic Processes
Like some other vitamins, folate in foods exists as a bound compound that must be split during digestion. It is both passively and actively absorbed, and then is transported in the blood circulation, usually attached to a protein. The major storage area is the liver. Once folate is absorbed, the body appears to retain most of it. Some is excreted into the bile, but then it is reabsorbed. The urine contains only a small amount.

Meeting Body Needs

Requirements/Recommendations
Requirements are related to its function in cell division, thus growth and development. The RDAs take into

consideration that, on average, about one-half of folate in food is digested and absorbed. (See table inside front cover.)

Sources

From its name you may have guessed that plant foods ("foliage"), such as vegetables, legumes, and fruits, are excellent sources. Actually, many animals foods supply folate, too.

The bioavailability and retention in foods is variable. Heating foods can destroy folate. Milling grains removes some folate. Some foods, like breakfast cereals and infant formulas, are fortified. Acids, including vitamin C with its antioxidant properties, protect folate. This makes fresh oranges and fresh or frozen orange juice, for instance, important commonly consumed foods as sources of folate.

Basic Diet

The Basic Diet Nutrient Profile (Table 2–3) shows the wide distribution of folate in foods.

Body Balance

Optimum/Adequate Signs and Symptoms

Evaluation of folate status is by the laboratory measure of folate in the blood. Sometimes folate metabolites in the urine are analyzed.

Insufficiency/Deficiency

The classic deficiency is **macrocytic (megaloblastic) anemia.** This is the anemia also caused by a deficiency of vitamin B_{12}, but this time the reason is different. When such a laboratory result is identified, the next step is to determine ("differential diagnosis") the true underlying cause. Why is this important? Folate can

TABLE 10–3
Summary of Water-Soluble Vitamins

Vitamins	Metabolism and Function	Deficiency	Meeting Body Needs

Vitamin C
Ascorbic Acid
	Form collagen
	Teeth firm in gums
	Hormone synthesis
	Resistance to infection
	Improve iron absorption
		Poor wound healing
		Poor bone, tooth development
		Scurvy
		  Bruising and hemorrhage
		  Bleeding gums
		  Loose teeth
			RDA: Adults, 60 mg
			Citrus fruits
			Strawberries, cantaloupe
			Tomatoes, broccoli
			Raw green vegetables

Thiamin
Vitamin B_1
	Coenzyme for breakdown of
	  glucose for energy
	Healthy nerves
	Good digestion
	Normal appetite
	Good mental outlook
		Fatigue
		Poor appetite
		Constipation ·
		Mental depression
		Neuritis of legs
		Beriberi
		  Polyneuritis
		  Edema
		  Heart failure
			RDA: 0.5 mg/1,000 kcal
			Pork, liver, other meats, poultry
			Dry beans and peas, peanut
			  butter
			Enriched and whole-grain
			  breads, cereals, rice
			Milk, eggs

Riboflavin
Vitamin B_2
	Coenzymes for protein and
	  glucose metabolism
	Fatty acid synthesis
	Healthy skin
	Normal vision in bright light
		Cheilosis
		  Cracking lips
		  Scaling skin
		Burning, itching, sensitive eyes
			RDA: 0.6 mg/1,000 kcal
			Milk, cheese
			Meat, poultry, fish
			Dark-green leafy vegetables
			Enriched and whole-grain
			  breads, cereals, rice

Niacin
Nicotinic acid
Niacinamide
	Coenzymes for energy
	  metabolism
	Normal digestion
	Healthy nervous system
	Healthy skin
	Tryptophan a precursor:
	  60 mg = 1 mg niacin
		Pellagra
		  Dermatitis
		  Sore mouth
		  Diarrhea
		  Mental depression
		  Disorientation
		  Delirium
			RDA: 6.6 mg NE/1,000 kcal
			Meat, poultry, fish
			Dark-green leafy vegetables
			Enriched or whole-grain
			  breads, cereals, rice
			Tryptophan in complete
			  proteins

appear to correct the anemia caused by a deficiency of vitamin B_{12}, yet allow the nervous system deterioration to continue. (See section on vitamin B_{12}.) By regulation of the Food and Drug Administration, over-the-counter folate supplements can contain no more than 400 µg—the RDA level for pregnant women. This restriction attempts to avoid the unfortunate situation of people self-medicating with folate, when, in fact, they have pernicious anemia from a deficiency of either extrinsic or intrinsic factor. Of course, people can take multiple quantities of supplements!

People at risk for folate deficiency, like other nutrients, are those who have inadequate intakes or an abnormality interfering with the body's processing of folate. This includes people eating a limited variety of foods or consuming foods where focus on vitamin retention was lacking, or people with gastrointestinal problems.

Several medications interfere with folate's metabolic processes. *Antacids* can increase the alkalinity of the stomach and small intestine, thereby reducing the amount of folate that is digested and absorbed. *Sulfasalazine, phenytoin,* and alcohol interfere with absorption and cellular utilization. *Methotrexate,* a powerful antitumor medication, is a folate antagonist; this is the basis for the medication's effectiveness against rapidly dividing cancer cells.

Needing special attention are healthy people during periods of rapid growth and development. Pregnant women are a classic example. Macrocytic anemia has long been observed as occurring during pregnancy. More recently, some research showed that babies with **neural tube defects,** a severe condition affecting the formation of the nerves of the spinal cord, can be born to some mothers with low folate levels. This observation prompted much debate in the scientific, health care, pharmaceutical, food processing, and political communities as to the best public health policy to establish. As a health care provider, you will want to recognize the importance of nutritionally evaluating and counseling women of childbearing age as to the importance of their preconceptual nutrition status. (See Chapters 3 and 12.)

Vitamins	Metabolism and Function	Deficiency	Meeting Body Needs
Vitamin B_6 Pyridoxine Pyridoxal Pyridoxamine	Coenzymes for protein metabolism Conversion of tryptophan to niacin Formation of heme	Gastrointestinal upsets Weak gait Irritability Nervousness Convulsions	**RDA:** Women, 1.6 mg; men, 2.0 mg Meat, whole-grain cereals, dark-green vegetables, potatoes
Vitamin B_{12} Cobalamin	Formation of mature red blood cells Synthesis of DNA, RNA Requires intrinsic factor from stomach for absorption	Pernicious anemia: lack of intrinsic factor, or after gastrectomy Macrocytic anemia: leads to neurologic degeneration	**RDA:** Adults, 2.0 µg Animal foods only: milk, eggs, meat, poultry, fish
Folate Folic acid Folacin	Maturation of red blood cells Synthesis of DNA, RNA Not a substitute for vitamin B_{12}	Macrocytic anemia: in pregnancy, sprue	**RDA:** Adults, 180–200 µg Dark-green leafy vegetables, meats, fish, poultry, eggs, whole-grain cereals
Biotin	Component of coenzymes in energy metabolism Some synthesis in intestine Avidin, a protein in raw egg white, interferes with absorption	Occurs only when large amounts of raw egg whites are eaten Dermatitis, loss of hair	**ESADDI:** Adults, 30–100 µg Organ meats, egg yolk, legumes, nuts
Pantothenic Acid	Component of coenzyme A Synthesis of sterols, fatty acids, heme	Occurs rarely Neuritis of arms, legs; burning sensation of feet	**ESADDI:** Adults, 4–7 mg Meat, poultry, fish, legumes, whole-grain cereals Lesser amounts in milk, fruits, and vegetables

The biggest concern about excess folate is its masking of pernicious anemia with its irreversible neurological deterioration, from a deficiency of vitamin B_{12}. Also, high intakes of folate appear to interfere with zinc absorption.

BIOTIN

Biotin is a sulfur-containing B-complex vitamin and is a constituent of enzymes involved in macronutrient metabolism. It is measured in micrograms (μg).

Studies concerning human biotin requirements are limited. Thus, instead of an RDA for biotin, an Estimated Safe and Adequate Daily Dietary Intake (ESADDI) is presented. (See table inside front cover.) Information on the biotin content of foods is incomplete, so this vitamin is not provided in tables of food composition. Generally, the best sources are liver, egg yolk, peanut butter, and soy flour. Fruit and meat are poor sources. Biotin is synthesized by intestinal microorganisms, but it remains unclear if it is absorbed.

Laboratory tests usually measure biotin levels in blood. Deficiencies have been produced experimentally. Symptoms are general and mimic those of other B-complex vitamin deficiencies, such as loss of appetite, nausea, vomiting, depression, and loss of hair. The classic study involved feeding raw egg whites (equal to about 60 eggs) to four healthy volunteers. Egg whites contain a protein (named *avidin*) that binds biotin and prevents its absorption. Heating the egg white inactivates the binding capacity of avidin by denaturing the protein. (See Chapter 7.) Toxic side effects of excessive doses have not been reported.

PANTOTHENIC ACID

Pantothenic acid is a B-complex vitamin involved in a key crossroads enzyme reaction of macronutrient metabolism. Its name means "everywhere."

As with biotin, studies concerning human pantothenic acid requirements are limited. Thus, instead of an RDA, an Estimated Safe and Adequate Daily Dietary Intake (ESADDI) is presented. (See table inside front cover.) Information on the pantothenic acid content of foods also is incomplete, so this vitamin is not provided in food composition tables. Pantothenic acid is widely distributed, especially in meats, whole grains, and legumes.

Pantothenic acid can be measured in blood or urine. Deficiency of pantothenic acid has been noted in people with severe malnutrition with symptoms common to those of other B-complex vitamins. For example, the neuritis sometimes seen in people with chronic alcoholism has been ascribed to pantothenic acid deficiency. However, when diets are extremely inadequate, several nutrient intakes are at levels that can produce such symptoms. It becomes exceedingly difficult to separate the symptoms and to ascribe them to a lack of one or another nutrient. Toxicity has not been reported.

Table 10–3 summarizes the water-soluble vitamins.

HEALTH AND NUTRITION CARE ISSUES

Vitamins are the "newest" group of food substances determined to be nutrients. From recorded time until the early twentieth century, the single cause and cure of many debilitating and deadly diseases slowly was shown to be something in food. During the first half of the twentieth century, that "something" was scientifically demonstrated to be what are known now as vitamins. Thus, these discoveries were made within the lifetimes of many people currently living! During the second half of the century, a focus of vitamin research turned to studying vitamins' role in providing lifelong optimal health and in preventing or delaying other diseases, or as a treatment once disease does occur. Emphasis for the vitamins has been on the body's reaction to decreased or increased intake or presence in a body fluid that can be measured, such as the blood or urine. Many fervent studies are being conducted, leading to some interesting results being published and highly publicized. However, the dramatic effects of old are not seen in these new experiments. Thus, new functions and practical uses of vitamins continue as theories, not proven facts. But as more sophisticated research instruments are developed, scientists should be able to study the different locations where vitamins actually function in healthy and diseased cells.

Vitamins seem to be the class of nutrients most intriguing to consumers. Surveys in the United States estimate that about one quarter of the people in some population groups take over-the-counter vitamin supplements, in addition to the vitamins supplied by their foods. The highest users are women, European-

Americans, people over the age of 45 years, and those with higher incomes and some college education who do not smoke cigarettes or drink alcohol.

Since vitamins exist in such minuscule amounts in foods and the body, scientists have needed highly sophisticated instruments to find and measure them. In clinical practice, we usually rely on calculating the amount shown in tables of food composition, on patients' symptoms, clinical signs of deficiency or toxicity, and laboratory measurements of blood and urine. (See Chapter 3.)

When screening a client's diet regarding vitamin content, five points to keep in mind are:

- Amount of the vitamin expected to be in the food.
- Quantity of the food that actually is eaten.
- Frequency with which the food is eaten.
- Stability of the vitamin during processing and cooking.
- Bioavailability of the vitamin, especially when ingested with other foods or dietary supplements, and upon body's need.

Based on history and current health/illness status, a patient may benefit from a decrease or increase in the administration of one or more vitamins. For these patients, implement the same procedures as for any other preventive or treatment method. Incorporate specific evaluation techniques, closely monitor important variables, and maintain consistent long-term follow-up.

PUTTING IT ALL TOGETHER

Vitamins, a diverse class of nutrients, are organic compounds with specific structures, properties, and food sources. They differ from the macronutrients inasmuch as they are needed in tiny amounts by the body, they cannot be manufactured in the body at all or in sufficient amounts, and they function mainly as coenzymes along the points of metabolic pathways.

? QUESTIONS CLIENTS ASK

1. If I do not get the RDA for all the vitamins each day, am I at risk of developing vitamin deficiency diseases?

Not necessarily. If your usual diet has been good, chances are that your body has sufficient stores of fat-soluble vitamins to last for several months and sufficient tissue saturation of water-soluble vitamins to protect you for several weeks. Children, because of their smaller body stores and rapid growth, are more susceptible to the effects of dietary deficiency. The chances of deficiency symptoms increase as your intake of a given vitamin falls lower and lower, especially if this continues over a period of time. So play it safe; choose a variety of foods so that over a week's time your intake will give you a daily intake that equals the RDAs.

2. Will a vitamin supplement give me extra energy?

Vitamins do not provide calories, and so they are not a direct source of energy. It is true that most of the

B-complex vitamins are needed in enzymes to change the carbohydrates, fats, and proteins of foods into forms that the body can use for energy. The amounts of vitamins needed are very small and are easily provided in the foods we eat. Extra B vitamins will do absolutely nothing to increase your energy level.

3. Are the foods in today's market practically depleted of vitamin value because of the poor soil on which they are grown?

No. Improvement of the soil with fertilizers, crop rotation, and soil testing are sound farming practices that increase crop yields of highly nutritious foods.

4. I am constantly meeting deadlines and become very frustrated and nervous when I have extra jobs to do. Should I take stress vitamins?

Contrary to popular ideas, emotional and psychological stress does not increase your need for vitamins. If you are eating a variety of foods from each of the groups in the Food Guide Pyramid, there is no need for additional vitamins. So-called stress vitamins vary greatly in the kinds and amounts of vitamins present in the mixture. People who are under physiological stress, as after surgery, burns, or other severe injuries, do require additional vitamins. These are determined for patients based on individual evaluation.

5. *Is vitamin C from natural sources such as rose hips better than pills from synthetic sources?*

Vitamin C has a specific chemical composition that is the same whether it comes from foods or pills. The only difference is in the greater cost of the rose hips. The body has no way of telling whether it was given a "natural" vitamin found in foods or supplements or a synthetic vitamin; either form is equally effective. Some vitamin products actually are synthetic vitamins with "added natural ingredients." Be sure to read the fine print!

✔ CASE STUDY: A Young Woman with Signs of Vitamin Toxicity

Irene F., 28, is convinced that the usual food supply is depleted of its vitamin content so that she needs to take vitamin supplements. She also read that large doses of vitamins can prevent colds and such chronic diseases as cancer and heart conditions. Besides a daily multiple vitamin tablet, she takes 8,000 µg RE vitamin A, 1 g vitamin C, and 200 mg α-TE vitamin E. She also uses her food processor to make carrot juice, which she drinks every day.

After several months Irene noticed that her skin was quite yellow and dry. Her hair was falling out and her joints ached. So she went to her HMO to discuss her observations and to get treatment for her symptoms.

Questions

1. What are the reasons for each of Irene's symptoms?

2. Why are the results of ingesting excesses of fat-soluble vitamins more likely to be obvious than the results of ingesting excess water-soluble vitamins?

3. What circumstances might indicate the need for a person to take vitamin A, vitamin C, or vitamin E supplements?

4. What are the recommended dietary allowances for vitamins A, C, and E for women Irene's age? How might Irene's need for vitamins be different from the RDAs?

5. What steps would Irene need to take to overcome the symptoms she observed?

6. The vitamin content of foods from the Food Guide Pyramid was explained to Irene. List five foods from this guide that are excellent sources of vitamin A, of vitamin C, and of vitamin E.

7. What food shopping, storage, and preparation techniques should Irene use to retain the vitamin value of foods?

8. From the viewpoint of your chosen career, describe the approaches you would take to examine the pros and cons of controversial health-related nutrition research.

9. Obtain a research article that might interest Irene. What points would you emphasize when interpreting the information with her?

FOR ADDITIONAL READING

American Dietetic Association: "Position Paper: Vitamin and Mineral Supplementation," *J Am Diet Assoc,* **96:**73–77, 1996.

Blank, S, et al: "An Outbreak of Hypervitaminosis D Associated with the Overfortification of Milk from a Home-Delivery Dairy," *Am J Pub Health,* **85:**656–659, 1995.

"Dietary Supplements: Recent Chronology and Legislation," *Nutr. Rev,* **53:**31–36, February 1995.

Hine, RJ: "What Practitioners Need to Know About Folic Acid," *J Am Diet Assoc,* **96:**451–52, 1996.

Slesinski, MJ, *et al.:* "Trends in the Use of Vitamin and Mineral Supplements in the United States: The 1987 and 1992 National Health Interview Surveys," *J Am Diet Assoc,* **95:**921–23, 1995.

Sydenstricker, VP, *et al.:* "Preliminary Observation in 'Egg White Injury' in Man and Its Cure with a Biotin Concentrate," *Science,* **95:**176–77, 1942.

Minerals

Chapter Outline

Increase calcium intake so at least 50 percent of youth aged 12 through 24 and 50 percent of pregnant and lactating women consume 3 or more servings daily of foods rich in calcium, and at least 50 percent of people aged 25 and older consume 2 or more servings daily. *

Decrease salt and sodium intake so at least 65 percent of home meal preparers prepare foods without adding salt, at least 80 percent of people avoid using salt at the table, and at least 40 percent of adults regularly purchase foods modified or lower in sodium. *

Reduce iron deficiency to less than 3 percent among children aged 1 through 4 and among women of childbearing age. *

✔ CASE STUDY PREVIEW

The marvelous roles minerals play in health and disease hold a fascination for researchers, health care providers, and consumers of all ages. Although interest in minerals has moved in esoteric directions, some classic mineral deficiency diseases continue to linger in the United States. What nutrition habits can youngsters develop so they experience a lifetime of optimal mineral health?

Healthy People 2000: National Health Promotion and Disease Prevention Objectives. Public Health Service, U.S. Department of Health and Human Services, Washington, DC, 1991, Objectives 2.8, 2.9, 2.10, p. 94.

Minerals are another fascinating group of nutrients. Today's media headlines together with widespread advertising emphasize the role of one or more minerals in promoting health or preventing and curing diseases. Life, health, and healing from diseases do depend on them, just like all the other nutrients.

Minerals, like vitamins, are termed micronutrients. The body contains small quantities of minerals and needs them in tiny amounts. As a class of nutrients, minerals are completely different from the others (except water). Minerals are *inorganic* elements, that is, they do not contain carbon, thus are not organic. Knowledge that minerals existed as part of living organisms has been recognized for centuries, since it is the mineral (inorganic) portion that remains after death or burning of plants and animals. But all the intricate functions of these elements did not begin to be discovered until the twentieth century. Thus, there remains a large "gray area" of tentative vs. actual knowledge about minerals. Today, with advanced laboratory instruments, minerals can be more closely studied in both health and disease, hence the realization they have numerous intriguing functions.

CHARACTERISTICS

Nomenclature

Minerals are a group of elements that are essential in small amounts. They must be supplied by the diet. Their names are found in the periodic table of elements.

Chemical Structure/Composition/Measurement

Minerals usually exist in three states: inorganic compounds (e.g., sodium chloride, hydrochloric acid); organic compounds (e.g., phospholipids, hemoglobin, thyroxine); and free ions (e.g., Cl^-, K^+, Ca^{2+}).

In foods, minerals are measured in milligrams (mg) or micrograms (µg). (Recall that 1 microgram means 0.000001 gram.) In body fluids, mineral concentrations are expressed in milliequivalents per liter, as metric Système International (SI) units. (See Table 4–8 and Appendix Table B–2.)

Distribution

As with vitamins, the other group of micronutrients, minerals are found in all foods except those that are highly refined. In the body, minerals are found in all tissues and fluids. About 4 percent of the body weight is minerals, about 5–6 pounds (2–3 kg) in the average adult.

Classification

Table 11–1 shows various ways minerals pertaining to the human body can be classified. The oldest method is by the total amount present in the body. Early scientists studied decayed and burned tissues to determine the remaining "ash." For example, they discovered that, in the human body, calcium and phosphorus comprise three-quarters of all mineral elements. For convenience, the minerals were divided arbitrarily into two groups, known as *macro* minerals (those greater than 0.005 percent of body weight) and *micro* minerals (the remainder).

A second method is according to the amount advised to be ingested in the diet. Traditionally, 100 mg is used as the dividing line between macro- versus microminerals.

TABLE 11-1
Mineral Composition of the Adult Body and Dietary Recommendations

	Approximate Amount in Adult Body 70 kg	Adult Dietary Recommendations
Minerals for Which a Recommended Dietary Allowance (RDA) Has Been Set		
Calcium	1,200 g	800 mg
Phosphorus	750 g	800 mg
Magnesium	30 g	280 mg (females) 350 mg (males)
Iron	4 g	15 mg (females) 10 mg (males)
Zinc	2 g	12 mg (females) 15 mg (males)
Iodine	30 mg	150 µg
Selenium	2 g	55 µg (females) 70 µg (males)
Minerals for Which an Estimated Safe and Adequate Daily Dietary Intake (ESADDI) Has Been Set		
Copper	150 mg	1.5–3.0 mg
Manganese	150 mg	2.0–5.0 mg
Fluoride	1 g	1.5–4.0 mg
Chromium	5 mg	50–200 µg
Molybdenum	9 g	75–250 µg
Minerals with Estimated Minimum Requirements		
Sodium	105 g	500 mg
Chloride	105 g	750 mg
Potassium	245 g	2,000 mg
Minerals for Which No Allowances Have Been Set		
Essential as parts of other nutrients:		
Sulfur	175 g	
Cobalt	5 mg	
Present in the body and possibly essential:		
Arsenic, Cadmium, Nickel, Silicon, Tin, Vanadium		
Present in the body but no known function:		
Aluminum, Barium, Boron, Bromine, Gold, Lead, Mercury, Strontium		

Properties

The properties of minerals are important both in foods and in the body. Minerals easily exist as ions and thus can be very reactive. (See Chapter 4.) Like vitamins, they cannot be interconverted one to another or to other nutrients. (Some people claim that transmutation of minerals can occur; this is false.)

Since they are elements, minerals are not affected by heat. Minerals have varying solubilities in water, acids, and bases. In water, minerals equilibrate between foods and liquid. For example, minerals can dissolve into the water portion of canned foods or cooking water, thus affecting the mineral nutritive values of these foods. Conversely, fresh vegetables canned in salted water will take up some salt. The pH of the surroundings can make minerals either more or less soluble. For example, calcium and iron are more soluble in acids and less in bases. This is especially important for their digestion and absorption in the stomach and small intestine. Like vitamins, minerals do not supply energy themselves, but are involved in all metabolic pathways of the macronutrients, the vitamins, and even of themselves. The human body truly is finely tuned.

FUNCTIONS

For convenience, minerals are discussed separately. Specific distinctive functions are listed under the headings for each mineral. (See succeeding sections.)

Overall, the first mineral functions were the most obvious to observe with the eye and measure with rudimentary scientific instruments. For example, the minerals that exist in largest quantities both in foods and in the body were easiest to measure. Thus, the presence of the minerals composing bones and teeth (high in calcium and phosphorus) led to the recommendation of consuming dairy foods (high in calcium and phosphorus) to supply those minerals.

Slowly, minerals' many other functions are being identified. Like most nutrients, minerals function in interrelated fashion with themselves or with other nutrients. Some categories of general functions include:

- *Part of structure of every cell.* The hard structures, like bones and teeth, have strength and rigidity from calcium, phosphorus, magnesium, and fluoride. The soft tissues, like organs and muscles, contain many minerals in their structures, including potassium, sulfur, phosphorus, and iron.

- *Part of vitamins, enzymes, and hormones.* Minerals are constituents of various regulatory compounds. For example, *vitamins:* sulfur is part of thiamin and biotin, and cobalt is part of vitamin B_{12}. *Enzymes:* selenium is part of glutathione peroxidase (part of the body's antioxidant system), and zinc is part of carbonic anhydrase (helps release carbon dioxide from red blood cells). *Hormones:* iodine is part of thyroxine. In many of these situations, minerals function as **cofactors,** that is, a mineral unites with the other compound in order for that substance to function. For example, calcium activates pancreatic lipase. In other cases, minerals act as catalysts to increase the speed of a reaction. For example, copper speeds up the reaction to incorporate iron into the hemoglobin molecule and zinc does the same in the formation of insulin by the pancreas.

- *Regulate response of nerves to stimuli and contraction of muscles.* Minerals control the passage of materials in and out of cells, thereby regulating the transmission of nerve impulses and muscle contractions. For example, working together, sodium, potassium, calcium, and magnesium all regulate the various cellular pumps and cellular membrane ion channels.

- *Maintain water and acid-base balances.* Water balance between the inside and outside of the cells depends in large part on the correct concentrations of potassium (inside) and sodium (outside). Acid-base regulation involves minerals, especially as buffer salts. (See Chapter 4.)

Requirements and Recommendations

Greater knowledge exists about some minerals than about others. Like other nutrients classified as "essential" (some fatty acids and amino acids), there are 15 minerals currently considered to be "essential"—a total of 15 out of 92 naturally occurring minerals. Sufficient scientific data exist for seven minerals to receive the classification of Recommended Dietary Allowances (RDAs). Five minerals have the designation Estimated Safe and Adequate Daily Dietary Intake (ESADDI). Three minerals that are main electrolytes in the body are grouped together under the designation "Estimated Minimum Requirements of Healthy Persons." (See Table 11-1, Chapter 1, and the inside front cover.)

The remaining minerals present in the body do not yet have a clearly identified human purpose sufficient to indicate a guideline for dietary intake. Research is exploring these minerals' possible roles. But some may have no purpose at all in the human body. They may be in the body because they happened to be in any of the many substances people ingest during the course of a lifetime, and the body just kept them.

Sources

Like all the other nutrients, no single food or group of foods supplies all minerals. Consuming a wide variety of wholesome foods within each group of the Food Guide Pyramid (Figure 1-5) usually ensures adequate and balanced mineral intake. Because of the tiny amounts involved and the expense of the laboratory tests, analysis of all the minerals contained in foods has not been done. This fact is very important because the general public and our clients and patients may be ingesting more, or less, minerals than may be superficially assumed. To locate minerals in foods eaten, a masterful sleuth is needed.

For instance, eating processed foods can decrease or increase intake. Some minerals are deleted in processing. For example, iron and other minerals are removed from whole grains during the refining process. Iron is the only mineral returned to refined grains bearing the label "enriched"; none of the others are. But whole grains contain more fiber, which can chelate some of the minerals. Of all the minerals, it is sodium, mainly as sodium chloride, that is most frequently added during food processing. It is added for its taste appeal and its role as a natural preservative. Several foods are fortified with calcium, for example, some breakfast cereals, breads, and orange juice. Iodide is added to table salt labeled "iodized," but this kind of salt is not routinely used in commercial food processing. Only some minerals are required to be listed on the Nutrition Facts panel of packaged foods, thus consumers cannot know the kinds and amounts of all minerals contained in the food. (See Figure 1-9.) Sometimes clues will be found in the ingredient list . . . but not always.

Water, except distilled water, contains various amounts of numerous minerals. Fluoride is present naturally in many water sources; some municipal water supplies are fluoridated. "Hard" water contains calcium and magnesium, with these minerals ion-exchanged with sodium to produce "soft" water. (See Chapter 4.) Water is a source of iron in some geographic regions. Bottled water or spring water contain minerals from the ground, which would be difficult to declare fully on the labels. Processed foods containing water also contain the minerals of the water supply where the foods were manufactured. For example, a canned, bottled, or boxed product manufactured by the same company but produced in two different areas will have different mineral contents reflecting the water supply in each area. The same situation occurs at home, in restaurants, and in institutional facilities.

Bioavailability The importance of bioavailability is highlighted in the discussion of each category of nutrients. Individual nutrient bioavailability varies widely, but is most obvious for minerals. Some bioavailable comparisons are:

Nutrient	*Amount Absorbed*
Carbohydrates, lipids, and proteins	About 90%
Sodium	About 95%
Calcium	About 10% to 40%
Iron	About 5% to 60%
Chromium	About 0.5% to 2%

Factors *favoring* the absorption of mineral elements include:

- Body need. *Example:* growing children and pregnant women require more calcium and so absorb a higher percentage of calcium from food.
- Chemical form. *Example:* heme iron in animal foods is more available than is the nonheme iron found in plant foods.
- Stomach and small intestine pH. *Example:* calcium and iron are better absorbed in an acid environment.
- Presence of other nutrients. *Example:* lactose in the intestinal tract improves the absorption of calcium; vitamin C enhances the absorption of calcium, iron, and zinc.

Factors *reducing* the absorption of mineral elements include:

- Excessive intake. *Example:* high intakes of any mineral above body needs may pass directly into the feces and be eliminated.
- Mineral-mineral interaction. *Example:* high intakes of zinc (as from supplements) reduce absorption of copper; high intakes of iron reduce absorption of zinc and copper; high intakes of calcium reduce absorption of iron.
- Chelating agents. *Example:* substances such as oxalic and phytic acids, fiber, fat, and some medications bind minerals, thus reducing absorption.
- Rapid digestive tract transit time. *Example:* fiber, mineral oil, laxatives, diarrhea, and food poisoning speed up the passage of foodstuffs so there is less time for absorption.
- Damage to digestive tract. *Example:* stomach and intestinal surgery, short bowel syndrome, sprue, cystic fibrosis, and intestinal parasites can cause short- and long-term mineral malabsorption.

Basic Diet

The Basic Diet Nutrient Profile (Table 2–3) identifies aggregate amounts for several essential minerals. For the minerals listed, the value can be considered representative of a food category, similar to the Exchange Lists for Meal Planning. (See Appendix Table A–2).

The actual amounts of all minerals present in most foods have not been determined. Thus, tables of food composition and computerized nutrient databases have incomplete tabulations. (Read the fine print accompanying them. See

Appendix Table A – 1.) The mineral content for the Nutrition Facts label (Figure 1 – 9) on food packages usually is calculated from food composition tables, rather than by laboratory chemical analysis of each food product. (Just imagine the cost of doing analyses for all the thousands of packaged foods available—with ever-changing ingredients.) Therefore, health care providers using such tools for calculating the mineral content of diets must interpret the results with caution. This is especially so when pronouncements about *intake deficiencies* are declared. More than likely, the mineral was not measured, but actually may be contained in the food eaten. Evidence from other assessment and monitoring procedures must corroborate suspected mineral intake deficiency findings. (See Chapters 3 and 16.)

MINERAL HIGHLIGHTS

Macrominerals

CALCIUM

Characteristics
Calcium is the most abundant cation in the body. Of the approximately 1,200 mg present in the adult, 99 percent is located in the skeleton.

Functions
The majority of calcium is part of the hard structure that forms bones and teeth. The remaining 1 percent of calcium has functions so crucial to life that they take first priority over bone mineralization. Calcium helps maintain permeability of cell membranes so that materials can enter and leave. For example, calcium is essential for normal muscle and nerve activity, such as heart contraction. Calcium aids in blood clotting. As a cofactor, calcium activates many enzymes, such as pancreatic lipase. Calcium is involved in chromosome movement prior to cell division. Calcium aids in the synthesis and release of neurotransmitters.

Metabolic Processes
Some calcium is absorbed in the stomach, but most is absorbed by active transport from the upper small intestine. Many factors enhance its absorption—an acid pH, vitamins C and D, some amino acids, lactose, and a greater body need. In blood circulation, most calcium is transported either as an ion or is bound to the protein albumin. Calcium blood levels are maintained within a narrow range by several opposing forces. For example, if blood levels begin to fall, two hormones, parathyroid hormone and calcitriol (vitamin D), intervene to increase calcium concentration. Opposing them is calcitonin (secreted by the thyroid gland). If calcium levels begin to rise, calcitonin is secreted, and the calcium level decreases back to the normal range. The chief storage deposit of calcium is bone trabeculae in the ends of the long bones. Bones are very active tissues. Each day from 250 to 1,000 mg of calcium enter and leave bones. Most calcium is excreted through the urine; calcium is filtered by the kidneys as blood circulates through them. Some calcium is found in the feces, especially the calcium that was not absorbed, or that which was present in digestive fluids and cells. (See Figure 11 – 1.)

Meeting Body Needs

Requirements/Recommendations
As the many roles of calcium become better known and as more people live longer, requirements and recommendations for calcium intake are adjusted. The current RDAs are listed in the table inside the front cover. Some professional organizations and governmental agencies recommend somewhat higher levels—1,500 mg for teenagers, pregnant/lactating women, and adults over 65 years; and 1,000 mg for women between adolescence and menopause.

No matter what the discrepancy is regarding calcium intake recommendations, consensus exists on several points. First, the importance of adequate tooth and bone mineralization during childhood is clear. Throughout childhood, adolescence, and young adulthood, bones grow in length and diameter and increase in hardness (mineral concentration) to achieve peak bone mass. If the diet is adequate in calcium and other nutrients, bones can continue to mineralize (positive calcium balance) until a person is about 30 years old. Second, people do not outgrow their need for calcium. Adults of all ages need to consume calcium throughout life. Third, nondietary factors can either increase or decrease bone mineralization; for example, genetic influences, hormones, physical activity, diseases, and medications.

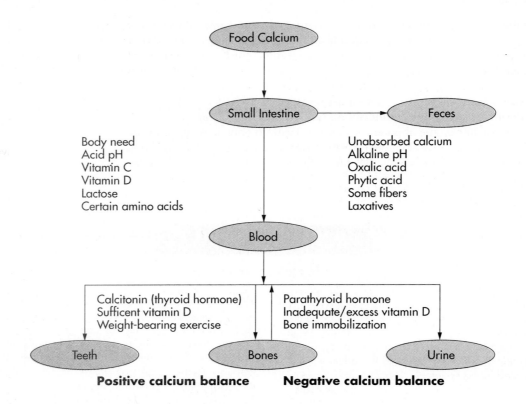

FIGURE 11–1
Some factors influencing the body's balance of calcium.

Sources and Basic Diet
As you look at the Basic Diet Nutrient Profile (Table 2–3) and Appendix Table A–1, you can see that calcium is widely distributed in foods. Without a doubt, though, is its high concentration in milk and foods made with milk. "Hard" water contains calcium, thereby contributing calcium whenever water is used as a beverage or added during food preparation.

Bioavailability of calcium must be considered. The most bioavailable food is milk. Several dark green leafy vegetables like kale, turnip greens, mustard greens, and collards are good sources of bioavailable calcium. Other greens like spinach, chard, and beet greens, as well as cocoa, contain oxalic acid. This is an organic acid that combines with some of the calcium to form an insoluble complex, reducing calcium absorption. Thus, these kinds of dark green leafy vegetables cannot be counted on for their calcium content. In similar fashion, phytic acid is an organic acid found in the outer layers of whole grains; it also can form a complex with some of the calcium. However, unless these vegetables and grains are the only ones expected to supply all the calcium, their intake poses no significant concern. For example, on food labels, the listed calcium content of both whole grain and enriched breads and cereals may be the calcium originally present in the grain, or from

milk used as an ingredient, or as intentional additives such as those added for fortification or for preservation, like calcium propionate. Legumes contain calcium. But the total amount of calcium absorbed from legumes is reduced because of the legumes' phytic acid and fiber contents. Making tofu from soybeans and treating cornmeal with lime water to make tortillas overcomes the chelation obstacle.

Canned fish with bones, like herring, salmon, and sardines, are good sources of calcium provided the bones are eaten. Sometimes calcium from bones is inadvertently included in processed fish, meat, and poultry foods. Automated deboning machinery cuts into soft, pliable bones as it removes the flesh portion.

Because of the public's interest in consuming non-food sources of calcium, several calcium supplements are marketed. Some, like calcium carbonate or citrate, are soluble in water and appear to be more bioavailable. Others, like those made from bonemeal and dolomite, present two problems. One is that the calcium is not as bioavailable. More importantly, they may contain harmful levels of toxic metals, like lead, that were stored in the bones of the marine or land animal. The FDA issued statements to alert consumers to this danger; health care providers and educators need to reinforce the message.

Body Balance

Optimum/Adequate Signs and Symptoms

Even with all that is known about calcium, body balance remains difficult to measure. This is because the body maintains tightly controlled homeostatic levels in blood. (See Metabolic Processes section.) To measure body stores, the bones must be measured. Currently, this can be done only with x rays. (See Figure 11–2.) As much as 30–40 percent of bone calcium is lost before changes can be detected by x rays. To examine bone calcium status with limited radiation exposure, improved instruments are being constructed. But so far, they are not suitable for screening or general monitoring purposes. The instruments are large, the techniques remain in research development, and the tests are costly.

Insufficiency/Deficiency

Calcium deficiency becomes evident only after years of insufficient intake. The negative long-term outcomes of inadequate intake are becoming more recognized. Greater numbers of people are reaching older adulthood; more people have shifted from ingesting milk and dairy foods to sodas and other beverages and foods that contain little or no calcium. This includes children and adults of all ages. Of special concern are younger people still mineralizing their bone storage deposits. Also, inadequate intake may place people with food allergies or lactase deficiency at long-term risk.

Many situations impair absorption of calcium, such as ingestion of alcohol, lowered acidity of stomach or small intestines, malabsorption of lipids, and surgical removal of part of the intestines. Anything that increases the rapidity of the food mass through the digestive tract can have a negative effect, like laxatives and excess fiber. Many medications form a complex with calcium, rendering either calcium or the medication, or both, nonabsorbable. (See Chapter 18.)

Immobilization and lack of weight-bearing physical activity increases calcium release from bone storage; this places nonambulatory and inactive people at risk for bone loss of calcium.

Conditions associated with inadequate calcium intake, impaired absorption, and increased loss include teeth and bone malformations; dental caries and periodontal disease; and osteoporosis. Currently, the best approach to prevent osteoporosis is for people to form

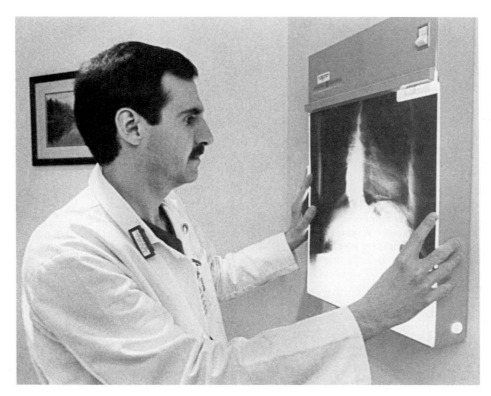

FIGURE 11–2
Calcium is stored in bones. A reduction in the amount of bone mass can have many causes. Vertebra, hip, and wrist fractures are common outcomes. (*Source:* Tom Wilcox/Merrill.)

strong bones while young and to maintain adequate nutrition and physical activity throughout life. Colon cancer and some forms of hypertension have been observed in some groups of people with inadequate intakes; however, there are many other factors associated with these diseases, too.

Excess/Toxicity

Conditions of excess and toxicity have not been observed in healthy people. Presumably, less calcium is absorbed as the dosage increases; the kidneys can filter and excrete excess calcium. However, since calcium is a cation, it may interfere with the absorption of other minerals. Researchers are investigating these mineral-mineral interactions. When toxic signs have been noted, the cause usually was related to excess vitamin D or to a disease state, such as parathyroid or kidney diseases.

PHOSPHORUS

Characteristics

About 85 percent of phosphorus is located in bones and teeth. The remaining amount is found in muscles, organs, blood, and other fluids. This mineral exists either as an ion, or in bound form, such as with phospholipids and phosphoproteins.

Functions

Phosphorus has numerous functions. Besides its role in the formation of bones and teeth, it is a major buffer for the body, a component of many enzymes, and necessary for muscle contraction and nerve activity. Importantly, it is part of the compounds adenosine triphosphate (ATP) and adenosine diphosphate (ADP), which store and release energy according to body needs. Phosphorus is part of deoxyribonucleic acid (DNA) and ribonucleic acid (RNA).

Metabolic Processes

Nearly all ingested phosphorus is absorbed. Less is absorbed in the presence of high levels of other cations like magnesium, aluminum, and calcium. Absorption is by both active and passive transport mechanisms, with vitamin D playing a role. The kidneys maintain body homeostasis by excreting more or less phosphorus, according to body needs.

Meeting Body Needs

Requirements/Recommendations

Very little research has been done. The RDAs advise the same amount of phosphorus as calcium (1:1 ratio), except during the first year of life when phosphorus is slightly lower compared to calcium. (See table inside front cover.)

Sources and Basic Diet

The Basic Diet Nutrition Profile (Table 2–3) illustrates phosphorus' wide distribution in the food groups. Like

calcium, dairy foods are the foremost source. Legumes, nuts, and animal foods are excellent sources. Thus, if the diet supplies enough calcium and protein, it will furnish ample amounts of phosphorus. Many processed foods contain phosphate as ingredients, such as carbonated beverages.

Body Balance

Neither deficiency nor toxicity appears to occur in healthy people. Since the kidneys play a pivotal role in body balance, their dysfunction can lead to serious phosphorus imbalances. (See Chapter 28.)

MAGNESIUM

Characteristics

The entire adult body contains about one ounce (20–40 g) of magnesium. About 60 percent is located in bones and teeth, about 40 percent in muscles and organs, and the remaining (about 1 percent) in the extracellular fluids.

Functions

The majority of magnesium is involved in bone and tooth formation, but its other functions are far-reaching. For example, it activates more than 300 enzymes, especially those storing and releasing energy. (It forms complexes with ATP.) Together with calcium, sodium, and potassium, magnesium regulates the transmission of nerve impulses and the contraction of muscles. Magnesium and calcium seem to work in pairs. Calcium is necessary for muscles to contract, and magnesium is necessary for them to relax. Magnesium is important for proper DNA and RNA formation and function, parathyroid hormone secretion, vitamin D conversion in the liver, and blood clotting.

Metabolic Processes

Magnesium absorption occurs along the entire length of the small intestine. Absorption is enhanced by vitamin D and decreased by high amounts of calcium, phosphorus, and fat. Body need influences absorption, which occurs by both active and passive transport. Bones are magnesium's major long-term storage site. The kidneys excrete magnesium.

Meeting Body Needs

Requirements/Recommendations

Knowledge of body requirements of magnesium is limited because of technical constraints and by the body being able to maintain close regulation of blood levels. The RDAs are based on a few studies and the amount of magnesium estimated to be available in the United States food supply. (See table inside front cover.)

Sources and Basic Diet

The Basic Diet Nutrient Profile (Table 2–3) lists magnesium's wide distribution in foods. Dairy foods are not a major source. Since magnesium is a constituent of chlorophyll in plants, dark green vegetables are a particularly rich source. Whole grains contain magnesium, but it is removed during refining and not returned. Legumes and nuts are other rich sources. "Hard" water is a reliable source as a beverage and in foods prepared with water.

Body Balance

Optimum/Adequate Signs and Symptoms

Methods to determine precise body balance have yet to be developed. Laboratory determination of blood and urine levels of magnesium can be performed. Such tests measure recent intake, rather than long-term, body status.

Insufficiency/Deficiency; Excess/Toxicity

Because of magnesium's abundant distribution in foods and the ability of the body to absorb, store, and excrete it based on body needs, observation of deficiency or toxicity is rare in healthy people. Several groups of people are at risk. For example, insufficiency or deficiency can occur with prolonged vomiting; extended suctioning of gastric contents; severe malabsorption; intestinal surgery; cirrhosis of the liver; and kwashiorkor. Malnourished patients given high glucose nutrient solutions with minimum magnesium may exhibit refeeding syndrome because of magnesium's role in ATP function. (See Chapter 17.) Many over-the-counter and prescription medications can interfere with magnesium's absorption or excretion. For example, calcium gluconate, corticosteroids, and diuretics can lower blood levels; magnesium-containing antacids can elevate blood levels. Kidney diseases severely impair magnesium balance. (See Chapter 28.)

SULFUR

Characteristics

Sulfur is present in three amino acids—methionine, cystine, and cysteine—and so is present in all proteins in the body. Connective tissue, skin, hair, and nails are especially rich in sulfur. Two B-complex vitamins, thiamin and biotin, as well as coenzyme A, contain sulfur in their molecules.

Functions

As part of proteins, sulfur provides cell structures. Sulfur is one of the intracellular electrolytes and thus maintains body acid-base homeostasis. (See Chapter 4.)

Metabolic Processes

Sulfur is absorbed either as part of amino acids or as the ion. It is excreted in the feces and urine.

Meeting Body Needs

Currently, there is no recommended level of intake. Diets adequate in protein supply liberal amounts of sulfur.

Body Balance

In healthy human beings, neither deficiency nor toxicity has been observed. A rare hereditary absence of a sulfur-containing enzyme causes a form of mental retardation.

Sulfite, a food preservative containing sulfur, had been commonly used to prevent food discoloration and spoilage. In 1986, the FDA removed it from the GRAS list. (See Chapter 2.) Although the exact mechanism remains unclear, some people with asthma reacted adversely to the preservative by exhibiting severe wheezing; in some people the asthmatic reaction was fatal. Sulfite also was commonly added to medications as a preservative. Now, food and medication labels must state its presence. Health care providers need to alert their clients.

SODIUM

Characteristics

Of the approximately 100 g of sodium in the adult body, half is present in the extracellular fluid, 40 percent in bone, and about 10 percent in the intracellular fluid.

Functions

Sodium, potassium, and chloride are the most abundant electrolytes in the body. These three are the main ions involved in maintaining the body's osmotic pressure and water balance. (See Chapter 4.) Other major functions of sodium include cell permeability maintenance, muscle contraction, nerve impulse transmission, and acid-base balance. Sodium facilitates nutrient absorption, like glucose and amino acids.

Metabolic Processes

Nearly all ingested sodium is absorbed. The most common absorption mechanism is active transport with glucose. Sodium homeostasis is maintained inside the body mainly by adrenal hormone control on the amount of sodium excreted by the kidneys. When sodium intake is high, aldosterone levels decrease, causing an increase in urinary excretion of sodium. The reverse reactions occur when sodium intake is below the body's needs—hormone levels increase, resulting in less sodium being excreted. Varying amounts of sodium are excreted through the skin (sweat) and feces.

Meeting Body Needs

Requirements/Recommendations

Sodium requirement depends mostly on the amount lost and that needed to maintain extracellular fluid volume. For example, fluid volume increases during times of growth, like infancy and pregnancy. (See inside front cover.)

Sources and Basic Diet

The Basic Diet Nutrient Profile (Table 2–3) identifies the wide distribution of sodium in foods. However, the major source of sodium is salt—usually sodium chloride. (See Appendix Table B–3.) Not only is it used in home food preparation and at the table, it is present in considerable amounts in many commercially processed foods such as ham, bacon, luncheon meats, smoked poultry and fish, cheese; pickles, olives, catsup, meat sauces, and relishes; and snack foods like potato chips, pretzels, crackers, and some sodas. Sodium is present in other compounds often used in food preparation, such as monosodium glutamate (MSG), baking powder, and baking soda. Appendix Table A–1 illustrates the contrast in sodium content between processed and unprocessed foods. With foods chosen according to the Food Guide Pyramid (Figure 1–5), about three-fourths of the sodium present in fresh foods is from the milk and meat groups. With few exceptions, foods lowest in sodium are fresh fruits and vegetables, rice, pasta, and cooked cereals. "Softened" water, in which calcium and magnesium are removed and sodium is substituted, often is an overlooked contributor. So is the sodium content of certain prescription and over-the-counter medications. (See Chapter 18.)

Body Balance

Optimum/Adequate Signs and Symptoms
The healthy body is able to maintain sodium balance over a wide range of dietary intakes and environmental climates. The sodium content of a 24-hour urine collection is a good indicator of the previous day's dietary intake, compared to the amount needed by the body. Blood levels are easier to collect; serum sodium concentrations are measured routinely in health care settings.

Insufficiency/Deficiency
Dietary deficiency of sodium seldom occurs. Sodium deficiency can result from extremely heavy sweating, especially in people before they adapt to extremely hot and humid climates. Infants are especially vulnerable. Athletes such as long-distance runners and bikers, or people doing heavy work in hot environments, have increased sodium needs due to losses in perspiration. Such losses usually can be replaced by adding extra salt to food before and after the activity. (Water intake should be increased, too. See Chapter 4.) Salt tablets are not recommended. Other people at risk for sodium depletion are those experiencing severe vomiting or diarrhea, or who have conditions such as adrenal gland disease or cystic fibrosis.

Excess/Toxicity
Usually, healthy people can handle high intakes of sodium. However, when the body is unable to excrete excess sodium, additional water is held in the extracellular fluid. Edema or ascites is characteristic of several pathological conditions, including congestive heart failure, renal failure, excessive secretion of cortical hormones by tumors of the adrenal gland, and liver disease. Steroid medications used in the therapy of several conditions, such as arthritis, inflammatory diseases, and organ transplants, can lead to sodium and water retention. About one in four Americans has some elevation of blood pressure. Nearly half of these people are **salt sensitive,** that is, sodium intakes considered typical for most people lead to elevation of blood pressure for them. In all of these conditions, modification of sodium intake is included as part of therapy.

POTASSIUM

Characteristics
Potassium is a major electrolyte in the body. Nearly all is located inside cells.

Functions
One of potassium's essential roles is in the normal contractions of all muscles, including the heart. Other potassium functions include synthesis of protein and glycogen, maintenance of fluid balance and pH, and transmission of nerve impulses.

Metabolic Processes
Nearly all ingested potassium is rapidly absorbed through the small intestine and colon. In healthy people, balance is finely regulated. Excess potassium is readily excreted by the kidneys. Regulation of potassium excretion is opposite that of sodium—increased levels of aldosterone increase potassium's excretion. However, the kidneys have less control over potassium excretion than over sodium excretion. Even when a potassium deficiency exists, the kidneys continue to excrete some potassium. Some potassium is excreted in feces.

Meeting Body Needs

Requirements/Recommendations
Only a few studies have evaluated potassium status. The current Food and Nutrition Board's suggested intake levels are shown in the table inside the front cover of the text.

Sources and Basic Diet
The Basic Diet Nutrient Profile (Table 2–3) shows the extensive distribution of potassium throughout the various food categories. Appendix Table A–1 lists the potassium content of individual foods. Specific foods especially high in potassium are bananas, oranges, grapefruit, dried fruits, cantaloupe; potatoes, broccoli, carrots and celery; legumes; meat, poultry, fish; milk. Whole grains have more potassium than refined ones. "Salt substitutes" usually replace sodium with potassium.

Body Balance

Optimum/Adequate Signs and Symptoms
Normally, dietary deficiency of potassium does not occur. Determination of serum potassium concentrations is a routine laboratory test.

Insufficiency/Deficiency

Low intakes are unusual. But restricted food consumption places people at risk. People following severe weight loss programs or who have anorexia nervosa or alcoholism need to be evaluated. Potassium deficiency can be a side effect of many disorders. Vomiting, diarrhea, and diabetic ketoacidosis cause potassium loss. Medications such as thiazide diuretics and some antibiotics and steroids result in potassium loss. (See Chapter 18.) Nasogastric suctioning or intravenous (IV) hydration without potassium are common therapies that require monitoring of potassium levels. Some of the signs and symptoms of potassium deficit are common to many different situations. Examples include nausea, vomiting, apprehension, listlessness, muscle cramps and weakness—from leg cramps to constipation caused by decreased peristalsis. The most serious abnormalities are cardiac arrest and respiratory failure, which can result in death. Prevention of potassium depletion is accomplished by emphasis on potassium-rich foods. Treatment is with potassium-containing medications or IV solutions.

Excess/Toxicity

Potassium toxicity can be a complication in severe dehydration, adrenal insufficiency, and renal failure. Observations include numbness of the face, tongue, and extremities; muscle weakness; and cardiac arrhythmia. Cardiac failure may occur.

CHLORIDE

Characteristics

Chlorine exists nearly always as the ion chloride. It is the major anion of extracellular fluid, thus providing balance with the cation sodium.

Functions

Chloride's chief function is electrolyte balance regulation. It is the major anion of gastric juice (hydrochloric acid—HCl). Lung function depends on the chloride ion being exchanged for another anion, bicarbonate (HCO_3^-) during oxygen and carbon dioxide exchange. (See Chapter 4.)

Metabolic Processes

Chloride absorption occurs by passive transport, but follows along with the active transport absorption of sodium. Whether chloride is present in foods or gastric juice, it is absorbed into the blood circulation in the intestines. Like other ions, excess chloride is excreted by the kidneys, paralleling sodium. Chloride also is excreted through the skin along with sodium in sweat, and in feces.

Meeting Body Needs

Requirements/Recommendations

Determining exact chloride requirements is difficult because its balance is so closely interrelated with the other electrolytes and experimental techniques remain inadequate. The current Food and Nutrition Board's suggested intake levels are shown in the table inside the front cover of the text.

Sources and Basic Diet

Neither the Basic Diet Nutrient Profile (Table 2–3) nor Appendix Table A–1 lists the chloride content of food. The major source of chloride is sodium chloride. Thus, any food with added salt contains chloride. About 60 percent of the NaCl molecule is chloride. It also is present in high protein foods, like eggs, fish, and meat.

Body Balance

The concentration of chloride in serum is the standard test used to determine body balance.

Insufficiency/Deficiency; Excess/Toxicity

Chloride deficiency rarely occurs, except in situations of digestive tract upsets, such as vomiting, diarrhea, and gastric suctioning. Profuse sweating places people at risk, especially infants, children, and athletes.

Toxicity secondary to dehydration is the major example. Water is lost from the body, so the concentration of chloride increases in the remaining body fluids.

MINERAL HIGHLIGHTS

Microminerals

IRON

Characteristics

The total amount of iron in the adult body is only about 2 to 4 g, thus classifying iron as a micromineral. Most of this iron is present in blood as hemoglobin. This protein consists of heme (iron-containing compound) attached to globin (protein).

Functions

As hemoglobin, iron has the crucial function of carrying oxygen from the lungs to all cells and returning with carbon dioxide. (See Chapter 4.) A small amount of iron, as the heme molecule, is present in muscles as myoglobin, where it holds oxygen for the needs of muscles. Iron also is a cofactor for many enzymes, such as those required for the release of energy from the macronutrients, conversion of beta-carotene to preformed vitamin A, and formation of collagen. Iron is involved in the immune process.

Metabolic Processes

Iron absorption, like that of calcium and some other minerals, varies considerably because it is influenced by many factors. (See Table 11–2.) Once iron enters the blood circulation, it is transported bound to a protein, transferrin. Iron is stored chiefly as the compound ferritin, located in blood and cells. Hemosiderin is another storage form of iron, located in the liver. Since iron is part of hemoglobin, it is found in bone marrow and the spleen. The body uses its iron economically and recycles it. For example, the red blood cells, after their normal life span of about 120 days, release the iron from hemoglobin; then the body reuses the iron for the synthesis of new hemoglobin. Storage amounts are different in healthy adult men and women. Men have about 1,000 mg; women have about 300 mg. Excretion of iron is by daily perspiration, shed gastrointestinal cells, urine, skin, hair, and nails. Blood loss from any cause is a loss of iron—menstruation, bleeding ulcers, hemorrhage.

Meeting Body Needs

Requirements/Recommendations

Dietary iron is required to meet the changing needs of the body over the life span. Requirements include: maintaining adequate stores; replacing daily losses, including during menstruation in women; and providing sufficient amounts for muscle and blood expansion, especially during infancy, adolescence, and pregnancy. Recommended levels of dietary iron also consider the bioavailability of iron from the food supply. (See section below.) The RDAs for iron are listed in the table inside the front cover.

Sources and Basic Diet

The Basic Diet Nutrient Profile (Table 2–3) and Appendix Table A–1 list the chemical analysis of iron in foods. Total iron content varies, but is highest in animal muscle foods, legumes, whole grain and enriched grains, and dark-green vegetables. Iron is the only mineral added to grains labeled "enriched." Some cereals are "fortified" with additional levels of iron.

Iron bioavailability from food must be considered. First, the body has a great capacity to absorb greater amounts of iron from food during times of need. For example, rapidly growing children, pregnant women, and people with iron-deficiency anemia can absorb a greater proportion of iron from food compared to healthy adult men.

Second, the chemical form of iron (heme vs. nonheme) in food is important. (See Table 11–3.) Heme iron is part of the hemoglobin and myoglobin molecules and thus occurs only in animal foods. Nonheme iron is present in both animal and plant foods. About 40 percent of iron in meat, poultry, and fish is heme iron; the remaining 60 percent is nonheme iron. All iron in plant foods is nonheme iron. Importantly, about 25 percent of heme iron is available for absorption, but only 5 percent of nonheme iron is available.

Heme iron sources are lean muscle meats, poultry, and fish with liver, oysters, and clams being especially high. Nonheme iron is supplied by meat, poultry, fish; legumes; nuts; whole, enriched, or fortified grains; dried fruits; dark-green leafy vegetables; and dark molasses. Dairy foods are low in iron.

Third, methods of food preparation and food combination can increase or decrease absorption. Two methods of enhancing iron availability are cooking high-acid foods in iron skillets and providing a source of vitamin C at mealtimes. (See Table 11–2.)

TABLE 11–2
Factors That Influence Iron Absorption

Increase Absorption	Decrease Absorption
Body needs unmet	Body needs met
Growth	Ferric iron (Fe^{3+})
Pregnancy	Nonheme iron without meat or vitamin C
Acute blood loss	Lack of stomach acid
Anemia	Tannins in tea (with meals)
Heme iron	Coffee with meals (less effect than tea)
Ferrous iron (Fe^{2+})	Phytic acid; oxalic acid
Meat in same meal	Calcium, phosphorus, magnesium, zinc
Vitamin C in same meal	Antacids with meals
Gastric acidity	Malabsorption disorders

TABLE 11–3

Iron Content of Average Servings of Some Common Foods

	Serving Size	Iron Content (mg)		Serving Size	Iron Content (mg)
Heme Iron = 40%; *Nonheme Iron = 60%*			*Nonheme Iron = 100%*		
			Broccoli, spear	1 med.	2.1
Liver, beef	3 oz	5.3	Turnip greens	½ cup	1.6
Oysters	3 oz	5.0	Spinach, cooked	½ cup	1.5
Beef, ground, lean	3 oz	2.1	Peas, cooked	½ cup	1.3
Turkey, dark	3 oz	2.0	Tomato, raw	1 med.	0.6
Lamb, roast	3 oz	1.7	Potatoes, white	1 med.	0.5
Tuna, light	3 oz	1.6	Beets, cooked	½ cup	0.5
Pork, fresh, roast	3 oz	0.9	Beans, green	½ cup	0.5
Chicken, breast	3 oz	0.9	Carrots, cooked	½ cup	0.4
Salmon, pink	3 oz	0.5	Cabbage, raw	1 cup	0.4
			Lettuce, iceberg	1 cup	0.3
Nonheme Iron = 100%					
Beans, lima, dry, cooked	½ cup	3.0	Prunes	10	2.4
Beans, canned, with pork	½ cup	2.3	Dates	10	1.0
Lentils, cooked	½ cup	2.1	Raisins	1 oz	0.6
Peanut butter	2 tbsp	0.6	Banana	1 med.	0.4
Walnut halves	½ oz	0.5	Cantaloupe	¼	0.3
Almonds, shelled	½ oz	0.5	Apple	½ oz	0.2
			Orange	1 med.	0.1
Cereals					
branflakes (40%) with added iron	1 oz	8.1	Molasses		
			black strap	1 tbsp	5.0
cornflakes, plain	1 oz	1.8	light	1 tbsp	0.9
oatmeal, cooked	1 cup	1.6			
wheat, shredded	1 oz	1.2	Sugar		
			brown	1 tbsp	0.3
Breads			white	1 tbsp	0.0
whole wheat	1 slice	1.0			
white, enriched	1 slice	0.7	Cheese, cheddar	1 oz	0.2
rye, enriched	1 slice	0.7	Milk	1 cup	0.1

Body Balance

Optimum/Adequate Signs and Symptoms

The major pivotal point for body balance appears to be intestinal iron absorption determined by body iron stores. Thus, if the body needs less iron, ingested iron may be immediately excreted directly in the feces; the opposite occurs when the body needs more—more is absorbed. Assessment of iron adequacy depends on an evaluation of dietary intake, clinical signs and symptoms, and laboratory blood tests, including tests for hemoglobin, hematocrit, transferrin, and ferritin. (See Chapter 3.)

Insufficiency/Deficiency

As iron stores are depleted, the serum transferrin level rises so that more iron can be absorbed from the intestinal tract. With continuing deficiency the serum ferritin and serum iron levels fall, as does the transferrin saturation. Once available iron stores are used up, less and less hemoglobin is produced even though the body can reuse its own iron over and over again. Iron-deficiency anemia is described as **microcytic** (small cell) and **hypochromic** (low color). With the fall in hemoglobin levels, the amount of oxygen carried to the tissues is reduced. Pallor, headache, dizziness, and fatigue even with little physical effort are common symptoms as the anemia becomes more severe. Progressively, there is shortness of breath and increased heart rate as the lungs and heart attempt to compensate for reduced oxygen levels.

Anemias are caused by many factors other than iron deficiency. Among the nutrient deficiencies that can lead to anemias are folate, vitamin B$_6$, vitamin B$_{12}$, vitamin C, vitamin E, and copper. Thus, the kind of anemia and its cause must be determined before appropriate therapy can be provided.

An important public health example is lead poisoning. Lead interferes with the formation of hemoglobin and can result in the diagnosis of "anemia." The amount of lead absorbed is greater in people whose diets are deficient in iron, calcium, and zinc. (Recall that there is competition or interference among many minerals for absorption. In lead poisoning, lead absorption is greater. A nutritionally adequate diet is essential for reducing the amount of lead absorbed.) At special risk are children living in poverty where exposure to lead paint is highest and who are least likely to be well nourished. Children with lead poisoning exhibit poor appetites, vomiting, weight loss, hyperactivity, and learning disabilities. In severe cases, convulsions and death occur. (See Chapter 13.) Lead poisoning can occur after ingesting acidic food or beverages (like orange juice or wine) stored in pottery with lead glazes or in lead crystal carafes.

Iron insufficiency may be corrected with attention to the factors influencing iron absorption. By the time iron deficiency is obvious, iron supplementation usually is necessary. In severe iron-deficiency anemia, blood loss from the gastrointestinal tract should be suspected. Instances include bleeding ulcers, ulcerative colitis, or the blood losses that accompany parasite infestation such as hookworm. Also, take into account the frequency with which people donate blood or women experience extraordinary menstrual losses.

Excess/Toxicity

Iron excess usually does not occur under normal circumstances since the body finely regulates absorption. However, iron toxicity can occur. Too much absorbed iron can lead to large deposits of iron in the liver, lungs, pancreas, and other tissues. A rare example is the condition of **hemochromatosis,** a genetic defect characterized by excessive absorption of iron leading to organ damage, skin pigmentation, and cirrhosis of the liver. More commonly seen is **hemosiderosis,** in which excessive iron storage does not lead to tissue damage. It can occur when iron therapy is continued long after it is needed. Also, in the United States, every year about 2,000 children suffer iron poisoning from accidently swallowing iron supplements formulated for adults. Iron overload can be a complication of any abnormal breakdown of red blood cells, such as **hemolytic anemia.**

ZINC

Characteristics

Zinc is widely but unevenly distributed in plants and animals, including humans. In people, the highest concentrations are found in bones, liver, hair, skin, and the prostate gland.

Functions

Zinc appears to have numerous functions, some of which have yet to be identified. It appears to be involved in numerous enzymes (about 70). Examples are alcohol dehydrogenase (required for the metabolism of alcohol by the liver), carbonic anhydrase (transfer of carbon dioxide from the tissues to the lungs), and carboxypeptidase (splits off amino acids during protein digestion). Zinc is essential for the synthesis of DNA and RNA and thus of proteins. Consequently, zinc is involved in normal growth, development, wound healing, and sexual maturation. Zinc is necessary for the release of vitamin A from the liver, for normal taste sensitivity, for cell membrane structure, and for normal immune responses.

Metabolic Processes

Methods to study the intricacies of zinc metabolic processes continued to be researched. Zinc is absorbed actively, passively, and by carrier-facilitated transport. Several factors seem to be involved in zinc's absorption. Like many other minerals, zinc appears to be absorbed according to body need. A zinc-binding compound found in pancreatic secretions is necessary. Small amounts of zinc are better absorbed than zinc supplied in large doses. Zinc is transported in the blood, usually attached to the protein albumin. Although zinc is found throughout the body, no special storage spaces have been identified. Zinc is excreted through the feces from unabsorbed zinc and from that present in digestive tract secretions; from losses of skin cells, sweat, and hair; and a small amount in urine.

Meeting Body Needs

Requirements/Recommendations

Since the body so efficiently regulates zinc and there is no precise test to determine zinc status, exact requirements currently are unknown. Knowledge is being learned from observations of people on tightly controlled nutrient intakes, such as those on total parenteral nutrition (TPN) feedings. (See Chapter 17.) The RDAs consider the estimated zinc content of usual dietary intakes in the United States, along with findings from human research studies. (See table inside the front cover.)

Sources and Basic Diet

The Basic Diet Nutrient Profile (Table 2–3) lists zinc. Zinc is widely distributed in animal foods and plant foods that are good sources of protein. Especially excellent are oysters, organ meats, muscle meats, and dark meat of poultry; eggs; legumes; peanuts and peanut butter. Dairy foods and whole-grain breads, cereals, rice, and pasta; dark-green and deep-yellow vegetables make important contributions. Fruits and other vegetables are low in zinc. Bioavailability must be considered. For example, under research conditions, zinc in animal foods like meat is better absorbed than from plant foods containing fiber and phytate.

Body Balance

Optimum/Adequate Signs and Symptoms
The balance of zinc is carefully controlled by the healthy body. Many laboratory tests have been tried in an effort to find one that is reliable. The one most commonly used is blood tests. Hair analysis, once popular because hair is easy to obtain, is not considered indicative of zinc levels in the body.

Insufficiency/Deficiency
The first descriptions of zinc deficiency in human beings were documented in the 1960s. First, in Iran and Egypt, young boys who had been eating a diet in which more than half the calories were provided by unleavened whole-grain bread failed to grow and did not mature sexually. Although the bread was a good source of zinc, almost all of the zinc was chelated by the phytate. Second, in the United States, observation of zinc deficiency was among the pioneer recipients of parenteral nutrition because zinc had not been added to the solutions. (See Chapter 17.) Third, a hereditary disorder causes the inability to absorb zinc, resulting in severe zinc deficiency known as **acrodermatitis enteropathica.** Other people at risk include anyone who is malnourished, especially rapidly growing children.

Zinc deficiency has many ramifications, including impaired immunity, slow wound healing, hair loss, and skin lesions. Patients complain of changes in taste sensitivity, such as **hypogeusia** (diminished taste sensitivity) and **hyposmia** (diminished odor sensitivity); altered tastes of saltiness, sweetness, or bitterness of foods; and lessened appetites. Because of zinc's interrelationship with vitamin A metabolism, night blindness is possible. (See Chapter 10.)

Excess/Toxicity
Ingestion of large doses (2 gm) of zinc has resulted in stomach irritation and vomiting. The subtle effects of long-term small amounts, seen most commonly in people who take dietary supplements, include copper deficiency, anemia, impaired immunity, and a decrease in high-density lipoproteins (HDLs). (See Chapters 19 and 24.) Thus, the chronic intake of zinc supplements greater than 15 mg/day by adults is not recommended, unless under close medical supervision. Zinc poisoning is possible after ingesting acid foods or beverages stored in galvanized containers.

IODINE

Characteristics
Iodine exists mainly as an ion, iodide. Although an element, it is not a metal. In the body, most iodine is located in the thyroid gland, with lesser amounts in salivary and stomach glands. However, all cells contain tiny amounts. In the environment, iodine is unevenly distributed, with highest concentrations in ocean water and nearby coastal areas.

Functions
Iodide's major function is as a constituent of two hormones, triiodothyronine (T_3) and thyroxine (T_4). These hormones influence most organ systems of the body. For example, the hormones regulate energy metabolism, including basal metabolism, and the metabolic processing of the macronutrients. The hormones are necessary for normal physical and mental development.

Metabolic Processes
Iodine is ingested in foods mostly as the inorganic ion (iodide) and as organic compounds (attached to amino acids). Absorption occurs rapidly and throughout the digestive tract. The ionic form is better absorbed than the organic form. Iodide is taken up by the thyroid gland from blood circulation. About three-quarters of the iodide content of the body is located in this gland. The kidneys promptly excrete iodide. A small amount is excreted in feces and sweat.

Meeting Body Needs

Requirements/Recommendations
Requirements relate to the formation of the proper amount of the thyroid hormones. The RDAs take into account the wide distribution of iodide in regular and processed foods. (See table inside front cover.)

Sources
Three major categories of foods contribute to iodide intake. First, the greatest natural source of iodide is seafood. Examples include saltwater fish, shellfish, and seaweeds. Plants grown along the seacoast are good sources; foods grown inland and freshwater fish contribute little iodide. Thus, iodide content of plants, unlike some other minerals, depends on the iodide content of the soil in which the plant grows. Second, the most reliable way to ensure an adequate iodide intake is to use iodized table salt. Its use is a major public health intervention in the eradication of iodine deficiency diseases. However, sodium chloride used in commercially processed foods usually is not iodized. The third source is iodide that becomes an unintentional additive to foods. (See Chapter 2.) For example, dough conditioners used in commercial bread-making may contain iodide, which then is incorporated into the finished baked product. Iodide-containing disinfectants have been used, especially in milk processing. Iodide can be present in animal feeds. These multiple sources of iodide caused the food industry to find alternatives whenever possible. The outcome has been a steady decline in extra sources of iodide in the total food supply.

Some foods contain substances (**goitrogens**) that interfere with the body's use of thyroid hormones. Certain plants of the cabbage family, including cabbage, brussels sprouts, cauliflower, radishes, rutabagas, and turnips, contain the natural chemical goitrin. Likewise, goitrogens are found in peanuts, cassava, and some oilseeds such as rape seed. Fortunately, heating inactivates the goitrogens. Also, only when the raw vegetables form a substantial part of the diet is the goitrogen of concern. The typical consumption of these foods in American diets presents no problem.

Body Balance

Optimum/Adequate Signs and Symptoms
Proper functioning of the thyroid gland is the hallmark for optimal iodide balance. Laboratory tests to measure adequacy include measurement of iodide in urine, blood levels of T_3 and T_4, and radioactive iodide uptake by the thyroid gland.

Insufficiency/Deficiency
An insufficiency of iodide leads to decreased hormone production, followed by lower energy metabolism rates. In an effort to produce more hormones, the thyroid gland enlarges. The condition is known as **simple goiter,** or, if widespread in a particular geographical area, **endemic goiter.** In mild deficiency, the only outward sign is slight enlargement of the thyroid gland, visible at the neckline. However, in a woman who is pregnant, irreversible birth defects can occur in her baby. The mother's body is unable to supply the fetus with enough iodide. Thus, because this deficiency occurs during such a critical period of development, the baby is more severely affected than the mother. The condition, **cretinism,** is marked by low basal metabolism, dry skin, thick lips, enlarged tongue, arrested skeletal and muscle development, and severe mental retardation.

Excess/Toxicity
Because the kidneys rapidly excrete iodide, toxicity is extremely rare.

SELENIUM

Characteristics
Selenium, like iodine, is an element but is not a metal. In the body, selenium is present in all cells, especially in the kidney, liver, and pancreas. The selenium content of soil varies widely throughout the world, a fact that enabled researchers to study its essential role in human health. (See section, Meeting Body Needs.)

Functions
Many functions of selenium remain unknown. The major function appears to be its presence in the enzyme glu-

tathione peroxidase. This is one of the enzymes involved in the body's antioxidant defense system, thus classifying selenium as an "antioxidant." Selenium is part of other enzyme systems. Selenium is important for the proper functioning of the pancreas and the immune system.

Metabolic Processes
Selenium is ingested in food in both the inorganic and organic forms. In contrast to iodide, the organic form of selenium is more bioavailable and is absorbed mainly from the duodenum. No special storage spaces are evident. The kidneys can excrete varying amounts of selenium. Up to about two-thirds of excreted selenium is through the urine, with the remainder mostly in the feces. However, with excessive intakes, selenium also is excreted through the sweat and lungs.

Meeting Body Needs

Requirements/Recommendations
That selenium was required by humans was not proven until 1979, when scientists in China discovered selenium was valuable in helping to prevent the condition known as **Keshan disease.** This was prevalent in children and women of childbearing age. In the United States, clinicians also observed the necessity of adding selenium to the parenteral solutions fed to their patients. (See Chapter 17.)

Although other scientific studies have been conducted, there is limited information about selenium across the life span. The RDAs are estimated from worldwide human and animal studies. (See inside front cover.)

Sources
Limited food analysis data exist, thus selenium is not listed in the Basic Diet Nutrient Profile (Table 2–3) nor in Appendix Table A–1. In foods, selenium is found combined with sulfur-containing amino acids (methionine and cysteine). Thus, animal protein foods such as seafood, kidney, liver, and meat are the highest sources. Like iodine, the selenium content of the soil is directly related to the amount found in the plants grown in it. Thus, the region where individual plant foods are grown dictate their selenium content. Of the plant foods, grains are higher in selenium than are vegetables and fruits.

Body Balance

Optimum/Adequate Signs and Symptoms
Body homeostasis seems to be maintained even with wide differences in intake. The most common laboratory tests measure glutathione peroxidase levels of platelets, blood selenium levels, or urinary selenium.

Insufficiency/Deficiency
Signs of selenium deficiency overlap other nutrient deficiency diseases. Examples are muscle weakness, joint discomfort, and disorders of the heart muscle or pan-

creas. People potentially at risk are those being fed tightly controlled nutrient solutions and vegans, especially growing youngsters, who eat plant foods grown in soil low in selenium.

Excess/Toxicity

Selenium toxicity in animals was known about before it was recognized in humans. Cattle that grazed in areas with high amounts of selenium in the soil classically developed stiffness, blindness, hoof deformity, and hair loss. Death sometimes occurred. In humans, signs of toxicity are evident in the areas of the world where constant exposure to slightly elevated amounts of selenium exists. Abnormal observations include hair loss; fingernail discoloration, tenderness, and loss; and a "garlic" breath odor. When excess selenium was ingested during a short period of time, usually from dietary supplements, abnormalities included vomiting, hair loss, increasing fatigue, and a "sour-milk" breath odor.

COPPER

Characteristics

At the beginning of the twentieth century, copper was noted to be a dietary essential element because of its role in hemoglobin formation. Copper is found throughout the body, with the largest concentrations in the liver, brain, heart, and kidneys. In the environment, copper is widely but unevenly distributed throughout the world.

Functions

Copper is essential for the correct functioning of many enzyme systems. Some examples are: ceruloplasmin (involved in iron transport in blood); cytochrome c oxidase (essential for release of energy from the macronutrients); superoxide dismutase (plays a role in the body's antioxidant system).

Metabolic Processes

Absorption of copper appears to occur throughout the digestive tract, with most from the stomach or duodenum, where an acid pH enhances its absorption. Several other factors enhance or inhibit copper's absorption. For example, large amounts of zinc, calcium, magnesium, or vitamin C can decrease its absorption. In blood circulation, about 95 percent of the copper is transported as ceruloplasmin. The liver is the main storage organ. Feces are the major excretion route. Absorbed but unneeded copper is excreted into bile, which then becomes part of the feces that also contain unabsorbed copper.

Meeting Body Needs

Requirements/Recommendations

Research to determine requirements for copper is ongoing. Because only imprecise data are available, the Food and Nutrition Board placed the recommended copper intake in the category designated as Estimated Safe and Adequate Daily Dietary Intake (ESADDI). (See table inside front cover.)

Sources and Basic Diet

The Basic Diet Nutrient Profile (Table 2–3) lists the copper content of groups of food. Copper content depends on the amount in the soil. Thus, the amount of copper in any specific food ingested will vary. Among foods from similar geography, the highest sources are organ meats, shellfish, whole grains, legumes, nuts, mushrooms, and chocolate. Milk is a poor source. Water may be a source, especially household or institutional sources that travel through copper pipes.

Body Balance

Optimum/Adequate Signs and Symptoms

Measurement of blood components are commonly used laboratory tests. Some examples are superoxide dismutase (SOD), copper levels, and amounts of ceruloplasmin. Each test must be interpreted with caution. For example, ceruloplasmin is influenced by inflammation or hormonal changes.

Insufficiency/Deficiency

Copper insufficiency is unusual in healthy people. People at risk include those with malabsorption or those taking high doses of zinc, vitamin C, or antacids. Infants at risk are those born prematurely (did not build up adequate stores) or with persistent diarrhea. Children at risk are those fed diets composed only of milk. A rare hereditary disorder, **Menkes' disease,** is characterized by low levels of the copper-containing proteins, severe mental retardation, and sparse, steely (kinky) hair.

Excess/Toxicity

Copper excess is unusual in healthy people. Instances have occurred from ingesting acidic food or beverages stored in copper or brass containers; or ingesting abnormally acidic water that travels through copper tubing and valves. Symptoms include vomiting and diarrhea. A rare hereditary disorder, **Wilson's disease,** is characterized by slowly increasing amounts of copper stored throughout the body, such as in the liver, brain, and eyes. The skin, hair, and eyes take on a golden or greenish-golden hue.

MANGANESE

Characteristics and Functions

In the body, manganese is found in all cells, mostly in the mitochondria. Of the organs, highest concentrations are in the pancreas, bone, and liver. Manganese is part of several enzyme systems. Functions of the enzymes include energy metabolism, fatty acid and cholesterol synthesis, urea formation, and antioxidant protection. Unique to manganese is that magnesium can substitute for the manganese in many of the enzymes.

Metabolic Processes
Manganese is absorbed poorly from the small intestine. Its absorption and transport in the blood is shared with iron. For example, increased levels of iron inhibit absorption of manganese; manganese is transported by transferrin. No particular storage location is known. Manganese is excreted mainly through the bile, then into the feces. Little is present in urine.

Meeting Body Needs

Requirements/Recommendations
Research to determine requirements for manganese is ongoing. Because imprecise data are available, the Food and Nutrition Board placed the recommended manganese intake in the category designated as Estimated Safe and Adequate Daily Dietary Intake (ESADDI). (See table inside front cover.)

Sources
Manganese is found mainly in plant foods rather than animal foods. Soil concentration is indicative of the amount in foods. Major food sources are whole grains, legumes, vegetables, and fruits. Although tea has especially high levels, the manganese appears to be unabsorbed.

Body Balance
Reliable laboratory tests for manganese have yet to be established. However, serum levels sometimes are monitored. In humans, neither deficiency nor toxicity has been seen, except in extreme research conditions. Ability to maintain body balance may be due to adequate dietary sources, strong body homeostatic controls, and interchangeability of manganese and magnesium in enzyme systems.

FLUORIDE

Characteristics and Functions
Fluorine exists in the body as compounds called fluorides. Small amounts cause striking reductions in dental caries (decay). By combining with calcium compounds, fluoride enables the tooth enamel to be more resistant to acids produced by mouth bacteria when they act on certain carbohydrates. (See Chapter 5.) Fluorides also may be useful in maintaining bone structure, thereby reducing the incidence of osteoporosis.

Metabolic Processes
Ingested fluoride is better absorbed from drinking water than from food. Absorption occurs rapidly, directly from the stomach. From the blood circulation, fluoride is taken up by bones and teeth. Nearly all fluoride is stored in bone. The mineral is excreted in urine.

Meeting Body Needs

Requirements/Recommendations
Requirements for fluoride are based on its uptake by bones and teeth. The Food and Nutrition Board placed the recommended fluoride intake in the category designated as Estimated Safe and Adequate Daily Dietary Intake (ESADDI). (See table inside front cover.)

Sources
The most reliable source is fluoridated water. For municipal water supplies, the recommendation is a fluoride concentration of 1 part fluoride to 1 million parts water (about 1 mg per quart). Besides drinking water, the water used in commercial and household food preparation is a sometimes overlooked source. Examples include fluoridated water added to make pastas, rice, and cooked cereals; soups; frozen juices; powdered beverages; ice cubes; and sodas.

Alternatives to the use of fluoridated community water are bottled fluoridated water; topical application of stannous fluoride by a dentist or dental hygienist together with the use of fluoride toothpaste; or the use of fluoride drops or tablets. (Fluoride-containing infant and children's vitamins are available by prescription from a physician or dentist.)

Food sources are highly variable, depending on the region's fluoride content of water and soil. Ocean fish eaten with bones, such as salmon, herring, and sardines, and tea are excellent sources. Bone meal and dolomite may be high in fluoride, but also may contain toxic levels of undesirable minerals like lead and arsenic. Thus, these supplements are to be avoided.

Body Balance
Caries-free teeth may be one sign of optimal body balance. Research studies are exploring fluoride's potential role in preventing or treating osteoporosis.

Excess accumulation of fluoride is known as **fluorosis.** Mottled teeth may occur in children when the concentration of overall ingested fluoride is 2 to 8 mg per kilogram of body weight. The appearance of irregularly distributed patches, first chalky-white, then yellow, grey, brown, or black, are evident on the teeth's enamel. Long-term daily ingestion of 20–80 mg fluoride has resulted in skeletal fluorosis.

CHROMIUM

Characteristics and Functions
Chromium is a metal that exists in both the inorganic and organic forms. The adult body contains about 5 mg of chromium, with the highest concentrations in muscle,

fat, and skin. More information is known about chromium's role in animals than in human beings. Chromium is a constituent of glucose tolerance factor (GTF). This compound also contains amino acids and nicotinic acid. GTF is involved in the efficient uptake of insulin by cell membranes. Thus, chromium appears to be important for the optimal metabolism of glucose. Other possible functions of chromium are in the synthesis of fatty acids, cholesterol, and RNA.

Metabolic Processes

The metabolic processing of chromium is under investigation. Chromium is absorbed from the small intestine, especially the jejunum. The organic form appears better absorbed. About 0.5 percent of chromium is absorbed from the diet. In blood circulation, chromium is transported by transferrin and possibly albumin, globulins, and lipoproteins. No specific storage site has been located. Chromium is excreted by the kidneys.

Meeting Body Needs

Requirements/Recommendations
Limited knowledge exists regarding human requirements for chromium. The Food and Nutrition Board placed the recommended chromium intake in the category designated as Estimated Safe and Adequate Daily Dietary Intake (ESADDI). (See table inside front cover.)

Sources
Detection of chromium content of foods and determination of chromium bioavailability are under investigation. Current information suggests chromium is present especially in whole grains and meats. (Refining of grains removes the chromium.) Other sources include yeast, cheese, mushrooms, and prunes. Commercial and home preparation of foods may be important factors. For example, stainless steel contains chromium. Acid foods heated in such containers can leach out chromium. This may explain the higher levels found in some beers and wines fermented in stainless steel vats.

Body Balance

Methods to study chromium balance in human beings are limited. Because chromium is so widely distributed in the environment, states of insufficiency or deficiency are uncommon. People potentially at risk include those with narrow, highly refined food selections; those with long-term malabsorption; and those on parenteral nutrition. Some people who have insulin resistance have benefited from increased chromium. Chromium cannot replace insulin therapy in people with insulin-dependent diabetes mellitus (IDDM). (See Chapter 26.)

Excessive intake of chromium from food or dietary supplements may not lead to excessive levels of chromium in the body since such tiny amounts are absorbed.

More knowledge may be learned as people continue to take self-prescribed, unmonitored chromium dietary supplements. The chemical form of chromium in food and supplements does not appear to be toxic.

MOLYBDENUM

Characteristics and Functions

Molybdenum is an element that exists in the inorganic and organic forms. The adult body contains about 9 mg of molybdenum, with the highest concentrations in the liver, adrenal glands, and kidneys. Molybdenum is widely distributed in the environment; its content in plants and animals reflects soil content. Conducting research has required artificial means of creating deficiencies and excesses. More information is known about molybdenum's role in plants and animals than in human beings. Molybdenum is involved in enzyme systems, including xanthine dehydrogenase and xanthine oxidase (both involved in the formation of uric acid), and sulfite oxidase (involved in catabolism of the sulfur-containing amino acids, methionine and cysteine).

Metabolic Processes

The metabolic processing of molybdenum is under investigation. Molybdenum appears to be absorbed from the stomach and small intestine. About 25–80 percent is absorbed. Absorption is influenced by several factors. For example, sulfur and copper appear to decrease its absorption. In blood circulation, molybdenum seems to be transported in both its inorganic and organic forms. No specific storage site has been located. Molybdenum is quickly excreted by the kidneys. The body also excretes molybdenum via bile, sweat, and hair.

Meeting Body Needs

Requirements/Recommendations
Limited knowledge exists regarding human requirements for molybdenum. The Food and Nutrition Board placed the recommended molybdenum intake in the category designated as Estimated Safe and Adequate Daily Dietary Intake (ESADDI). (See table inside front cover.)

Sources
Detection of molybdenum content of foods and determination of molybdenum bioavailability are under investigation. Current information suggests molybdenum is present especially in legumes, whole grains, meats, fish, and poultry. Refining of whole grains removes the molybdenum. Other sources include nuts, milk, yogurt, and cheese. Vegetables and fruits are low in molybdenum content.

Body Balance

Methods to study molybdenum balance in human beings are limited. Because molybdenum is so widely distributed in the environment, neither insufficiency nor deficiency states under normal conditions have been documented. One example of an artificially induced deficiency was a patient fed parenterally, without molybdenum, for 18 months. Abnormalities included nausea, vomiting, disorientation, and decreased uric acid and sulfate in the urine. Another example is a rare hereditary disorder that leads to a defect of an enzyme-required molybdenum cofactor (molybdopterin). Among other abnormalities, the disorder results in mental retardation.

Excessive molybdenum intake appears to increase the urinary loss of copper. Toxic levels can do the same. In addition, blood and urine levels of uric acid are increased, leading to goutlike symptoms. (See Chapter 27.)

A summary of minerals appears in Table 11–4.

TABLE 11–4
Summary of Selected Mineral Elements

Element	Location/Function	Utilization	Adult Daily Allowances and Food Sources
Calcium	99% in bones, teeth Nervous stimulation Muscle contraction Blood clotting Activates enzymes	10 to 40% absorbed Aided by vitamin D and lactose; hindered by oxalic acid, fiber, basic pH Parathyroid hormone regulates blood levels *Deficiency:* fragile bones; osteoporosis	**RDA:** 1,200 mg up through age 24; 800 mg age 25 and after Milk, cheese, ice cream Mustard and turnip greens Cabbage, broccoli Clams, oysters, salmon
Copper	Utilization of iron for hemoglobin formation Pigment formation Myelin sheath of nerves	In form of ceruloplasmin in blood Abnormal storage in Wilson's disease *Deficiency:* rare	**ESADDI:** 1.5–3.0 g Liver, shellfish, meats, nuts, legumes, whole-grain cereals
Fluoride	Prevent tooth decay	Storage in bones and teeth Excess leads to tooth mottling	**ESADDI:** 1.5–4.0 mg Fluoridated water
Iodine	Form thyroxine for energy metabolism	Chiefly in thyroid gland *Deficiency:* endemic goiter	**RDA:** 150 µg Iodized salt Shellfish, saltwater fish
Iron	Mostly in hemoglobin Muscle myoglobin Carries oxygen to cells Oxidizing enzymes for release of energy	5–20% absorption Acid and vitamin C aid absorption Daily losses in urine and feces Menstrual loss *Deficiency:* anemia is common in infants, children, young women	**RDA:** men, 10 mg; women, 15 mg Organ meats, meat, fish, poultry Whole-grain and enriched cereal Green vegetables; dried fruits
Magnesium	60% in bones, teeth Transmit nerve impulses Muscle contraction Enzymes for energy metabolism	Salts relatively insoluble Acid favors absorption Dietary deficiency unlikely; occurs in alcoholism, renal failure	**RDA:** 280–350 mg Milk, meat, green leafy vegetables, legumes, whole-grain cereals
Phosphorus	80–90% in bones, teeth Acid-base balance Transport of fats Enzymes for energy metabolism; protein synthesis	Vitamin D favors absorption and use by bones Dietary deficiency unlikely	**RDA:** 800 mg Milk, cheese, ice cream Meat, poultry, fish Whole-grain cereals, nuts, legumes

TABLE 11–4, continued

Element	Location/Function	Utilization	Adult Daily Allowances and Food Sources
Potassium	Intracellular fluid Protein and glycogen synthesis Water balance Transmit nerve impulse Muscle contraction	Almost completely absorbed Body levels regulated by adrenal glands; excess excreted in urine *Deficiency:* starvation, diuretic therapy	**Minimum:** 2,000 mg Ample amounts in meat, cereals, fruits, fruit juices, vegetables
Sodium	Extracellular fluid Water balance Acid-base balance Nervous stimulation Muscle contraction	Almost completely absorbed Body levels regulated by adrenal glands; excess excreted in urine and by skin *Deficiency:* rare, occurs with excessive perspiration	**Minimum:** 500 mg Table salt Baking soda and powder Milk, meat, poultry, fish, eggs
Zinc	Enzymes for transfer of carbon dioxide Taste, protein synthesis	*Deficiency:* growth retardation; altered taste	**RDA:** men, 15 mg; women, 12 mg Plant and animal proteins

HEALTH AND NUTRITION CARE ISSUES

Minerals are the "oldest" substances known to remain in foods and human beings, although this "ash" was not recognized as a group of essential nutrients until the twentieth century.

In the United States, mineral insufficiencies or deficiencies receiving the most health policy attention are those with severe disfiguring or physical and mental impairments as consequences: calcium (promotes bone and tooth formation and prevents osteoporosis), fluoride (promotes tooth strength and prevents dental caries), iron (promotes mental and physical well-being and prevents iron-deficiency anemia), and iodine (promotes well-being in current and future generations and prevents goiter). These four minerals are the ones most commonly added to foods or water as a health measure for the general population. Starting in the 1960s, a burst of renewed interest was sparked as the need to know more about mineral metabolism became obvious for patients receiving total parenteral nutrition (TPN) solutions. Much of what is known now about various minerals originated with these "pioneer" patients, clinicians, and researchers. (See Chapter 17.)

Since the body is able to store and reuse minerals, deficiencies may not occur for years. Factors reducing mineral status are primary deficiency from inadequate intake of mineral-containing foods; or secondary deficiency from decreased release of minerals from foods during digestion, decreased absorption from numerous reasons, or increased losses, ranging from perspiration to blood loss to kidney disease. Over-the-counter and prescription medications can affect minerals, usually more than one at a time. For example, *digoxin* (a heart medication) increases urinary losses of calcium, magnesium, potassium, and zinc. Although the desired effect of *furosemide* (a diuretic) is to increase sodium and potassium excretion, it causes increased excretion of calcium, magnesium, and zinc. (See Chapter 18.)

Mineral excesses do occur, but the nuances may be undiscernible. For example, overly ingesting one or two minerals over all others may cause a severe reduction in the absorption of some of the others. The person does not feel or notice anything, so does not realize possible negative effects may be lurking, either in the near or distant future. Mineral toxicities have been reported from excessive intake of, notably, selenium, iron, fluoride, iodide, zinc, and copper. Thus, unless medically indicated and closely monitored, megadoses of minerals are to be avoided, especially as doses of single nutrients. Levels of minerals found in a diet of a wide variety of wholesome foods continue to furnish human requirements as a balanced mixture and are below levels of toxicity seen by the indiscriminate use of concentrated amounts of certain minerals in pill form.

When screening a client's diet regarding mineral content, five points to keep in mind are:

- Minerals are widely scattered among nearly all foods.
- Bioavailability of minerals varies considerably, especially when ingested with other foods or dietary supplements, and upon body's need.
- Minerals contained in animal foods generally are more bioavailable than minerals in plant foods. This may be of consequence for diets composed chiefly of plant foods.
- Mineral content of water added to foods, such as to pasta, rice, cooked cereals, frozen beverages, and soups, and mineral content of water consumed as ice cubes or as beverages must be highlighted. These may be of consequence to overall amount consumed and stored in the body, especially over time.
- Fats, oils, and white sugar do not contain minerals.

PUTTING IT ALL TOGETHER

Minerals are inorganic elements. Varying amounts are found in nearly all foods. Overall, minerals function in body anabolism, catabolism, and regulation.

? QUESTIONS CLIENTS ASK

1. I am a 34-year-old woman. Should I take a calcium supplement?

The best source of calcium is foods. If you drink 2–3 cups of milk a day and eat a variety of foods from the Food Guide Pyramid, you will get about 800 mg calcium. Some nutrition scientists now recommend a daily calcium intake of 1,000 mg for women your age to help maintain good bone mineralization. An additional cup of milk, yogurt, or some cheese would be desirable.

2. I am allergic to milk. What is the best calcium supplement for me?

Calcium carbonate is 40 percent calcium, so a 500 mg tablet provides 200 mg calcium. You can select the least expensive brand. However, avoid any with bone meal and dolomite since such natural products often

are contaminated with mercury or lead, both of which are toxic.

3. Can too much calcium cause kidney stones?

A healthy person can have a calcium intake of about 2,500 mg (more than three times the RDA) without any appreciable effect on calcium excreted in the urine. People who have kidney stones or who are at risk for them may excrete more calcium when they have high intakes. About 80 percent of kidney stones contain calcium, chiefly as calcium oxalate. Severely cutting down on calcium intake not only increases the loss of calcium from the bones but increases the absorption of oxalate. Thus, the calcium from bone could combine with oxalate to form stones, even though calcium intake is reduced. To minimize the problem, a susceptible person should drink plenty of fluids.

4. *I cook soups and stews in a heavy iron pot. Does this have any effect on the iron in the food?*

This is one way to increase your iron intake. Acid foods such as tomatoes are especially good since they dissolve the iron from the cookware and thus increase the amount of iron in the food.

5. *I had a hair analysis that showed I was deficient in zinc. What is a good supplement for me?*

Currently, hair analyses are not considered to be reliable for making a diagnosis of any mineral deficiency. The composition of hair changes not only by the supply of nutrients to it, but also by treatments that hair undergoes, such as shampoos, conditioners, and sprays, which might add to, or subtract from, the trace minerals in the hair. So hair analysis is a waste of money. Besides, taking a zinc supplement on the basis of any one test might interfere with the absorption of other minerals such as copper. It is not a good idea to self-prescribe such supplements.

✔ CASE STUDY: A Teenager with Iron-Deficiency Anemia

Ellie M., 15, is not doing well in school. She is frequently absent because of headaches. When she is there, her teachers notice she is not concentrating in class. At home, her parents find her sleeping instead of doing her homework. She is irritable to her family; she never wants to do anything with them any more by complaining she is too tired. Even her friends have noticed she is moody and can't keep up with their high-energy activities.

At a health checkup the following screening information was obtained. Ellie is 63 in. (160 cm) tall, without shoes, and weighs 108 lb (49 kg) in light indoor clothing. Blood test results were: hemoglobin = 10 g/dl (6.2 mmol/L); hematocrit = 30%. By history, Ellie started her menses at age 12; her monthly period lasts about one week, with heavy flow the first three days. Ellie's eating habits are erratic and she is afraid of "getting fat." A typical food intake is:

Breakfast: Cold cereal with sliced banana; skim milk.

Lunch: Eats school lunch, depending on the menu. Her favorite meals are pizza, spaghetti with tomato sauce, and grilled cheese sandwiches. On other days, she just eats the foods she likes, usually the salad and applesauce or peaches. Sometimes she drinks all the milk, but usually only about half.

After school: Low-fat cookies; low-calorie soda.

Dinner: Poultry, fish, or pasta; green vegetable, like peas; salad with low-calorie dressing; low-fat cake; skim milk.

Evening snack: Pretzels; low-calorie soda.

Questions

1. Using Appendix Figure D–2, plot Ellie's height and weight. What is the comparison between her height and weight? What other information would be helpful?

2. What are normal levels for hemoglobin and hematocrit? (See Appendix Table D–2.) What is your interpretation?

3. What are some reasons that teenage girls are especially at risk for iron-deficiency anemia? What factors apply to Ellie?

4. Why is fatigue a common symptom of anemia? What are other possible causes of Ellie's symptoms that might be misleading of the correct diagnosis?

5. An iron supplement, *ferrous sulfate,* was prescribed for Ellie. Could a high-iron diet achieve the same results? Explain.

6. In her health class at school Ellie learned that some foods are good sources of heme iron. What is meant by heme iron? What foods supply heme iron? What are such foods in Ellie's typical day? What other foods in Ellie's current diet are good sources of nonheme iron?

7. What nutrients enhance or interfere with iron absorption? What nutrients are necessary for

the synthesis of hemoglobin? With Ellie, plan revised meal patterns and menu structures for a typical day to improve the food combinations at meals and snacks.

8. Some of the foods in Ellie's current diet are good sources of minerals other than iron. What are they? What other minerals may be low in Ellie's overall diet? What are some potential short-term or long-term effects? How would you counsel Ellie?

FOR ADDITIONAL READING

American Dietetic Association: "Position Paper: The Impact of Fluoride on Dental Health," *J Am Diet Assoc,* **94:**1428–30, 1994.

Bove, LA: "Sodium & Chloride," *RN,* **59:**25–29, January 1996.

Calcium Requirements of Children," *Nutr Rev,* **53:**37–40, 1995.

Hoppe, B: "Taking the Confusion out of Calcium Levels," *Nurs95,* **25:**32KK–32MM, July 1995.

Hu, H, et al, "The Relationship Between Bone Lead and Hemoglobin," *JAMA,* **272:**1512–17, 1994.

Mertz, W: "Risk Assessment of Essential Trace Elements: New Approaches to Setting Recommended Dietary Allowances and Safety Limits," *Nutr Rev,* **53:**179–85, 1995.

Mushak, P, and Crocetti, AF: "Lead and Nutrition: I. Biologic Interactions of Lead with Nutrients," *Nutr Today,* **31:**12–17, 1996.

"Optimal Calcium Intake." *NIH Consens Statement,* June 6–8; 12(4):1–31.

Perez, A: "Hyperkalemia," *RN,* **58:**33–37, November, 1995.

Perez, A: "Hypokalemia," *RN,* **58:**33–36, December, 1995.

Raimer, F: "Identifying Magnesium and Phosphorus Imbalances," *Nurs95,* **25:**32CC–32EE, February 1995.

Wardlaw, GM and Weese, N: "Putting Calcium into Perspective for Your Clients," *Top Clin Nutr,* **11:**23–35, December 1995.

PART II

Food and Nutrition Applications Throughout Life

CHAPTERS IN PART II

Part II examines the nutritional needs of people at various stages throughout the life span. This part of the textbook brings forward the knowledge and skills learned in the preceding eleven chapters. Practical pointers are incorporated into each chapter.

Chapter 12 covers pregnancy and lactation. It looks at childbearing women's nutritional changes. Food and nutrition issues specific to new mothers and fathers are emphasized.

Childhood is the focus of Chapter 13. The chapter starts at birth and ends with adolescence. This chapter traces children's growth, development, nutritional needs, and food selections. Community nutrition resources available to children and their families are described.

Keeping young and middle-aged adults nutritionally healthy is the purpose of Chapter 14. The various influences and stressors faced by such adults are addressed. Since most readers of the textbook are in this age range, some topics may spark debate on individual approaches to the multiple issues presented.

Chapter 15 identifies the interplay between ever-evolving adult maturation and modifications in foods and nutrients. The concluding focal points are examples of special health concerns and available nutrition programs.

Nutrition During Pregnancy and Lactation

Chapter Outline

PREGNANCY
Fetal and Maternal Changes
Meeting Nutrition Needs
Special Food and Nutrition Issues
High-Risk Pregnancy
LACTATION

Maternal and Paternal Decisions
Meeting Nutrition Needs
Special Food and Nutrition Issues
Putting It All Together
Questions Clients Ask
Case Study: A First Pregnancy

Reduce the incidence of fetal alcohol syndrome to no more than 0.12 per 1,000 live births.*

Increase to at least 85 percent the proportion of mothers who achieve the minimum recommended weight gain during their pregnancies.*

Increase to at least 75 percent the proportion of mothers who breastfeed their babies in the early postpartum period and to at least 50 percent the proportion who continue breastfeeding until their babies are 5 to 6 months old.*

 CASE STUDY PREVIEW

Many people look forward to having children. Pregnancy and the early months after the baby is born are times of anticipation and change. What nutritional and lifestyle habits are of special focus for parents?

**Healthy People 2000: National Health Promotion and Disease Prevention Objectives.* Public Health Service, U.S. Department of Health and Human Services, Washington, DC, 1991, Objectives 14.4, 14.6, and 14.9, pp. 110–11.

Ahealthy, fully developed, and well-nourished baby is the hope of all parents. Most parents realize this dream. The woman who is in good health and who has maintained good nutrition *prior* to conception as well as during pregnancy has the best chance for a pregnancy without complications, a healthy baby, and the ability to breast-feed.

For the expectant mother, father, and their families, the time surrounding pregnancy is a unique period. There is much to think about and do. Each member of the obstetric team, including physicians, nurses/midwives, dietitians/dietetic technicians, health educators, social workers, and psychologists, plays a crucial role in the positive outcome of the pregnancy. In the area of nutrition, the mother-to-be's nutritional status and dietary intakes and practices need to be evaluated and monitored. Suitable individualized counseling and referral, for example, to WIC (Special Supplemental Food Program for Women, Infants, and Children), should be provided in a coordinated manner.

Pregnancy

FETAL AND MATERNAL CHANGES

Fetal Growth and Development

During the 9 months of pregnancy the healthy fetus will achieve a weight of about 3 to 4 kg (about 6½ to 8¾ lbs) and a length of about 50 cm (20–21 in.). The skin, skeleton, digestive tract, heart, liver, lungs, nervous system, and kidneys begin to develop during the second to the eighth weeks. If any problem, such as a nutrient deficiency or toxicity or a harmful environmental substance, interferes with development during this critical period, the organs may not reach their full potential or some permanent damage may result. Once a critical period has passed, any abnormality cannot be fully corrected.

Maternal Changes

The size of the uterus, placenta, amniotic fluid, and blood volume increase to serve the developing fetus. The breasts enlarge in preparation for lactation. Energy (fat) and protein reserves are built up for the time of delivery and lactation.

Weight Gain The weight of the various components that make up these fetal and maternal changes accounts for about 9.5 kg (20.9 lb). Obstetric health-care providers should adopt specific and reliable procedures to routinely measure the height and weight of the pregnant woman and record the results at each prenatal visit.

Weight gain is divided into three recommendations, based on the woman's prepregnancy weight for height or BMI (Body Mass Index). (See Chapter 3 for calculation and Table 12–1 for interpretation.) Weight gain for women of normal BMI should be between 11.5 and 16 kg (25 and 35 lbs). Women with a lower BMI should gain more; women with a higher BMI should gain less. Short women in each category should attempt to reach the lower end of their range. A young woman or a woman with a history of previous inadequate weight gain for herself or another infant should aim for the higher end of the range.

TABLE 12–1

Recommended Total Weight Gain Ranges for Pregnant Women,[a] by Prepregnancy Body Mass Index (BMI)[b]

Weight-for-Height Category		Recommended Total Gain	
		kg	lb
Low	(BMI < 19.8)	12.5–18	28–40
Normal	(BMI of 19.8 to 26.0)	11.5–16	25–35
High[c]	(BMI > 26.0 to 29.0)	7.0–11.5	15–25

Source: Subcommittee on Nutritional Status and Weight Gain During Pregnancy: *Nutrition During Pregnancy* Part I: Weight Gain, Food and Nutrition Board, Institute of Medicine. National Academy of Sciences. National Academy Press, Washington DC, 1990.

[a] Young adolescents and black women should strive for gains at the upper end of the recommended range. Short women (< 157 cm. or 62 in.) should strive for gains at the lower end of the range.

[b] BMI is calculated using metric units.

[c] The recommended target weight gain for obese women (BMI > 29.0) is at least 6.8 kg (15 lb).

Pregnancy is not the time to correct obesity. Although obesity increases the risks of pregnancy, the loss of weight poses an even greater threat to the fetus. The lower limit of recommended weight gain for extremely obese women is 6.8 kg (15 lb).

The rate of weight gain is as important as the total amount of weight gained. (See Figure 12–1.) A rate of weight gain of about 0.4 kg (1 lb) each week during the second and third trimesters is recommended for women with a normal BMI. For underweight women, the recommended rate is higher (0.5 kg) per week, with a slower rate (0.3 kg) urged for overweight women. The rate of weight gain for extremely obese women needs to be individualized.

If the pregnant woman gains rapidly during the first trimester, she should not try to compensate for this by slowing her rate of gain during the remainder of her pregnancy. Rather, she should maintain a steady weight gain. On the other hand, a low weight gain during the first two trimesters cannot be fully made up during the final trimester.

Throughout pregnancy, the pregnant woman needs to receive nutrition education and personalized counseling. Also, her physical activities should be assessed and she should be encouraged to maintain suitable levels and patterns of exercise.

MEETING NUTRITION NEEDS

Recommended Dietary Allowances

The Recommended Dietary Allowances (RDAs) are increased during pregnancy and even more so during lactation. (See tables inside front and back covers.) Although the individual needs of each woman should be determined, there are limited available techniques. One approach is obtaining a food history or food frequency, along with asking pertinent questions about special conditions such as restricted food choices, an erratic work schedule, or limited income. (See Appendix Figure C–2 for a sample.) In addition, routine laboratory tests should be assessed, especially hemoglobin, hematocrit, or ferritin.

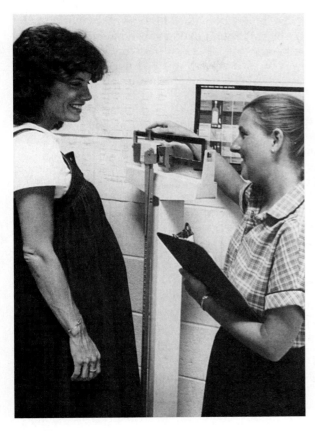

FIGURE 12–1
A steady rate of weight gain should be maintained throughout pregnancy. (*Source:* School
of Nursing, University of Pennsylvania, and Gates Rhodes, photographer.)

Nutrient Highlights

Water Water need is increased during pregnancy, but only by about 30 ml
(1 oz) each day! Water chiefly is needed for the fluid in the extracellular space,
amniotic sac, and fetus. Water, though, is important for proper bowel function
and to replace fluid loss from vomiting, if morning sickness occurs.

Energy The rate of weight gain is the best indicator of caloric adequacy or
inadequacy. Since the majority of the weight gain of the fetus happens during the
second and third trimesters, increased energy needs by the woman occur during
those months. Of utmost importance to keep in mind, however, is the recognition that the foods chosen to provide the energy should also provide sufficient
water, fiber, essential fatty acids, protein, vitamins, and minerals. For example,
while a daily lunch of a soda and potato chips may provide energy for weight
gain, it supplies negligible amounts of the other nutrients required by the
pregnant woman and her rapidly growing fetus.

Protein During the first trimester protein synthesis is small in comparison to
that during the remainder of the pregnancy. Besides fetal development, additional

protein is necessary for increased blood volume and growth of the uterus, placenta, breasts, and skeletal muscles. Throughout her pregnancy, a woman is in positive nitrogen balance. (See Chapter 7.)

Vitamins The need for nearly all vitamins is increased during pregnancy, some vitamins more than others. (See table inside front cover.) Survey data show that actual intake is less than the RDAs for vitamins D, E, B_6, and folate. However, the dietary intake recommendations usually can be met by food.

Exceptions occur for women in special circumstances. For example, women with lactose intolerance may need to direct extra attention toward obtaining sufficient amounts of vitamin D and riboflavin. Women who are strict vegetarians (vegans) may need supplementation with vitamin B_{12}.

Research is continuing regarding the role of folate in preventing neural tube defects such as spina bifida. (See Chapter 10.) Except in situations where a folate deficiency is suspected, the folate needs of the woman prior to conception and during pregnancy can be met by well-selected foods. When this is impossible, supplementation is necessary and monitoring of blood indicators is to be done. This points out the crucial need for preconception health and nutrition care— something many women do not have.

Evidence does exist that preformed vitamin A has toxic effects on the fetus and thus supplementation with this vitamin should be avoided.

Minerals The need for most minerals appears to increase during pregnancy. Despite research limitations, survey data indicate that dietary intake is less than the RDAs for iron, zinc, calcium, and magnesium. However, the RDAs do include a large safety range, and during pregnancy, as a response to increased body need, the rate of absorption of minerals increases and the rate of excretion decreases. Also, evidence from animal and human studies indicates that some minerals, for example, selenium and zinc, may be toxic in high doses. Consideration also must be given to nutrient-nutrient interactions during absorption and metabolism by the body. For example, excessive intakes of calcium and vitamin D interfere with the absorption of zinc. (See Chapter 11.)

The largest dietary increase is in response to the iron requirement. Because of many women's limited stores of iron, the demands of the growing fetus (especially during the last trimester), and the blood losses at delivery, the iron requirement is increased beyond that which generally could be met by even very wise food selections. Of all the nutrients, only iron is advised to be supplemented (e.g., a low-dose iron preparation of 30 mg of ferrous sulfate daily) during the second and third trimesters.

Unless there are reasons why a woman may be deficient in certain minerals, the dietary intake for minerals in general should be able to be supplied by suitable food choices. (See RDAs Table and Figure 1–5.) Such reasons include the woman whose daily dietary intake of calcium is below 600 mg from foods that also supply phosphorus and magnesium, or who is lactase deficient, or whose selection of foods restrict the intake of targeted minerals. Improving nutrient intake from foods is preferable to adding specific nutrients here and there. Overall dietary balance of all nutrients provides the best opportunity for body homeostasis.

In the past, sodium was restricted. Since the fetus, maternal tissues, and expanding blood volume all require sodium, its restriction is no longer recommended, unless intake is excessive for body needs.

TABLE 12–2

Examples of Food Selections During Pregnancy and Lactation

	Pregnant Woman	Pregnant Teenage Girl	Lactating Woman
Cereal, whole-grain or enriched	1 serving	1 serving	1 serving
Bread, Rice, Pasta, whole-grain or enriched	5 servings	5 servings	5 servings
Vegetables, including			
Dark-green leafy or deep yellow	½ cup	½ cup	½ cup
Potato	1 medium	1 medium	1 medium
Other vegetables	½–1 cup	½–1 cup	½–1 cup
(One vegetable to be raw each day)			
Fruits, including			
Citrus	1 serving	1 serving	1 serving
Other fruit	3 servings	3 servings	3 servings
Milk, whole, low-fat, or skim	3–4 cups	5–5½ cups	5–5½ cups
Meat, Fish, Poultry, cooked weight	4 oz	4 oz	4 oz
Eggs	3 to 4 per week	3 to 4 per week	3 to 4 per week
Butter or Fortified margarine			
Additional Servings of Above Foods	To meet caloric needs	To meet caloric needs	To meet caloric needs
Desserts, Cooking Fats, Sugar, Sweets			

Food Choices

Because of increased energy, protein, vitamin, and mineral needs, wholesome foods should be chosen from the Food Guide Pyramid. (See Table 12–2 for examples of food selections for pregnant women. For evaluation of nutrient intake, use the Basic Diet Nutrient Profile—Tables 2–3 and 2–4.) The addition of 1½ to 2 cups of milk supplies necessary nutrients besides calcium. Pregnant teenagers need to consume 5 to 5½ cups of milk to meet their calcium requirements. Many culturally acceptable foods can be substituted for the nutrients supplied in fluid milk, and dairy products are widely incorporated as ingredients in numerous packaged foods.

In recent years, the importance of the sanitation and safety of food choices made by pregnant women has become better recognized. The danger of microorganisms and mercury poisoning top the list. All foods should be selected with utmost care from reliable purveyors and handled consistently with good hygiene practices. (See Chapter 2.) Of all groups of people, the most concern is for pregnant women. For example, pregnant women are at high risk for mercury poisoning, mainly caused by industrial pollution. For the fetus, exposure during the first trimester is the most critical period. In the United States, public health advice is that pregnant women are to be informed to eat no more than 7 oz a week of predatory fish, like shark or swordfish. Posted notices in fishing areas, supermarkets, and as menu attachments are commonly found in regions along sea coasts, but not as commonly at inland establishments.

SPECIAL FOOD AND NUTRITION ISSUES

Common Dietary Adjustments

During the first trimester, early morning nausea and vomiting (morning sickness) usually can be alleviated by eating a high-carbohydrate food before arising, such as dry toast, crackers, or hard candy. Fatty and fried foods should be restricted. Several small meals a day, rather than three large meals, are desirable. Fluids should be taken between meals instead of at mealtime. Odors can be a factor, so strong aromas from foods or cooking should be eliminated as much as possible.

During the latter part of pregnancy, constipation is a common occurrence. It is caused, in part, by the pressure of the uterus as it enlarges against the colon. Also, hormones produced by the placenta have a relaxant effect on the muscles of the gastrointestinal tract. Constipation usually can be avoided by placing more emphasis on a liberal intake of fluids, raw fruits and vegetables, and whole-grain breads and cereals, along with a regular program of physical activity.

Multivitamin and Mineral Supplementation

For women whose dietary intakes are inadequate or who are at high medical risk, such as those carrying multiple fetuses or who smoke cigarettes, nutrient supplementation is recommended.

The suggested nutrients include the vitamins C, D, B_6, and folate and the minerals calcium, iron, copper, and zinc. To promote optimal absorption of these nutrients, it is advised that the pregnant woman take the supplement between meals or at bedtime rather than with meals or snacks.

Food Cravings

During pregnancy some women have cravings for certain foods—pickles and ice cream being the focus of cartoonists. There is no evidence that such cravings indicate a particular nutrient need. Provided they do not interfere with the intake of a balance of a variety of foods, there is no harm in humoring these cravings.

Pica

Some pregnant women, especially in some ethnic and geographic regional groups, eat a variety of nonfood substances such as clay, chalk, cigarette ashes, ice, and laundry starch. This abnormal craving for substances with little or no nutritional value is known as **pica.** The reasons for these unusual cravings are not known. As little as a handful of clay or laundry starch or as much as a quart of clay or a box of laundry starch is consumed. Pica raises several concerns: the nonfood substances may be eaten instead of nutritious foods; these nonfood substances usually interfere with absorption of nutrients; substances such as clay could lead to intestinal blockage; and eating plaster or air freshener could produce toxic reactions. A woman who eats as much as a box of dry laundry starch is obtaining about 1,600 kcal, but with no fluid, protein, vitamins, or minerals.

HIGH-RISK PREGNANCY
What Constitutes High Risk?

A high-risk pregnancy is one in which there is a greater likelihood of spontaneous abortion, stillbirth, a low-birth-weight infant, and/or an infant who has congenital abnormalities. High-risk pregnancy also includes the mother who is more likely to give birth prematurely or is herself at any other health or nutritional risk.

The principal factors that increase the risks are summarized in Figure 12–2. Not every woman who falls in the high-risk category will have a low-birth-weight infant or one who has malformations. Nor is every woman susceptible to maternal

PREPREGNANCY STATUS

Age:

- Immaturity with pregnancy under 17 years of age
- Risk especially high before 15 years
- First baby after 35 years

Inadequate diet:

- Lack of essential nutrients
- Crash dieting, especially by teenagers
- Strict vegetarianism without supplements of calcium, vitamin B_{12}

Inadequate nutritional status:

- Anemia
- Stunted growth
- Low weight for height
- High weight for height

Inadequate socioeconomic environment:

- Lack of income to purchase foods
- Lack of education
- Inability to improve economic status
- Inadequate understanding of nutrient requirements

Prior unsuccessful pregnancies:

- Miscarriage, stillbirth, low-birth-weight infant
- Eclampsia
- Three or more pregnancies within 2 years

Alcohol, tobacco, or drug abuse

Chronic disease:

- Diabetes mellitus, cardiovascular disease, renal disease, cancer, cystic fibrosis, HIV/AIDS, inborn errors of metabolism

DURING PREGNANCY

Continuation of conditions existing prior to pregnancy plus:

- Too little weight gain; excessive weight gain
- Inappropriate pattern of weight gain
- Pica
- Multiple births

FIGURE 12–2
Factors that increase the risks of pregnancy.

complications. However, prudence suggests that health professionals identify as early as possible any existing risks and provide counseling toward improvement of the diet as well as change in lifestyle.

Teenage Pregnancy

About one million teenagers become pregnant each year. For her own growth and development toward adulthood, a girl's nutritional requirements are considerable. To these requirements must be added nutrients to meet the increasing needs of the fetus. Unfortunately, prior to conception and during pregnancy, teenage girls often skip breakfast, snack on foods of low nutrient content, and restrict food intake to achieve a slim figure.

Teenagers facing pregnancy, both girls and boys, often drop out of school, leaving them to face a life of limited opportunity. Add to these problems parental disapproval, ostracism by peers, and poor self-image.

In especially high-risk circumstances are pregnant girls under 15 years of age. They have more frequent complications of anemia, eclampsia, and long, difficult labor. Babies born to them are more often low-birth-weight and have a higher rate of **neonatal mortality;** that is, the infants die in the period immediately after birth or during the first month of life. Repeated pregnancies before the age of 20 place both the young woman and the unborn child in an extremely high-risk category.

Anemias

Microcytic, hypochromic iron-deficiency anemia is seen in many pregnant women. Macrocytic anemia caused by folate deficiency is less common, but with potential serious side effects, such as neural tube defects in the fetus. Macrocytic anemia in pregnant women who are vegans may be due to deficiency of vitamin B_{12}. All these anemias are preventable by dietary improvements and/or supplements of iron, folate, or vitamin B_{12}, respectively. Optimal preconception nutritional status, along with optimal nutrient intake during pregnancy, are essential for the prevention of these three forms of anemia. (See Chapters 10 and 11.)

Alcohol Ingestion

Following ingestion of an alcoholic beverage, the alcohol content of the maternal circulation rapidly rises and crosses over into the fetal circulation. The alcohol is removed much more slowly from the fetal blood than from the maternal blood. In the meantime, the oxygen supply to the fetus is reduced and fewer fetal cells are produced.

FETAL ALCOHOL SYNDROME

Small head circumference
Small mid-face
Small eyes
Thin upper lip; poorly developed groove between tip of nose and upper lip
Poor swallowing reflexes
Retarded growth
Impaired nervous system development
Learning disabilities; mental retardation

Fetal alcohol syndrome (FAS) occurs in one of every 1,000 births and is a leading cause of mental retardation. The full syndrome is most prevalent in infants born to women who chronically abuse alcohol or consume about five drinks daily. But as few as two drinks daily has been associated with learning disorders, attention deficits, and prematurity. Although an occasional drink has been taken by a pregnant woman without any apparent harm, the safest course is not to drink at all. Fetal alcohol syndrome is one condition that is entirely preventable!

Illicit Drugs

The use of illicit drugs such as *cocaine* and its derivatives places the woman and her fetus at severe medical high risk. Evidence continues to accumulate about the metabolic side effects and subsequent nutritional ramifications of acute and chronic drug abuse. The best approach, although not always possible, is immediate drug rehabilitation.

Hyperemesis Gravidarum

About 2 percent of all pregnant women experience vomiting during the first trimester that is so severe as to be life threatening. Dehydration, high blood levels of sodium, low blood levels of potassium, and alkalosis result when vomiting is prolonged. Hospital care is essential to correct the fluid and electrolyte imbalances and to gradually initiate oral feeding.

Pregnancy-Induced Hypertension

This condition, also called **preeclampsia,** is characterized by increased blood pressure, swelling of the hands, face, and ankles, and proteinuria. A sudden gain in weight about the 20th week of pregnancy indicates water retention and the presence of preeclampsia. At worst, preeclampsia leads to convulsions and coma, a condition known as **eclampsia.** The causes of the condition are poorly understood, but lack of prenatal care and poverty are associated. Restriction of calories, protein, and/or sodium is no longer considered to be useful, and is potentially dangerous.

HIV Infection and AIDS

Depending on the stage of the disease at the time of conception and throughout pregnancy, malnutrition is a threat to both the woman's health and her baby's. Malnutrition is due largely to inadequate food supply, inappropriate food choices, increased nutrient demands caused by the primary and opportunistic infections, and various digestive tract complications, including malabsorption of nearly all nutrients. (See Chapter 29.)

Diabetes Mellitus

At the initial preconception or prenatal visit, a woman should be screened for possible gestational diabetes mellitus. Questions include: whether there is a family history of diabetes, whether there have been any complications such as miscarriages or stillbirths in earlier pregnancies, and whether there have been any babies weighing more than 9 pounds. The existence of any or all of these factors necessitates further study to determine whether the woman might develop gestational diabetes mellitus.

A woman who has previously diagnosed diabetes mellitus, especially if she required insulin, can have a successful pregnancy and a healthy baby. Prior to and throughout her pregnancy she and her obstetric caregivers must form a highly interactive team for constant monitoring and support.

The fundamental nutritional needs of a pregnant woman with diabetes are the same as those of other pregnant women. During early pregnancy when nausea and inadequate food intake are problems, it may be necessary to lower the insulin intake to avoid hypoglycemia. Later in the pregnancy the gradual increase in weight increases the insulin requirement slightly. The maintenance of a gradual weight gain is extremely important, as is the avoidance of fluid retention. Mealtimes and snacks need to be consistent and correlated with physical activity and sleep patterns.

Lactation

MATERNAL AND PATERNAL DECISIONS

To breast-feed her infant is a matter of choice for a woman in the United States. Often the father plays a pivotal role—whether for or against it. Usually the decision to breast-feed is made during pregnancy. During the past 100 years, breast-feeding first declined, then increased during the 1960s and 1970s, and again declined during the 1980s and 1990s. About 45 percent of all newborns and 80 percent of five- to six-month-old infants are fed a commercial formula as either their sole source of nourishment or as a supplement to breast milk. (Review Objective 14.9 from "Healthy People 2000" at the beginning of this chapter.) Surveys indicate breast-feeding is more widely practiced by older and better educated women with higher incomes. The nine months of pregnancy slowly build up the mother's nutrient reserves of energy (body fat), proteins, vitamins, and minerals for release during the lactation period.

MEETING NUTRITION NEEDS

Recommended Dietary Allowances

The recommended allowances during lactation equal or exceed those for pregnancy. The RDAs are subdivided into recommendations for months 1 through 6 and for months 7 through 12 of lactation. (See tables inside front and back covers.)

Nutrient Highlights

Water About one *additional* liter (about 1 quart) of fluid—in the form of water, food, and other beverages (e.g., soups, milk, fruit juices)—is necessary, based on milk production, environmental conditions, and maternal physical activity.

Energy During the first six months, the amount of milk produced each day is about 750–800 ml (25–27 oz). Human milk furnishes about 67–77 kcal/dl. Thus, the total milk production accounts for 500–615 kcal. However, converting the energy value of food to the energy value of milk is only 80–90 percent efficient. Therefore, each day's milk production requires about 600–740 kcal.

With satisfactory weight gain during pregnancy the woman has stored body fat to furnish 200–300 kcal per day for milk production during the first three months. Allowing for the use of these body fat reserves, the diet should be increased about 500 kcal daily. (See table inside back cover.) As body fat stores are used for milk production, there naturally is a steady small loss of body weight. Deliberate attempts to lose more weight should not be attempted during lactation.

Protein Each 100 ml human milk supplies about 1.2 g protein. In 800 ml, the protein content is 9.6 g. Changing food protein into human milk protein also is not 100 percent efficient. Each woman differs in her nutritional needs to produce milk. The RDAs are 65 g for the first six months and 62 g for the second six months. (See table inside front cover.)

Vitamins and Minerals The nutrient content of some vitamins (vitamins D, B_6, and B_{12}) and some minerals (selenium and iodine) of human milk may reflect either maternal intake or her body stores. The levels of some nutrients, especially calcium and folate, are maintained in the milk at the expense of the mother's stores if she does not consume sufficient amounts.

Food Choices

A variety of foods chosen in adequate amounts from each of the food groups in the Food Guide Pyramid satisfactorily meets the nutrient requirements of the new mother. (See Figure 1–5 and Table 12–2.) Incorporating the fluid increases may take some planning. Otherwise, only in unusual circumstances are additional dietary supplements necessary during lactation.

Unless they have been resolved, if the mother was at high risk during pregnancy, similar issues face her after her child is born. Under normal conditions, examples of unique issues for the mother include adjusting to new roles, returning to work or school, and having the stamina to awaken at various hours to feed the baby.

Many women are unclear and uncertain about their breast-feeding skills. They need the guidance of their health care team and the support of the father, relatives, friends, and employer.

PUTTING IT ALL TOGETHER

Pregnancy sometimes is called "expecting," as in "I'm expecting a baby!" Truly, this is a period of great anticipation. Nutrition is important prior to, during, and after pregnancy for both the mother and the baby.

? QUESTIONS CLIENTS ASK

1. We are planning to have a baby but I am 30 lb overweight. Should I lose this weight before becoming pregnant?

Yes, if you lose weight gradually at a rate of about 1½ to 2 lb a week. By losing weight you may avoid some of the complications that occur in some overweight women. Your low-calorie diet must be nutritionally adequate so that important reserves of nutrients are not depleted prior to your pregnancy.

2. I drink about 3 cups of coffee a day. Will this harm my baby?

Caffeine is a stimulant. It crosses from the maternal to the fetal circulation where it is removed more slowly. There probably are no adverse effects from drinking 2 to 3 cups daily. It is important to avoid the accumulative effect from many sources of caffeine such as tea, many soft drinks, cocoa and chocolate products, and medications. Read the labels carefully! (Refer to Table 18-2 for amounts of caffeine.)

3. Should I stop using aspartame?

Moderate use of this alternative sweetener in coffee or on cereal is not known to be harmful. Women who have phenylketonuria should refrain from its use since one of the amino acids that make up this sweetener is phenylalanine. Persons with this metabolic condition tolerate phenylalanine only in carefully measured amounts. (See Chapter 27.)

4. I have been taking 5 g of vitamin C daily to prevent colds. Should I continue to do so during my pregnancy?

Large amounts of vitamin C can lead to vitamin C dependency in the baby. This means that ordinary intakes of the vitamin, such as the baby would receive from breast milk or formula, will not meet daily needs. Also, the evidence about the value of megadoses of vitamin C in preventing colds is not strong. Vitamin C intake should be restricted to that provided by the diet.

✔ CASE STUDY: A First Pregnancy

Anna and Paul, both 32 years old, just found out that they are going to have their first child. Anna is one month pregnant.

Results of Anna's routine assessment studies are: Height = 67 in. (170 cm), without shoes; prepregnancy weight = 126 lbs (62 kg), in indoor clothing; frame size = medium. Hematocrit,

hemoglobin, and serum ferritin levels = within normal limits.

The health history revealed: both parents-to-be are interested in natural childbirth; Paul plans to attend all classes and be in the delivery room; both are health conscious; they run about 2 miles, 4 days a week; neither of them smokes;

they have a glass of wine with dinner; Anna plans to breast-feed the baby.

Anna always was interested in maintaining an attractive appearance and is concerned about gaining too much weight. She is worried that additives in foods may harm the fetus and wonders whether she should continue to shop at the supermarket or start buying all foods at an all-natural specialty store. She is taking 5 g vitamin C daily since she believes that this cuts down on the incidence of colds, but no other dietary supplements.

A typical day's meal pattern and menu choices are:

When she runs:	Bottled water.
Breakfast:	Orange juice; cereal with skim milk; espresso.
Lunch:	Pita bread sandwich with chicken or tuna salad; apple, orange, or banana; packaged square of carrot cake made without preservatives; bottled fruit juice drink.
Dinner:	Broiled, roasted, or stir-fried meat, poultry, or fish; rice or potato; green or yellow vegetable; tossed salad with low-calorie salad dressing; low-calorie gelatin dessert; wine.
Snacks:	Dried fruit, like raisins; or air-popped popcorn.

Questions

1. How does Anna's weight compare with standards for her age and height? (See Appendix Table D–1 and Figures D–5 and D–6.) Calculate Anna's Body Mass Index (BMI). (See Chapter 3 and Appendix Figure B–1.)

2. What is the recommended total weight gain for Anna? (See Table 12–1.) How much should she gain during the first trimester? During the second trimester? During the third trimester?

3. What factors in Anna's history are favorable for a successful pregnancy? Which are unfavorable? How might you counsel her and Paul about strengthening the successful ones and altering the unfavorable ones?

4. Evaluate Anna's food choices for her new status of mother-to-be. (See the Food Guide Pyramid, Figure 1–5; or the Basic Diet, Tables 2–3 and 2–4; or Nutritive Value of the Edible Part of Food, Appendix Table A–1.) What food changes or additions might Anna make to her present intake to meet her nutritional needs during pregnancy?

5. Anna does not like to drink milk. According to the Food Guide Pyramid, what is the recommended number of servings from the dairy group? (See also Table 12–2.) How would you explain to her the importance of including sufficient foods from this group each day? What other foods could she include as comparable nutritious substitutes for milk? List several ways Anna could increase her intake of milk and these foods.

6. Anna's friends told her she should take a variety of dietary supplements. What are some questions to ask Anna before giving her an answer?

7. How would you advise Anna about her concerns regarding food shopping? About food additives? Identify two additives and discuss their impact on pregnant and/or lactating women. Also, what are some food safety issues for pregnant women? (See also Chapter 2.)

8. What changes in diet will help ensure successful lactation when the baby arrives? What suggestions can you give about breast-feeding to provide guidance for both Anna and Paul?

FOR ADDITIONAL READING Foulke, JE: "Mercury in Fish: Cause for Concern?" *FDA Consumer,* **28:**5–8, September 1994.

Story M, and Alton, I: "Nutrition Issues and Adolescent Pregnancy," *Nutr Today,* **30:** 142–51, 1995.

Yu, SM, and Jackson, RT: "Need for Nutrition Advice in Prenatal Care," *J Am Diet Assoc,* **95:**1027–29, 1995.

Nutrition for Infants, Children, and Teenagers

Chapter Outline

Increase to at least 75 percent the proportion of parents and caregivers who use feeding practices that prevent baby bottle tooth decay.*

Increase to at least 90 percent the proportion of school lunch and breakfast services and child care food services with menus that are consistent with the nutrition principles in the Dietary Guidelines for Americans.*

Increase to at least 75 percent the proportion of the Nation's schools that provide nutrition education from preschool through 12th grade, preferably as part of quality school health education.*

✔ CASE STUDY PREVIEW

Each child is an unique individual whose heredity and environment shape the course of his or her life. Woven into daily life are aspects of food and nutrition. What childhood memories of food do you have?

**Healthy People 2000: National Health Promotion and Disease Prevention Objectives.* Public Health Service, U.S. Department of Health and Human Services, Washington, DC, 1991, Objectives 2.12, 2.17, and 2.19, pp. 94–95.

Foods and their nutrients are essential to life. In the early years of life the child's nutritional health depends on the family unit. The parents must have knowledge of the changing food needs of the child and must also have sufficient resources to provide food, shelter, and clothing for the family. Equally important, parents create the cultural and psychological environment that influences the development of food habits, setting the pattern for later years.

During the preschool years some children depend solely on family caregivers for their nutritional needs. For other preschool children the responsibility for meeting nutritional needs is shared by the family and others, such as caregivers in child-care centers and babysitters. The child entering school becomes influenced by teachers and peers, and learns to broaden his or her experiences with food. For many people, the adolescent years often are turbulent as the teenager seeks independence and freedom from adult rules and standards.

Infants

GROWTH AND DEVELOPMENT

Body Size

Infants vary widely in their growth patterns, so it is unwise to compare one infant with another. Yet there is value in being familiar with typical patterns of growth and development. On average, infants double their birth weight during the first four months and triple their birth weight by the age of 10–12 months. The baby is born with a proportionally large head and short arms and legs. During the first year, the body proportions change, and continue to do so in early childhood. The trunk, arms, and legs become longer.

A series of growth charts have been developed for normal ranges of weight, length, and head circumference. (See Appendix Figures D-1 to D-4.) With these charts, parents and the pediatric health care team plot an infant's measurements and growth progress compared with infants of the same gender and age. If the infant moves along a given percentile rating, individual progress is likely to be satisfactory for that infant—although a short deviation into another percentile channel usually is not important. However, an infant who dramatically or steadily moves into a different channel, either higher or lower, and remains there, prompts identification of the cause. For example, if the infant suddenly drops into a lower channel, eating problems, negative environmental situations, or perhaps an illness that is interfering with growth, should be explored.

Body Composition

At birth about 75 percent of the infant's weight is water and 12–15 percent is fat. The relatively high amount of water, low amount of subcutaneous fat, and the proportionately large surface area explain why precautions must be taken to keep the infant well hydrated and warm. By the end of the first year, the body water content is about 60 percent and fat is 24 percent.

Bones are quite "soft" at birth; mineralization of the bone protein matrix continues throughout the childhood and teenage years into young adulthood. The body calcium content is three times as great at the end of the first year as it was at birth.

At birth, full-term healthy infants have a hemoglobin level of 14–27 g/dl (8.7–16.8 mmol/L). This high level provides sufficient iron stores for the first several months when it is expected that the baby is breast-fed human milk containing small amounts of iron. By three months of age, the hemoglobin level is about 10–17 g/dl (6.2–10.6 mmol/L).

Organ Development

Most brain development occurs during fetal life and the first five to six months. The brain reaches about 90 percent of its adult size by four years. Severe malnutrition during pregnancy and the first months of life leads to inadequate development of the central nervous system. The poorly nourished infant and child may never reach full mental potential.

The newborn's stomach has a capacity of about 30 ml, and at one year can hold about 240 ml. The ability to digest simple sugars (like lactose), emulsified fats, and protein is present at birth in the full-term infant. The pancreas gradually increases its secretion of amylase and lipase during the first year so that complex carbohydrates and fats can be given.

The kidneys achieve their full functional capacity by the end of the first year. Young infants are unable to excrete high levels of waste that results from feeding undiluted cow's milk, or concentrated infant formulas, or inadequate water.

Feeding Abilities—Psychological and Physiological

Parent-infant bonding (the close mother-infant—or father-infant—relationship that is established when the baby is cuddled while being fed) begins shortly after birth. This close relationship influences desirable feeding behaviors and also sets the stage for developing social, emotional, and psychological values that continue from infancy into childhood. Mother-infant bonding is developed equally well whether the baby is breast- or bottle-fed. Formula-fed babies can also develop a close bond when fed and cuddled by the father. When parents are unavailable, a consistent surrogate caregiver to feed the baby must be found.

Physical development of feeding behaviors depends on the nervous control of the muscles of the mouth, the coordination of eyes and hands, and the ability to hold and grasp objects, and to sit up. Infants vary widely in the ages at which each stage of development occurs. (See Figure 13–1.)

At birth, infants are able to coordinate breathing, suckling, and swallowing. Eyes do not yet focus, but the baby is able to "root" for food. This **rooting reflex** is a response to touch that causes infants to turn toward the nipple to find it. If the mother strokes the baby's cheek or lip closest to the nipple, the baby will turn in that direction and seek the nipple. The **extrusion reflex** is exhibited during the first two or three months when solid food is placed on the tongue and the tongue pushes it out of the mouth. Clearly the baby is not yet ready for solid foods. This is a natural response and has nothing to do with the baby's like or dislike for a food.

At three to four months the tongue movements change so that gradually solid food can be given. Most pediatric team members recommend that solid foods be introduced between four to six months, depending on each infant's nutrient needs and feeding abilities.

At about six months babies begin to grasp objects and put them into their mouths. They sit up and open their mouths when they see food approaching. They can clean food off their spoons and move the food from the front to the

BIRTH TO 6 MONTHS	SEVEN TO 10 MONTHS
Roots for nipple	Bite matures
Suckles	Rotary chewing
	Transfers food from side to side
FIVE TO 7 MONTHS	Curves lips around cup
Begins to sit	
Follows food with eyes	**TWELVE MONTHS**
Lips close over spoon	Sociability increases
Begins to swallow solid foods	Interest in solids increases
	Cup drinking improves
SIX TO 8 MONTHS	
Tongue moves laterally	**SECOND YEAR AND BEYOND**
Controls position of food in mouth	Circular rotary chewing
Controls swallow	No pause in side-to-side transfer
Munches	Begins to use utensils

FIGURE 13–1
Feeding capabilities in the infant and toddler. (*Source:* Satter, E.: "The Feeding Relationship: Implications for Obesity." Reprinted with permission from *Food & Nutr. News,* 56(4):24, 1984, published by National Livestock & Meat Board.)

back of the tongue. Teeth begin to erupt about the fifth to sixth months. Iron-fortified cereals mixed with a little formula or vitamin C–containing fruit juices are good choices at this time to replenish the baby's iron stores.

During the second six months babies become familiar with the taste, aroma, and texture of a variety of fruits, vegetables, and meats. Babies progress from semisolid to more solid foods, then from strained to mashed to chopped foods. As manual dexterity develops, foods are picked up with the hands and fingers, put into the mouth, munched or chewed, and swallowed. By the end of the first year feeding becomes increasingly messy. Parents need to be taught that infants learn by feeling their food, and that bringing food from hand to mouth is not yet a fully coordinated skill.

MEETING NUTRITION NEEDS

Recommended Dietary Allowances

The Recommended Dietary Allowances (RDAs) for the first six months are set at the level that a healthy, well-nourished infant would receive from breast-feeding. For the second six months, the levels of nutrients are based on satisfactory growth on a formula plus solid foods. (See tables inside front and back covers.)

Nutrient Highlights

Water The surface area of infants is large in proportion to the total weight, and considerable amounts of water are lost daily through the skin. Young infants need additional water to excrete wastes, especially by the immature kidneys. About 150 ml/100 kcal is the recommended fluid intake. Breast milk supplies the fluid requirement as does a formula diluted to give about 20 kcal/oz. To avoid the ever-present risk of dehydration, plain water should be given as dictated by climate and whenever babies have fevers, vomiting, or diarrhea.

Energy The energy requirement is high during the first year in order to support a high metabolic rate, rapid growth, and vigorous activity. The requirements vary widely from infant to infant. Under normal circumstances, infants are able to self-regulate their intakes; this innate ability must be allowed to flourish.

Carbohydrate and Fat No RDAs have been set for carbohydrate and fat. Almost half of the calories from human milk are supplied by fat. Human milk has about 6–9 percent linoleic acid, an essential fatty acid. With emphasis on lower fat intakes by the American public, there have been misguided efforts to lower the fat intake of infants. A lower fat intake by infants would necessitate increasing the protein and carbohydrate intakes to supply sufficient calories to support growth. A substantial increase in the protein intake results in an additional burden to excrete nitrogenous wastes—a burden infants' kidneys may not be able to take on. Nonfat, low-fat, or reduced-fat milks are not recommended for the first two years.

Protein Each day infants add from 3.0 to 3.5 g protein to their bodies for a total of about 1,200 g during the first year. At birth the protein content of the body is about 11 percent; by the end of the first year this has risen to 15 percent. Based on per kilogram body weight, the infant's requirement is about twice the adult allowance.

Vitamins All infants should receive a single dose of vitamin K immediately upon birth. Aside from that, human milk supplies the vitamins needed by infants, except possibly vitamin D. If exposure to sunshine is inadequate, a vitamin D supplement usually is recommended. Commercial formulas provide recommended levels of all vitamins.

Minerals Iron is the element requiring special attention during the first year. Full-term infants are born with adequate iron stores for the first four to six months. Human milk is low in iron but about 50 percent of the iron in breast milk is absorbed—it is highly bioavailable. Healthy breast-fed infants rarely develop iron-deficiency anemia. After four to six months the daily iron need is supplied by iron-fortified formula or iron-fortified cereal. Starting at birth, infants not breast-fed should have a daily iron intake of about 1 mg/kg of body weight, depending on the iron bioavailability in the formula selected. Human milk or formulas supply the recommended allowances for calcium, phosphorus, magnesium, and other minerals. Some pediatricians and pedodontists recommend a fluoride supplement for both breast- and formula-fed infants in areas where the water fluoride content is low.

SPECIAL FOOD AND NUTRITION ISSUES

Breast-Feeding

"Breast-feeding is best" is a teaching slogan used by health care providers. Currently, the number of infants *started* on breast-feeding is about the same number as those who are bottle-fed. (About 50 percent each. See Chapter 12.) Some cultures expect new mothers to breast-feed and actively support their efforts, but others do not. Although many new mothers start breast-feeding, for a number of reasons they discontinue it within a few weeks. This fact is troublesome. Many efforts are being made by members of the health and education disciplines to teach and support families in their breast-feeding endeavors.

Advantages of Breast-Feeding Human milk is free of contamination by disease-producing organisms, is instantly available at the right temperature for the infant, is nutritionally correct for healthy infants, and is usually less costly to produce even though the mother requires additional foods to support lactation. Importantly, human milk supplies immune factors that protect the infant against infections. Human milk contains the **bifidus factor** that promotes growth of desirable bacteria in the intestinal tract.

Contraindications From the standpoint of the mother, breast-feeding is contraindicated when she is afflicted with diseases that can impair her production of milk, or when it places an undue burden on her own health status, or if a substance can be passed from her to the baby through her milk. Each mother, infant, and situation must be assessed individually. Such conditions include tuberculosis, hepatitis B, and, in the United States, HIV/AIDS. Drugs used by drug abusers as well as many prescription medications are secreted into the woman's milk and have undesirable effects on the infant—including causing drug addition in the baby. A woman taking any drugs/medications whatsoever needs to be alerted about any possible harm that can come to her infant through such use.

Mothers employed outside the home often discontinue breast-feeding because there are no child-care facilities at the workplace. In this case, sometimes it is possible for breast milk to be expressed manually or by a breast pump, safely stored, and given later to the infant when under the care of others. Sometimes family members, including fathers, do not encourage the continuance of lactation. Many young mothers are unable to breast-feed because their own nutrition status has already been compromised or they do not see its relevancy. Individual solutions to each of these problems need to be sought so that breast-feeding can be achieved—for the sake of the infant and the mother.

From the standpoint of the infant, some genetic disorders interfere with the normal metabolism of either human or milk-based formula, for example, galactosemia. (See Chapter 27.) Some conditions seen in infancy, such as phenylketonuria, cystic fibrosis, bronchopulmonary dysplasia, diabetes mellitus, and cerebral palsy, now are treated by breast-feeding augmented by specialized formulas or adaptive feeding devices.

Initiation of Breast-Feeding So that mother-infant bonding can occur immediately after birth, the infant and mother usually remain close by each other for the first 24 hours. Much depends on the location (home or institutional setting) of the birth and the length of stay in a hospital or birthing center.

The nurse or lactation specialist can help the mother and the infant as breast-feeding is initiated by positioning the two of them and by showing the mother how to stroke the little cheek to encourage the baby to seek the breast. The infant should grasp the **areola** (the pigmented area that surrounds the nipple), not just the nipple, in its mouth. At first the baby will receive 15–45 ml of colostrum. **Colostrum** (the thin, yellowish fluid first secreted by the breast) supplies small amounts of nutrients but is important for the immune factors it supplies. This early stimulation helps to initiate milk flow—the so-called let-down reflex. (See Figure 13–2.) Self-demand feeding permits the baby to breast-feed when hungry, rather than according to an arbitrary adult-set time schedule. The mother soon learns to recognize when the baby is hungry and can tell the difference between crying about hunger or some other discomfort. Infants breast-feed as often as every two hours during the first weeks, but soon self-

FIGURE 13–2
The strong bond developed between mother and infant is one of many advantages of breast-feeding. Here the older child has become part of a warm relationship. (*Source:* Health Education Associates, Glenside, Pennsylvania, and Charles M. Cadwell, photographer.)

regulate to an approximate three- or four-hour schedule. About the second month, the baby begins to sleep through the 2 or 3 A.M. feeding. By five months, the baby no longer awakens for a feeding at 10 to 11 P.M.

Weaning Some infants breast-feed for up to a year; others wean to one or two bottles of formula within a few months if the mother's milk supply is insufficient or if the mother is returning to employment or school. Weaning from the breast is accomplished gradually. A bottle or cup feeding is substituted at a convenient feeding time. When the baby has become accustomed to this—after about a week or two—a second bottle or cup is offered. As long as two to three months is needed for full weaning.

Formula Feeding

Human and Cow's Milk Compared The many differences in the composition of human and cow's milk are shown in Table 13–1. Cow's milk contains almost three times as much protein, more than three times as much calcium, and more than six times as much phosphorus. Most of the protein in cow's milk is casein, while that in human milk is, for the most part, lactalbumin. Casein produces large, rubbery curds in the stomach, whereas lactalbumin curds in the

TABLE 13–1

Comparison of Human Milk, Cow's Milk, and Milk-Based Formula

Nutrient	Human Milk per 1,000 ml[a]	Cow's Milk (Whole) per 1,000 ml	Milk-Based Formula per 1,000 ml
Fluid			
Water, ml	897	894	875
Energy and Macronutrients			
Energy, kcal	718	620	670
Protein, g	10.6	33.4	15–16
Fat, g	44.9	33.9	33–37
Carbohydrate, g	70.6	47.3	70–72
Vitamins			
Vitamin A, RE	656	315	340–500
Vitamin A, IU	2,470	1,279	1,700–2,500
Vitamin D, µg		10[b]	10
Vitamin E, mg TE	1.3–3.3	5.7	5.7–8.5
Vitamin C, mg	51	10	55
Thiamin, mg	0.14	0.39	0.4–0.7
Riboflavin, mg	0.37	1.65	0.6–1.0
Niacin, mg NE	2.0	0.85	7–9
Vitamin B_6, mg	0.11	0.43	0.3–0.4
Vitamin B_{12}, µg	0.46	3.63	1.5–2.0
Folate, µg	51	51	50–100
Minerals			
Calcium, mg	328	1,208	550–600
Phosphorus, mg	144	945	440–460
Sodium, mg[c]	141	498	250–390
Potassium, mg[c]	523	1,544	620–1,000
Magnesium, mg	31	132	40–50
Iodine, µg	30–100		40–70
Iron, mg	0.3	0.5	1.4–12.5[d]
Zinc, mg	1.8	3.9	2.0–4.0

Source: Robinson, CH, et al: *Normal and Therapeutic Nutrition,* 17th ed. New York: Macmillan Publishing Company, Inc., 1986, p. 283.

[a] One liter of human milk = 1.025 g; 1 liter of cow's milk = 1.017 g.

[b] Assumes fortification of cow's milk with 10 µg vitamin D.

[c] Allowances for sodium and potassium ranges are considered to be safe and adequate.

[d] Values for formula not fortified and fortified with iron.

stomach are smaller and softer. Thus, lactalbumin curds are more accessible to the digestive enzymes and more completely digested. Cow's milk furnishes slightly less fat and lactose. Ounce for ounce, both milks supply about 20 kcal.

Proprietary Formulas Commercial formulas comprise about 90 percent of all formulas used in infant feeding. Cow's milk is the usual base ingredient for these formulas. Formulas are developed to resemble the composition of human milk. For example, the vitamin and mineral levels are adjusted to meet the RDAs for different age groups of infants and vegetable oil is substituted for the butterfat of cow's milk to furnish a higher intake of linoleic acid.

Special formulas are available for therapeutic purposes. For infants who are allergic to milk, nutritionally adequate formulas of soybean or protein hydrolysates are used. Enzyme disorders, such as galactosemia or lactase deficiency, require soybean or predigested formulas. Formulas are available for each specific inborn error of metabolism, such as phenylketonuria and maple syrup urine disease. A variety of formulas are marketed for preterm infants.

Formulas are available in several forms and concentrations: single strength, ready-to-feed, in quart cans, or in 4-oz or 8-oz disposable ready-to-feed bottles; concentrated liquid, which is measured into sterilized bottles and diluted with boiled or clean tap water; and powdered formula, to be diluted with water or added to other formulations as a nutrient/energy booster. (See Figure 13–3.)

The Food and Drug Administration (FDA) regulates the composition of infant formulas. All formulas must provide minimum levels of all essential nutrients, and must be produced under strict standards so that formulas are wholesome and safe.

Solid Foods

A sufficient supply of human milk or formula meets the nutrition needs of the full-term infant for four to six months. At this age, the infant has better control of the tongue movements and swallowing mechanisms for handling solid foods. Moreover, there is less likelihood of allergic reactions to food.

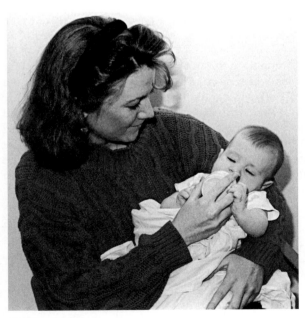

FIGURE 13–3
Numerous formulas are available to meet the nutrient needs of all babies. (*Source:* Drexel University and The Children's Hospital of Philadelphia, and Peter Groesbeck, photographer.)

Iron Iron supplementation is required during the second half of the first year. This can be supplied as iron-fortified formula used until the end of the first year, or as iron-fortified cereal.

Sugar Infants, like adults, happily respond to sweet foods. Parents and other caregivers may be tempted to bring this pleasure to the baby by offering sweet foods. Too much sugar, however, could displace needed foods supplying proteins, vitamins, and minerals. It can upset water balance (osmotic load) in the intestines and lead to diarrhea. It has been speculated to lead to a habit that demands sweet foods in later years. The sugar content of commercially prepared infant foods has been reduced. Fruit *juices* should be given to infants and not "fruit punch" or "fruit drinks" or other sweetened beverages like iced tea or soda.

Salt Babies require some sodium and, also like adults, they respond to the taste of salt. During the early months breast milk or formula supplies the sodium. Thereafter, supplementary foods provide increasing amounts of sodium. Far too often, parents and caregivers provide salt in excess of the baby's needs. The concerns are that unnecessary use increases the excretory load on the immature kidneys to excrete it and may accustom the infant to ingesting high levels that are continued throughout life.

Commercial vs. Home-Prepared Baby Foods Commercial baby foods are widely used. They save time, are consistent in nutritive value from jar to jar, and are as nourishing or more so than similar home-prepared foods. Many options are available, some better than others. Parents and other caregivers need to read labels for ingredient and nutrient listings. The more sugar, corn syrup, starch, and flour contained, the less nutrients will be provided per calorie. For example, strained or chopped meats, poultry, and vegetables have more nutritive value than "dinners"; plain fruits more than "desserts." Cost per jar is no indicator of nourishment.

Many parents derive satisfaction from the home preparation of foods for their baby. These foods usually cost less, offer more variations in texture, and accustom babies to foods of their culture. If parents choose to prepare their own baby foods, some education is necessary, especially regarding food sanitation and preservation of nutritive value. For example, foods should be quickly refrigerated or frozen, not left out at room temperature; fruits and vegetables should be cooked in minimum water, steamed, or microwaved. Although adult foods can be made into a consistency an infant is able to take, the ingredients used, preparation methods, and sanitation standards certainly may not be appropriate for a baby.

Sequence of Feeding Practices vary widely in the sequence with which foods are added and the age at which they are introduced. A typical sequence and suggested daily schedule are shown in Tables 13 – 2 and 13 – 3.

Food Habits

Nutrition and eating habits are developed in infancy and early childhood. Sound food habits foster flexible eating habits and enjoyment of a wide variety of foods, then and later in life. Parents have a wonderful opportunity and a tremendous responsibility for laying the foundation. Some guidance you might provide includes the following advice:

TABLE 13–2
Typical Sequence for Food Additions to Infant's Diet

Age	Food Additions
4–6 months (bottle-fed) 5–6 months (breast-fed)	Iron-fortified rice cereal. Mix precooked cereal with formula for thin consistency at first. Increase to 2 to 5 tablespoons by 7–8 months. Orange juice. Increase gradually to ½ cup. May be diluted with water at first. May be given from a cup. Mashed ripe banana, applesauce, strained pears, apricots, prunes, or peaches. Start with 1 teaspoon and increase to 3 to 4 tablespoons by 1 year. Strained vegetables: asparagus, green beans, carrots, peas, spinach, squash, or tomatoes. Start with 1 teaspoon and increase to 3 to 4 tablespoons by end of the year.
6–7 months	Strained meats and egg yolks. Mash hard-cooked egg yolk with a little formula, using ¼ teaspoon at first. Increase gradually. Plain yogurt. Avoid egg white until 10 months or so since it often gives allergic reactions.
5–8 months	Crisp toast, zwieback, teething biscuits; potato.
8–10 months	Chopped vegetables and fruits; pasta, rice.
10–12 months	Whole egg; plain puddings such as custard, Junket; cooked and mashed dried beans or peas; cottage cheese or soft cheese.

- Hold and soothe the baby while breast- or bottle-feeding. This provides feelings of satisfaction, security, and warmth. Propping up the baby in the crib to go to sleep with the bottle in the mouth increases the chances of aspiration. This feeding style likely leads to rampant dental caries in the developing teeth, a condition known as baby bottle tooth decay. Fruit juices or sweetened beverages cause the same problem. Mix cereals with a little formula and serve from a spoon as solid food, instead of feeding by bottle. After breast milk or plain formula, water is the fluid of choice.

- Regulate the feeding schedule according to the baby, not the clock. Give new foods at the beginning of the meal when the baby is hungry. Babies, like adults, are hungrier at some times more than others. Do not expect babies to finish every bottle or every food at every meal.

- Introduce only one new food at a time. Serve only small portions of a new food; a taste is enough. Wait four to five days before trying a second new food. In this way, detection of food intolerance or allergic reactions is possible.

- Express your enjoyment about food. Do not show your own dislike of a food by the expression on your face, remarks you say, or refusal to eat the food yourself.

- Expect the baby and toddler to touch and feel most foods and be messy. Spilling accidents will occur. No need to scold about developmental skills not yet practiced, nor negate learning that occurs by mistakes. Use a cup that doesn't tip over easily, a deep bowl with rounded edges, and a spoon that can be managed better by babies and toddlers. Provide safe and comfortable seating. (See Figure 13–4.)

TABLE 13–3

Typical Schedules for Six-Month-Old and Year-Old Infants

Six-Month-Old Infant		Year-Old Infant	
6 A.M.	Human milk or formula,[a] 7–8 oz	*Breakfast*	Cereal, 2–5 tablespoons Milk, 8 oz Chopped fruit, 2–4 tablespoons
8 A.M.	Orange juice, 3 oz		
10 A.M.	Human milk or formula,[a] 7–8 oz Cereal, 2–3 tablespoons	*Lunch*	Chopped meat, ½–1 oz *or* Egg, 1 Potato, 2–4 tablespoons Chopped vegetable, 2–4 tablespoons Milk, 8 oz
2 P.M.	Human milk or formula,[a] 7–8 oz Egg yolk, ½–1 Vegetable, ¼ to 2 tablespoons		
6 P.M.	Human milk or formula,[a] 7–8 oz Cereal, 2–3 tablespoons	*Snack*	Milk, ½ cup Zwieback *or* Teething crackers
10 P.M.	Human milk or formula,[a] 6–7 oz	*Dinner*	Cereal *or* Potato, 2–5 tablespoons Milk, ½ cup Chopped fruit, 2–3 tablespoons Toast *or* Zwieback
		Bedtime	Milk, 8 oz

[a] Human milk is recommended throughout the first year; otherwise iron-fortified formula is recommended as a substitute.

FIGURE 13–4

This toddler enjoys feeding herself. Touching food is important at this stage. (*Source:* Barbara C. Schwartz.)

GROWTH AND DEVELOPMENT

During these years, children's growth patterns vary widely. Some children, by heredity, are destined to be short and stocky, others tall and slender. Some children will have growth spurts at an earlier age than others. As with infants, satisfactory progress is best determined by following the child's own measurements from month to month, year to year. (See Appendix Figures D-1 to D-4.)

During the second year, the toddler's weight gain is about 3.5–4.5 kg (8–10 lb). From the second birthday to the ninth year, the increase in height and weight is at a much slower rate. For instance, the annual gain in weight drops to 2–3 kg (4–7 lb).

Many physical and behavioral changes set the stage for maturation in the nutritional area. By two years, children have lost most of their "baby fat." The next years show muscles increasing in size and firmness, and bones becoming stronger. Changes in body proportions, motor coordination, and mental development progress in orderly fashion for each healthy child, although the chronologic ages at which these changes occur vary from child to child.

MEETING NUTRITION NEEDS

Recommended Dietary Allowances

During these years, nutrient requirements vary from child to child, depending on individual variability. Interestingly, based on body weight, nutrient requirements of children are higher than those of adults. This is accounted for mainly because of growth—children are in a state of positive nitrogen balance. Refer to the inside front and back covers of the text for the RDAs of specific nutrients and energy.

Food Choices

Table 13–4 demonstrates possible food choices modified to provide the nutrients needed by children (and teenagers). Observe the two columns designated for children. (See also Tables 2–3 and 2–4.) Children require snacks to provide for their relatively high energy needs and to avoid excessive hunger at mealtimes. Meal patterns and menu structures usually entail minimeals and well-balanced snacks.

Young children prefer plain, simply flavored foods that are only lightly seasoned. Mixtures, as in casseroles, are well accepted as the child becomes older. Foods requiring chewing are essential for oral development of strong bones, teeth, and mouth muscles. These body structures are crucial for proper speech development. Foods too tough for primary teeth to chew, like meats, are better chopped for the preschool child and for the child whose secondary teeth are erupting. Lukewarm rather than steaming hot foods are preferred. Strongly flavored fruits and vegetables may be more acceptable in later school years.

Children sometimes go on food jags. They will eat only certain foods, for instance, peanut butter and jelly sandwiches. Usually these diversions of appetite do not last too long if the parents (and grandparents and other adults) make no particular point of them. If milk is refused as a beverage, it can be served as cooked cereals or puddings. Often it will be accepted again if it is poured into a decorated mug, or occasionally is flavored, or is colored with bright fruit purees

TABLE 13–4

Examples of Food Selections for Children and Teenagers

	CHILDREN		TEENAGERS	
Foods	*1–6 years*	*7–10 years*	*11–14 years*	*15–18 years*
Bread, rice, pasta, whole-grain or enriched	1–3 servings[a]	3 servings[a] or more	3 servings[a] or more	4–6 servings[a] or more
Cereal, whole-grain or enriched	⅓–½ cup	½ cup	¾ cup	1 cup or more
Vegetables				
Raw	Small serving[b]	Small serving[b]	⅓ cup	½ cup
Cooked (green leafy or yellow 3 to 4 times a week)	1–4 tbsp	¼ to ½ cup	½ to 1 cup	1 cup or more
Potato, white or sweet, or other starchy vegetable	2–4 tbsp	Small, or ⅓ cup	Medium, or ½ cup	Large, or ¾ cup
Fruits				
Citrus (or other good source of vitamin C)	⅓–½ cup	½ cup	½ cup	½ cup
Other	¼–½ cup	2–3 portions such as apple, banana, peach, pear		
Milk, reduced-fat (vitamins A and D fortified) (not skim)	2–3 cups	2–3 cups	3–4 cups	4 cups or more
Lean meat, Poultry, Fish	1–4 tbsp, chopped[b]	2–3 oz	3–4 oz	4 oz
Dried beans, Peas, Peanut butter	2–3 servings a week or more. Use ½ cup cooked dried beans or peas or 2 tablespoons peanut butter for 1 oz meat			
Eggs	2–3 weekly	2–3 weekly	2–3 weekly	2–3 weekly

[a] One serving equals 1 slice of bread or ½ cup rice or pasta.

[b] Young children should not be given chunks of meat, raw carrots, nuts, or popcorn on which they may choke.

or juices. Plain yogurt and mild cheeses are good substitutes. Children enjoy some (not always a lot of) variety to encourage wider experiences, to be recognized as being special, or to prevent boredom. As desire for independence increases and skills become better, children relish helping to buy, prepare, serve, and then *eat* those personalized foods.

A child who has reached school age may have developed new food likes, but may face other problems related to maintaining good nutrition. Mornings in many homes are too often rushed, so that breakfast is a hurried meal or may be skipped entirely. A child who is apprehensive about school may eat poorly at lunch. A short lunch period may be upsetting to the slow eater. Extremely active children may become unduly tired before meals.

Food Habits

The childhood years strengthen the development of health-promoting food habits started in infancy. During childhood, many food behavior problems are just the result of adult inattention to normal growth and development milestones. Plus, adults often fall into the trap of using food for non-nutritional purposes. For example, a child may be given a sweet as a reward or a bribe. Or dessert is withheld because the child was not a "clean-plater." Or the child is sent away from the table for misbehavior. Children associate cake, candy, soda, and snack foods with happy occasions, but are told that "soda is bad for you" or "candy will rot your teeth." The child is rightly confused when told that a particular food is bad, but is given it as a reward. Other times children even refuse a favorite food or dawdle over meals to gain attention.

FIGURE 13–5
Preschool children learn to taste new foods. They also imitate their peers in food acceptance. (*Source:* U.S. Department of Agriculture.)

Some tips you might provide to adults caring for children are (see Figure 13 - 5):

■ Serve meals in a pleasant place and in a calm unhurried atmosphere. Allow for a period of quiet and rest before meals if the child has become excited and fatigued. Allow sufficient time for meals in pleasant surroundings.

■ Encourage table manners and interaction with others that are appropriate for the child's age. Expect that children occasionally will spill beverages or knock dishes off the table. Avoid letting mealtimes slip into a battleground between children and adults, or other children.

■ Use foods that are colorful, varied in texture and flavor, and attractively served. Introduce small servings of new foods—gradually. Provide eating utensils and dishes that are safe and easy to hold and to use. Many vegetables, fruits, meats, and bread work out better as finger foods.

■ Remember that appetites decrease as growth rates slow between infancy and adolescence. Remember, too, that developmental milestones often dictate children's food and eating habits. For example, children refuse foods as a way of asserting their independence; these stages start in infancy and continue onward. Try to accept that much of these refusals just means that your child is getting older!

SPECIAL FOOD AND NUTRITION ISSUES

Most children grow up healthy and well nourished. But along the path of childhood, some children fail to achieve their full potential because of inappropriate food habits, an inadequate supply of certain nutrients, and/or an unhealthy environment.

Food Asphyxiation

Food asphyxiation can be fatal. Choking is caused when round, hard, or sharp pieces of food are inhaled into the airway and get lodged so that the supply of air to the lungs is cut off. Children under three years are at greatest risk. Some have delayed tooth development, so chewing and swallowing foods are difficult.

Round pieces of food, such as hot dogs, sausages, grapes, jelly beans, gumdrops, popcorn, nuts, and raw carrots are among the foods that have caused choking. Cooked carrots, hot dogs, and meats cut up lengthwise in small pieces, and grapes cut in quarters are less likely to cause difficulty.

Adults should supervise children when they are eating so that any incident can be attended to promptly. When eating, children need to sit upright and not run around.

Lead Poisoning

Children who live in old houses where interior surfaces were covered with lead-based paint are most at risk. Toddlers often chew on the painted surfaces, on paint chips, and on plaster, thus ingesting harmful amounts of lead. Lead also leaches out of lead plumbing joints in water pipes in homes, schools, and worksites; all children and adults are susceptible (See Chapter 11.)

Iron-Deficiency Anemia

About 9 percent of one- to two-year-old children and 4 percent of three- to ten-year-old children have iron-deficiency anemia. Most at risk are children and teenagers, especially girls, who have restricted food preferences. When screening food intake records, look for excessive intakes of milk and limited intakes of iron-rich foods. (See Chapters 10 and 11.)

Growth Failure

Because of children's high metabolic needs, growth failure is a common problem—albeit often preventable—seen by people in the health disciplines. In these situations, complex socioeconomic, cultural, and psychological problems— often combined with political upheaval and imperfect health care and educational systems—prevent the optimal nourishment, growth, and development of children. (See Figure 13–6.)

The most obvious problem is the protein-energy malnutrition (PEM) experienced by children as a consequence of famine, war, political unrest, and poverty. (see Chapter 8.)

Some children fail to grow because their parents or other caregivers do not know what nutrients children require or what foods children need. Research shows that children living in emotionally tense family environments in which there is little expression of affection often have poor intakes of food. Sadly, inadequate nutrition and growth failure are outcomes of child neglect and abuse.

Children who have acute or chronic medical conditions are at special risk. Missing the critical physiological periods of growth and development because of one or more abnormalities in nutrition metabolism results in permanent growth failure. The earlier it occurs and the longer it lasts, the more impossible it is for "catch up" growth to occur. Examples abound. The taste and smell of food are altered in children with upper respiratory infections; the child with recurring infections usually has a poor intake of food. Children who are physically or

FIGURE 13–6
The child on the right is 3 years younger than the child on the left. The younger child appears to be better nourished and cared for. (*Source:* Chavez/UNICEF.)

mentally challenged sometimes need modification of the eating environment. Other children have chronic health problems such as malabsorption disorders, diabetes mellitus, and liver, heart, or renal diseases that interfere with digestion, absorption, utilization, storage, or excretion. Other children harbor intestinal parasites. For children with diseases having a nutritional component, coordinated skills are required of the family, interdisciplinary pediatric health care teams, teachers, and community child advocates.

Vegetarian Diets A special circumstance of growth failure, which can lead to death, sometimes is seen in children being fed vegetarian diets. Young children have a limited stomach capacity and are unable to consume enough bulky foods to meet their energy and protein needs. Also, the high-fiber content of plant food diets reduces the absorption of minerals. When strict vegetarian (vegan) diets are given to children, soy milk fortified with calcium, vitamin D, and vitamin B_{12} should be used. Depending on the age of the child and food choices, among other factors, perhaps riboflavin, magnesium, zinc, and other vitamins and minerals need to be supplemented, also.

Sometimes insufficient lipids are included; addition of suitable amounts of vegetable oils containing sufficient total fat and essential fatty acids is necessary. Legumes, nuts, nut butters, tofu, and textured vegetable proteins provide relatively concentrated sources of energy and are also important for their other nutrient contributions. (See Chapter 7.)

Obesity

Research indicates that about 14 percent of obese infants, 40 percent of obese 7-year-olds, and 70 percent of obese 10- to 14-year-olds become obese adults.

From children's points of view obesity is a serious problem pertaining to their sense of self-worth and acceptance by their peers. Obese children are teased by their playmates. They are left out of games and athletic activities, and become increasingly inactive. Far too often, parents chide their children about their fatness and inactivity. Some children spend hours watching television or playing computer games, and as they do so, they make frequent trips to the kitchen.

The treatment of obesity is a family and neighborhood matter. The young obese child is dependent upon parents, other adults, and sometimes siblings for foods. Safe and enjoyable physical activities not requiring competition with peers are best to emphasize, such as walking, jumping rope, playing ball, and bike riding. Given the chance, most children enthusiastically "just like to play." (See Chapters 8 and 9.) One important aspect about pediatric intervention is that children are growing and thus must have nutrients to meet all requirements. (See RDA tables inside front and back covers.)

Hyperactivity

The diagnosis of "hyperactivity" (attention deficit disorder [ADD] or attention deficit hyperactivity deficit [ADHD]) occasionally is made when a child enters preschool or elementary school and is characterized by the teacher as a behavior problem. Some researchers claim that a diet free of artificial colorings and flavorings, salicylates, refined foods, and/or sucrose leads to marked improvement in the abnormal activity level or sleeping patterns of the child. Over the years, controlled research studies have failed to substantiate the proposed theory. Based on each child's medical, family, and nutritional history, suitable interventions should be determined. Whatever regimen is selected, it must be nutritionally and developmentally child-appropriate.

Preadolescents and Adolescents

GROWTH AND DEVELOPMENT

A third and final period of rapid growth and metabolism occurs during the teenage years. (The first was during fetal development and the second was during the first six months of life.) Teenagers have many interests in their physical and emotional development. For example, important topics are the size and shape of their bodies; their complexion; their overall appearance; and physical activity—but now more energized, peer-acceptable, and structured. Girls express a particular need for a good figure, healthy skin, and beautiful hair. Boys are more likely to be interested in tall stature, muscular development, and athletic vigor and stamina.

MEETING NUTRITION NEEDS

Recommended Dietary Allowances

The Recommended Dietary Allowances (RDAs) for calories for teenage boys and girls are somewhat higher than those for adult men and women, with corresponding higher allowances for most vitamins and minerals. These increases are to satisfy the increased nutrient needs caused by their adolescent growth spurt. (See tables inside front and back covers.)

Teenage girls who become pregnant must meet their own high nutritional requirements for a maturing body as well as those for a successful pregnancy. (See Chapter 12.)

Food Choices

Table 13–4 (p. 242) demonstrates possible food choices modified to provide the nutrients needed by teenagers (and children). This time observe the two columns designated for teenagers. (See Tables 2–3 and 2–4, also.) Surveys repeatedly indicate that diets of teenage boys and girls most frequently fail to meet the RDAs for vitamins A and C and the mineral calcium. Girls often do not receive enough iron. Of the Food Guide Pyramid, the dairy group needs special emphasis because of its great contribution to meeting teenagers' requirements for calcium. This time period is of immense importance to providing sufficient calcium storage in the bones. Milk consumption among girls sharply decreases after age 12 years. (Nearly half of American women over age 35 have completely stopped drinking milk.) Vitamins A and C intake would be substantially improved if leafy dark-green, and deep-yellow vegetables and citrus fruits were consumed more frequently. All these foods have high nutrient densities.

Food Habits

From the previous paragraph, the impression may be that all teenagers are poorly nourished and always eat great quantities of high-calorie, nutrient-sparse snacks. In fact, many teenagers have good food habits, are well nourished, and might serve as role models for others in their age group who need to improve their food habits. Perhaps we have not sufficiently appealed to teenagers themselves to be those peers!

SPECIAL FOOD AND NUTRITION ISSUES

Continuation of Previous Issues

Many of the same issues faced by children continue into adolescence. For example, teenagers are vulnerable to anemia because of their rapid growth, especially if they eat limited iron-containing diets. Teenage girls are especially vulnerable to loss of iron during menstruation. Chronic diseases sometimes become less stable because of hormonal changes or progressive deterioration. Family situations may improve or become more chaotic.

Some teenagers voluntarily restrict their food intake. Eating disorders such as anorexia nervosa and bulimia can interrupt normal growth processes in teenagers. Some, in the attempt to acquire the "right" figure or physique, adhere to overly strenuous exercise regimens. Others combine rigid exercise programs with crash diets to "meet weight" prior to an athletic event. For sports team hopefuls, bizarre eating habits, food choices, and nutrient supplementation are used in the search for a competitive edge. Moreover, attempting to fit in nourishing foods and regular eating times around the hectic schedules of practice, traveling, games—and school—is nearly an impossible goal for even the best intentioned teen!

At a time when teenagers are experiencing doubts about their body image, the presence of acne can be particularly distressing. The restriction or omission of fried foods, chocolate, nuts, candy, cola beverages, rich desserts, and iodized salt often has been recommended, although improvement usually has not been remarkable.

Appropriate advice to the teenager includes emphasis on a variety of foods from the Food Guide Pyramid. In addition to regular meals, a liberal intake of water, regular elimination, adequate sleep, and skin cleanliness are useful. Excluding fried foods, candy, and fatty desserts usually improves the nutrient quality of the diet.

CHILD NUTRITION PROGRAMS

The U.S. Department of Agriculture (USDA) provides substantial assistance through cash and distribution of food commodities to programs designed to improve the nutritional well-being of children of all ages. These programs have a nutrition education component. Most programs are administered through states and counties/cities.

The largest of these programs is the National School Lunch Program, which serves foods designed to furnish about ⅓ of the RDAs for each age group and meet the *Dietary Guidelines for Americans.* (See Figure 13-7.) Flexibility in foods provided encourages school districts to accommodate preferred ethnic foods of the children in their communities and to serve foods lower in lipids and salt. Children with special health care needs are to have appropriate foods and feeding devices.

The School Breakfast Program is an expansion of the school lunch law and is especially important in situations where children might not receive breakfast at home.

Child-care centers, summer school programs, and camps also are aided by cash and commodity distribution. The Special Milk Program provides reimbursement to defray the cost of milk. Children in school may purchase the milk below cost, or may receive it without charge if they are from families with low incomes.

The Special Supplemental Feeding Program for Women, Infants, and Children (WIC) provides nutrition education and vouchers for specific foods. The groups this program is intended to reach are pregnant and lactating women who have limited incomes and are at high risk, and their infants and children under five years who also are at high risk. (See Chapter 12.)

PUTTING IT ALL TOGETHER

The years from birth through adolescence are distinguished by their nutritional importance for the continuing health of the youngster and for laying the groundwork for future health and well-being throughout adulthood. Overall recommendations for the pediatric population include:

- Eat a variety of enjoyable foods that are nutrient dense. Drink enough water.
- Maintain appropriate growth rates and development of bone mineral density and lean muscle mass.

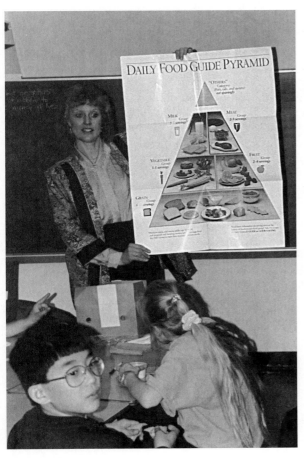

FIGURE 13–7
A dietetic technician registered (DTR) provides classroom nutrition education. Through provision of nourishing foods and nutrition information, school staff help children develop lifelong healthy food choices. (*Source:* Courtesy of Dallastown, Pennsylvania Area School District.)

- Obtain adequate sleep and regular outdoor physical activity.
- Avoid alcohol, drugs, and tobacco.
- Develop behavioral and self-esteem skills for coping with life.
- Obtain a diverse educational foundation to permit flexibility in the workplace.

? QUESTIONS PARENTS ASK

1. My baby seems to be constipated. What should I do?

Breast-fed babies usually have two or three bowel movements a day, while formula-fed babies have only one. If the stools are hard and dry, try increasing the plain water intake or giving ½–1 oz prune juice or strained prunes daily until the constipation is corrected.

2. If my child has diarrhea, what should I do?

Diarrhea can be a serious problem since it leads to loss of fluids and electrolytes. Your pediatric health care professional should be consulted if this lasts more than a day. If the diarrhea is mild, a glucose-electrolyte solution is usually recommended. As soon as this solution is tolerated, depending on the age of the child, return to breast milk, formula, or milk, given at half strength. Gradually increase the concentration to full strength. Fruit juices are not recommended since the high natural sugar content causes further loosening of the stools. Severe diarrhea requires hospitalization and intravenous feedings.

3. I use raw honey for sweetening foods. Isn't this better than table sugar for my children?

Honey, especially raw honey, should *never* be used during the first year of life. As it turns out, honey often carries the *botulinum* bacteria that can cause digestive tract upsets, lethargy, and feeding difficulties. This foodborne illness could require hospitalization and sometimes has been fatal. Limit the use of all sweeteners since they provide only calories, lead to enhancement of the "sweet tooth," and usually displace more nourishing foods. Encourage enjoyment of foods' own flavors and aromas.

✔ CASE STUDY: An Infant Has a Six-Month Checkup

Barry is scheduled for his six-month health checkup. At birth, he weighed 6.6 lbs (3 kg) and now weighs 14.3 lbs (6.5 kg). He was breast-fed until four months of age. Until then he received no supplements. At four months, he was weaned gradually to an iron-fortified formula. At five months, rice cereal fortified with iron was started. His mother, Debby, often props Barry in his crib with his bottle because she has much to do, and Barry's two-year-old brother, Jimmy, is always getting into things since he is in the "terrible two's."

Questions

1. Using Appendix Figure D–3, plot Barry's weight. About how much should Barry gain in the next six months? What other nutritional assessment information would you like to have? (See also Chapter 3.)

2. What are the important differences in the composition of human milk and cow's milk–based infant formulas?

3. What supplements, if any, might have been advisable during the months Barry was breast-fed? Explain.

4. What alternatives might be possible for Barry's family so that he does not need to be propped in his crib with his formula in a bottle? What potential hazards exist with the current practice?

5. What are some suggestions for successfully introducing Barry to new solid foods?

6. When could some finger foods be started? List three that would be appropriate.

7. What might be a sample daily feeding schedule that would be appropriate for Barry at eight months?

8. Soon Debby will return to work. Barry and Jimmy will attend a child-care center. What considerations related to nutritional practices should Debby inquire about when she interviews for a suitable center for her two sons?

✔ CASE STUDY: A Young Schoolgirl with Behavioral Eating Problems

Mrs. M. is worried about six-year-old Cindy's failure to grow. Cindy weighs 35 lb (16 kg) and is 42¼ in. (108 cm) tall. There is no evidence of infection or other disease. You ask Cindy what

foods she likes, and she replies: milk, ice cream, mashed potatoes, chocolate pudding, hot dogs, and peanut butter and jelly sandwiches. She says she does not like "icky" vegetables like spinach, squash, carrots, broccoli, and asparagus. But a spoonful of peas or corn is OK. Cindy participates in the school lunch program, but her teachers report to Mrs. M. that she runs around a lot and will not sit down with the other children.

Until she was five, Cindy had been fed early since Mr. M. comes home from work about 6:30 P.M. Now Cindy eats with her parents, but mealtime has become a battle. Mrs. M. described last night's dinnertime as typical of Cindy's behavior. She ate all her mashed potatoes, picked at her roast beef, but refused to touch the broccoli. Mr. M. insisted Cindy eat everything on her plate. Again and again Mr. M. scolded her: "Eat up. You don't know how lucky you are to have this good food." The dessert was chocolate pudding, one of Cindy's favorites. But her father told her she could not have any since she had not eaten all the food on her plate.

Long after Mr. and Mrs. M. had left the table, Cindy still sat there, staring at her food but not eating any of it. Finally her mother allowed her to leave the table. Just before bedtime, Mrs. M. gave Cindy ice cream, thinking to herself: "At least I'll get *something* into her."

Questions

1. From Appendix Figure D-2, determine Cindy's height and weight percentiles. About how much should Cindy gain in height and weight during the coming year?

2. About how many calories and how much protein should Cindy's diet include? What other nutrients are especially important for school-age children? Why?

3. Would mineral and vitamin supplements be a good idea? Why or why not?

4. List possible problems contributing to Cindy's inadequate food intake and failure to grow.

5. What additional information would be useful to have in planning for improvement of Cindy's food habits?

6. Suggest some behavioral changes that could lead to a happier mealtime for the family, as well as improved nutrition for Cindy.

7. Cindy's teacher suggested a parent conference with Mr. and Mrs. M., the school food service director, and the school nurse. What are some possible interactions among these adults? What positive outcomes might be planned?

FOR ADDITIONAL READING

American Dietetic Association: "Position Paper: Dietary Guidance for Healthy Children," *J Am Diet Assoc,* **95:**370, 1995.

American Dietetic Association: "Position Paper: Nutrition Services for Children with Special Health Needs," *J Am Diet Assoc,* **95:**809-12, 1995.

American Dietetic Association, Society for Nutrition Education, and American School Food Service Association: "Position Paper: School-Based Nutrition Programs and Services," *J Am Diet Assoc,* **95:**367-69, 1995.

Olson, RE: "The Folly of Restricting Fat in the Diet of Children," *Nutr Today,* 30:234-45, 1995.

Pollitt, E: "Does Breakfast Make a Difference in School?" *J Am Diet Assoc,* **95:**1134-39, 1995.

Prentice, A: "Calcium Requirements of Children," *Nutr Rev,* 53:37-45, 1995.

Rickard, KA, *et al:* "The Play Approach to Learning in the Context of Families and Schools: An Alternative Paradigm for Nutrition and Fitness Education in the 21st Century," *J Am Diet Assoc,* **95:**1121-26, 1995.

Nutrition for Young and Middle-Aged Adults

Chapter Outline

Increase years of healthy life to at least 65 years.*

Increase to at least 50 percent the proportion of postsecondary institutions with institutionwide health promotion programs for students, faculty, and staff.*

Increase to at least 50 percent the proportion of worksites with 50 or more employees that offer nutrition education and/or weight management programs for employees.*

 CASE STUDY PREVIEW

Society is changing. The workplace is changing. Families are changing. These changes require individuals to make adjustments. What changes related to your food and nutrition lifestyle have you experienced?

Healthy People 2000: National Health Promotion and Disease Prevention Objectives. Public Health Service, U.S. Department of Health and Human Services, Washington, DC, 1991, Objectives 8.1, 8.5, 2.20, pp. 101, 102, 95.

Rapid changes are occurring in society worldwide. Many of them relate to health, nutrition, and food choices. One issue adults face is choosing the "right" foods—for optimal health, for preventing chronic diseases, or for slowing down the aging process. Young adulthood may be seen as a time for potential chronic disease initiation and middle-aged adulthood as chronic disease expression.

Another issue is knowing how to judge the safety of biotechnologically developed foods, such as genetically altered foods. Producing convenience foods without creating tons of paper, metal, and other kinds of packaging materials that need to be recycled is an issue in many communities. Even with the abundance of foods produced in the United States, concerns still exist that hunger persists in this country as well as throughout the world. Some issues can be resolved by individual and family lifestyle changes; others require the collective efforts of everyone and every nation.

Young Adults

MEETING NUTRITION NEEDS

Age Range

Young adulthood may be considered the time span covering people in their 20s and 30s.

Recommendations and Nutrients

From a nutritional standpoint, not everything is completely "adult" in young adults. Physical growth actually continues to about the age of 21 years, and bone growth continues to the mid-to-late twenties. Because of this, the RDAs for vitamin D and calcium are set at the adolescent levels until young adults celebrate their 25th birthday. (See tables inside front and back covers.)

By the time young adults are in their twenties, **cellular hyperplasia** (additional growth in cell numbers) ceases. At that point, the RDAs are based on the nutrient requirements for continual maintenance and repair of the body, rather than the further need to account for the nutrient demands of growth. Exceptions occur for women during times of pregnancy and lactation. (See Chapter 12.)

IMPACTS ON NUTRITIONAL HEALTH

Lifestyle Influences

Most children and teenagers hope to reach the goal of a healthy, productive, successful, and personally fulfilling life as adults. After graduation from high school, young people usually seek independence by obtaining a job, entering college, or joining military service. Some move out of their parents' home into their own apartments, either by themselves or with friends. Consequently, they are likely to think, "I can finally do what I want."

Rapidly, young adults realize that food can be costly to buy, whether at a restaurant or just at a supermarket. Also, a lot of expensive equipment is necessary—a refrigerator, stove, pots and pans, utensils, dishes, dishwasher, table and chairs, let alone the microwave and imported computerized coffee maker. Plus, shopping, cooking, and cleaning up after meals all take a great deal of time. Added to this complexity is the realization that skill is required to produce all the courses of a meal, done at the right time. No wonder young adults relish "care packages" sent from home to the college dorm, or weekly visits back home to check in on the family, which just happen to coincide with dinner time.

Family Factors

For the majority of men and women in their 20s and 30s, this is the time of tremendous interest in dating, at first perhaps just casual dating, but as years pass, interests change to committed relationships, usually leading to marriage. Up to that point, food and nutrition interests may revolve principally around which club to be seen in, which restaurant to dine in, which parties to attend on weekends, or which resort to go to on vacations.

Cooking for one can become monotonous, so various group activities may be planned that include some sort of meal, such as a group of coworkers going out for drinks and snacks on the way to a sports event, several friends getting together in someone's kitchen for an informal pot-luck dinner, and close friends meeting for conversation and coffee at an intimate cafe or at a local fast-food restaurant.

Although not always, marriage usually changes all this. (We say "usually" because only about 30 percent of adults—even married adults—are estimated to live in traditional households, that is, households where at least one adult stays at home.) At first, couples generally plan to eat every meal together, with images of candlelight dinners every night. They soon find out that doing this takes tremendous efforts at juggling schedules. Other commitments soon may interfere with these romantic plans. Sharing shopping, food preparation, and clean-up doesn't always go as envisioned, either. Comedians still seem to get laughs when they talk about who is expected to take out the trash after dinner, as opposed to who actually does it!

Commonly, women have their children during these years. (See Figure 14-1.) Most times, children dramatically change a woman's lifestyle. Typically, the children's needs come before the mother's and her schedule revolves around theirs. For example, she may skip breakfast to make certain they leave on time for school. Working outside the home might influence her to purchase convenience foods she knows are not in the best nutritional or health interests of her children. Mothers with child-care arrangements have the extra burden of making certain the children's food and nutrition requirements are met by the day care center or caregiver. Sometimes guilt can become overwhelming. She may feel terrible about not being home when the children return after school, as her own mother had been, to greet them and listen to their day's school stories over mugs of hot chocolate and jars full of homemade cookies. Fathers many times assume greater responsibilities for food shopping and preparation and feeding of children—often without any formal or informal education for this role.

Today, many young children know how to heat frozen foods for meals and snacks in the microwave. Older children and teenagers meet their friends after school and on weekends at the food court of the local shopping mall. Teaching children how to food shop, set the table, cook or bake, and share food experiences as a family unit is a lost art in many families. Realizing this, it has become popular for some parents to send their teenagers to special classes to learn social graces and the techniques of fine dining and sophisticated entertaining.

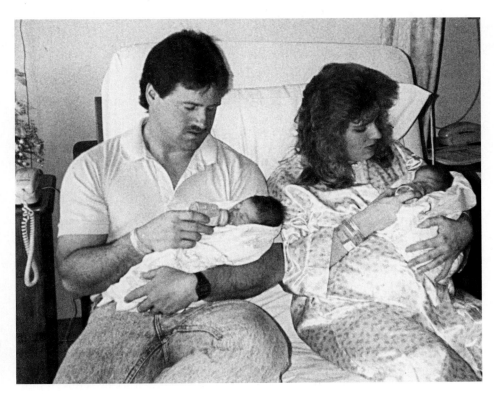

FIGURE 14-1
The ideal situation occurs when both parents participate in the nutritional nurturing of their children. (*Source:* Doylestown Hospital, Doylestown, PA, and Donald Lenz, photographer.)

Career Considerations

The decades of the 20s and 30s are the years when most adults believe they are getting prepared and then really launching their careers. Most look forward to advancing within their company and occupation; often they must be willing to relocate to obtain a sought-after position—just when they finally discover all the out-of-the-way markets for the best food buys in their old neighborhood.

For those who started living away from home at college, becoming accustomed to dorm food or "college mystery meat" is a trying experience. Drinking coffee to keep oneself awake all night to study, eating too many meals at the fast-food eateries on and off campus, and going to too many parties that serve too many fat- and sodium-laden foods—all these changes usually play havoc with one's appearance. Freshmen complain of gaining 5 to 25 pounds by the time they go home for the winter holiday break. Even adolescent complexion problems seem to resurface. One of the most serious social problems on campus is alcohol consumption. No matter what message is given to college students, for many, this is a time of excessive drinking. Another serious problem for many students, mostly women but also men, is eating disorders like anorexia nervosa or bulimia. (See Chapter 9.)

For young adults who start employment after high school, having regular paychecks is a real boon. Frequently eating out, though, can consume a large portion of a paycheck. Some employed young adults start college in evening programs. This kind of schedule—working all day and attending classes most evenings—usually puts a strain on any kind of regular eating pattern. The majority of meals may be composed of convenience and fast foods. Eating dinner late at night right before going to bed can become the norm rather than the exception.

As young adults advance in their careers, food can take on different roles. Sometimes attendance at dinner parties, luncheons, or banquets is required. Another issue closely related to food is money. Especially in the earlier years before other responsibilities are assumed, some young adults may have extra money to spend on food. Conversely, some young adults need to carefully watch their food budgets since they have so many other expenses like payments for rent or mortgages, cars, credit cards, and insurance policies, along with the expenses of raising young children.

Health and Fitness Interests

Because most young adults are healthy, many of their interests relate to fitness. Chronic diseases, unless present in their families, are not so ominous to young adults. Instead, fitness for the sake of one's appearance, endurance, stamina, and as a way to meet other people plays a larger role in the young adult's life. Sometimes appearances are deceiving, however; for with the exception of the diseases that are ongoing from childhood, the initiation of chronic diseases starts in or continues silently throughout young adulthood. Examples of nutritionally related diseases include cardiovascular diseases, obesity, cancer, alcoholism, drug abuse, periodontal disease, and osteoporosis. Signs or symptoms of most of these diseases are not obvious to people during young adulthood. Thus, often symptom-free young adults unknowingly continue with lifestyles that are not conducive to optimal nutritional well-being or long-term health.

Middle-Aged Adults

MEETING NUTRITION NEEDS

Age Range

Middle-aged adulthood may be considered to start in a person's 40s and continue into the late 50s and early 60s. But keep in mind that some people consider themselves "young" or "middle-aged" until well into retirement—and even beyond!

Recommendations and Nutrients

The RDAs for middle-aged adults remain at the levels established for their younger and their slightly older counterparts. (See tables inside front and back covers.) Surprisingly, little research has been conducted regarding maintaining the nutritional status of middle-aged adults who are healthy. Healthy people normally do not seek out health care. Thus, they do not participate in the research projects that study similar-aged people who do have a family history of, or who do exhibit signs and symptoms of chronic diseases. Much of the knowledge about nutrition and health for middle-aged adults is based on research with at-risk individuals rather than on "healthy" people, or on general surveys of statistically selected groups, or on extrapolations from results of studies conducted with people of other age categories.

Lifestyle Influences

Although independence, dreams, exploring, planning, and launching out into life may be considered the hallmarks of young adulthood, achieving that success, power, financial reward, and personal fulfillment by middle age has become the expectation of many people for themselves or for other family members or friends. This rush to success in 20 to 25 years belies the fact that many great accomplishments are made much later in life by creative, inventive, and business-minded men and women. Such an obsession as this can destroy the self-respect, health, and nutritional status of many middle-aged adults.

Nevertheless, the early days of struggle are behind most people by this point. They are ready to enjoy life to the fullest. Many have the time, money, and material possessions to do this. Unless the house was refinanced, the mortgage has been paid off. Better quality home furnishings, as well as finer quality china, have replaced the secondhand ones, or those given as shower and wedding gifts, or the ones ruined by children and pets. Vacations to exclusive resorts, foreign travel, or luxurious ocean cruises now are possible, if not every year, then at least for special celebrations such as a 25th wedding anniversary. Culinary delectables are much enjoyed and described as one of the highlights of these occasions. Routine entertainment is at more prestigious restaurants. The better quality home decor leads to more elegant dinner parties, perhaps catered. No more pot-luck dinners around a friend's kitchen table! Food definitely has taken on the symbol of one's status.

Family Factors

Increasingly, middle-aged adults live in an altered family unit, with nearly half having separated, divorced, or become widowed. Many have remarried, but others live alone, or with their children as a single head of household, or with another family, or in a variety of other family arrangements. Many middle-aged adults begin to realize their own parents and relatives are aging; thus, middle-aged adults must spend more time helping their older relatives with daily routines, including food shopping and preparation.

These middle-aged family units change the meal patterns, menu structures, and food selections set in young adulthood. For example, with huge-appetite teenagers in the household, mealtimes can last all day—and into the night. (See Figure 14-2.) For single people, cooking for one can seem like déjà vu. In fact, millions of middle-aged adults are living alone today. For adults in "second family" units, babies and small children again may become part of their family lives. Others have grandchildren to dote on and to give special food treats—foods they know are not in the best nutritional interests of children's health. By middle age most adults experience the "empty nest syndrome," slowly sending their children to college or watching them leave home to start careers and families of their own. In many households, once again only couples are present. They spend their time sending food packages to children living in the dorm, paying for wedding feasts, and inviting their children to "stop by and have dinner with us some evening." The life cycle continues.

Career Considerations

Typically, middle-aged adults are in their peak earning capacity years. Yet some completely change careers, while others leave corporate employment to become entrepreneurs. For some the decision is voluntary; for others, their jobs were abolished or their employers went out of business. Suddenly finding themselves

FIGURE 14-2
Wholesome, delicious foods can be prepared easily for a complete meal or quick snack. (*Source:* Anthony Magnacca/Merrill.)

standing in line for food stamps is not what these people had ever expected! Worrying about financial problems causes many to overindulge in alcohol and eat incessantly. Others lose their appetites and neglect to eat.

Usually in middle age both men and women have reached their top educational levels, although many are returning to school for additional studies to enable them to start a "second career." Others take adult or continuing-education classes for self-interest or for professional advancement. Nowadays middle-aged adults who have reached their career potential must keep up to date if they want to remain competitive with younger coworkers. So many are returning to the classroom evenings and weekends (and consequently eating late at night before going to bed).

Health and Fitness Interests

Middle-aged adults can experience the first pangs of chronic diseases. For some, the symptoms send them to the physician, dentist, or other health professional in sufficient time for appropriate intervention. (See Figure 14-3.) Unfortunately, some people are too busy to heed (or they choose to deny) their body's warning signals until extensive treatment is required. Some wait too long, whether it is a woman with a lump in her breast, or a man with chest pains that he calls "indigestion," or either of them for a dentist's visit because of bleeding gums.

Middle-aged adults often take a renewed interest in health and fitness, this time predominantly from a wellness standpoint. They rejoin health clubs, gyms, and spas. They take up power walking, jogging, cross-country skiing, and swimming. They ride exercise bikes in front of the television set. Many take a new look at their appearances and decide to do something about that "middle-age spread." Enrollment in weight-loss programs by middle-aged women continues at a brisk pace.

FIGURE 14–3
People are learning to shop differently in supermarkets. Reading labels is important
when people want to select healthier foods. Asking questions and requesting that
certain foods be available places people in command of their food purchases.
(*Source:* Courtesy of *FDA Consumer.*)

SPECIAL FOOD AND NUTRITION ISSUES

During young- and middle-aged adulthood, a continuum of food and nutrition
issues ebb and flow. For example, women often are more cognizant of such issues
because of their experiences with pregnancy and lactation. Men may recognize
issues when a similar-aged coworker suffers a heart attack.

Oral Contraceptives

Varying proportions of estrogens and progestogens are used in oral contraceptives. In many ways these preparations produce changes that are similar to those
in pregnancy. The blood levels of some nutrients and their metabolites are
increased after a woman takes oral contraceptives for several months. Serum
cholesterol, triglycerides, iron, copper, and retinol-binding protein are increased
somewhat. The levels of other nutrients and their metabolites are decreased:
vitamin B_6 , folate, vitamin B_{12}, zinc, vitamin C in leukocytes, and riboflavin in red
blood cells.

For women whose nutritional status is good it is generally believed that the
biochemical changes are not of practical significance. However, for women
whose nutritional status is already borderline, the use of oral contraceptives might
increase the deficiency. Improvement of borderline diets is the best approach.
Vitamin and mineral supplements are not necessary.

Premenstrual Syndrome

Premenstrual syndrome (PMS) is a group of physical and emotional changes that precede the onset of menstruation. From two weeks to a few days prior to the onset of menstruation many women notice food cravings, irritability, depression, headache, breast tenderness, backache, bloating, and weight gain. Symptoms of hypoglycemia are sometimes present. The causes of this syndrome are not well understood.

The treatment of PMS remains controversial. Many supplements, including vitamin B_6, vitamin E, and magnesium, have been tried. Although some women have reported that such supplements reduced the symptoms, most of the reports have not been based on controlled studies. The value of dietary changes has not been proven. Most of the suggested dietary changes also enhance nutritional status and are generally recommended for reducing the risks of chronic disease. Some suggested interventions, though, are unbalanced, and result in negative side effects.

Oral and Dental Health

Most adult Americans visit their dentist and hygienist more frequently than other health care providers. Before age 35, dental caries is prevalent, but gingivitis and **periodontal diseases** become increasingly common after age 35. These oral diseases are characterized by bleeding gums and decreased bone mass in the jaw especially, leading to loosening and loss of teeth. Many factors contribute to this condition, including inadequate oral hygiene. Nutritionally, a person's diet should be evaluated for overall nutrient adequacy. Special focus should be given to the intake of water, protein, vitamins A, D, K, C, and folate, and the minerals calcium, magnesium, fluoride, iron, and zinc; to the form of carbohydrate-containing foods; to the texture of the diet; and to the sequence of eating followed by tooth brushing and dental flossing. The importance of oral and dental health to ongoing positive nutritional status cannot be overemphasized.

PUTTING IT ALL TOGETHER

Because most young and middle-aged adults are healthy, they usually do not receive the personalized food and nutrition assessment and counseling that health care professionals routinely provide for people with a diagnosed illness. Yet these two groups of adults *do* receive nutrition education. What happens most often is that generic advice or admonishments are provided, based on at-risk groups of people, which very likely are dissimilar in genetic background and environmental conditions. This easily could lead to erroneous conclusions and interventions. As we enter the twenty-first century, keeping healthy adults healthy certainly is worthy of nutrition research and heightened *individualized* attention in health practice settings.

? QUESTIONS CLIENTS ASK

1. I'm a student serving on our campus' food services committee. What can I suggest to improve the nutritional status of our students?

Food service areas should provide a wide variety of nutritious foods from which students could choose; delicious, nutritious foods could be made available at midnight during finals week; vending machines could contain nutrient-dense foods, like reduced-fat milk, fruit juices, fresh fruits, fresh vegetables, and whole-grain breads and cereals; nutrition videos could be aired on the campus cable TV station; and students in your nutrition course could write a "Nutrition Q & A" column for the campus newspaper.

2. I've just started my first job. How can I keep "on the run" and healthy, too?

Start your day with nutritious foods—at home, on the way to work, or once you arrive at work. Keep tasty, sealed snacks in your briefcase, desk, or locker—bottles, boxes, or cans of fruit *juice;* boxes of milk; and low-fat and low-salt nuts and popcorn, whole-grain crackers and pretzels. Have a drink of water each time you pass the water fountain, or drink your own water. If there is a community refrigerator, keep yogurt, cheese, milk, fresh fruit, and salad fixings. Take time for lunch, a walk outside, and socialization.

✔ CASE STUDY: An Executive at Risk for Chronic Diseases

Mrs. Dana Z., 43, is an executive in a large firm that is having financial problems. Consequently, Mrs. Z. is putting in longer hours, coming home late, and going to work on weekends. Her two preadolescent children complain they never see her.

She has trouble sleeping. She worries about the future of her job. Her husband's employment also is in jeopardy and both of them are concerned about keeping up payments on their home and continuing to save for their children's college education. Although Mrs. Z. tried to quit smoking a year ago, her stress level is too high; she smokes ½–1 pack of cigarettes each day. Her father died of a heart attack at age 50 and her mother has diabetes mellitus.

Her hectic work and home schedules allow no time for exercise; she cannot recall the last time she and her husband had some relaxation time together. She skips breakfast so she can get the children off to school on time. She stops at a convenience store on the way to work and buys coffee and a breakfast sandwich. By midmorning she is hungry, and gets a soda and candy bar from the vending machines in her building on the way outside to have a cigarette. She usually sends out for lunch and dinner—a sandwich, chips, coffee or soda, and pie or cake, which she eats at her desk. Sometimes she goes out for a business lunch or dinner where she has a full-course meal and wine. On the rare occasions she can get home for dinner, she stops at a fast-food restaurant to pick up foods she knows her children will like. Into the late hours of the night, she works at her computer in her home office and to stay alert, she drinks coffee, eats cookies, and smokes.

The time arrived for the annual physical examination provided by her company. Screening anthropometric and laboratory assessment data revealed: Height = 64 in. (163 cm) without shoes; weight = 135 lb (61 kg) in indoor clothing; frame size = small; total serum cholesterol = 225 mg/dl (5.85 mmol/L).

Questions

1. What is an appropriate weight for Mrs. Z.? What is your evaluation of the blood study result? What other nutritional assessment data would be helpful to have?

2. What risk factors for various chronic diseases are present in Mrs. Z.'s life?

3. How could Mrs. Z. modify her diet (for example, meal patterns, menu structures, and food selections) to make it more health promoting? (See also Chapter 2; Food Guide Pyramid and Nutrition Label, Figures 1–5 and 1–9; Basic Diet, Tables 2–3 and 2–4; and Nutritive Value of the Edible Part of Food, Appendix Table A–1.)

4. What are some practical suggestions for ways Mrs. Z might reduce her risk of chronic diseases, in addition to modifying her diet?

5. Develop a plan for implementing dietary and lifestyle modifications in a gradual and progressive manner over the next few months with Mrs. Z.

6. Mrs. Z. wonders what she can order when she (a) goes to a restaurant for a business meal; (b) goes on a transcontinental flight; (c) stays at a hotel during a business trip. What might be suggested?

7. What nutrition and health concerns do you have for her husband? For their two children?

8. What are some nutrition education techniques that might be helpful when providing dietary counseling with Mrs. Z. and her family?

FOR ADDITIONAL READING

American Dietetic Association and Canadian Dietetic Association: "Position Paper: Women's Health and Nutrition," *J Am Diet Assoc,* **95:**362–66, 1995.

Finn, SC: "Women's Health: Anatomy of an Issue," *Top Clin Nutr,* **11:**1–7, December 1995.

Gerrior, SA, *et al:* "Differences in the Dietary Quality of Adults Living in Single versus Multiperson Households," *J Nutr Educ,* **27:**113–19, May/June 1995.

McNutt, K: "Putting Nutrition Priorities into Perspective," *Nutr Today,* **30:**168–71, 1995.

Nutrition for Older Adults

Chapter Outline

Maturation
Meeting Nutrition Needs
Special Food and Nutrition Issues
Older Adult Nutrition Programs

Putting It All Together
Questions Clients Ask
Case Study: An Older Man Has Nutritional
   Concerns

Increase to at least 80 percent the receipt of home food services by people aged 65 and older who have difficulty in preparing their own meals or are otherwise in need of home-delivered meals.*

Extend to all long-term institutional facilities the requirement that oral examinations and services be provided no later than 90 days after entry into these facilities.*

✔ CASE STUDY PREVIEW

Life is a progression, with nutrition and food choices important along each step of the way. Have conversations with people older than yourself. Learn about changes they experienced in the kinds of foods available in the marketplace, food preparation in the home or institutional setting, breakthroughs in the science of nutrition, changes in the health care system, and changes in health care practice in your major discipline. What fascinating facts did you learn?

Healthy People 2000: National Health Promotion and Disease Prevention Objectives. Public Health Service, U.S. Department of Health and Human Services, Washington, DC, 1991, Objectives 2.18, 13.13, pp. 95, 109.

Aging is the continuum of change that occurs from birth to death. It includes growth and decline. The changes are gradual, starting in a person's early 20s. The rates of change differ at various stages of life and from one individual to another. These changes determine the nutritional guidance or services provided. In research centers throughout the world gerontologists are studying factors, including nutrition, associated with **aging. Gerontology** is the study of aging, and **geriatrics** is the science dealing with the abilities, diseases, and care of aging persons.

The older adult age category is the most complex because older adults are the most heterogeneous. They bring their genetic makeups, all their life experiences, and the results of their environmental exposures. Older adults vary in their physical wholeness, their feelings, and their capacities. Most older persons cope very well. They have managed a lifetime of problems that younger people often cannot appreciate. Older men and women have dealt with changes in the family structure—having children, seeing children leave home, enjoying grandchildren, and perhaps suffering the death of a spouse and other loved ones. Income changes often have been considerable over their lifetimes. Social mores and values likewise have changed. Many have participated in community and world affairs and are a valuable resource of experience, talents, and skills. Older adults have perspective and wisdom.

Income and living arrangements have a profound effect on the nutritional status of older Americans. Sufficient or inadequate income often leads to the choice of using money for bills or for food, with food often being last. About four of every five older adults live in their own houses; they shop and cook for themselves, eat with others at neighborhood restaurants, or participate in government-sponsored nutrition programs. Retirement, personal-care, and life-care communities are an attractive option for many older couples and individuals, with one major amenity being one or more served daily meals. Some of the more frail reside in institutions with varying quality levels of nutrition services.

MATURATION

Age Range

As time goes by, it becomes evident that there is no age that clearly designates who is "older" or "old." Sometimes the classification is: 65–74 years, *young-old;* 75–84 years, *middle-old;* and 85 years and beyond, *old-old.* Indeed, one person may have reached the age of 95 and be enjoying vigorous health and productive activity, while another, 30 years younger, is living a life of confusion at age 65!

Body Size and Composition Changes

At the cellular level, no body cells except those of the central nervous system are as old as the whole person. For example, the cells of the digestive tract and liver rapidly turn over in a few days; the red cells have a life span of 90–120 days. With aging, there is a gradual diminution of cell reproduction rate and there are fewer functioning cells of some organs such as the kidneys and liver, and of some tissues like muscle.

Lean Body Mass and Fat Even with the same level of physical activity, older adults have less lean body mass than younger adults. The loss is gradual. Compared to a 25-year-old person, by age 65 a man has lost about 12 kg of lean body mass and a woman has lost about 5 kg. The lean mass has been replaced by fat, and the organs gradually become infiltrated with fat. This decline in lean body mass means less endurance in vigorous activity. Energy basal metabolism slowly decreases.

Body Water With a reduction in lean body mass and an increase in body fat, the amount of body water decreases. Water imbalances are more common and severe; for example, the edema that accompanies cardiac and renal failure, and the dehydration that is an all too common complication in fevers, infections, and diarrhea. Since the thirst mechanism is less pronounced in older adults, many fail to drink as much water as they require.

Collagen The collagen content of the body increases considerably with aging, leading to less elasticity of cells. The blood vessels rupture more easily and bruise marks become more evident. Cells of the digestive and respiratory tracts are affected, as are the esophageal, anal, and bladder sphincters. Joints are less flexible, often painful, and arthritic. Wrinkling of the skin becomes evident in spite of the best applications of skin creams.

Bone Mass Older people often remark when they are measured for height: "That can't be right. I am 5 feet 6, not 5 feet 4." Or listen to the person who says "I can't reach the top shelves in cabinets any more." Indeed, much bone mass loss has occurred to account for these observations.

Up to a person's early 30s, the deposit of minerals into bone is greater than the loss. After that age, bone loss begins to exceed bone gain; it is greater and faster in women than in men. On average, loss in women is about 0.5 percent per year up to menopause. Then for three to five years after menopause, the yearly loss triples. Following that, loss drops down to about 1 percent a year. Thus, by ages 70–80 years, in both women and men, a great deal of bone mass loss has occurred, and the possibility of osteoporosis and fractures has increased.

Sensory Changes

A gradual loss of taste buds is common, but for the most part, food seasoning can compensate for this. Some people may eat more highly seasoned, spicy foods only to complain of dyspepsia an hour or two after meals. Numerous medications contribute to altered taste perceptions. (See Chapter 18.) A deficiency of zinc also has been associated with reduced taste appeal.

The sense of smell is as important to the enjoyment of food as is the sense of taste. Those who cannot smell the foods eaten find them to be lacking their familiar distinctive characteristics. This situation is aggravated when foods are prepared elsewhere and then served, as in restaurants, residential dining areas, and hospital rooms. Importantly, when the sense of smell is diminished, an important guide to determining the safety of food is lost.

Sensitivity to touch becomes less acute. Picking up and holding small objects, like cooking and eating utensils, are more difficult with reduced finger dexterity.

Dimming of vision is common as are conditions of the eyes such as cataracts. Brighter, but not harsh lighting is preferred so that menus can be read; visually challenged individuals need assistance to know the placement of foods on plates. With dimming vision many people no longer can read the print on food labels—ingredient and nutrition listings, and expiration dates.

Hearing loss is common, with the upshot being that some older adults find participating in conversation at mealtimes difficult. Because they lose so much of what is being said, they may isolate themselves from the company of others.

Gastrointestinal Function Changes

Perhaps the most common problem pertaining to the digestive tract is the inability to chew. Loss of teeth, poorly fitting dentures, or sore gums give rise to the all too common stereotype that all elderly persons eat soft, mushy food. Multiple nutritional deficiencies often occur because the person is unable to chew meat, raw vegetables, nuts, grains, seeds, and some fruits. Oral hygiene and dental care often is the key to improving the nutritional status of the individual as well as contributing to the enjoyment of food and mealtimes.

Reduced salivary flow has a number of implications: first, some dry foods are too difficult to swallow, and second, the reduced flow of saliva increases dental plaque development. Numerous medications also increase the dryness of the mouth.

Reduced elasticity of the esophageal muscles also contributes to swallowing difficulties. Movement of the food bolus decreases, and esophageal contractions are less forceful. Regurgitation of stomach contents occurs when the esophageal sphincter functions poorly. The esophagus is irritated by acid from the stomach and further aggravated by acid foods, spicy foods, and fiber.

For individuals producing less hydrochloric acid in the stomach, or none at all, several adverse effects happen. Bacterial overgrowth, fermentation, and bloating increase; iron absorption decreases, and foodborne pathogens linger. If intrinsic factor is not produced, vitamin B_{12} is not absorbed.

Some persons experience decreased tone of the gastrointestinal tract muscles. Impaired circulation to the intestinal tract further aggravates this problem. With less tone, there also is slower food transit through the tract. Common complaints include heartburn, abdominal cramping, diarrhea, or constipation.

The net amount of nutrient absorption usually is very good. Most older adults absorb water, simple sugars, fatty acids and triglycerides, amino acids, vitamins, and minerals as well as younger persons.

Metabolic Changes

The basal metabolic rate decreases about 2 percent per decade after age 25, or about 10–12 percent by age 75. In addition, the lifestyles of many older adults become more sedentary, leading to a significantly lower total energy expenditure. (See Chapter 8.)

Vitamin D synthesis in the skin is about half as great as that for younger adults. For a number of reasons, many people stay indoors more, away from the sun. Often, older adults drink little vitamin D–fortified milk, further compromising the absorption of calcium and its uptake by the bones.

Hormonal changes occur, affecting the body's metabolic regulation. For example, fasting blood glucose values likely are within normal range, but a high test dose of glucose shows that the recovery rate to the fasting level is slower than that for younger persons given the same challenge. Sometimes this is mistakenly diagnosed and treated as diabetes mellitus. (See Chapter 26.)

When anemia is present, the correct diagnosis is necessary for proper treatment. For example, *macrocytic* anemia may suggest either a deficient intake of folate or vitamin B_{12}, or both; *microcytic* anemia suggests lack of iron. (See Chapters 10 and 11.) Also essential for proper diagnosis is the determination of any possible hemorrhage, especially small daily amounts lost from the bowel that frequently are unnoticed.

Many changes taking place in the later years of life can have major consequences on a person's mental outlook. Retirement not only affects one's income, but to some, represents the loss of purposeful activity and association with coworkers. A likely result is a diminished sense of worth and self-esteem. The death of a spouse probably is the most stressful event a married person encounters. This loss is exaggerated when other family members and friends live at a distance from the bereaved. Grief, loneliness, and apathy can lead to disinclination to food shop, omission of meals, and restricted food choices; others find solace in overeating. Illness itself, or worry about it, often becomes overwhelming, resulting in fear of eating certain foods or favoring other ones.

Multiple diseases affect the brain and nerves, thereby affecting mental factors related to food choices, nutritional well-being, and health. Numerous medications have neurological side effects that impose negative nutrition responses. On the other hand, some nutrient deficiencies or toxicities can mimic neurological symptoms that result in misdiagnosis and treatment. Thus, careful nutritional assessment and monitoring by health care providers requires alertness to possible nutrition-neurologic interactions.

MEETING NUTRITION NEEDS

Nutritional Assessment

Most older adults enjoy good nutritional health. A small percentage are at increased risk for malnutrition. Routine screening is imperative to uncover small problems before they escalate. Some factors that increase the risk of malnutrition in older adults are listed in Figure 15-1. Also, Appendix Figure C-1 was devised to be a screening tool for older adults. Any individual at risk should be referred for further measurement of nutritional status.

- Socially isolated; living alone
- Below poverty level
- Over 75 years
- Unintentional weight loss
- Fad diets; erroneous beliefs about foods and needs for them
- Physical impairment—unable to shop and cook food
- Interference with eating; poor appetite; dry mouth; lack of teeth or ill-fitting dentures; poor sense of taste and smell
- Common gastrointestinal complaints: heartburn, dyspepsia, diarrhea, flatulence, constipation
- Medications that interfere with food intake or absorption
- Chronic illness that interferes with food intake
- Draining wounds, bedsores
- Grief, apathy, depression, suicidal tendencies
- Alcoholism
- Mental illness

FIGURE 15-1
Factors that increase the risks of malnutrition in older adults.

Recommended Dietary Allowances

As noted in Chapter 14, relatively little research has been conducted to determine the nutritional requirements of adults, especially older adults. Presently, it is assumed that the RDAs for persons over 51 years are the same as those for younger persons except in the instance of energy; for nutrients based on energy, like thiamin; and for iron in nonmenstruating women. (See tables inside front and back covers.)

Nutrient Highlights

Water A liberal fluid intake is especially important for older adults. First, as noted above, the amount of total water in the body is decreased and risk of dehydration is ever present. Second, the number of functioning nephrons is considerably lower, requiring the kidneys to work harder to eliminate wastes. Third, foods with higher water content, like fresh vegetables and fruits, and milk are eaten less often. Sample interventions are: remain cognizant of fluid intake and excretion; alter food preparations, like increasing consumption of soups made with vegetables and tender meat, poultry, and fish, or drinking fruit juices diluted with seltzer water; or substituting low-calorie sodas or flavored bottled beverages for regular soft drinks if energy intake must be limited and water is unavailable.

Energy The average energy allowance for men 51 years and over is 2,300 kcal and for women, 1,900 kcal. For many sedentary older adults these allowances are too high to maintain energy balance. Generally the requirement drops sharply after age 51. Many men maintain healthy body weight at 1,700 – 1,900 kcal, while women who are sedentary may need as little as 1,400 – 1,600 kcal. These lower energy requirements necessitate careful choice of foods so that nutrient requirements can be met within caloric ranges. Each increase in physical activity increases the caloric expenditure, thus improving the likelihood that nutrient needs will be reached.

Fiber Older adults often exclude fiber-rich foods in the belief that such foods are irritating to the gastrointestinal tract. Health care providers can explain how fiber helps both digestion and elimination by its ability to stimulate peristalsis, hold water to form a soft stool, and improve transit time. Menus prepared by food service operations must reflect these special needs. Additions of fiber to the diet, whether for an individual client or for groups, should always be made gradually since a sudden increase can lead to bloating, gas, and even diarrhea.

Vitamins The diets of some older adults do not meet the RDAs for vitamins A, C, B_6, and folate. Yet there is little physical or biochemical evidence of vitamin deficiencies. However, since it is difficult and costly to determine when an individual is at risk, it seems prudent to aim for a diet that meets the RDAs. Some older persons are at risk for vitamin D deficiency if they do not drink milk and if they have very limited exposure of the skin to sunlight.

Minerals Calcium is the mineral most likely to be inadequate in the diet. The average intake by women is less than 500 mg. The RDA for calcium is 800 mg, but many researchers and clinicians now recommend 1,000 – 1,500 mg in order to minimize bone mass loss. Other factors are involved and must be assessed, such as: adequacy of diets during childhood and younger adulthood; current calcium, vitamin D, and magnesium status; and current weight-bearing physical activity.

With a well-balanced diet, each 1,000 kcal will furnish about 6 mg iron. This necessitates a caloric intake of about 1,650 to 1,700 kcal daily to satisfy the RDA of 10 mg. If caloric needs are lower, individuals must choose foods that are especially rich in bioavailable iron. Some may be advised to take an iron supplement if there is risk of, or documented laboratory indication of, inadequate iron nutriture.

Zinc intake requires special attention since the mineral is primarily in foods often neglected by older people because of inability to chew meat and nuts or digestive problems after eating legumes. Adjusting the texture of such foods may be one solution, such as using ground meat items like hamburger, meatloaf, and meatballs or using ground nuts, like smooth peanut butter.

Meal Patterns and Menu Structures

The Food Guide Pyramid (Figure 1–5) continues as the basis for planning nourishing meals and menus. Routine evaluation and monitoring can be accomplished by using the Basic Diet Nutrient Profile (Tables 2–3 and 2–4). A little extra effort in meal planning, preparation, and service carries a big bonus in enjoyment and well-being.

Social isolation is a major risk factor for malnutrition. Those who live alone often need practical, easy-to-prepare recipes for one. If there is a freezer in the home, a family-size casserole can be prepared and divided into individual portions for freezing and later use. Meat, poultry, fish, legumes, or vegetable soups can be prepared in quantity and frozen in small portions. Setting an appealing table brings pleasure to the meal. A meal eaten by a window or while listening to or viewing a news program provides variety and maintains involvement in community and world events. Sharing a meal with a friend or neighbor avoids isolation for two persons! (See Figure 15–2.)

Those who live in residential facilities or who are admitted to hospitals need to remain as independent as possible. Dining room decor should be pleasant, well lighted, and easy to traverse. Whether in the dining room or the individual's room, staff members need to be attentive, speak at an acceptable speed and loudness, and assist only to accommodate. Mealtimes, including the bedtime snack, need to be arranged for the convenience of the people eating, not of the administrators and staff. Menus must be planned to be nourishing, include traditional food preferences, and printed with larger, bolder type styles. Foods presented must be wholesome, attractive, and tasty. Without a doubt, mealtime is a perceived highlight of the day. On the whole, creativity, commitment, competence, and caring are more important resources for achieving this outcome than are money and time.

SPECIAL FOOD AND NUTRITION ISSUES

Energy Balance

Since measurement of an older person's actual height becomes more difficult to ascertain, using height-weight standards is less valid and reliable than for younger people. Weight itself also is less informative, since adult body composition changes—an older adult, even at the same *weight* as earlier in life, has a different weight proportion of muscles, adipose tissues, bones, and fluids. Cautiously interpret assessment data. (See Chapter 3.)

Higher levels of body weight appear to provide greater protection against illness and can reduce the severity and length of illness. On the other hand, those

FIGURE 15–2
A meal with friends in an attractive environment helps to ensure satisfactory nutrition.
(*Source:* Dunwoody Village, Newtown Square, PA, and Peter Zinner, photographer.)

who are 10–15 percent under standard weight are more prone to infections, longer illnesses, and somewhat higher mortality rates.

Obesity in older adults means greater stress on joints that are often fragile and arthritic; greater demand on the respiratory and cardiovascular systems; higher association with increased hypertension; and greater risk if surgery is needed. However, older adults usually find losing weight less easy to achieve because of their lower basal metabolism and lower physical activity. A daily intake of 1,000 kcal or less often is needed to bring about even a modest weight loss. At such low caloric levels, intakes of essential fatty acids, vitamins, and minerals are inadequate. A supplement that furnishes 100 percent of the RDA levels of minerals and vitamins is needed. The person also may exhibit negative nitrogen balance. Thus, sustained weight loss in older adults should be attempted only with close nutrition and other health guidance and observation.

Physical Activity

Of persons over 65 years, less than one-third walk on a regular basis and only 10 percent engage in vigorous exercise. Exercise improves cardiovascular and respiratory fitness and general well-being. Each 100 kcal expended in exercise is just as effective as each 100 kcal removed from a diet. Engaging in moderate exercise such as walking, bike riding, and mild "stretch" calisthenics improves overall vigor and emotional outlook, along with a corresponding increase in calorie intake. In turn, these improvements yield greater physical and mental health and greater intake of all essential nutrients, provided that food choices are made from nutrient-dense foods. Thus, physical activity is crucial at all ages—

babies crawl, children run, teenagers race, young adults jog, middle-aged adults walk, and older adults stroll. (See Figure 15–3.)

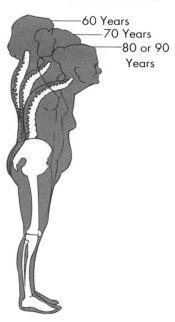

60 Years
70 Years
80 or 90 Years

Osteoporosis

Osteoporosis is recognized as common in older adults. Loss of height, back pain, the "dowager's hump," greater difficulty breathing and eating because of the compression of the vertebrae making the thoracic area smaller and misformed, and hip fractures all become more obvious as both women and men grow older. Osteoporosis is seen in women earlier; men are protected longer chiefly because of their larger, more mineralized bones. Especially for women, lifelong inadequate calcium intake, a drop in estrogen production after menopause, and too little exercise are important factors in the development of osteoporosis. Cigarette smoking and excessive consumption of alcohol also increase the risks. As with many chronic conditions, genetics and medications are influences.

Since the condition becomes progressively worse as a person ages, intervention is aimed at halting its course and is multipronged. Nutritionally, the focus is on adequate calcium from foods, and supplements as indicated; vitamin D; and limited intake of iron, phytate, and wheat bran at the same meal as calcium. Again, individually assess and monitor each person to find the best approach for health care.

FIGURE 15–3
Establishing health-promoting habits throughout life yields benefits. People enjoy longer, more productive older years. Routine outdoor physical activity, coupled with wise food choices and social connections, form a solid foundation. (*Source*: Anthony Magnacca/Merrill.)

Medications

Many older persons swallow as many as 10–15 pills a day. Sometimes they have more than one health care practitioner and do not reveal to all of them the other medications they are taking. Also, older adults may take a number of over-the-counter medications and vitamin/mineral and other dietary supplements. Confusion and even death can occur when alcohol and commonly prescribed anti-depressants, sleeping pills, or tranquilizers mistakenly are taken together. When screening older adults for nutritional health, determine which prescribed medications, over-the-counter preparations, or supplements are being taken, and the timing of administration. Advise, verbally and in writing, exactly when a given medication is to be taken, whether it should be taken with or without food and beverages, and what side effects such as dry mouth, nausea, diarrhea, and constipation may be expected. (See Chapter 18.)

Alternative Therapies

The later years of life often bring, or threaten to bring, serious health problems like heart disease, cancer, and arthritis. Thus, it is not surprising that many older Americans try bizarre diets, herbal remedies, and supplements of all kinds. For example, popular so-called cures for arthritis are found in magazines, circulars, and on television infomercials—ranging from honey and vinegar, to copper bracelets, to megadoses of vitamins. There is no known cure for arthritis, nor do any of these highly advertised remedies help the individual. From a nutritional point of view, those who have arthritis are best served by consuming a well-balanced diet and by losing weight if obese since joints are benefited by less weight bearing upon them, or by gaining weight if thin to provide a cushion of adipose tissue in case of falling down. As an ancient form of practice, alternative therapies periodically receive great renewed fanfare and support. (See Chapter 2.)

OLDER ADULT NUTRITION PROGRAMS

The U.S. Department of Agriculture provides programs specially designed to improve the nutrition of older adults. Through amendments to the federal Older Americans Act, regional agencies on aging throughout the United States provide a variety of services to persons over 60 years.

A major nutrition program is the Congregate Dining Program, held at senior-citizen centers. This program provides a noontime meal each day for at least five days a week. Each meal furnishes at least one-third of the RDAs. Provisions for modified diets can be made. Nutrition education, periodic health checks, and recreational and social activities are important components of the program. For persons unable to walk or drive to the center for meals, transportation is provided.

A second major nutrition program is the Home-Delivered Meals Program, sometimes known as "Meals-on-Wheels." This program provides meals for homebound older adults who cannot shop or cook meals for themselves. Usually, one hot meal is delivered for immediate consumption at midday, along with a second cold meal for the evening. In some regions, a week's supply of meals are provided at one time. In either situation, drivers of the vehicles have the opportunity to check in with the recipient on a face-to-face basis. Meals are paid for on a sliding fee scale according to the person's ability to pay. Many civic groups, hospitals, and churches are involved with this program.

A third program is the Food Stamp Program. For those who qualify on the basis of their incomes, this program helps provide opportunities to obtain foods. Nutrition guidance and targeted shopping tips, like buying food for one, make the most of stretching the older adult's buying power.

PUTTING IT ALL TOGETHER

Although health-promoting measures taken in the early years of life yield the greatest protection against chronic diseases in the older years, changes in lifestyle at any age can have positive effects. Important elements for improving the quality of life for older Americans are choosing foods according to the Food Guide Pyramid, engaging in outdoor physical activity, avoiding substance abuse, and maintaining social connections.

? QUESTIONS CLIENTS ASK

1. Would frozen meals be a good idea for me to eat?

Many kinds of frozen meals are available for breakfast, lunch, dinner, and snacks. Since the label lists the ingredients and some nutrition information, you could tell if the contents might be right for you. Check prices; they can vary for the same kind of food product. Rounding out the meal with something fresh, like fresh fruit at breakfast or a salad at dinner, along with some milk, helps balance the nutrients.

2. What can I do about constipation?

Constipation (and diarrhea) can be a problem because of several factors, such as reduced tone of the intestines and changes in diet. Some factors to consider are: making certain you are drinking enough water and other fluids; eating enough fiber, like fresh fruits, vegetables, and whole grains; getting enough physical activity; and having regular mealtimes and habits of elimination.

✔ CASE STUDY: An Older Man Has Nutritional Concerns

Mr. Joseph S. is an alert 88-year-old man who lives alone in his home. Two months ago he fell, had a hip fracture, and currently uses a walker to get around his house. He used to do all the repairs on his home, but now can't use a ladder.

Mr. S. just received a large utility bill. He says his living expenses keep going up while he has to live on a fixed income, mostly Social Security and a small amount of interest from savings. By the time all the bills are paid, there sometimes is not quite enough money for food. He is afraid he will be a burden on his son, who lives 15 miles away.

Mr. S. notices his appetite isn't the same as it used to be and he does not have the energy to cook for himself. He forces himself to eat a little bit, but he finds he has trouble chewing food. His clothes do not seem to fit anymore; they just hang—a big difference from how he used to dress. He also has been experiencing cramping, gas, and diarrhea. These symptoms appear to be like those describing cancer of the bowel that he read about in a magazine for senior citizens. He wonders if this could be the underlying cause of all his problems.

Questions

1. Using Appendix Figure C–1, evaluate Mr. S.'s nutritional health.

2. Medical studies did not find any evidence of cancer. What are some factors that may explain Mr. S.'s gastrointestinal symptoms? Hip fracture? Chewing difficulties? Lack of energy? Alterations in appearance?

3. What nutritional assessment evaluations are especially pertinent to perform with Mr. S.? (See Chapter 3.)

4. What nutrients are especially needed by a man like Mr. S.? What foods supply them? (See also Food Guide Pyramid and Nutrition Label, Figures 1-5 and 1-9; Basic Diet, Tables 2-3 and 2-4; and Nutritive Value of the Edible Part of Food, Appendix Table A-1.)

5. Plan a sample meal pattern and menus for one day with Mr. S. (See also Chapter 2.)

6. Mr. S. is thinking about ordering dietary supplements designed for "active seniors" advertised in his magazine because he wants his former pep and energy. How would you respond?

7. What role, if any, might Mr. S's son have regarding his father's foods and nutrition situation?

8. If Mr. S. lived in a neighborhood near where you attend school, or live, or work, what are some voluntary or governmental resources related to foods and nutrition that might be available for Mr. S.?

FOR ADDITIONAL READING

Gray-Donald, K: "The Frail Elderly: Meeting the Nutritional Challenges," *J Am Diet Assoc,* **95**:538-40, 1995.

Nutrition Screening Initiative's Technical Review Committee: "Appropriate and Effective Use of the NSI Checklist and Screens," *J Am Diet Assoc,* **95**:647-48, 1995.

O'Donnell, ME: "Assessing Fluid and Electrolyte Balance in Elders," *Am J Nurs,* **95**:40-46, November 1995.

Ryan, C, Eleazer, P, and Egbert, J: "Vitamin D in the Elderly: An Overlooked Nutrient," *Nutr Today,* 30:228-33, 1995.

Saunders, MJ: "Incorporating the Nutrition Screening Initiative into the Dental Practice," *Spec Care Dentist,* **15**:26-37, January/February 1995.

PART III

Food and Nutrition Therapies

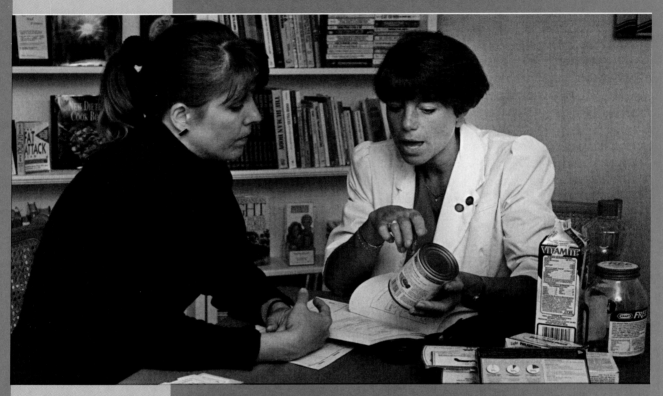

In Part III you will learn how to apply the principles of nutrition to the therapy of specific pathological conditions. The maintenance of good nutrition is crucial to all patients with both acute and chronic diseases.

Satisfactory nutrition of patients depends on the team approach with the patient and his or her family and support system, physician, nurse, dietitian, pharmacist, social worker and other health care providers participating. Nutritional screening identifies patients who are at risk, and comprehensive assessment is made for patients who are found to be at increased nutritional risk. Dietary counseling is an essential aspect of therapy so that the patient and his or her family understand the rationale for any diet modifications and will be able to implement any changes indicated.

The adequate normal diet is the foundation upon which modified diets are planned. In some instances the basic diet is modified for texture and consistency. Other conditions require alteration in energy and/or one or more nutrients; for example, high- and reduced-calorie diets, sodium-restricted diets, modified-fat diets, and so on. Enteral and parenteral nutrition under carefully controlled modifications are essential for complex metabolic changes such as those that occur in burns, renal failure, AIDS, and others. Infections, fevers, the immune response, stress, trauma, metabolic changes, and surgery influence nutritional requirements.

Numerous medications used in the treatment of diseases often lead to changes in appetite or taste, or may interact with nutrients so that the absorption and metabolism of the nutrient is reduced or increased. On the other hand, foods and their nutrients may alter the action of medications. Some medications are taken with meals or with specific foods, while others are more efficacious when taken away from mealtime. Mineral and vitamin supplements may be indicated when medications reduce the utilization of nutrients. Nutritional products named in this section are intended only as examples and do not constitute endorsement by the authors nor suggest that choices are limited to the products listed.

Nutritional Care in the Clinical Setting

Chapter Outline

Illness and Nutrition
The Team Approach
Nutritional Assessment in the Clinical
  Setting
Estimation of Energy and Protein
  Requirements

Practical Applications for Nutritional Care
The Patient's Meals
Putting It All Together
Questions Clients Ask
Case Study: An Older Woman with Severe
  Arthritis

Increase to at least 95 percent the proportion of people who have a specific source of ongoing primary care for coordination of their preventive and episodic health care. *

✔ CASE STUDY PREVIEW

In this chapter you will conduct a nutritional assessment of an older woman with limited income and various health problems. You will plan for improved nutrition and to enable her to remain in her own home.

Healthy People 2000: National Health Promotion and Disease Prevention Objectives. Public Health Service, U.S. Department of Health and Human Services, Washington, DC, 1991, Objective 21.3, p. 124.

All health care including nutrition is preventive in nature and is divided into three levels. **Primary prevention** focuses on people who currently have no health problems but might be at risk for future problems. **Secondary prevention** involves education, counseling, and/or treatment of individuals experiencing health problems to prevent complications. **Tertiary prevention** focuses on rehabilitation of people with short-term or chronic problems, disabilities, or functional difficulties that have compromised their health.

ILLNESS AND NUTRITION

Illness and Food Acceptance

The many physiological, cultural, economic, and emotional factors affecting food acceptance were discussed in Chapter 2. People who are ill face added problems related to their meals. Sometimes it is necessary to take therapeutic measures that distress rather than provide immediate comfort. Members of the health care team play an essential role in bridging the gap between comfort and therapy. (See Figure 16–1.)

Illness itself often reduces interest in food because of anorexia, gastrointestinal distention, or discomfort following meals. Inactivity and some medications also reduce the desire for food.

A hospital patient is subject to many stressors. Numerous questions are asked, some of which are very personal. The patient is escorted to various parts of the hospital for numerous tests, examinations, and therapies that sometimes involve what seem to be endless waits. Sometimes the purpose of the procedure is not adequately explained, or if it is, one is too distracted to fully comprehend. The

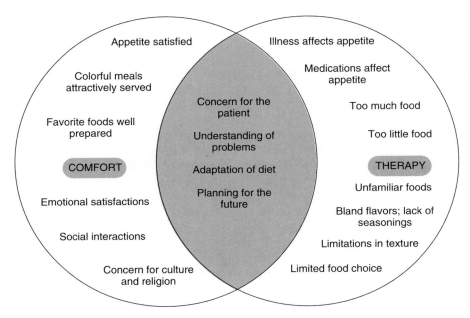

FIGURE 16–1
Patients seek comfort through food. Modified diets often present problems regarding food acceptance. Through counseling and encouragement the nurse, dietitian, and dietetic technician can help the patient to resolve these problems.

patient may view the frequent entering of the hospital room by staff as an invasion of privacy. Being addressed by a first name may be considered demeaning by an older person who prefers being called Mr., Miss, or Mrs. Roommates are sometimes incompatible.

Times of awakening and meals often vary greatly from the customary schedule. Control over one's life and future seemingly has been lost. The patient may feel identified not as a person, but as "the ulcer in room 213." Numerous health care professionals, staff, and other hospital employees pass through the room; care seems to be provided by a cast of thousands. Yet, unless there is a primary care nurse or nursing case manager, care is so fragmented that no one person has responsibility for the patient.

If the diet is modified, the patient may be getting less or more food than usual. The change in flavor or texture of some diets may not be appealing. Some modified diets make the patient feel deprived and punished. Often there is unwillingness to accept any change. The patient may react with irritation or even anger. Complaints about food, indifference to diet, poor eating, or ignoring suggestions made by doctors, nurses, or dietitians may occur.

Illness and Nutrition

Illness has many effects on nutrient requirements and utilization. Lack of appetite, vomiting, and pain often prevent sufficient intake of food. Severe diarrhea reduces nutrient absorption, so that weight loss, dehydration, and signs of malnutrition may occur. By raising the metabolic rate, fever increases the need for calories, protein, and vitamins. In metabolic diseases, nutrients often are not utilized fully. The patient who must remain in bed or in a wheelchair for a long time usually loses increased amounts of nitrogen and calcium from the body. As you become more experienced in the care of patients, numerous examples of the effects of illness upon nutrition will become apparent.

Medical Nutrition Therapy

POSSIBLE AIMS OF MEDICAL NUTRITION THERAPY

Maintain normal nutrition
Correct nutritional deficiency, e.g., high-protein diet
Change body weight
Adjust to body's ability to use a nutrient, e.g., diet for diabetes mellitus
Permit the body or an organ to rest

An appropriate diet is an integral part of the care plan developed for each patient. The normal diet is the basis on which all therapeutic or modified diets are planned. The diet may supplement medical or surgical care or be the specific treatment for a disease.

Medical nutrition therapy is the use of specific nutrition services to treat an illness, injury, or condition. Medical nutrition therapy begins with assessing the nutritional status of patients deemed to be at risk. A personal diet plan to reach desired outcomes is developed. It may include diet modification, supplementation, and/or enteral or parenteral nutrition. Medical nutrition therapy can occur in acute care, long-term care, home health care, or outpatient care settings. It can be preventive to reduce the incidence and severity of chronic diseases. An essential component of comprehensive health care, it will result in both health benefits and reduced health care costs.

Most hospitals, extended care facilities, and clinics have a diet manual that describes diets served or used in the facility, as well as procedures and policies for ordering diets or dietary changes, arranging conferences with a dietitian, channeling patients' requests, answering patients' questions, and so on. All members of the health care team should acquaint themselves with this important resource.

THE TEAM APPROACH

Providing optimal nutrition for any patient requires a coordinated team approach. The basic team members are the physician, dietitian, dietetic technician, and nurse. Other members may include the pharmacist, social worker, speech language pathologist, and occupational and physical therapist. All can and should serve as nutrition advocates for the patient. (See Figure 1–1.) The central person is the patient, who must be involved and should participate as much as possible in planning for and implementing all care.

The physician and dietitian work closely together in prescribing the appropriate diet for each patient. Each diet prescription is the result of an evaluation of the patient's nutritional needs, symptoms, and laboratory tests if they have been done. The physician should explain the purpose of the diet to the patient.

The dietitian translates the prescribed diet into foods or nutritional products. Assessment of the patient's nutritional status, writing of a nutritional care plan, creation of meal plans incorporating both prescribed dietary modification and the patient's preferences, and counseling the patient and the family are among the dietitian's responsibilities. The staff of the dietary department is responsible for preparation and serving of food and is also involved in the patient's response to the diet. The dietitian may be assisted by a dietetic technician.

The nurse has the most continual and direct contact with the patient. After assessing the patient's status, the nurse makes a **nursing diagnosis.** A nursing diagnosis is a statement of a health-related problem or potential problem resulting from identification of signs and symptoms, and is based on objective and subjective data that can be verified. It differs from the medical diagnosis in that it includes factors such as assistance, hygiene, prevention, and safety. Nursing diagnoses involve conditions that can be treated legally and independently by nurses. Nursing interventions and evaluations can then be planned.

A tentative list of nursing diagnoses has been developed of which three mention nutrition specifically. Many nursing diagnoses have nutrition implications. A list of nursing interventions coordinated with the nursing diagnoses has also been developed.

In primary nursing, one nurse is responsible for the care of a patient from admission to discharge. Through coordination, management, and communication, the nurse is accountable for the care, even when not physically present.

Case managers oversee, manage, and account for the complete health care of a group of patients over a period of time. Case management can take place in the hospital, extended care facility, and/or home and follow patients as they move from one to another. Case managers may be nurses, social workers, dietitians, or possibly other members of the health care team. The objectives are to provide quality care, prevent fragmentation of services, and contain costs.

NUTRITIONAL ASSESSMENT IN THE CLINICAL SETTING

No single test exists that unfailingly predicts nutritional risk, nor is the ultimate combination of tests known. Few patients present a completely normal profile when assessed by a variety of methods. Although numerous techniques exist to identify individuals at risk or frankly malnourished, assessment is still an evolving science and art.

Assessment of nutritional status should be a routine aspect of the care of all patients. Since in-depth assessment is time consuming, costly, and not required by

---

NURSING DIAGNOSES MENTIONING NUTRITION

Nutrition, alteration in: less than body requirements
Nutrition, alteration in: more than body requirements
Nutrition, alteration in: potential for more than body requirements
(Carpenito, 1995)

SOME NURSING DIAGNOSES WITH NUTRITIONAL IMPLICATIONS RELATE TO

body temperature, bowel elimination, breast feeding, fluid volume, growth and development, infant feeding patterns, knowledge deficit, impaired mobility, noncompliance, parenting, self-care deficit, skin integrity, and swallowing (Carpenito, 1995)

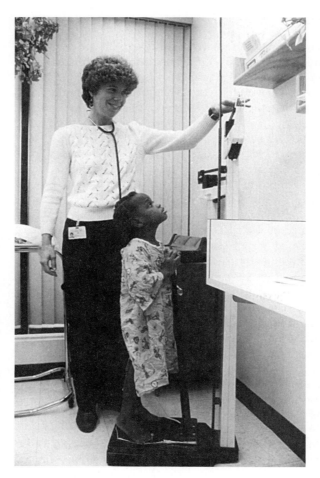

FIGURE 16–2
A patient's weight should be taken at the first visit to an ambulatory care setting or admission to the hospital. Periodic weights should be taken to determine changes from time to time. (*Source:* Department of Health and Human Services, Office of Human Development Services.)

all patients, a screening protocol, or description of steps to be taken, is needed to identify individuals at risk for malnutrition. Only patients at risk will have a comprehensive nutritional assessment.

Prospective audits have shown that malnourished patients have longer hospital stays, a higher rate of morbidity and mortality, and decreased response to therapies than those who are not malnourished. If at-risk patients can be recognized early and measures taken to prevent or reverse malnutrition, morbidity and mortality can be reduced and hospital stays shortened.

Screening

Nutrition screening is the process of identifying characteristics known to be associated with nutrition problems. The screening protocol will vary from facility to facility depending on the nature of the institution and its patient population.

Every patient should be weighed and measured upon admission to the health care system to provide baseline information. In addition, weight should be monitored periodically, whether or not weighing is part of the physician's order. (See Figure 16-2.) Patients who cannot stand or who are immobile can be weighed in chair scales or bed scales. (See Chapter 3 for a discussion of weighing and measuring techniques.) Weights significantly above or below accepted standards or recent unplanned marked weight gains or losses may be indicative of nutritional risk.

Depressed serum albumin and total lymphocyte counts are thought to have as great prognostic value for nutrition problems as more elaborate testing batteries. Nutrition histories and clinical evaluations (see Chapter 3) can adequately single out many individuals at risk nutritionally. Functional assessment of skeletal and respiratory muscles, heart, and liver are also good predictors of malnutrition. In a long-term care facility, additional indicators of risk include chronic disease, decubitus ulcers, and low hemoglobin and hematocrit values. Use of medications that interact with nutrients increase risk of poor nutritional status. Multiple medications can increase this potential. (See Chapter 18.) Aspects of the patient's history that may alert the assessor to increased risk of malnutrition are presented in Figure 16-3.

Concern about the connection between malnutrition and poor patient outcomes has led several national organizations to create the Nutrition Screening Initiative. Its objective is to promote nutrition screening as an essential element in the nation's health care system, focusing initially on older people who are at high risk. For a nutrition screening instrument developed by this initiative, see Appendix Figure C-1.

Comprehensive Assessment

Patients who have been shown through screening to be at risk of malnutrition require more detailed nutrition assessments including nutrition, medical, and medication histories. Once the nutritional status and the nature, extent, and cause of any problems have been determined, a nutrition care plan can be developed.

Anthropometric Measurements In the hospital setting, weight changes are of greater relevance than comparing the weight to some standard. For example, a patient may have an unexplained significant weight change before seeking health care, yet still be overweight. Determining the cause of the unplanned weight loss (or gain) is more important at this point than comparing the present weight to a height/weight chart. If no record of past weight(s) is available, learning the

Decubitus ulcers: also known as bedsores or pressure sores

- Recent unplanned loss or gain of 10% or more of usual body weight
- Inadequate or inappropriate oral intake of nutrients
- Difficulty with chewing or swallowing and/or poor appetite
- Multiple medications
- Increased needs due to infection, fever, trauma, burns, recent surgery
- Chronic disease(s) or impairment of any major body system
- Nutrient losses due to malabsorption, draining wounds or abscesses, prolonged vomiting or diarrhea
- Depression or mental or cognitive impairment
- Assistance needed for activities of daily living (ADL)
- Alcohol and/or substance abuse

FIGURE 16-3
Some factors associated with risk for malnutrition.

% OF USUAL BODY WEIGHT

$$\frac{\text{present weight}}{\text{usual weight}} \times 100\%$$

<90% or >110% indicates
need for further assessment

ASSESSMENT OF VISCERAL PROTEINS

Serum albumin, prealbumin,
Transferrin
Dietary protein intake
Blood urea nitrogen
Immune function

ASSESSMENT OF SOMATIC PROTEINS

Anthropometric measures
Lean body mass measures

CREATININE HEIGHT INDEX

80% = moderate depletion

Recall Antigens Commonly Used:

Candida albicans, mumps,
streptokinase-streptodornase
(SK-SD), trichophytin,
coccidioidin, and tuberculin
skin test antigen (purified
protein derivative, or PPD)

patient's usual weight and comparing it to the present weight can be valuable. Recent unplanned weight loss is a critical factor in the potential for malnutrition. Standards for determining risk appear in Appendix Table D-2.

Abnormal triceps skinfold (TSF) and midarm muscle circumference (MAMC), discussed in Chapter 3, are also indicators of malnutrition.

Laboratory Measurements Measurement of serum albumin and serum transferrin is thought by many to have the greatest potential for identifying changes in visceral proteins. Serum albumin is most frequently measured and is thought by some to be the single best predictor of the course of the illness. Depressed values are associated with delayed wound healing, decreased resistance to infection, increases in duration of hospital stay, morbidity, and mortality. Nonnutritional factors such as stress, infection, and decreased synthesis also affect serum albumin levels. Prealbumin is sometimes measured because it has a very short half-life, decreasing rapidly during malnutrition and returning promptly to normal on refeeding.

Transferrin may be analyzed directly from the serum or calculated from the total iron-binding capacity (TIBC). Transferrin has a shorter half-life than albumin, and therefore provides an earlier indication of protein deficiency. Levels are also influenced by factors such as stress and trauma. Reference standards for serum albumin and transferrin appear in Appendix Table D-2.

Somatic protein, found in muscle, can be estimated by measuring urinary creatinine, a breakdown product of creatine that is found in muscle. Creatinine excretion for a 24-hour period is compared to reference standards for men and women based on body weight or by computing the creatinine height index, a 24-hour urinary creatinine compared to the projected output of a healthy adult of the same gender and height. The collecting of all urine for a 24-hour period is not always practical.

Nitrogen balance is approximated by comparing urinary urea nitrogen (UUN) in a 24-hour sample of urine to nitrogen intake. A factor of 4 is added to UUN to cover nonurea nitrogen losses such as those from the skin, nails, hair, and feces. (See also Chapter 7 and Appendix Figure B-1.)

Serum enzymes are low in healthy individuals, but in some disease states are released into the blood. Enzyme evaluations allow the location, extent, and course of a disease to be determined and followed.

Vitamin levels in plasma reflect recent intake while intracellular levels mirror overall vitamin status. Individuals with depressed vitamin levels are at risk for infections and delayed wound healing.

Immunological Measurements The total lymphocyte count is a commonly measured indicator of immune response. Decreased levels are associated with a reduced ability to fight infections. (See Appendix Figure B-1 and Table D-2.)

Delayed cutaneous hypersensitivity involves the injection of a panel of recall antigens into the skin with response evaluated after 24, 48, and/or 72 hours, depending on the test protocol. The normal individual responds to one or more of the antigens with development of an **erythema** (redness) and **induration** (hardening) at the injection site. Sometimes the size of the response area is evaluated.

Seriously undernourished patients might display **anergy,** a failure of the immune system to recognize and respond to foreign proteins. Mildly depleted individuals would show a response intermediate between normal and anergy. Skin test responses would be impaired in malignancy, stress, infectious diseases caused by some pathogenic bacteria, fungi, parasites, and viruses, metabolic diseases, surgery, trauma, burns, and with some medications such as steroids, chemotherapy agents, and immunosuppressants. (See also Chapter 18.)

Classification and Criteria for Malnutrition Malnutrition is sometimes classified as follows:

- **Kwashiorkor malnutrition** is characterized by compromised protein status. Serum albumin and transferrin and immune response are depressed. The patient may appear well nourished. (See also Table 7–2.)
- **Marasmic malnutrition** involves depletion of lean body mass and energy reserves. Body weight, triceps skinfold (TSF), and/or midarm muscle circumference (MAMC) are below standard. Serum albumin and transferrin may be normal. (See also Table 7–2.)
- **Combined kwashiorkor-marasmic malnutrition** involves both protein-energy malnutrition and visceral protein depletion.

Ongoing Assessment

When assessment indicates that nutritional intervention is needed, periodic follow-up assessments should be conducted. Protocols will vary, but the patient should be weighed several times a week. Other anthropometric measurements, serum albumin and transferrin, total lymphocyte count, and cell-mediated immunity (see Chapter 19) may be evaluated periodically. The effectiveness of the nutritional care plan can thus be evaluated. Members of the health care team should be alert to potential changes in the nutritional status of all patients, even those shown to be normal by the initial screening process.

Patients who are hospitalized for two weeks or longer are at particular risk of developing **iatrogenic** (doctor- or hospital-induced) **malnutrition** due to inadequate nutrient intake, stress, and blood loss through surgery and blood tests. Especially for such patients, but for other patients also, nutritional assessment, surveillance, and vigilance on the part of the care team must be ongoing.

ESTIMATION OF ENERGY AND PROTEIN REQUIREMENTS

When the patient is judged to be malnourished and nutrition intervention is indicated, means of determining energy and protein needs are helpful.

Energy Needs

To set goals for energy intake, the sum of the basal needs plus activity is viewed in light of the nutritional goal for the patient—rehabilitation, maintenance, or weight loss. (See Chapter 8 for factors influencing energy requirements.)

One way to determine the energy needs of hospitalized patients is to measure the resting energy expenditure (REE). The volumes of oxygen consumed (VO_2) and carbon dioxide expired (VCO_2) are measured with portable equipment that can be brought to the patient's bedside. The REE is generally elevated by burns, fever, infection, multiple fractures, trauma, and hypermetabolic states, and lowered in malnutrition. The Harris Benedict equations derived from indirect calorimetry measurements with increments for injury factors are also used to determine energy needs, as are simpler formulas. (See Chapter 8.)

Protein Needs

Stress increases nitrogen loss. If urinary nitrogen losses are known and there are no abnormal exudates or malabsorption factors, the protein intake can be planned to correct for these losses. When urinary losses are not known, protein intake can be proportional to energy intake. (See Chapter 7.)

PRACTICAL APPLICATIONS FOR NUTRITIONAL CARE

Omnibus Budget Reconciliation Act (OBRA)

This legislation provides for extended care facility reforms, some of which relate to nutrition and diet. Residents must have nutritional assessments upon admission and periodically thereafter, a dining setting to optimize quality of life, an adequate palatable diet and sufficient fluids, plus a choice in where, when, and what food is eaten. A qualified dietitian must be employed on a full-time, part-time, or consultant basis.

Joint Commission on Accreditation of Healthcare Organizations (JCAHO)

The JCAHO is a private nonprofit organization that accredits hospitals, long-term care facilities, and organizations offering acute care, ambulatory care, mental health services, and/or home care. The standards it measures include functions that affect patient outcomes and are interdisciplinary. Facilities are encouraged to develop plans to meet these standards that are consistent with the organization's resources and patient needs.

All patients are to have a nutrition screening and, if indicated, a nutrition assessment. A plan for nutrition therapy is developed, ordered, implemented, and monitored. Nutrition care teams include physicians, registered dietitians, nurses, pharmacists, and other health care professionals.

Medical Record

The medical record is an information-sharing tool that assists in communication and in coordinating activities of the health care team. Nutrition information is an important part of the medical record. It is a legal document.

A dietitian or dietetic technician records the nutrition information. It might include a care plan with diet orders; a summary of the diet history, medical nutrition therapy, and patient outcomes to be achieved; a report of the patient's acceptance and tolerance of the diet; communications between the dietary staff and other members of the health care team concerning the patient's nutrition; requests for diet consultations or instructions; the patient's progress toward achieving objectives; and a nutritional care discharge plan.

A Kardex system or notebook may be used to organize information about nutrition and/or nursing care plans for a patient in a concise form where it can be found and read quickly without paging through the medical record. This information is gathered from the medical record and the patient.

Health care facilities have their own policies concerning the format of their medical records. Some are organized according to sources of information, some use a body systems approach, others are structured around the patient's health problems. Many facilities have computerized their medical record keeping. (See Figure 16–4.)

FIGURE 16–4
The patient's medical record is a primary means of communication for the interdisciplinary team. On reviewing this medical record, the nurse extracts information on the patient's food intake and on tests that relate to nutritional status. (*Source:* Dunwoody Village, Newtown Square, PA, and Peter Zinner, photographer.)

Problem-Oriented Medical Record (POMR) Problem-oriented medical records are organized according to the patient's major problems. They include a database, problem list, and progress notes. Entries are organized according to the SOAP format. The problem list may change as the care plan is implemented. Nursing diagnoses may also be recorded according to the SOAP format. Additional formats in use include Charting by Exception (CBE), in which only changes are recorded, FOCUS on problems, and others.

Home Health Services

Home health services are urgently needed. It has been estimated that 20–40 percent of extended care facility residents could be at home if adequate services were available.

Home Health Care The Diagnosis-Related Group (DRG) system of reimbursement has resulted in earlier discharge from hospitals and a greater emphasis on home care. This trend has many nutrition ramifications, for more than 60 percent of patients referred to home health care are on modified diets. More patients are going home with tube feedings and parenteral nutrition. (See Chapter 17.) Many have chronic diseases that affect food intake, digestion, and metabolism. Multiple medication therapies may compromise nutritional status.

Home health agencies provide nursing services, home health aides, occupational and physical therapy, speech therapy, and social services. Dietitians with

SOAP FORMAT EXAMPLE

S: subjective; e.g., lack of appetite
O: objective; e.g., 10 lb weight loss
A: assessment; e.g., inadequate intake of calories, protein, iron
P: plan; e.g., contact physician about iron supplement, have patient eat meals with others to increase social interaction

Diagnosis Related Groups (DRGs):
classification system that groups patients according to principal diagnosis, surgical procedure, age, significant comorbidities or complications, and other relevant criteria

these agencies provide nutrition screening and assessment, consultation with other team members, care conferences, and home visits. They prepare nutrition education materials for staff and clients and provide staff development workshops.

Home Delivered Meals Many individuals with limited mobility can remain in their homes if they have home delivered meals. The meal consists of a dinner delivered at midday. Often, a cold meal for the evening is also provided. This service is provided through senior centers or Meals on Wheels programs through volunteers. (See Chapter 15.)

Hospice

Advance Medical Directive: also known as Living Will or Durable Medical Power of Attorney

Hospice care for terminally ill patients is a philosophy, not a facility. The aim is to provide continuity of care, taking the patient's preferences and the needs of the family and caregivers into account. The patient remains at home as long as is desired and feasible. The hospice may be a special unit of a hospital, extended care facility, or boarding home. Pain alleviation is an important factor in the care. A patient's right to self-determination should be respected. For example, the patient's advance medical directive might specify that no intravenous or tube feeding is desired.

THE PATIENT'S MEALS

Ensuring Adequate Nutrition and Pleasant Mealtimes

A selective menu helps to provide the patient with preferred foods and also gives him or her some sense of control over the situation. The nurse or dietetic technician, hospital volunteers, or family members may assist the patient in completing the day's menu. The menu choices should be reviewed to ensure that the meals are nutritionally adequate. (Refer to Figure 17–1.)

A pleasant environment is essential to the greatest enjoyment of food. Every effort should be made to ensure that surroundings are clean, orderly, and well ventilated. Patients should have time to rest before and after meals and treatments, and physical therapy should be scheduled away from mealtime. Activities not associated with meal service should be avoided while patients are eating.

If there is a delay because of a laboratory test, arrangements should be made to keep food hot or chilled in a nearby kitchen. A microwave oven is helpful for reheating food. Before the tray arrives, the patient's hands should be washed, and the patient should be positioned for comfort, safety, and ease of eating. If the patient wears dentures or glasses, they should be in place. The tray should be checked to be certain that everything needed is there and is conveniently placed. Sometimes it is necessary to open food containers, cut the meat, or spread the bread topping. The patient should be encouraged to eat but should not be made to feel rushed. The tray should be removed promptly when the patient is finished. In some circumstances, patients may eat in a dining room or otherwise be brought together to provide social interaction.

The nurse is more likely than anyone else to observe how well the patient eats, what kinds and amounts of food are refused, and attitudes toward food. Problems such as poorly fitting dentures and inability to chew, portions that are too large for some older persons or too small for teenagers, and excessive numbers of missed meals because of an order for nothing by mouth (NPO)—required for some tests—should be observed, reported, and resolved.

Good Tray Service

Meals are often the high point of the day for patients. An attractive tray of well-prepared food presented cheerfully goes a long way toward ensuring acceptance. The essentials of good tray service are

- A tray of sufficient size for uncrowded arrangement of dishes
- An immaculately clean, unwrinkled tray cover and napkin
- An attractive pattern of spotless china, sparkling glassware, shining silverware, and attractive, clean paper goods
- Convenient, orderly arrangement of all items on the tray so that the patient can reach everything easily
- Portions of food suitable for the patient's appetite and needs
- Food attractively arranged with appropriate garnishes
- Hot foods served on warm plates and kept warm with food cover; cold foods served on chilled dishes
- Trays promptly served to the patient so that food is at its best

Feeding the Patient

Acutely ill, disabled, and elderly infirm patients must sometimes be fed. Patience and understanding are required, especially if the patient eats slowly.

Proper patient positioning will assist swallowing and prevent reflux and aspiration. A sitting position is preferable, but if this is not feasible elevate the head. If lying down is unavoidable, it should be on the side with the head turned. Eating while laying flat on the back increases the possibility of aspiration.

Protect the patient's clothes with a napkin. Before starting to feed the patient, make sure you have everything you need. Feed the patient at a rate appropriate for the condition. Encourage the patient to indicate the order in which foods are preferred and to participate in any way possible. Sit rather than stand, so that you and the patient will be more comfortable and relaxed. If the patient is visually challenged, describing the food items and their placement on the tray and possibly rearranging foods for the patient's convenience will be helpful.

Focus on the patient, rather than the feeding process. Avoid giving the impression that feeding is a bother and you must hurry. Talk about pleasant things and give the patient a chance to talk. Listen to what the patient has to say. Give encouragement for any progress. (See Figure 16–5.)

Family members may wish to participate in feeding the patient. They may also wish to bring favorite foods consistent with any dietary modification and, of course, food sanitation standards. The family can thus feel that they are helping with the care.

PUTTING IT ALL TOGETHER

Nutrition is an essential part of patient care in any health facility. The health care team working together with each member providing expertise in a specialty can help provide optimal nutrition for patients. Assessment to determine nutritional status, careful planning, and documentation will accomplish this goal.

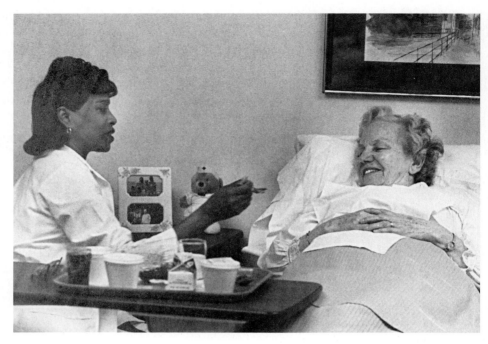

FIGURE 16–5
When feeding a patient, the care provider should be seated and unhurried. Pleasant conversation between patient and assistant helps to provide a congenial environment. (*Source:* Dunwoody Village, Newtown Square, PA, and Peter Zinner, photographer.)

? QUESTIONS CLIENTS ASK

1. I've heard and read so much about identifying the genes associated with chronic diseases and about gene therapy. How does this affect nutrition and nutrition therapy?

As we learn more about genes, their involvement with disease, and their reaction with nutrients, it will be possible to plan diets specific to individual genotypes rather than broad-based diets that may not apply to all.

2. Why do you weight me? My weight seldom changes.

It is important to know your weight for baseline information. Unplanned weight loss or gain can be an indication of some health problem, so we monitor your weight periodically to determine whether there is any change.

3. Why did you do a nutrition screening with my mother? I think her diet is fine.

The screening will address various factors that can contribute to risk of poor nutrition—medications, problems with chewing, and so on.

✔ An Older Woman with Severe Arthritis

A volunteer who delivers meals to the home-bound noticed that Ruth A. was not eating her meat or her raw fruits and vegetables. She reported this to her supervisor, and a visit and assessment were made by the social worker and

home health nurse. The nurse assembled the data recorded in the diet history in preparation for a visit by the consultant dietitian.

Mrs. A. is a 76-year-old widow who lives alone. Her income is from Social Security. She is

63 in. (160 cm) tall and weighs 92 lb (42 kg). Her usual weight is 103 lb (47 kg). She had a recent unplanned weight loss of 11 lb (5 kg). Her hair is dry and thin, her skin pale, her lips inflamed, and she has fissures around her lips. Her posture is stooped.

She usually rises at 8 A.M. and has meals alone at approximately 9 A.M., noon, and 5 P.M. She seldom snacks and does not smoke or drink alcoholic beverages. Mrs. A. has poorly fitting full dentures and her appetite is poor. She has constipation.

Her hemoglobin is 9.4 g/dl, hematocrit 33% and serum albumin 2.8 g/dl. She takes buffered aspirin twice a day for relief from arthritis.

Her favorite foods are bread and canned fruit, which she eats twice daily. She does not eat meat, raw fruits, or raw vegetables, which she likes but cannot chew. She doesn't drink milk because she thinks that at her age she does not need it.

Complete the nutrition screening form (Appendix Figure C–1) for Mrs. A. Using the diet history form in Appendix Figure C–2, fill in as much of the information as you can for Mrs. A. From information provided, write a possible 24-hour food intake.

Questions

1. Which of the data about Mrs. A. are anthropometric? Clinical? Biochemical? Dietary? Point out data from one or more of these areas that reinforce information from another area. Explain.

2. Compare results from each category of nutritional assessment to standards. (See Figure 1–5, Table 3–1 and Appendix Table D–2.) Which results are normal? Low? High?

3. How may her arthritis affect her food intake?

4. What role does dental health play in Mrs. A.'s food intake? How might this situation be dealt with?

5. What may be the relationship between Mrs. A.'s food intake and her constipation? How might her food intake be modified to relieve this condition?

6. What are the nutritional implications of aspirin? (See Chapter 18.)

7. In the counseling process, what are some goals and outcomes toward which Mrs. A. and the dietitian might plan?

8. Using the Food Guide Pyramid (see Figure 1–5) and the Basic Diet (see Table 2–4), help Mrs. A. plan a nutritious one-day menu based on her resources, daily routine, food preferences, and health assessment.

9. Should Mrs. A. receive vitamin/mineral supplements? Explain your answer.

10. What suggestions might you make to ensure that her mealtimes are as pleasant as possible?

11. Prepare a POMR problem list for Mrs. A.

12. Write SOAP notes for any of these problems.

13. Identify community resources to assist Mrs. A. in remaining in her home.

14. How might you and Mrs. A. evaluate her progress toward the goals in question 7?

FOR ADDITIONAL READING

"Appropriate and Effective Use of the NSI Checklist and Screens," *J Am Diet Assoc,* **95:**647–48, 1995.

Bélanger, M-C, and Dubé, L: "The Emotional Experience of Hospitalization: Its Moderators and Its Role in Patient Satisfaction with Foodservices," *J Am Diet Assoc,* **96:**354–60, 1996.

Breslow, RA, and Bergstrom, N: "Nutritional Prediction of Pressure Ulcers," *J Am Diet Assoc,* **94:**1301–06, 1994.

Bulechek, GM, *et al:* "Nursing Interventions Used in Practice," *Am J Nurs,* **94:**59–66, October 1994.

Calianno, C, and Pino, T: "Getting a Reaction to Anergy Panel Testing," *Nurs 95,* **25:**58–61, January 1995.

Carpenito, LJ, ed: *Nursing Diagnosis: Application to Clinical Practice,* 6th ed, Philadelphia: JB Lippincott, 1995.

Corti, M-C, *et al:* "Serum Albumin Level and Physical Disability as Predictors of Mortality in Older Persons," *JAMA,* **272:**1036–42, 1994.

Gallagher-Allred, CR, *et al:* "Malnutrition and Clinical Outcomes: The Case for Medical Nutrition Therapy," *J Am Diet Assoc,* **96:**361–69, 1996.

Grace-Farfaglia, P, and Rosow, P: "Automating Clinical Dietetics Documentation," *J Am Diet Assoc,* **95:**687–90, 1995.

Herbelin, K, *et al:* "A Blueprint for Success with OBRA Inspections of Nursing Facilities," *Dietet Curr,* **21**(4):17–20, 1994.

"How to Write Outcome Statements," *Nurs 95,* **25:**32Q, January 1995.

Krasker, GD, and Balogun, LB: "1995 JCAHO Standards: Development and Relevance to Dietetics Practice," *J Am Diet Assoc,* **95:**240–43, 1995.

McGinn, N: "Patient-Centered Care: Opportunities for Health Care Professionals to Work Together," *Dietet Curr,* **20**(3):11–14, 1993.

Mears, E: "Prealbumin and Nutrition Assessment," *Dietet Curr,* **21**(1):1–4, 1994.

Pachter, LM: "Culture and Clinical Care: Folk Illness Beliefs and Behaviors and Their Implications for Health Care Delivery," *JAMA,* **271:**690–94, 1994.

Phaneuf, C: "Screening Elders for Nutritional Deficits," *Am J Nurs,* **96:**58–60, March 1996.

Position of the American Dietetic Association: "Cost-Effectiveness of Medical Nutrition Therapy," *J Am Diet Assoc,* **95:**88–91, 1995.

Position of the American Dietetic Association: "Nutrition Services in Health Maintenance Organizations and Other Forms of Managed Care," *J Am Diet Assoc,* **93:**1171–72, 1993.

Snyder, B: "An Easy Way to Document Patient Ed," *RN,* **59:**43–45, March 1996.

Enteral and Parenteral Nutrition

Chapter Outline

Normal Diet
Liquid Diets
Soft Fiber-Restricted Diets
Enteral Nutrition Support
Parenteral Nutrition Support

Oral or Tube Feeding or TPN?
Putting It All Together
Questions Clients Ask
Case Study: A Visually Impaired Patient
  Has Surgery

*For the two out of three adult Americans who do not smoke and do not
drink excessively, one personal choice seems to influence long-term health
prospects more than any other: what we eat.**

✔ Case Study Preview

Having a patient well nourished on a diet as similar to normal as the situation permits is an essential part of therapy for any patient. You will follow
a patient who is visually impaired through the postoperative progression
from nothing-by-mouth (NPO) to a regular diet.

**The Surgeon General's Report on Nutrition and Health.* U.S. Department of Health and Human
Services, 1988, p. 1.

The objective of all patient feedings is maintaining or restoring good nutrition. Feeding may be by enteral or parenteral routes; sometimes both are used simultaneously. Although technically **enteral nutrition** means any feeding involving the gastrointestinal tract, this term is often used specifically to refer to tube feeding into the esophagus, stomach, or small intestines. **Parenteral nutrition** provides nutrients intravenously.

The name of a diet should indicate any modification as specifically as possible. Low-calorie diet or low-protein diet has little meaning without a frame of reference; 1,500-kcal diet or 40-g protein diet is more specific. Naming a diet for the clinician who developed it may help immortalize that person, but gives no hint as to the nature of the modification. Naming a diet for a disease condition leads to needless duplication of diets since a certain modification may be appropriate for several disease conditions.

NORMAL DIET

Regular, General, Full, or House Diet: other names for normal diet

The most frequently used of all diets is the normal diet. Like a modified diet, it is a very important part of the patient's therapy. The normal diet in hospital usage follows the principles outlined in the preceding units and is planned to provide the Recommended Dietary Allowances. (See the Food Guide Pyramid (Figure 1–5) and the Basic Diet (Table 2–4.) The normal diet in the hospital places no restrictions on food choice. Strongly flavored vegetables, fried foods,

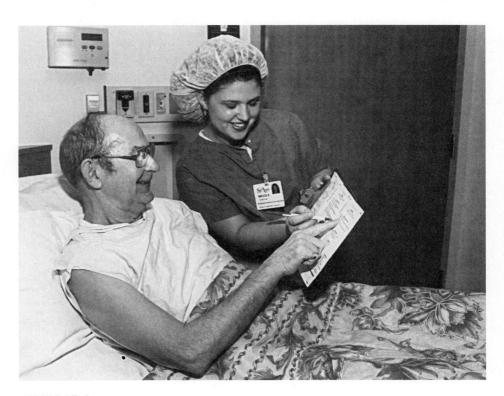

FIGURE 17–1
A dietetic technician assists a visually impaired patient with filling in his selective menu, thereby assuring that he will receive preferred foods. (*Source:* Anthony Magnacca/ Merrill.)

cakes, pies, pastries, and spicy foods all have a place on the menu, but they should be used with discretion. Many hospitals and nursing homes provide a selective menu from which the patient may make food choices, thus avoiding items that cannot be tolerated or are strongly disliked. Choosing foods from a menu also provides the patient with some sense of control over the situation. (See Figure 17–1.)

LIQUID DIETS

Clear-Liquid Diet

The clear-liquid diet features food and beverages both clear and liquid at body temperature. This diet includes tea, coffee, or coffee substitute, fat-free broth, carbonated beverages, clear fruit juices, fruit ices, and plain flavored gelatin. No milk is included. Low-residue, high-protein, high-calorie oral supplements may be given to increase nutritive value.* Small amounts (30–60 ml) are offered every hour or two with the volume increased as the patient is better able to tolerate them.

The diet relieves thirst, helps maintain fluid and electrolyte balance, and helps stimulate peristalsis in the postoperative patient. Depending on the choices of fluids, the diet will provide 200–500 kcal, some sodium and potassium, and some vitamin C. It is inadequate for most nutrients, and its use for more than a day or so will lead to weight loss, wasting of tissues, and multiple nutrient deficiencies.

Full-Liquid Diet

This diet consists of liquid and foods that liquefy at body temperature. (Food allowances and a sample menu appear in Table 17–1). It can be nutritionally adequate. The diet is offered in six feedings or more. Initially, amounts smaller than those stated in Table 17–1 are given. The protein level may be increased by adding nonfat dry milk to milk-base items. Formulas such as Ensure®, Meritene®, and Sustacal® may be included.† Because they are fortified with iron and other minerals and vitamins, such beverages can increase the adequacy of the intake.

This diet is high in lactose, which makes it inappropriate for anyone with decreased lactase activity. Suggestions for foods with reduced lactose appear in Figure 22–2. Ensure® and Sustacal® are lactose free. A full-liquid diet is also high in fat and calcium and low in fiber.

SOFT FIBER-RESTRICTED DIETS

Mechanical Soft Diet

The mechanical soft diet is planned for those who have difficulty in chewing because of no teeth or poorly fitting dentures. It may also be ordered for a person who has chewing problems after a stroke or for patients who have had head or neck surgery. The modification from the normal diet is in texture only and there is no restriction on seasonings or methods of preparation. People lacking teeth or well-fitting dentures vary greatly in their ability to chew and diets should not be

Clear-Fluid Diet: another name for clear-liquid diet

INDICATIONS FOR A CLEAR-LIQUID DIET

Anorexia, nausea, vomiting
Acute diarrhea
Preparation for bowel surgery
Postoperative progression
Transition to oral feeding after tube or parenteral feeding

Full-Fluid Diet: another name for full-liquid diet

INDICATIONS FOR A FULL-LIQUID DIET

Postoperative progression from clear liquids to solid foods
Acute infections of short duration
Acute gastrointestinal upset
Inability to chew

Dental Soft Diet: another name for mechanical soft diet

---

*CitriSource®, Citrotein®, Sandoz Nutrition, Minneapolis, MN.

†Ensure®, Ross Laboratories, Columbus, OH; Meritene®, Sandoz Nutrition, Minneapolis, MN; Sustacal®, Mead Johnson & Company, Evansville, IN.

TABLE 17–1
Food Allowances and Sample Menu in Full-Liquid Diet

Food Allowance for One Day	Sample Menu
6 cups lowfat milk 2 eggs (in custards or pasteurized eggnog) 1–2 oz strained meat or poultry 1 cup strained citrus juice ½ cup tomato or vegetable juice ½ cup vegetable purée ½ cup strained cereal 2 servings dessert: soft custard, plain ice cream, sherbet, plain gelatin, plain pudding, plain yogurt, fruit ice 3 tablespoons sugar, syrups 1 tablespoon margarine Broth, bouillon All beverages including high-protein, high-calorie supplements Flavoring extracts, salt, syrups Protein: 85 g Fat: 55 g Calories: 1,800	*Breakfast* Grapefruit juice Strained oatmeal with margarine, hot lowfat milk, and sugar Milk Coffee with lowfat milk and sugar *Midmorning* Orange juice Custard *Lunch* Broth with strained beef Tomato juice Vanilla ice cream Milk *Midafternoon* Milk shake *Dinner* Cream of asparagus soup Eggnog, commercial Strawberry gelatin with whipped cream Tea with lemon and sugar *Evening* High-protein, high-calorie beverage

unnecessarily restricted. Some can masticate whole foods very satisfactorily with their gums; others cannot. Composition of this diet varies from institution to institution. Common modifications are as follows:

Meat and poultry are minced or ground; fish usually is sufficiently tender without further treatment.

Vegetables are cooked. They may be cooked a little longer than usual to be sure they are soft, and may be diced or chopped.

Chopped raw tomatoes and chopped lettuce are sometimes used.

Soft raw fruits: banana, citrus sections, berries, grapes, diced soft pear, peach, apple, apricots, melons. All cooked canned and frozen fruits.

Soft rolls, bread, and biscuits instead of crisp rolls, crusty breads.

All desserts on a normal diet that are soft, including pies with tender crusts, cakes, puddings. Finely chopped nuts and dried fruits.

Soft Diet

The soft diet is used intermediately between the full-liquid and the normal diet in any situation in which the full-liquid diet was indicated. It may be used for patients with mouth sores or mild gastrointestinal problems. It is nutritionally adequate and includes soft foods that are easy to chew and digest and are moderately low in fiber, stimulating flavorings, and excessive fats. (See Table 17–2.)

TABLE 17-2

Food Allowances and Sample Menu for Soft Diet

Foods Allowed	Sample Menu[a]
Beverages—all	*Breakfast*
Bread—white, fine whole-wheat, rye without seeds; white crackers	Orange sections
Cereal foods—dry (except bran) and well-cooked breakfast cereals; hominy	Oatmeal
grits, macaroni, noodles, rice, spaghetti	Lowfat milk
Cheese—mild, soft, such as cottage and cream; cheddar; Swiss	Sugar
Desserts—plain cake, cookies; custards; plain gelatin or with allowed fruit;	Whole-wheat toast
plain ice cream, ices, sherbets; plain puddings, such as bread, cornstarch,	Margarine
rice, tapioca; plain yogurt	Coffee with milk, sugar
Eggs—all except fried	
Fats—cream, margarine, vegetable oils and fats in cooking	*Lunch*
Fruits—raw: ripe avocado, banana, grapefruit or orange sections without mem-	Tomato bouillon
brane; canned or cooked: apples, apricots, fruit cocktail, peaches, pears,	Melba toast
plums—all without skins; Royal Anne cherries; strained prunes and other fruits	Broiled flounder
with skins; all juices	Asparagus tips
Meat—very tender, minced, or ground; baked, broiled, creamed, roasted, or	Soft roll
stewed: beef, lamb, veal, poultry, fish, bacon, liver, sweetbreads	Margarine
Lowfat milk—in any form	Golden cake with fluffy white icing
Soups—broth, strained cream or vegetable	Lowfat milk; tea, if desired
Sweets—all sugars, syrup, jelly, honey, plain sugar candy without fruit or nuts,	
molasses; use in moderation	*Dinner*
Vegetables—white or sweet potato without skin, any way except fried; young	Grapefruit juice
and tender asparagus, beets, carrots, green beans, peas, pumpkin, squash	Broiled chicken
without seeds; tender chopped greens; strained cooked vegetables if not ten-	Baked potato without skin, margarine
der; tomato juice	Julienne green beans, margarine
Miscellaneous—salt, seasonings and spices in moderation, gravy, cream	Dinner roll
sauces	Margarine
	Apple crisp
	Lowfat milk
	Tea or coffee, if desired

[a] May be given in six feedings by saving part of the food from the preceding meal.

Pureed Diet

The pureed diet features foods that are smooth and soft, require little if any chewing, and are easy to swallow. It is used for patients who have difficulty chewing and swallowing, such as some stroke patients.

All foods are pureed or blenderized unless they are already smooth. Liquids can be added to create a consistency appropriate for the needs of the individual. Additional fat or sweeteners may be added to increase calories. Pureeing each food separately and assembling in layers or shaping by putting through a pastry tube makes foods more attractive and may increase intake.

Controlled-Consistency, Modified-Fluid Diet (for Dysphagia)

With **dysphagia** there is a problem in swallowing or in the transfer of food from the mouth to the stomach. No two patients are identical. A swallowing evaluation is conducted, usually by a speech-language pathologist, and an individualized diet of consistency appropriate to the patient's condition is planned. Depending on the patient's status, foods may need to be thick or thin in viscosity, light or heavy in density, smooth or textured. Commercial thickeners may be used. Thin liquids are usually the most problematic items.

TABLE 17–3

Indications, Contraindications for, and Possible Disadvantages of Tube Feeding

Indications	Contraindications	Possible Disadvantages
Inability to chew or swallow	Intestinal obstruction	Diarrhea—bacterial, osmotic, or lactose-induced
Paralysis	Paralytic ileus	Dehydration or fluid overload
Unconscious or comatose	Bowel sounds absent	Aspiration pneumonia
Esophageal paralysis, trauma, or obstruction	Severe gastrointestinal bleeding	
Head or neck surgery or irradiation	Uncontrollable diarrhea	
Unwillingness to eat as in anorexia nervosa	Unmanageable vomiting	
Hypermetabolic state as cancer or severe burns	Fistula in gastrointestinal tract	
Need for continuous or nighttime feeding		

ENTERAL NUTRITION SUPPORT

Enteral feeding by tube is used for the patient with a functioning gastrointestinal tract who cannot be adequately nourished by oral intake. (See Table 17–3). The formula must be nutritionally adequate and well tolerated. It should be similar to the normal diet in nutrient composition unless the patient's condition dictates otherwise. Many types of formulas are available.

Composition and Characteristics

Proteins may be in the form of intact proteins such as milk, egg, or beef; protein hydrolysates that have been partly broken down to peptides and amino acids; or free amino acids. Intact proteins require normal digestive enzyme function and absorption. Protein hydrolysates are used when there is reduced ability to digest foods or absorb nutrients. Peptides or free amino acids are sometimes used when there is severe enzyme lack or malabsorption.

Part of the carbohydrate is often in the form of glucose polymers that are easily digested and absorbed. Starch or a disaccharide such as sucrose or lactose may be used. Many formulas are lactose free because of lactose intolerance in some malabsorption disorders.

Fat sources in formulas include vegetable oils, butterfat, lecithin, mono- and diglycerides, and **medium-chain triglycerides (MCTs).** Medium-chain triglycerides are fats with 8–12 carbon chains that do not exist in nature and are manufactured. MCTs are digested and absorbed more rapidly than longer chain fats and are transported by the portal vein. Commercially prepared formulas that are intended to be adequate nutritionally are fortified with vitamins and minerals. They do not contain essential fatty acids.

At full strength, formulas commonly provide 1 kcal/ml. More concentrated formulas are available to meet elevated caloric needs without excessive volume. When these are used, care must be taken to provide adequate fluids. Formulas tend to be low in fiber, but fiber-enriched formulas are available.*

Osmolality

The **osmolality** of the formula, the number and size of particles per kilogram of water, is an important factor in patient tolerance. **Isotonic** formulas have the

*Jevity™, Ross Laboratories, Columbus, OH; FiberSource™, Sandoz Nutrition, Minneapolis, MN.

same osmolality as body fluids. **Hyperosmolar** formulas have an osmolality greater than that of the body's extracellular fluids and may cause rapid fluid and electrolyte shifts, resulting in diarrhea, especially if they are introduced into the small intestine.

Since carbohydrates are digested most rapidly, they have the greatest influence on osmolality. Glucose and sucrose have a higher osmolality than more complex carbohydrates. Free amino acids have a higher osmolality than intact proteins. A high level of electrolytes increases osmolality; fat has little effect.

Types

Formulas may be categorized as incomplete supplements, modular feedings, balanced complete, milk base, lactose free, chemically defined, and specialized. Incomplete supplements are intended to add nutrients to a diet but not to provide complete nourishment. Examples are hydrolyzed proteins, vitamin and mineral supplements, glucose polymers,* and medium-chain triglycerides (MCT).† **Modular feedings** involve the mixing of protein, fat, carbohydrate, vitamin, and mineral components into a formula tailored to the patient's needs.

Blenderized formulas, which are made by blending ordinary foods, or baby foods can be made at home. Most institutions use commercially prepared, **polymeric** (more than two molecules) formulas. These can be standard, which are similar to a regular diet in nutrient composition, usually are lactose free, and have low osmolality. High-nitrogen formulas are designed to meet increased protein needs. They are lactose free and have low to moderate osmolality. Concentrated formulas provide high energy and protein in a smaller volume. They are lactose free and moderate to high in osmolality.

Chemically defined formulas were tested for use in the space program, but lack of palatability led to use of other modes of feeding. They contain amino acids or hydrolyzed protein, vegetable or MCT oil, and oligosaccharides or hydrolyzed cornstarch. See Table 17–4.

Specialty formulas have the nutrient content adjusted for specific conditions.

Administration

The patient's condition and the length of time the tube feeding is likely to be given determine the site of the feeding tube. For short-term feedings a nasogastric tube is often used.

For long-term tube feeding or when a nasogastric tube is not appropriate, as in injury to the esophagus, a gastrostomy tube may be inserted surgically into the stomach. Neither the nasogastric nor the gastrostomy tube is suitable for gastroesophageal reflux, intractable vomiting, or inadequate gastric emptying. When the stomach must be bypassed, a jejunostomy tube may be used.

The feeding may be continuous or intermittent. A continuous feeding permits a larger daily volume and decreases the chances of diarrhea and bacterial growth. It is usually necessary for jejunal feedings. Intermittent feeding allows greater patient mobility. Because the stomach acts as a reservoir and diarrhea is therefore less likely, bolus feeding is satisfactory for feedings that enter the stomach. Formulas may be delivered by gravity drip or infusion pump.

Formulas are usually diluted at first and gradually increased to full strength. Rate of administration is also begun slowly and increased, although concentration

---

Hyperosmolar: also called hypertonic

Balanced Complete Formula: also known as blenderized formula

Elemental Formula, Hydrolyzed Monomeric Formula, Predigested Formula, Space Diet: other names for chemically defined formulas

EXAMPLES OF SPECIALTY FORMULAS

High in branched chain amino acids and low in aromatic amino acids; for trauma and liver disease
High fat and low carbohydrate; for pulmonary conditions
Low or devoid of phenylalanine: for phenylketonuria

Bolus Feeding: another name for intermittent feeding

---

*Polycose®, Ross Laboratories, Columbus, OH; Moducal®, Mead Johnson & Company, Evansville, IN.
†MCT, Mead Johnson & Company, Evansville, IN.

TABLE 17–4

Comparison of Balanced Complete and Chemically Defined Formulas

	Balanced Complete	Chemically Defined
Advantages	Seldom cause diarrhea Can use regular foods Palatable, can be taken orally	Minimal digestion Rapid absorption Low-residue
Disadvantages	Thick, can clog a fine feeding tube	Hyperosmolar Unpalatable Expensive

and rate are not increased simultaneously. Gradual advancement to full strength and rate decreases the possibility of diarrhea. Elevating the head of the patient's bed during and after feeding reduces the chance of aspiration.

Home Tube Feeding

Long-term tube feeding may be done at home with health care team members providing instruction and ongoing support and assistance to make sure it is carried out in a safe and sanitary manner. (See Figure 17–2.)

PARENTERAL NUTRITION SUPPORT

Peripheral Parenteral Nutrition (PPN)

A 5% dextrose in water solution, D_5W, is commonly given by a peripheral vein to provide fluids and a small number of calories. Electrolyte solutions can also be provided in this manner. Sometimes higher concentrations of dextrose plus amino acids and lipids are given to foster improved nutrition over a short period of time.

Total Parenteral Nutrition (TPN)

When feeding by any enteral route is not feasible and the patient is debilitated or hypermetabolic, total parenteral nutrition (TPN) may be used. (See Table 17–5.) Enteral feeding is preferable, and there is a saying, "If the gut works, use it."

Prospective TPN recipients are thoroughly assessed nutritionally and metabolically. While they receive TPN, blood levels of various nutrients are frequently monitored and adjustments made in the solution if indicated. An interdisciplinary nutrition support team monitors the patient's status.

Composition of Solutions

Proteins are provided as crystalline amino acids that can be adjusted in type and amount according to patient needs.

Carbohydrate in hypertonic dextrose solutions provides calories and spares amino acids for protein synthesis. The high dextrose load can produce hyperglycemia and, for patients with pulmonary problems, cause respiratory distress. To prevent or correct these problems, varying proportions of calories are given as fat.

Emulsions of soy or safflower oil, glycerol, or egg yolk to provide energy and prevent essential fatty acid deficiencies were in the past given separately from

Total Parenteral Nutrition (TPN): also known as hyperalimentation or intravenous hyper-alimentation; central parenteral nutrition

ADVANTAGES OF TUBE FEEDING OVER TPN

Maintains integrity of gastrointestinal tract
Preserves normal sequence of intestinal and liver metabolism
Water-soluble nutrients better handled
Fewer complications
Equipment and solutions need not be sterilized
Less expensive

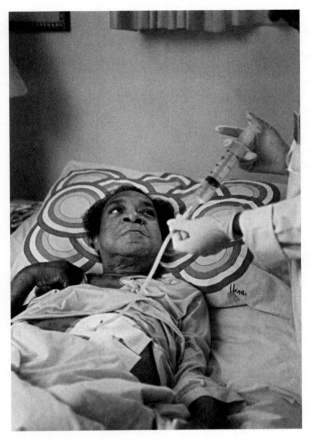

FIGURE 17–2
A community health nurse assists with the delivery of formula to a patient receiving tube feeding at home. (*Source:* Courtesy of LifeCare Alliance.)

the amino acids and dextrose. Lipid emulsions are isotonic and can be given by peripheral or central vein. Total nutrient admixtures that modify the macronutrients into a relatively stable solution have become available. Vitamins, electrolytes, and other minerals are also provided, some in the solution and some by injection.

**Total Nutrient
Admixtures:** also known as
3-in-1 formulas

Administration

TPN solutions are usually prepared to specification in the hospital pharmacy. Since they are hypertonic, for delivery a catheter is inserted into a large-caliber central vein such as the subclavian vein into the vena cava. The high rate of blood flow dilutes the solution quickly. For intermediate-term feeding of several months, a peripherally inserted central catheter (PICC) may be used.

The solution is infused slowly at first and gradually increased to the desired intake. This permits the patient to adjust and the health care team to evaluate the patient's tolerance. Infusion may be continuous or cyclic. Prolonged continuous infusion favors deposit of fat and **hepatomegaly,** or enlarged liver. Cyclic TPN provides amino acids plus glucose part of the day and amino acids alone during the remainder, which promotes mobilization of glycogen and fat. Possible complications of TPN are outlined in Table 17–5.

TABLE 17-5

Possible Indications for and Disadvantages of Total Parenteral Nutrition

Indications	Disadvantages
Uncontrollable diarrhea	Infection through catheter or around catheter site
Unmanageable vomiting	Venous thrombosis caused by irritation from hypertonic fluid
Peritonitis	Catheter insertion complications—pneumothorax, hydrothorax,
Paralytic ileus	arterial injury, hematoma
Short bowel syndrome	Metabolic complications—hyperglycemia, osmotic diuresis,
Gastrointestinal fistula	hyperosmolar coma, metabolic bone disease
Radiation enteritis	Atrophy of gastrointestinal cells
Intestinal obstruction	Cost
Hypermetabolic states like severe burns or cancer	
Severe inflammatory bowel disease	
Chronic pancreatitis	
Multiple organ failure	
Nutritional rehabilitation pre- or postoperatively	

Home Parenteral Nutrition

Individuals with little likelihood of ever achieving adequate intake by any enteral route may be candidates for home TPN. The patient and family must be carefully evaluated, taught, and demonstrate ability to carry out the procedure. Practical details such as obtaining equipment must be arranged and emotional support provided.

Patients on home TPN, usually use cyclic delivery and can plan their infusion over a 12-hour period, most of it while they are sleeping. The schedule can be individualized and can allow the patient to assume many former activities. For those who prefer or require daytime infusion, portable pump systems allow mobility during TPN. (See Figure 17-3.)

ORAL OR TUBE FEEDING OR TPN?

In many situations these modes of feeding are not mutually exclusive. For example, some tube-fed patients can take some food orally and should be encouraged to do so.

The ultimate goals are to have the patient well nourished on a food intake as close to normal as possible. When a patient is moving from one form of feeding to another, an abrupt changeover is seldom appropriate. Rather, a transition is planned in which one method is decreased while the other is increased. The former usually is not terminated until the patient can be adequately nourished on the latter.

PUTTING IT ALL TOGETHER

Many are surprised to learn that the normal diet is the most frequently used diet in a health care facility. Diets can be modified in texture and consistency to meet patient needs while remaining nutritionally adequate. Complete nourishment by tube feeding or intravenously, in an institution or at home, is also possible.

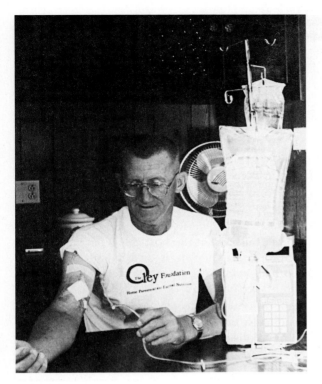

FIGURE 17–3
This man is completely nourished by total parenteral nutrition in his home with health care team members providing education and support. (*Source:* Courtesy of the Oley Foundation.)

? QUESTIONS CLIENTS ASK

1. I have an aunt who is in a nursing home. What can the family do to make sure she eats well?

If she receives a selective menu, help her fill it in when you or other family or friends visit. If she can't respond, report favorite foods (and dislikes) to the staff. If you wish to bring in a favorite food, check to make sure it is consistent with any diet modifications. If you visit at mealtime, help provide assistance with eating if needed and do anything you can to provide a pleasant surrounding for your aunt and her roommates or friends at the facility.

2. I have a neighbor who is in the hospital and who has a lot of trouble swallowing. They give him thickened pureed foods. I don't understand. I'd think liquids would be easier to swallow; they'd just slide right down.

A swallowing evaluation was probably performed with your neighbor since people with swallowing difficulties have different needs and responses. Thickened pureed foods were found to be best and safest for him. Actually, liquids seem to cause the most problems in people with swallowing problems.

3. My neighbor is receiving a tube feeding at home. How do they choose which formula to give her?

Your neighbor's nutritional needs and medical condition would determine the type of formula. The first objective is to provide adequate nutrition, which might influence the caloric concentration. If she had a condition such as a gastrectomy (removal of the stomach) and the formula is delivered to the small intestine, nutrients would be in a form that would compensate for the lack of digestion in the stomach. If she had a lactase deficiency, a lactose-free formula would be appropriate. There are many types of formulas, so one appropriate to her condition would be available.

Mr. Charles R., age 63 and visually impaired, is admitted to the hospital for abdominal surgery. His weight is normal as are his blood and urine tests. Mr. R. and his wife both work, have adequate incomes, and enjoy good health. Medical insurance will cover the hospital bills.

After surgery, as soon as peristalsis returned, Mr. R. was given clear liquids and then briefly, full liquids. Currently he is receiving a soft diet; and since he is progressing satisfactorily, he will receive a normal diet tomorrow.

On a visit to her husband, Mrs. R. tells the dietetic technician that she has some questions about nutrition and her husband's diet. How might they be answered?

Questions

1. Why did Mr. R. receive a clear-liquid diet after surgery? What foods are allowed on it? What nutrients did it provide?

2. Why was he next on a full-liquid diet? Did this diet provide all the nutrients he needs to recover from his surgery?

3. Mrs. R. says her husband has good teeth and can chew, so she does not understand why he is on a soft diet. What explanation might be given?

4. When he is on the normal diet, she asks, may he eat all kinds of foods? Since he is less active than usual, should he omit any foods?

5. Mrs. R. would like to bring some favorite food to her husband in the hospital. What would be appropriate when he is on the soft diet? The normal diet? What are some sanitation precautions she should observe?

6. You are involved in Mr. R.'s care. How might you assist with his meal selection and at mealtimes, considering his visual impairment?

FOR ADDITIONAL READING

American Dietetic Association: *Handbook of Clinical Dietetics,* 2nd ed. New Haven, CT: Yale University Press, 1992.

Bockus, S: "When Your Patient Needs Tube Feedings: Making the Right Decisions," *Nurs 93,* **23:**34–43, July 1993.

Cassens, D, *et al:* "Enhancing Taste, Texture, Appearance, and Presentation of Pureed Food Improved Resident Quality of Life and Weight Status," *Nutr Rev,* **54:**S51–S54, 1996.

Chicago Dietetic Association and South Suburban Dietetic Association: *Manual of Clinical Dietetics,* 4th ed. Chicago: American Dietetic Association, 1992.

Cole-Arvin, D, *et al:* "Identifying & Managing Dysphagia," *Nurs 94,* **24:**48–49, January 1994.

Eisenberg, PG: "Feeding Formulas," *RN,* **57:**46–53, December 1994.

Forloines-Lynn, S: "How to Smooth the Way for Cyclic Tube Feedings," *Nurs 96,* **26:**57–60, March 1996.

Gauwitz, DF: "How to Protect the Dysphagic Stroke Patient," *Am J Nurs,* **95:**34–38, August 1995.

Gianino, S, *et al:* "The ABCs of TPN," *RN,* **59:**42–48, February 1996.

Hodgkin, G, and Maloney, S, eds: *Diet Manual Including a Vegetarian Meal Plan,* 7th ed. Loma Linda, CA: Seventh-Day Adventist Dietetic Association, 1990.

McCrae, JD, *et al:* "Parenteral Nutrition: Hospital to Home," *J Am Diet Assoc,* **93:**664–73, 1993.

Pardoe, EM: "Development of a Multistage Diet for Dysphagia," *J Am Diet Assoc,* **93:**568–71, 1993.

"Position of the American Dietetic Association: Nutrition Monitoring of the Home Parenteral and Enteral Patient," *J Am Diet Assoc,* **94:**664–66, 1994.

"Position of the American Dietetic Association: Legal and Ethical Issues in Feeding Permanently Unconscious Patients," *J Am Diet Assoc,* **95:**231–35, 1995.

Sansivero, GE: "Why Pick a PICC?" *Nurs 95,* **25:**34–42, July 1995.

Viall, C: "Taking the Mystery out of TPN," Part I. *Nurs 95,* **25:**34–43, April 1995; Part II. *Nurs 95,* **25:**56–59, May 1995.

Food, Nutrient, and Medication Interactions

Chapter Outline

Increase to at least 75 percent the proportion of primary care providers who routinely review with their patients aged 65 and older all prescribed and over-the-counter medicines taken by their patients each time a new medication is prescribed.*

✔ Case Study Preview

Many medications affect nutrition, some in minor ways and others in ways that are profound and possibly dangerous. You will assist a woman who is being treated for depression with an MAOI—a medication with serious and even potentially fatal complications if consumed with inappropriate foods.

*Healthy People 2000: National Health Promotion and Disease Prevention Objectives. Public Health Service, U.S. Department of Health and Human Services, Washington, DC, 1991, Objective 12.6, p. 107.

The purpose of this chapter is to provide an overview of the many possible food, nutrient, and medication interrelationships that can occur. Medications widely used in various disease states will be discussed in chapters that follow.

PHARMACOLOGIC THERAPY

Pharmacologic therapy refers to treatment with medications. It may be the sole treatment or used in conjunction with other therapies. The pharmacist is the health care team member most directly involved with medications, but such medications and their effects are relevant to the practice of all health care professionals.

Medication therapy is an integral part of the treatment of many disease conditions. When medications and food are taken together, the availability of either or both may be affected. Medication-nutrient interactions are of greatest potential significance to those in poor nutritional status, with decreased metabolic or excretory capacity, and/or on long-term therapies often involving multiple medications. (See Figure 18-1.)

Over-the-counter and/or mail order medications, when used concurrently with prescription medications, can reduce or even trigger an incompatibility reaction. Taking another person's medications, consuming alcohol with medications, and misunderstanding, forgetfulness, or noncompliance concerning prescription medication usage can trigger or intensify medication-nutrient interactions. Older adults, who use at least 50 percent of medications consumed and frequently take several medications, are especially at risk. Older people may have reduced ability to absorb medications, metabolize them less efficiently and, because of decreased kidney function, have more difficulty in eliminating them. Some older individuals have inappropriate medications prescribed.

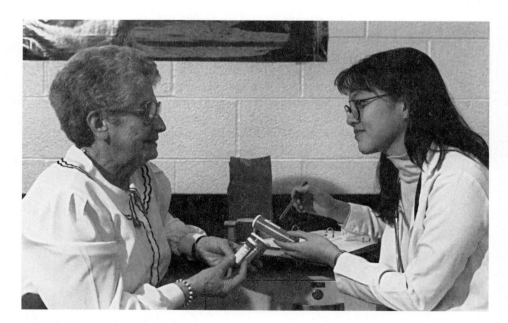

FIGURE 18-1
A client brings her medications to be checked by a nurse for possible medication-nutrient interactions and other incompatibilities. (*Source:* School of Nursing, University of Pennsylvania, and Denise Angelini Kosman, photographer.)

Medications most commonly utilized in office practice have been identified. Since these medications are used by millions of people, it is helpful to know what food-nutrient-medication interactions take place, and their dietary implications. (See Table 18–1.) The section that follows describes in more detail the many ways in which medications interfere with nutrition, and the ways in which food and nutrients may modify medication effectiveness.

TABLE 18–1
The 20 Medications Most Frequently Used in Office Practice

Class of Medication and Generic Name	Possible Nutritional Effects	Dietary Strategies
ANALGESIC		
Acetaminophen	Gastrointestinal disturbance	Take with food to ↓ gastric irritation
Aspirin	Nausea and vomiting, gastric pain or bleeding	Take on empty stomach for rapid analgesia; with food to ↓ gastric irritation
ANTIANXIETY		
Alprazolam	Anorexia, dry mouth, nausea and vomiting, constipation, diarrhea	Take with food or water; avoid alcohol
ANTIBIOTIC		
Amoxicillin	Nausea and vomiting, diarrhea	Take with food
Cefaclor	Anorexia, nausea and vomiting, sore mouth, diarrhea	Avoid alcohol
ANTIHISTAMINE		
Terfenadine	↑ appetite, dry mouth, nausea and vomiting, diarrhea	Take with food, avoid alcohol
ANTIHYPERTENSIVE Ace Inhibitor		
Captopril	Anorexia, transient hypogeusia, sore mouth, nausea and vomiting, diarrhea, ↑ blood: potassium, ↓ blood: sodium	Take 1 hour before meals, avoid potassium-containing salt substitutes, maintain adequate hydration, adequate salt intake
Enalpril	Anorexia, impaired taste, dry mouth, nausea and vomiting, diarrhea, constipation	Avoid potassium-containing salt substitutes
Calcium Channel Blocker *Diltiazem*	Anorexia, dry mouth, nausea, constipation	Take 1 hour before or 2 hours after meals
BRONCHODILATOR		
Albuterol	↑ appetite, altered taste, nausea and vomiting, diarrhea	Take with foods, limit caffeine intake
CARDIOTONIC		
Digoxin	Anorexia, nausea and vomiting, diarrhea, ↑ urinary: calcium, magnesium, potassium	Take on empty stomach, avoid ↑ sodium, take adequate potassium, avoid bran
DIURETIC		
Furosemide	Anorexia, dry mouth, ↑ thirst, nausea and vomiting, abdominal cramping, constipation, diarrhea, ↓ blood: potassium, sodium, ↑ urine: potassium, sodium	Take with food, ↓ sodium intake, adequate potassium intake

TABLE 18–1, *continued*

Class of Medication and Generic Name	Possible Nutritional Effects	Dietary Strategies
Triamterene	Dry mouth, nausea and vomiting, gastric upset, constipation, diarrhea, electrolyte imbalance	Take with or after meals, avoid ↑ potassium foods, avoid potassium-containing salt substitutes
ESTROGEN REPLACEMENT		
Estrogens	Nausea and vomiting, abdominal cramping, bloating, edema	Take with food, possible ↓ sodium
HISTAMINE₂ RECEPTOR ANTAGONIST		
Ranitidine	Nausea, abdominal pain, constipation, diarrhea	Take with food, limit caffeine, avoid alcohol
NONSTEROIDAL ANTI-INFLAMMATORY		
Diclofenac sodium	Nausea, dyspepsia, flatulence, heartburn, gastrointestinal ulceration and bleeding, constipation, diarrhea	Take with food or milk
Ibuprofen	↓ appetite, dry mouth, nausea and vomiting, dyspepsia, flatulence, gastrointestinal ulceration and bleeding, constipation, diarrhea	Take with food, avoid alcohol
Naproxen	Dry mouth, sore mouth, nausea, dyspepsia, heartburn, gastrointestinal pain, ulceration and bleeding, constipation, diarrhea	Take with food or milk, avoid alcohol
STEROIDAL ANTI-INFLAMMATORY		
Prednisone	Nausea and vomiting, indigestion, abdominal distention, peptic ulcer, fluid and sodium retention, fluid and electrolyte imbalance, negative nitrogen balance, ↓ carbohydrate tolerance, ↓ vitamin D activity	Take with food, ↓ sodium intake, ↑ potassium and protein intake, ↑ vitamin D intake, possible ↓ calorie intake, avoid alcohol
THYROID HORMONE REPLACEMENT		
Levothyroxine	Appetite change, nausea and vomiting, diarrhea, ↓ weight	Take on empty stomach, do not take iron supplement within 4 hours

Source: Adapted from Schappert, S.M.: "National Ambulatory Medical Care Survey: 1992 Summary," *Advance Data from Vital and Health Statistics,* No. 253, DHHS Pub. No. (PHS) 94–1250. Public Health Service, Hyattsville, MD, August 18, 1994, Table 22, p. 15.

MEDICATIONS AND FOOD INTAKE

Medications may affect food intake by decreasing or increasing appetite, depressing or elevating taste acuity or leaving an unpleasant aftertaste, or causing **stomatitis** (sore mouth), **xerostomia** (dry mouth), nausea, and/or vomiting.

Anorectic medications such as amphetamines are sometimes used to promote weight loss. Tolerance to these medications develops rapidly, so their value in weight control programs is limited. *Phenylpropanolamine,* a sympathomimetic, is an ingredient of many over-the-counter reducing aids and decongestants. Appetite suppressants currently available include *dexfenfluramine, diethylpropion, fenfluramine, mazindol, phenylpropanolamine,* and *phentermine.* (See Chapter 9.) Antacids, antihistamines, caffeine, cough medicines, digitalis, and narcotics can also decrease appetite.

Dysgeusia (altered taste sensation) or **hypogeusia** (diminished taste sensation) may be caused by *captopril, clofibrate, fluorouracil, griseofulvin* (antifungal), *levodopa* (Parkinson's disease management), and *penicillamine* (chelator). Others such as *allopurinal, penicillin,* and *quinidine* can leave unpleasant aftertastes. Many cancer chemotherapy agents cause dry or sore mouth, upset stomach, nausea, and/or vomiting. Some antibiotics can produce similar gastrointestinal disturbances. All these medicines have the potential for decreasing food intake.

Medicines can be a source of nutrients inappropriate for modified diets. For example, **antitussives** to relieve or prevent coughing and expectorants may contain sugar, while antacids and antibiotics are potential sources of sodium.

MEDICATION-NUTRIENT INTERACTIONS AND ABSORPTION

Nutrient Absorption

The small intestine, where most medications and nutrients are absorbed, is the major site of medication-nutrient interactions. Various modes of action are possible.

Transit Time Laxatives decrease nutrient absorption by reducing gastrointestinal transit time. **Anticholinergics,** which reduce acid secretion, used in the management of peptic ulcer, and ganglionic blocking agents used to treat hypertension, interfere with transmission of nerve impulses in the autonomic nervous system. Transit time is thereby increased with constipation a possible result.

pH Alteration Antacids raise the pH of the stomach contents, thereby reducing the absorption of iron, calcium, zinc, and folate.

Malabsorption *Cholestyramine, clofibrate,* and *colestipol,* used to reduce blood cholesterol, and *neomycin,* an antibiotic, may lead to malabsorption of fats and fat-soluble vitamins. *Methotrexate,* a folate antagonist, causes malabsorption of that B vitamin.

Mucosal Damage Medications such as *neomycin* and *colchicine* cause changes in the intestinal mucosa leading to malabsorption of nutrients including fat, nitrogen, lactose, calcium, iron, sodium, potassium, and vitamin B_{12}.

Medication Absorption

The presence of food in the stomach both lowers gastric pH and delays emptying of the stomach. Some medications are less well absorbed in an acid medium, and delayed gastric emptying can decrease the absorption rate of the medication. When rapid activity is desired as with **analgesics** (pain relievers), antiasthmatics, **antipyretics** (fever-reducing agents), hypnotics, or anti-infectives for acute infections, it is best to take the medication on an empty stomach. Antibiotics, aspirin, and *theophylline* are better absorbed in a fasting state.

BRAND X COUGH SYRUP

Ingredients: sucrose, high fructose corn syrup, phenylpropanolamine, sodium citrate, sodium saccharine

Medications Better Absorbed When Taken with Food: estrogens, *griseofulvin* (high-fat meal), *hydralazine, nitrofurantoin, propranolol, carbamazepine, metoprolol*

Specific nutrients or food components also have the potential to decrease medication absorption. The amino acids from a high-protein diet can inhibit the absorption of *levodopa, methyldopa,* and *theophylline.* It has been suggested that individuals taking *levodopa* for Parkinson's disease consume a breakfast and lunch low in protein to better control symptoms during the day. Most of the dietary protein should be included in the evening meal. Calcium, magnesium, and iron can form insoluble chelates with *tetracycline,* so dairy products and iron supplements should be spaced at least three hours away from tetracycline intake. A nutritionally unacceptable practice is the elimination of dairy items and iron-containing foods by patients taking tetracycline. The team of dietitian, nurse, pharmacist, and physician can help plan food and medication schedules to optimize nutrient intake and medication effectiveness.

MEDICATION-NUTRIENT INTERACTIONS AND METABOLISM

Nutrient Metabolism

Antimetabolites and antivitamins have chemical structures similar to the actual metabolites and vitamins. They block enzymatic reactions required for DNA synthesis, thus leading to the death of a cell. Cancer chemotherapy agents such as *fluorouracil* and *methotrexate* act in this manner.

Cycloserine, hydralazine, isoniazid, and *penicillamine* are vitamin B_6 antagonists. The anticonvulsants *phenytoin* and *phenobarbital* cause calcium loss so that rickets and osteomalacia can result from long-term therapy unless supplements are given. Folate deficiency can also occur in extended anticonvulsant therapy.

Oral contraceptives are associated with increased blood levels of lipids, vitamins A and E, iron, and copper, and decreased blood levels of vitamin C, riboflavin, folate, and vitamins B_6 and B_{12}. Carbohydrate and tryptophan metabolism are also affected. The significance of these changes is not clear as tissue levels of nutrients are frequently normal in spite of altered blood values.

Medication Metabolism

Damage to the intestinal mucosa can decrease the rate and amount of the medication that is absorbed. Medication metabolism in the liver may be affected by dietary composition. For example, there is some evidence that *theophylline* is best utilized when accompanied by a high-carbohydrate, low-protein diet. *Indoles,* found in vegetables of the Brassica family such as brussels sprouts and also in charcoal-broiled meat increase the rate of medication metabolism, thereby decreasing the effective time of the medicine.

Vitamin supplements may interfere with medication effectiveness. Pyridoxine has been shown to decrease the utilization of *levodopa* while vitamin K is an antagonist of *warfarin.*

MEDICATION-NUTRIENT INTERACTIONS AND EXCRETION

Nutrient Excretion

Furosemide and *thiazide* are potassium depleting, and a sodium-restricted diet, frequently 2,400 mg, emphasizing foods high in potassium is usually prescribed with them. Sodium and fluid retention may accompany use of adrenal corticosteroids, *clonidine*, estrogens, *guanethidine*, *hydralizine*, *methyldopa*, *indomethacin*, and *phenylbutazone*. Dietary sodium restriction and sometimes diuretic therapy are prescribed as adjuncts to the use of these medications.

Medication Excretion

Food and nutrients affect medication excretion by changing the pH of the urine. Acidic urine favors excretion of tricyclic antidepressants. Alkaline urine may prolong the effective time of amphetamines, *imipramine* (antidepressant), and *quinidine*.

Medication excretion can be affected by the mineral content of the diet. If sodium restriction accompanies therapy with the antidepressant *lithium*, increased reabsorption of the medication can result in lithium toxicity.

Natural licorice containing glycyrrhizic acid causes sodium and fluid retention, hypokalemia, and hypertension, potentially complicating treatment of patients already receiving therapy for hypertension. Most licorice sold in the United States is made with artificial flavorings.

SPECIFIC DRUG REACTIONS

Alcohol

Although alcohol is not a prescription drug, it has many qualities of other medications and can affect nutrition in numerous ways. Alcohol may also be viewed as a food since it yields 7 kcal/g, but it does not provide other nutrients in significant amounts. Moderate use of alcohol need not affect the nutritional status of the individual. Heavy alcohol intake decreases the appetite, however, and drinkers may substitute alcohol for more nutritious foods, with resulting nutrient deficiencies.

Metabolism of alcohol requires thiamin, niacin, and pantothenic acid, which may already be in short supply. Protein, fat, and carbohydrate metabolism may be altered both by decreased intake and by functional changes leading to hypoglycemia, triglyceride accumulation, fatty liver, impaired protein synthesis, and possible acidosis, among other complications.

Long-term use of alcohol is linked to increased risk of cancer of the esophagus. Alcohol may cause chronic pancreatitis resulting in interference with the digestion of fats and proteins. About 10–20 percent of the 10 million Americans who may be considered chronic alcohol abusers develop progressive liver disease. This disease is due to the effect of alcohol and is not prevented by a nutritious diet, although malnutrition can intensify the course of the disease.

Wernicke-Korsakoff syndrome is sometimes seen in people who abuse alcohol. Symptoms of Wernicke's encephalopathy are mental confusion, **ataxia** (uncoordinated gait), **nystagmus** (rolling eyeball movement), and **ophthalmoplegia** (eye muscle paralysis). Korsakoff's psychosis appears later in the course of the disease and features memory loss and delirium. A severe thiamin deficiency is involved and the symptoms respond to thiamin and other B-complex therapy.

High alcohol intake can decrease the rate of metabolism of other medications, causing toxicity. For example, alcohol intoxication inhibits the metabolism of some barbiturates causing a synergistic central nervous system depression that can result in unconsciousness, coma, and even death. Alcohol enhances the effect of oral hypoglycemic agents, sometimes to the point of loss of consciousness. Moderate consumption of alcohol does appear to have a protective effect against coronary heart disease.

Avoid Alcohol with:
anticonvulsants, antigout agents, antihistamines, anti-inflammatory agents, barbiturates, oral hypoglycemic agents, sedatives, tranquilizers

Disulfiram: also known as *Antabuse*

Disulfiram, when taken with alcohol, leads to abdominal and chest pain, headache, nausea, and vomiting. For this reason disulfiram is sometimes used under careful medical supervision and with supportive therapy as an alcohol aversion medication.

Caffeine

Caffeine is consumed by all age groups since it is found in chocolate and many soft drinks as well as coffee and tea. (See Table 18-2.) There is no evidence that moderate amounts of caffeine are harmful to the average healthy person. Although individual tolerances vary, intakes in excess of 1,000 mg/day can cause diarrhea, headache, heart palpitations, and sleeplessness.

Caffeine can increase the heart rate and basal metabolic rate and acts as a diuretic. The half-life of caffeine is increased in women taking oral contraceptives. Coffee increases acid secretion by stimulating the gastric mucosa; the same effect is noted with decaffeinated coffee. It appears, therefore, that caffeine is not the substance responsible for this action.

Caffeine is found in medications used as alertness or stay-awake agents, cold remedies, diuretics, headache and pain relievers, stimulants, and weight control medications. More than 1,000 over-the-counter medications contain caffeine.

Large amounts of caffeine taken with *theophylline* increase the risk of side effects of this drug, probably resulting from similar metabolic actions of the substances. Caffeine taken with neuroleptic medications such as *fluphenazine* and *haloperidol* can precipitate the medications, thereby reducing their effective-

TABLE 18–2
Caffeine Content of Some Beverages, Foods, and Over-the-Counter Medications

Item	Amount	Caffeine (mg)
Beverages		
Coffee, brewed	6 oz	80–175
Coffee, instant	6 oz	60–100
Coffee, decaffeinated	6 oz	1–5
Coffee, espresso	2 oz	90–110
Tea, steeped	6 oz	20–100
Cocoa, beverage	6 oz	2–20
Coca-Cola (regular and diet)	12 oz	45
Foods		
Milk chocolate	1 oz	1–10
Dark chocolate	1 oz	5–35
Chocolate cake	1 piece	20–30
Over-the-Counter Medications		
Anacin or *Midol*	2 pills	64
Excedrin	2 pills	130
NoDoz	2 pills	200

Source: Adapted from "Java Jolt," *University of California at Berkeley Wellness Letter,* **11**:8, August 1995.

ness. Caffeine does not reverse the depressant effects of alcohol on the central nervous system; therefore it is of no value in ''sobering up.''

Monoamine Oxidase Inhibitors (MAOI)

Probably the best known and potentially one of the most harmful medication-nutrient interactions is that between monoamine oxidase inhibitors and **pressor** (blood pressure elevating) amines, especially *tyramine.* Foods high in pressor amines stimulate the release of norepinephrine. Ordinarily the enzyme monoamine oxidase (MAO) inactivates the pressor agents, preventing the release of excess norepinephrine. Monoamine oxidase inhibitors block this reaction and the norepinephrine levels rise unchecked.

Consuming foods high in tyramine, a pressor amine with action similar to epinephrine, while on MAOI therapy can lead to headaches, palpitations, hypertension, and **hypertensive crisis,** an abrupt elevation of systolic blood pressure to 210 mm Hg or more. (See Chapter 24.) Major cerebrovascular accidents (strokes) have been documented. Both medication dosage and level of dietary tyramine influence the severity of the reaction.

Many foods contain tyramine in small quantities, but large amounts have been reported only in foods that are aged, fermented, or spoiled. Although more than 200 foods have been implicated, many of the reports involved spoiled foods or

Monoamine Oxidase Inhibitors: include antidepressants, *tranylcypromine sulfate, phenelzine sulfate, isocarboxazid;* antimicrobial *furazolidone;* antineoplastic *precarbazine;* antihypertensive *pargyline hydrochloride*

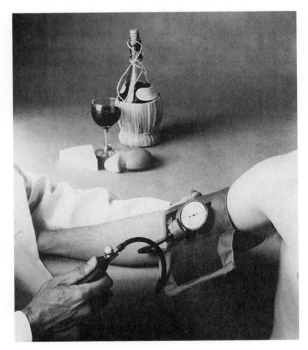

FIGURE 18-2
Chianti wine and aged cheeses are among foods that are high in tyramine content. High tyramine intakes by patients who are being treated with monoamine oxidase inhibitors may provoke sharp, sometimes dangerous, increases in blood pressure. (*Source:* Food and Drug Administration, U.S. Department of Health and Human Services, Washington, DC)

were difficult to verify. Almost 80 percent of reported cases involved aged cheeses. Few food analyses for tyramine are available.

Foods that must be eliminated by individuals on MAOI therapy are those that cause reactions when consumed in ordinary or even small servings. (See Figure 18–2.) Current therapeutic philosophy excludes such foods while allowing limited quantities of foods that may cause hypertensive crisis only when taken in large amounts. Such an approach improves patient compliance while eliminating unnecessary restrictions. Lists of foods to be avoided and those to be used in controlled amounts appear in Figure 18–3. Presently there is no justification for excluding other foods.

Emphasis should be on fresh foods. Any foods subject to bacterial contamination, spoiled foods, and foods that have been stored for prolonged periods should be avoided. Diet counseling should begin before MAOI therapy is initiated. In some instances a tyramine-restricted diet is begun several days before MAOI therapy. Since MAO production returns slowly after MAOI withdrawal, the diet should be continued for four weeks after the medication use ends.

Illicit Drugs

Illicit drugs have considerable potential for influencing nutritional status. Marijuana appears to increase the appetite, especially for sweets, with the attendant possibility of weight gain. Heroin affects the insulin response to glucose. Cocaine or crack depress appetite with weight loss common. There is also the possibility that available money will be spent on drugs instead of food.

Members of the health care team can cooperate to assist in reducing side effects from medication-nutrient interactions. In some cases smaller, more frequent doses of the medication or alternate medications may be feasible. Some strategies for controlling side effects such as anorexia and nausea appear in Table 25–1. The Food and Drug Administration (FDA) has proposed a program to

CHEDDAR CHEESE

Ingredients: milk, cheese culture, salt.
Aged over six months.

AVOID

Aged cheeses such as blue, brick, Brie, Camembert, Cheddar, Emmentaler, Gorgonzola, Gruyere, Mozzarella, Parmesan, Roquefort, Stilton
Aged meat, liver, dry sausage, meat extracts, pickled or smoked fish
Fermented soybean products (miso, some tofu)
Alcoholic beverages: ale, beer, chianti, vermouth
Banana peel, Italian broad beans, sauerkraut
Yeast extracts including yeast supplements
Any high-protein food that has been aged, fermented, pickled, smoked, contaminated, spoiled, or stored for a long time

ALLOW SMALL SERVINGS

Alcoholic beverages: distilled spirits, red wine, white wine, port
Avocado, raspberries
Chocolate
Peanuts
Soy sauce
Unpasteurized cream or yogurt

FIGURE 18–3
Tyramine-restricted diet.

provide patients receiving new prescriptions with useful, adequate, and easily understood written information about the medication.

PUTTING IT ALL TOGETHER

Diet and medications are both important in a patient's therapy. Unfortunately, they can interact to change the effectiveness of either or both. Strategies such as timing of medications, changing medications, and/or modifying food intake can ensure that the patient receives maximum benefit from both diet and medications.

? QUESTIONS CLIENTS ASK

1. I find my pills hard to swallow and am thinking of mixing them in coffee or tea at mealtime. Would this be all right?

No; the heat of the beverages can reduce the effectiveness of some medications. The tannin in the tea may also interfere with the absorption of your medications.

2. I enjoy going to a local coffee bar. Does espresso have more caffeine than regular coffee?

Ounce for ounce, espresso has two to three times as much caffeine as regular coffee. However, because of the difference in the size of the cups in which they are served, an average serving of espresso and of regular coffee have about the same amount of caffeine, around 100 mg.

3. My mother is a senior citizen and takes several over-the-counter medications on a regular basis. Should her doses be different from doses for a younger person?

Your mother should discuss her use of these medications with her health care provider. Age-related liver changes may alter the rate of metabolism of these medications so that smaller doses may be indicated. Also, there may be potential for medication-nutrient interactions for which the health care provider could provide counseling.

4. I see that histamine$_2$ receptor antagonists are now available in over-the-counter form. My uncle takes these in prescription form for peptic ulcers. I don't understand why anyone can walk into a store and buy these. Couldn't it be dangerous?

The lower dose, nonprescription form has been approved for heartburn and indigestion only. It is all right to use them as directed for occasional heartburn or indigestion, but if the condition persists, see your health care provider. Do not use this medication for other symptoms.

✔ CASE STUDY: A Secretary with Depression

Jennifer P., 30 years old, shares an apartment with two friends. For some time she has become increasingly depressed. As her depression deepened she found it harder to perform activities of daily living such as getting ready for work and carrying out her share of the household responsibilities. Recently she has been discharged from her position as a secretary because of poor job performance.

Since she has been home she has remained in her nightclothes all day, lying on the sofa and watching television. Jennifer's apartment mates have become increasingly concerned about her

condition and accompany her to a physician who refers her to a psychiatrist. *Imipramine* (Tofranil) is first tried, but Jennifer's depression does not respond, so *phenelzine* (Nardil), an MAOI, is prescribed. Before therapy is begun, a dietitian who works with the psychiatrist counsels Jennifer and her apartment mates about a tyramine-restricted diet.

Questions

1. What is the function of MAO? What is tyramine? Explain how MAOI works.

2. What are the symptoms of hypertensive crisis?

3. Jennifer ordinarily has low blood pressure and wonders if foods high in pressor amines present any danger to her when she is taking *phenelzine.*

4. Jennifer and her friends like to have an alcoholic beverage in the evening. Suggest some beverages that would be appropriate.

5. One of Jennifer's roommates heard that cola beverages should be omitted during MAOI therapy. What would an appropriate response be?

6. Jennifer likes to have a bagel with cream cheese for breakfast. She wonders if she must eliminate the cream cheese. Must she?

7. Jennifer is surprised to learn that she is to follow a tyramine-restricted diet for several weeks after MAOI therapy is completed. Explain why this is so.

FOR ADDITIONAL READING

"All About Drugs: Adverse Effects to Watch For," *Am J Nurs,* **95:**32–60, October 1995.

Burk-Shull, KA: "Protein Redistribution to Improve Quality of Life for Those with Parkinson's Disease: A Review," *Top Clin Nutr,* **10:**65–70, December 1994.

Cerrato, PL: "OTC Interactions: Vitamins and Minerals," *RN,* **56:**28–33, June 1993.

Drake, AC, and Romano, E: "How to Protect Your Older Patient from the Hazards of Polypharmacy," *Nurs 95,* **25:**34–41, June 1995.

Farley, D: "FDA's Rx For Better Medication Information," *FDA Consumer,* **29:**5–10, November 1995.

Flieger, K: "Aspirin: A New Look at an Old Drug," *FDA Consumer,* **28:**19–21, January/February 1994.

Friedman, GD, and Klatsky, AL: "Is Alcohol Good for Your Health?" *New Engl J Med,* **329:**1882–83, 1993.

Howser, RL: "What You Need to Know About Corticosteroid Therapy," *Am J Nurs,* **95:**44–49, August 1995.

Hussar, DA: "Reviewing Drug Interactions," *Nurs 93,* **23:**50–57, September 1993.

Hussar, DA: "Helping Your Patient Follow His Drug Regimen," *Nurs 95,* **25:**62–64, October 1995.

Hussar, DA: "New Drugs," *Nurs 96,* **26:**52–56, May 1996.

Lewis, CW, *et al:* "Drug-Nutrient Interactions in Three Long-Term-Care Facilities," *J Am Diet Assoc,* **95:**309–15, 1995.

Lilley, LL, and Guanci, R: "Adverse Effects of NSAIDs," *Am J Nurs,* **95:**17, August 1995.

Pelican, S, *et al:* "Nutrition Services for Alcohol/Substance Abuse Clients," *J Am Diet Assoc,* **94:**835–36, 1994.

Physician's Desk Reference. 51st ed. Montvale, NJ: Medical Economics Company, 1997. (published annually)

Physician's Desk Reference for Nonprescription Drugs. 18th ed. Montvale, NJ: Medical Economics Data, 1997. (published annually)

Wilcox, SM, *et al:* "Inappropriate Drug Prescribing for the Community-Dwelling Elderly," *JAMA,* **272:**292–96, 1994.

Nutrition in Immunity and Allergic Reactions

Chapter Outline

Immunity
Nutrition and Immune Function
Disorders of the Immune System
Allergic Reactions

Putting It All Together
Questions Clients Ask
Case Study: A Girl with Allergies

Throughout the world, generalized malnutrition is a common cause of acquired, correctable immune system dysfunction.*

✔ CASE STUDY PREVIEW

Wheat and eggs are widely used in food processing and may appear unexpectedly in foods. Ten-year-old Vicki is allergic to both. Can she have a cake for her birthday party? Can she eat things her friends enjoy? You will explore some of her problems in this chapter's case study.

**The Surgeon General's Report on Nutrition and Health.* U.S. Department of Health and Human Services, 1988, p. 435.

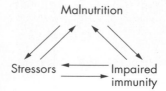

Malnutrition

Stressors ← → Impaired immunity

Compromised nutritional status depresses immune function, resulting in reduced resistance to stressors such as trauma, infection, and fever. Since true allergies are mediated by the immune system, increased knowledge of immune function has led to expanded understanding of allergy.

IMMUNITY

Impaired immune response may be primary (congenital) or secondary to nutritional imbalances, infections, trauma, cancer, AIDS, or immunosuppressive medications. The simplest and most basic defense against invasion of antigens is the physical barrier of healthy, intact skin and mucous membranes. In addition, systemic immunity is provided by specialized cells that respond to exogenous factors such as bacteria and viruses, and endogenous factors such as malignant cells. Cells of the immune system may be "fixed" in certain tissues such as bone marrow, lymph nodes, spleen, and thymus, or carried by the blood to the site where they are needed.

All types of immunity are closely interrelated. (See Figure 19–1.) The nature of the **antigen,** any substance that when introduced into the body is recognized by the immune system as foreign, determines which component of immune response assumes the greatest importance.

Nonspecific Immunity

Nonspecific Immunity: also known as nonadaptive immunity

Generally the first resistance to microbial invaders is nonspecific, an innate, generalized reaction to a foreign substance that does not require recognition of a particular antigen. Included are the skin and mucous membrane barriers, mucus, ciliary and visceral movements, the complement system, interferon, and iron-binding proteins such as lactoferrin and transferrin, phagocytes, and febrile and metabolic responses.

Interferon: protein formed when cells are exposed to a virus. Noninfected cells exposed to interferon are protected against viral infection.

Complement System The complement system is a group of proteins that interact with each other in a stepwise, cascade progression and mediate or intensify antigen-antibody reactions. Complement pierces the cell membrane, causing fluid to seep out. A defective complement system is associated with increased susceptibility to bacterial infections.

Antibody: protein synthesized in the presence of an antigen to counteract the toxic effects of the antigen

Phagocytosis **Phagocytosis,** the ingestion and destruction of microorganisms and cellular debris by phagocytes, is the main defense against bacteria. Circulating phagocytes include granulocytes and large scavenger cells called **macrophages.** Macrophages are directly involved in killing bacteria and viruses and also in readying antigens for recognition by T and B lymphocytes. **Lysosomes,** containing digestive enzymes, are released by the phagocytes and also by complement and are involved in the destruction of the invading microorganisms.

Specific Immunity

Specific Immunity: also known as adaptive immunity

Specific immunity, an immune response to a specific antigen, is an acquired, adaptive response. Lymphocytes are involved in a specific reaction to each type of antigen that, on first encounter, is slight. A memory system produces a more intense response on subsequent encounters.

There are two types of specific immunity, **cell-mediated,** which occurs inside or on the surface of the cell, and **humoral,** which occurs in blood and

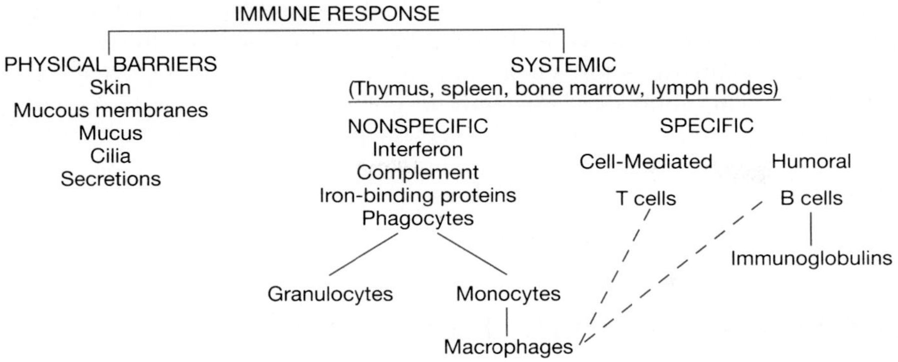

IMMUNE RESPONSE

PHYSICAL BARRIERS
Skin
Mucous membranes
Mucus
Cilia
Secretions

SYSTEMIC
(Thymus, spleen, bone marrow, lymph nodes)

NONSPECIFIC
Interferon
Complement
Iron-binding proteins
Phagocytes

SPECIFIC

Cell-Mediated   Humoral

T cells   B cells

Granulocytes   Monocytes

Immunoglobulins

Macrophages

FIGURE 19–1
A summary of immune response.

fluid outside the cell. For unknown reasons, some antigens produce a cell-mediated response and others a humoral reaction. The two systems have separate but interdependent functions. Furthermore, both react synergistically with nonspecific factors such as complement and phagocytosis to eliminate microorganisms.

Cell-Mediated Immunity Central to the cell-mediated immune system are lymphocytes called T cells. Before entering the blood or lymph, the cells differentiate in the thymus, where their surfaces develop receptor sites that are specific to one antigen. Upon initial contact with unique antigen, the T cells become activated. Receptors specific to the antigen are formed. The T cells are now primed for their immune function. When the T cells next come in contact with that antigen, proteins called **lymphokines** produced by antigen-activated lymphocytes are released. Lymphokines help to mobilize other immune factors and with the aid of other cells such as macrophages destroy the antigen.

Cell-mediated immunity is responsible for resistance to fungal and viral infections and transplant rejection, and plays a role in regulation of humoral immunity. It is measured by delayed cutaneous hypersensitivity (see Chapter 16). A positive reaction to the recall antigens indicates that T cells and phagocytes have gone to the injection site as an immune response to the antigen.

Several subsets of T cells have been identified, among them "helper," "killer," "lymphokine-producing," and "suppressor" cells. Helper T cells enhance the activity of B cells. Killer T cells can destroy bacteria, cancer cells, or cells activated by viruses.

Mounting knowledge of T cells holds promise for the prevention and treatment of disease. Patients who undergo organ transplants must receive immunosuppressive medications to prevent their own T cells from recognizing the transplanted tissue as foreign and rejecting it. Greater understanding of T cell function may enable researchers to develop ways of preventing such rejections. Decreased T cell count is one of the criteria for a diagnosis of AIDS (see Chapter 29).

Humoral Immunity The humoral immune system involves lymphocytes called B cells. The B cells do not react directly with antigens, but rather function through a more complex pathway. When activated by a particular antigen, B cells specific to that antigen differentiate into plasma cells. The plasma cells, in turn,

T Cells: also known as T lymphocytes or thymus-dependent lymphocytes

B Cells: also known as B lymphocytes or bursa-equivalent lymphocytes produce five types of antibodies or immunoglobulins (Ig), similar in basic structure but differing in amino acid chain, molecular weight, and function.

When the antibody is formed it binds to the antigen and the complex is phagocytosed. The antibodies also activate complement that binds with the antigen-antibody complex. As the complement proteins interact they produce enzymes to destroy cell walls and membranes. Macrophages then ingest the destroyed cells. A deficiency of B cells is associated with most viral infections as well as some bacterial infections.

NUTRITION AND IMMUNE FUNCTION

Nutritional deficiency is the most common cause of secondary defective immunocompetence. Fasts or inadequate intakes of short duration in healthy individuals appear to have little effect on immune function. In protein-energy malnutrition there is atrophy of the organs central to the immune system—thymus, lymph nodes, and spleen. T cell numbers are reduced as is phagocytosis. Infectious diseases are usually more virulent in malnourished persons. For example, measles, usually a mild disease in well-nourished individuals, frequently is life-threatening to the malnourished child. Persons with protein-energy malnutrition, hospitalized patients, drug users and alcohol abusers, some infants, older individuals, and food faddists are at especial risk for depressed immune function.

Infants

At birth immune response is low, but infants are protected by antibodies received from their mothers in fetal life. Breast milk also provides IgA. Infants small for gestational age may have decreased immune function, resulting in a higher incidence of infections.

Older Adults

Immunocompetence or ability to develop an immune response peaks in adolescence. With increasing age there is a progressive decline in immune function, especially cell-mediated immunity, while autoimmune manifestations increase. Older people have a higher incidence of cancer, degenerative diseases, and infection, all associated with decreased immune function and possibly autoimmune factors. In aging, lean body mass decreases, blood levels of many nutrients fall, and malnutrition is sometimes a problem, raising the question of whether dysfunction in immunocompetence is due, at least in part, to malnutrition.

Specific Nutrients

Any nutrient deficiency or excess that negatively affects complement, maturation of T cells or B cells, or that depresses DNA synthesis, cell division, or replication would decrease cell-mediated and humoral immunities. Macrophages require energy to release enzymes.

Energy Moderate caloric restriction has no known impact on immunocompetence. Indeed, animals fed diets adequate in all other nutrients but moderately restricted in energy live longer, have fewer tumors, and a slower decline of immune function with age. Whether these findings have relevance for humans is unknown. However, obese individuals have a greater incidence of respiratory infections and postoperative wound infections than normal weight people.

Protein Protein synthesis is central to immunocompetence and protein nutrition profoundly affects immune function. However, a protein deficit without other accompanying nutrient deficiencies is rare. Children with kwashiorkor (see Table 7–2) and hospital patients with kwashiorkor malnutrition (see Chapter 16) display depressed immunocompetence, especially the cell-mediated type.

Not only quantity but also quality of protein affects the immune response. Diets that fail to supply sufficient amounts of essential amino acids to synthesize the immune factors have been associated with increased incidence of infections.

Other Nutrients Single nutrient deficiencies in humans are rare and most information on the effects of nutrient lacks or excesses on immune function comes from animal studies. Practical implications for humans are unclear.

DISORDERS OF THE IMMUNE SYSTEM

Autoimmune Diseases

Sometimes the body's immune defenses turn on the body itself, and antibodies directed against its cells and organs are produced. These autoantibodies are factors in autoimmune diseases such as insulin dependent diabetes mellitus, rheumatoid arthritis, systemic lupus erythematosus, and myasthenia gravis. Nearly 80 autoimmune diseases affecting 50 million Americans have been identified. The cause of autoimmune conditions is unknown, but could involve viruses, chemicals, medications, and/or drugs that so change cells that the immune system responds as if they were foreign. Heredity, gender, and aging may be factors since autoimmune conditions are more common in some families, among women, and in older people. Nutrition therapy can help to control autoimmune diseases.

Immunodeficiency Diseases

When one or more components of the immune system is lacking or inadequate, immunodeficiency disorders result. Patients with advanced cancer may be affected because of the disease process and/or side effects of therapies. Some infants are born with immune system defects.

Acquired immunodeficiency syndrome (AIDS) is caused by the human immunodeficiency virus (HIV), which destroys a type of helper T cell, the CD_4 T cell. Affecting many body systems, AIDS involves **opportunistic infections,** rare cancers, and dementia, the latter through brain and spinal cord damage.

The infections are caused by ordinarily harmless organisms that "exploit the opportunity" of an incomplete immune system. Presently there is no cure for AIDS, although some medications will slow the replication of the virus. AIDS is further discussed in Chapter 29.

People with immunodeficiency disorders are at high risk of foodborne infections and should take especial precautions to prevent them. For additional information on foodborne conditions, see Chapter 2.

ALLERGIC REACTIONS

A problem central to any discussion of allergic reaction is terminology. Many people call any adverse reaction to food an allergy. Self-diagnosis and parental diagnosis are common. Some view "allergy" as a widespread problem, responsible for many physical and emotional difficulties. The terms *food intolerance* and *food allergy* are used with varying shades of meaning.

TABLE 19–1

Some Nonallergic Food Intolerances

Type	Cause	Example(s)
Metabolic food disorder	Metabolic defect	Lactose intolerance
Food additive	Monosodium glutamate (MSG)	Chinese Restaurant Syndrome
	Tartrazine (FD&C yellow No. 5)	Bronchospasm
	Sulfites	Bronchospasm
Pharmacological	Pressor amines	Headache, hypertension
	Caffeine	Headache, palpitation, panic attacks
Anaphylactoid (anaphylaxis type)	Chocolate, egg white, strawberries, shellfish, tomatoes may induce such reactions	Release of chemical mediators such as histamine without IgE or other immune mechanism
Food poisoning	Infectious agents	*Salmonella, Staphylococcus*

Food Intolerance

Food Intolerance: adverse reaction to a food or food component not involving the immune system

Some adverse reactions to food that do not involve the immune system and therefore are not true allergies are presented in Table 19-1. For most of these intolerances, therapy is similar to that of a true food allergy—avoidance of the offending food or additive. This type of reaction far outnumbers true allergic reactions.

Food Allergy

Food Allergy: adverse reaction to a food or food component that is immunologically mediated, reproducible in a food challenge, and causes functional change in the target organ

Food allergies are most common in infants and usually diminish as the child grows. Immaturity of the digestive and immune systems is a contributing factor. Allergies occurring before age three are more likely to disappear than those that originate later. Development of allergy is influenced by heredity, gastrointestinal permeability, immune response, and exposure to food. Children appear to inherit tendencies toward allergic reactions, not specific allergies. For example, a parent may have hay fever and the child a milk allergy.

Incidence

In the first months of life the gastrointestinal tract is permeable, and intake of cow's milk during this time may be related to increased incidence of allergy to the milk. Exclusive breast-feeding tends to reduce the incidence of allergy as does postponement of the introduction of solids, especially among infants at increased risk due to heredity or other factors.

The most common allergies, rhinitis, or hay fever, and to a lesser extent, asthma, are triggered by inhaling dust, pollen, and other allergens. Although the incidence of food allergies is hard to document, it has been estimated that 1–3 percent of adults have some allergic reaction to food. Individuals with other allergies have a higher incidence of food allergies than their nonallergic peers.

Mechanism

IgE-Mediated: also known as immediate hypersensitivity, reaginic sensitivity, or atopic sensitivity

The vast majority of food allergies involve an IgE-mediated response. Other immunoglobulins may be involved in some allergic reactions. Food proteins that act as antigens bind to IgE, which is attached to sensitized mast cells or basophils. This contact results in an antigen-antibody reaction with release of mediators

322

CUTANEOUS	SYSTEMIC ANAPHYLAXIS
Atopic dermatitis (eczema)	Mild
Urticaria (hives or welts)	Itching
	Redness of skin
GASTOINTESTINAL	Slight fever
	Urticaria
Nausea	Severe
Vomiting	Abdominal pain
Pruritis (severe itching) of lips, oral mucosa,	Cardiac arrhythmias
pharynx	Chest pain
Swelling of throat	Convulsions
Abdominal distention	Cyanosis
Cramping	Dyspnea
Diarrhea	Fever
	Hypotension
	Palpitation
RESPIRATORY	Urticaria
	Vomiting
Asthma	Throat constriction
Nasal congestion	Violent coughing
Rhinitis (hay fever)	Shock
Sneezing	Loss of consciousness
Wheezing	

FIGURE 19–2
Some symptoms of food allergy.

such as histamine and prostaglandins. These mediators cause the symptoms of food allergies. (See Figure 19-2.) Reactions can be immediate or delayed. Gastrointestinal symptoms are by far the most common, followed by cutaneous manifestations. Respiratory effects are less frequent. The role of T cells in allergic reaction is not clearly understood. They may be involved with celiac disease and protein enteropathies.

Clinical Findings

Evidence of IgE-mediated allergy can occur from within a few seconds to several hours after ingestion of the food. Gastrointestinal symptoms may begin in the area first exposed with burning, itching, and swelling of the mouth, gums, lips, tongue, and pharynx. Manifestations may end here or extend to other parts of the system as the allergen proceeds through the digestive tract. Food allergy may play a role in **otitis media** (inflammation of the middle ear) in children and joint pain and arthritis-like conditions in adults. There is controversy as to whether allergies are involved in migraine headaches and behavior changes.

Systemic anaphylaxis, severe allergic reaction, is the most dangerous allergic reaction. In especially severe episodes, emergency treatment with vaso-pressors, corticosteroids, oxygen, and artificial respiration may be necessary. Anaphylactic reactions have been reported when a food was consumed before heavy exercise that did not occur when the food was not followed by strenuous activity.

Allergens are usually acidic proteins or glycoproteins. Although almost every food, additive, dye, and preservative in the American diet has been implicated in allergic reactions, the vast majority of such reactions involve a few foods—cow's milk, eggs, fish and shellfish, legumes (soybeans and peanuts), tree nuts (cashews and filberts), and wheat.

Food Allergy Diagnosis

PROTOCOL FOR DIAGNOSING FOOD ALLERGY

Assessment
Food and symptom diary
Immunological testing
Elimination diet
Double blind food challenge

Many steps are involved in the assessment and diagnosis of a food allergy. Assessment should proceed in the following order.

Assessment In the history, symptoms and their timing, suspect foods and amounts required to produce a reaction, and a family history of allergies are obtained. Frequent episodes suggest a common food, multiple foods, or a nonfood cause. The physical examination includes anthropometric measurements and evaluation of growth and development in the child. Chronic disease conditions and allergic symptoms are checked. Nutrition screening (see Chapter 16) can pinpoint individuals at risk of malnutrition.

Food and Symptom Diary A detailed record of foods eaten, medications taken, and timing of symptoms is kept for two weeks. It also provides information on dietary adequacy since, in an attempt to avoid offending foods, some people have placed themselves on grossly inadequate diets.

Immunological Testing A drop of antigen is placed on the skin, which is then punctured. A small edematous area surrounded by erythema is considered positive. The test is sensitive but does overdiagnose as reactions are seen in 30 percent of tests. A positive response should be followed with a food challenge.

Other Tests The radioallergosorbent extract test (RAST) involves the use of radioactively labeled anti-IgE, and the enzyme-linked immunosorbent assay (ELISA) uses an enzyme for testing suspect foods. Both are more costly and no more accurate than skin testing, but might be used when the skin is especially sensitive, where there is skin disease, or when skin testing may produce an anaphylactic reaction. Any of these immunological tests yields a list of suspect foods.

Elimination Diet The patient begins with a simple elimination diet, omitting only foods that are under suspicion because of history, food diary, and immunological tests. Every effort should be made to ensure nutritional adequacy. If this regimen does not afford relief from symptoms, a more restricted elimination diet is taken. Numerous standard elimination diets exist. It may involve only one item from each of the food groups plus exotic foods the person has never eaten. In extreme cases, a hypoallergic chemically defined formula* is necessary if the symptoms continue. If symptoms still persist, the allergy probably is not to food.

Food Challenge After the individual has been symptom free for two to four weeks, suspect foods are reintroduced one at a time, beginning with very small amounts that are gradually increased. Foods known to cause life-threatening

---

*Vivonex®, Sandoz Nutrition, Minneapolis, MN; Peptamen®, Clintec Nutrition, Deerfield, IL.

anaphylaxis are excluded. If there is a positive reaction, the food may be tried again in six months, since food allergies sometimes have spontaneous remissions.

If there is uncertainty about the reaction or a psychological dimension, a double blind food challenge (DBFC) will be performed. In a double blind test neither the patient nor the clinician knows the nature of the allergen.

Many foods incriminated through histories and immunological tests do not yield similar results through double blind testing. For double blind challenges to be worthwhile, the patient must want the true answer, whether or not it confirms some preconceived opinion that is held.

Therapy

Compared to the complex diagnostic procedure, treatment of a food allergy is straightforward—eliminate the offending food. Some people may be able to consume small amounts of the food without problems, but when tolerance is very low or nonexistent, all traces of the food must be excluded. Where minute amounts may produce systemic anaphylaxis, the individual should wear medical alert jewelry indicating the allergen(s) and carry an epinephrine kit in case the offending food is inadvertently consumed. Desensitization therapy has not proven useful in food allergies.

When mild allergies to several foods exist, a **rotary** or **rotational diet** in which the offending food is eaten only once in five days and alternated with other allergy-causing foods may be used. This system provides a greater variety of foods and possibly greater nutritional adequacy.

In infants with an allergy to cow's milk for whom breast-feeding is not feasible, hypoallergenic formulas* are used. Some clinicians avoid soybean-base formulas because soybeans are almost as allergenic as cow's milk.

Individuals with proven allergies should be aware of **cross-reactivity,** allergy to foods of the same botanical or other classification. For example, a person with an allergy to soybeans is at risk of being allergic to other legumes such as peanuts.

Patient Counseling The meal pattern should fit in with that of the family and must ensure nutritional adequacy. An allergy to milk means that another source of calcium must be provided. Allergy to citrus fruits dictates that good sources of vitamin C must be emphasized.

Reading labels on food packages and interpreting this information are essential. Under current labeling regulations, any product with two or more ingredients must state all ingredients on the label. Since manufacturers sometimes change their formulations, it is prudent to read the label each time the food is purchased. Also, different brands of the same food may have different ingredients.

Foods that contain milk are detailed in the lactose-restricted diet (see Figure 22–2). Products likely to contain gluten, a protein found in some grains, are listed in Figure 22–3. Items that may cause reactions in individuals with egg, wheat, corn or soy allergies appear in Figures 19–3, 19–4, 19–5, and 19–6. See also Figure 19–7. Every effort must be made to ensure that diets are nutritionally adequate. See the Food Guide Pyramid (Figure 1–5) and the Basic Diet (Table 2–4).

---

*Nutramigen®, Pregestimil®, Mead Johnson & Company, Evansville, IN.

SOME RELATED ALLERGENS

Anacardiacae: cashew, mango, pistachio

Betulacae: filbert, hazelnut, wintergreen

Leguminosae: alfalfa, black-eyed pea, broad bean, chick pea, common bean, lentil, licorice, pea, peanut, soybean

CHILI WITH BEANS—Brand A

Ingredients: beans, water, beef, tomatoes, cornstarch, sugar, spices, salt

CHILI WITH BEANS—Brand B

Ingredients: beans, water, tomatoes, beef, wheat flour, chili powder, onion, garlic salt

CHILI WITH BEANS—Brand C

Ingredients: water, beans, beef, tomatoes, textured soy protein, monosodium glutamate

CREAMY CANDY CORN

Ingredients: sucrose, corn sweetener, egg albumin, salt, honey, confectioners' glaze, artificial flavor, artificial color (FD & C Yellow No. 5)

Eggs in any form, commercial egg substitutes

Beverages—eggnog, malted beverages, Ovaltine, root beer, wine

Breads and rolls containing eggs—crust glazed with egg, French toast, sweet rolls, pancakes, muffins, waffles, pretzels, zwieback

Desserts—cake, cookies, custard, doughnuts, some ice cream, some sherbet, meringue, cream-filled pies (coconut, cream, custard, lemon, pumpkin), puddings

Meats—meat loaf or meat balls, breaded meat dipped in egg

Egg noodles, pastas containing egg

Salad dressings—cooked dressing, mayonnaise

Sauces—Hollandaise, tartar

Soups—broth, bouillon, consommé (if cleared with egg), egg drop

Sweets—many cake icings, candies: cream, chocolate, fondant, marshmallow, nougat

Miscellaneous—baking mixes, dessert powders, fritters, pastries, souffles

Key label words—egg, egg white, egg yolk, dried egg, albumin, egg albumin

FIGURE 19–3
Diet without eggs—Examples of foods that may contain eggs.

Beverages—malted milk, instant coffee with added cereal grain, coffee substitutes, beer, gin, whiskey

Breads, crackers, rolls—all breads including wheat, hot breads and muffins; baking powder biscuits; gluten bread; matzo, pretzels, zwieback; white and graham crackers; griddle cakes, waffles. *Note:* bread, cracker, or wafers made of 100% rye, corn, rice, soy, or potato flours may be used

Cereals—bran, bran flakes, Cheerios, Cream of Wheat, farina, granola-type, Grape-Nuts, Grape-Nuts flakes, Kix, Maltex, Muffets, New Oats, Pablum, Pep, Pettijohn's, Puffed Wheat, Ralston cereals, Shredded Wheat, Special K, Total, Wheatena, wheat flakes, wheat germ, Wheaties, Wheat Chex, and others (read ingredient list on labels)

Desserts—cake or cookies, homemade, from mixes, or bakery; doughnuts, ice cream, ice cream cones, pies, puddings

Flour—white, whole wheat, graham, all purpose

Gravies and sauces thickened with flour. May use cornstarch thickening

Meats—prepared with flour, bread, or cracker crumbs such as croquettes and meat loaf; stews thickened with flour or made with dumplings; frankfurters, luncheon meats, or sausage in which wheat has been used as a filler; canned meat dishes such as stew, chili, stuffings and commercial stuffing mixes

Pastas—macaroni, noodles, spaghetti, vermicelli

Salad dressings—thickened with flour

Soups—commercially canned (read ingredient label), bouillon cubes

Key label words—flour, wheat, wheat starch, wheat germ, farina, seminola, bran (unless specified as corn bran, etc.)

FIGURE 19–4
Diet without wheat—Examples of foods that may contain wheat.

TASTY CORN MUFFIN MIX

Ingredients: cornmeal, enriched flour, sugar, shortening, leavening, salt

PUTTING IT ALL TOGETHER

Immune function, involving a complex interacting series of mechanisms, is essential for patient health and well-being. Individuals whose immune systems are compromised have reduced ability to mount a defense against infectious agents. True food allergies involve reactions of the immune system.

Beverages—ale, beer, carbonated, gin, grape juice, instant tea, milk substitutes, soy milk, whiskey

Breads and crackers—cornbread or corn muffins, enchiladas, English muffins, corn chips, tacos, tortillas, graham crackers

Cereals—cornmeal, hominy, cornflakes, Corn Chex, Grape-nuts, Kix, multigrain cereal, grits

Desserts—cakes, candied fruits, canned or frozen fruits or juices, cream pies, ice cream, pastries, pudding mixes, sherbet

Fats—corn oil, corn oil margarine, gravies, salad dressings thickened with cornstarch, mayonnaise, salad dressings, and shortenings—unless source of oil is specified

Flours—cornmeal, cornstarch

Meats—bacon, cured, tenderized ham, luncheon meats, sausage, scrapple

Soups—all commercial; homemade thickened with cornstarch

Sweets—candy, cane sugar, corn syrups, corn sugars, imitation maple syrup, imitation vanilla, jams, jellies, preserves

Vegetables—Harvard beets (thickened with cornstarch), corn, mixed vegetables containing corn, succotash

Miscellaneous—baking powder, batters for frying, catsup, cheese spreads, commercial mixes of all types, confectioner's sugar, monosodium glutamate, peanut butter, popcorn, sandwich spreads, vitamin capsules, yeast, chewing gum

Key label words—corn, corn syrup (dextrose), corn starch, cane sugar, cornmeal, corn bran

FIGURE 19–5
Diet without corn—Examples of foods that may contain corn.

KRISPY POTATO SNACKS

Ingredients: dehydrated potatoes, shortening, cornmeal, salt, BHA

Beverages—soy milk, beer, wine, chocolate and cocoa mixes, coffee creamers, fast-food shakes

Breads and cereals—high protein breads and cereals, English muffins, granolas, bread stuffing mixes

Desserts—some cakes and cake mixes

Fats—soybean oil, margarine, salad dressings, nonstick sprays, bacon bits, shortening—unless type of fat is indicated

Flour—soy flour

Gravies and sauces—soy sauce, Teriyaki sauce, Worcestershire sauce, gravy mixes

Meats—meat extenders, textured vegetable protein (TVP), soy protein, cold cuts, tofu, veggie burgers, miso, tempeh, some canned meats and fish

Pastas—casserole mixes

Sweets—candies, chocolate chips, pancake syrup

Vegetables—potato mixes, instant mashed potatoes

Miscellaneous—dip mixes, soy nuts

Key label words—soy, soya, soy fiber, soy flour, soy milk, soybean oil, soy protein, textured vegetable protein (TVP), modified food starch, tofu, vegetable gum, vegetable starch, lecithin, hydrolyzed vegetable protein, hydrolyzed soy protein. Soy is widely used in food manufacturing. Read labels carefully.

CHUNKY TUNA

Ingredients: tuna, water, hydrolyzed soy protein, salt

FIGURE 19–6
Diet without soy—Examples of foods that may contain soy.

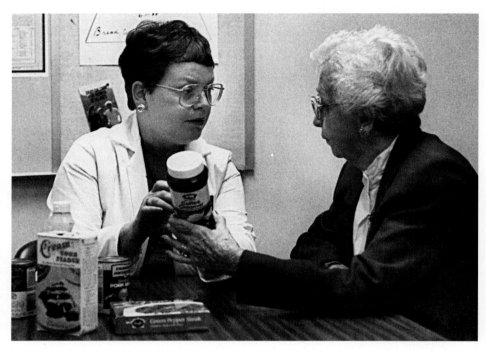

FIGURE 19–7
Having an allergy to corn, this patient must read ingredient labels carefully since numerous products contain corn or corn products. (*Source:* Dunwoody Village, Newtown Square, PA, and Peter Zinner, photographer.)

❓ QUESTIONS CLIENTS ASK

1. What is cytotoxic testing? Can it diagnose food allergies?

In cytotoxic testing, a food allergen is added to a sample of an individual's blood. The blood is examined under a microscope to see if the white blood cells change or die. This mode of testing has been studied by researchers who concluded that it does not diagnose food allergy. Cytotoxic testing is also called Bryan's test, leukocyte antigen testing, and food sensitivity testing.

2. I am pregnant. Will breast-feeding help to prevent food allergies in my baby?

Exclusive breast-feeding will help reduce the possibility of food allergies early in life. At that time the gastrointestinal system is immature. Breast-feeding does not guarantee that food allergies may not develop later, but it reduces the chances.

3. Many of my friends tell me they have food allergies. Are such allergies that common?

In one study, between 16 and 20 percent of people surveyed claimed that they or somebody in the household had a food allergy. It is believed that 1 to 2 percent of adults and 2 to 4 percent of children have true food allergies, which are reactions that involve the immune system. Some of your friends may also have food intolerances that do not involve the immune system.

FOR ADDITIONAL READING

Carroll, P: "Guidelines for Counseling Patients with Food Sensitivities," *Top Clin Nutr,* **9**:33–37, June 1994.

Chandra, R: "Nutrition and Immunity in the Elderly: Clinical Significance," *Nutr Rev,* **53**(Part II):S80–S85, 1995.

As an infant, Vicki A. displayed a sensitivity to cow's milk and was given a hypoallergenic formula. Her mother is allergic to shellfish, and her father has asthma. Vicki was able to tolerate cow's milk by the time she was two.

Throughout her childhood, Vicki has had a variety of nonspecific symptoms including stomachaches, headaches, listlessness, red watery eyes, cramps, and diarrhea. She is now ten years old and during the past year the symptoms have appeared more frequently. She has failed to gain weight during this period and currently is 52 in. (132 cm) tall and weighs 55 lb (25 kg). The symptoms led her parents to consult a physician, who referred them to an allergist.

A diet history, skin tests, and a food diary that Vicki kept led the allergist to suspect that wheat and egg were the allergens. Elimination diets and double blind food challenges confirmed this suspicion. Vicki is to follow an egg- and wheat-free diet.

Questions

1. Why was Vicki asked to keep a food diary?

2. Give possible reasons for Vicki's failure to gain weight.

3. What foods did Vicki avoid on the egg-free elimination diet?

4. What foods did Vicki avoid on the wheat-free elimination diet?

5. Suggest ways in which Vicki's menus may be made as much as possible like those of the rest of her family.

6. Write a day's wheat-free, egg-free menu for Vicki, keeping in mind her age and growth needs. (See Chapter 13.)

7. Vicki takes a packed lunch to school. Write three different lunch menus she might take.

8. Next month Vicki will be responsible for the snacks for her Girl Scout troop meetings. Suggest four different snacks she might provide.

9. Vicki wonders if she will outgrow her food allergies. What might you tell her?

10. Explain the relationship between food allergies and the immune response.

11. Visit a food market and/or look at the packages in your food cabinet and freezer. Read the ingredient labels on 10 items. Which would be acceptable for Vicki? Explain. Suggest substitutes for any items contraindicated for Vicki.

12. Vicki's birthday is coming up soon. She wants a birthday cake. What might you suggest?

Food Allergy and Intolerances. NIH Pub. No 93-3469, Public Health Service, Bethesda, MD, 1993.

Hanson, IÅ, *et al:* "Early Dietary Influence on Later Immunocompetence," *Nutr Rev,* **54:**S23–S30, 1996.

Hingley, AT: "Food Allergies: When Eating is Risky," *FDA Consumer,* **27:**27–31, December 1993.

Howard, BA: "Guiding Allergy Sufferers Through the Medication Maze," *RN,* **57:**26–31, April 1994.

Lavin, J, and Haidorfer, C: "Anergy Testing—A Vital Weapon," *RN,* **56:** 31–33, September 1993.

Mills, JA: "Systemic Lupus Erythematosus," *New Engl J Med,* **330:**1871–79, 1994.

Rosen, FS, *et al:* "Medical Progress: The Primary Immunodeficiencies," *New Engl J Med,* **333:**431–40, 1995.

Roubenoff, R, and Rall, LC: "Humoral Mediation of Changing Body Composition During Aging and Chronic Inflammation," *Nutr Rev,* **51:**1–11, 1993.

Shronts, EP: "Basic Concepts of Immunology and Its Application to Clinical Nutrition," *Nutr Clin Prac,* **8:**177–83, 1993.

Simons, FER, and Simons, KJ: "The Pharmacology and Use of H$_1$-Receptor-Antagonist Drugs," *New Engl J Med,* **330:**1663–70, 1994.

Nutrition in Stress, Infections, and Fevers

Chapter Outline

*Decrease to no more than 5 percent the proportion of people aged 18 and older who report experiencing significant levels of stress who do not take steps to reduce or control their stress.**

*Increase to at least 85 percent the proportion of people found to have tuberculosis infection who completed courses of preventive therapy.**

✔ **Case Study Preview**

After the number of new cases of tuberculosis declined over 60 years, a recent increase has concerned health care workers. You will plan care for a homeless woman with few skills and resources who has developed tuberculosis.

Healthy People 2000: National Health Promotion and Disease Prevention Objectives. Public Health Service, U.S. Department of Health and Human Services, Washington, DC, 1991, Objectives 6.9 and 20.18, pp. 100 and 123.

A number of unrelated pathological conditions interfere with the maintenance of good nutritional status. Among them are stress, infections, and fevers. Protein or protein-energy deficiencies may follow if dietary adjustments are not made promptly.

STRESS

Stress is defined as any event that threatens the body's steady state or homeostasis. Stress is a complex phenomenon involving both psychological and physiological factors. Stressors can be synergistic. For example, coexisting personal illness or injury with resulting lost income could cause the stress of personal debts to be greater than they would be to a healthy person.

Emotional Stress

Anxiety and tension are common stressors, and the reaction to them varies from person to person. An occurrence that one person finds stressful does not necessarily produce stress in another individual. Some stressors are positive, such as outstanding personal achievements or vacations. Change itself can be stressful. (See Figure 20–1.)

There is some evidence that emotional stress plays a role in the etiology of diseases such as cancer, cardiovascular disease, hypertension, and peptic ulcer. Some people eat more and others less while under emotional stress, with the potential for over- and undernutrition. No evidence exists to justify the need for the numerous multivitamin "stress" supplements marketed to increase resistance to stress or reverse nutritional deficiencies supposedly caused by stress.

Physiological Stress

Metabolic response to physiological stress is more specific.

Acute Phase In response to stressors the hypothalamus signals the release of the catabolic hormones: glucocorticoids, catecholamines (epinephrine and norepinephrine), and glucagon. These hormones cause increased protein catabolism, gluconeogenesis, glycogenolysis, and lipolysis, with resulting loss of adipose and lean body tissue. Catecholamines also mediate hypermetabolic states and depress insulin release, which may result in hyperglycemia independent of diabetes.

In these states, priority is given to supplying blood to the heart and skeletal muscles. The blood supply to the gastrointestinal tract is restricted and the catecholamines decrease motility. Anorexia, distention, or constipation are likely to follow. Stimulation of the parasympathetic nervous system may lead to nausea, vomiting, and diarrhea.

If the stress is of brief duration, the changes in stress hormone levels are short lived and unlikely to produce a catabolic response. On the other hand, if the stressor is intense and/or prolonged, a catabolic state with nitrogen wasting results.

Increased metabolic rate and nitrogen excretion are closely correlated. Protein catabolism is a normal response to stress and cannot be prevented, although adequate intake can help to replenish the lost protein. Nitrogen losses usually peak five to ten days after the stress. As the catabolic state is reduced through wound healing or moderation of the disease process, nitrogen losses decrease.

For the severely stressed patient in a critical care setting, assessment data obtained after the peak metabolic response are more valuable than those collected earlier, because the stress can alter many test results. Admission assessment need only identify patients at risk for malnutrition and give some direction about fluid and energy requirements.

SOME COMMON STRESSORS

death of family member, divorce, losing job, personal injury, illness, financial problems, pregnancy, new family member, new job, outstanding personal achievement, beginning or ending school, vacation, change in eating habits

Catecholamines: also known as the hormones of stress

FIGURE 20–1
What factors in this woman's work situation would you find stressful? (*Source:* School of Nursing, University of Pennsylvania, and Denise Angelini Kosman, photographer.)

Nutritional Requirements During Stress

The requirements are influenced by the extent and duration of the stress and the age and nutritional status of the individual. (See inside back cover for added energy needs imposed by different stressors.) Since food intake is frequently decreased during stressful situations, endogenous sources of glycogen, fat, and protein are important. An optimal nutritional state prior to stress will provide greater stores.

INFECTIONS

Infection is the invasion of the body by a pathogenic agent that multiplies and produces harmful effects. A synergistic relationship exists between malnutrition and infection (See Figure 20-2.) Poor nutritional status lowers resistance to infection by depressing the immune response; severe malnutrition can render common infectious diseases life threatening. Infection, on the other hand, involves protein catabolism, and frequently anorexia, nausea, vomiting, and diarrhea further decrease the food intake and intensify the nutrient losses. Enteric infections also interfere with absorption and reduce nutrient utilization. Fever, a frequent accompaniment, increases the energy needs.

Metabolic Response

Response to infection is similar to response to any other physiological stress, except that lipolysis and ketone formation are suppressed, probably because of high insulin levels. This decreased ability to utilize fat renders muscle protein

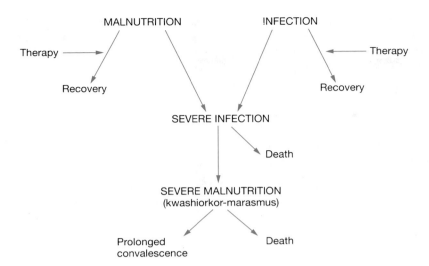

FIGURE 20–2
Malnutrition alone or an infection alone when subjected to therapy lead to recovery. But when malnutrition and infection are present at the same time there is a synergistic relationship which makes therapy difficult and the final outcome questionable.

losses even greater. Available nitrogen is used to synthesize immune response components. If the infection is prolonged, nitrogen conservation eventually begins, but not until considerable body weight loss and protein wasting have occurred.

Nutritional Factors

The most critical nutritional problem in infection is protein catabolism, which will occur even in mild infection without fever. As with other severe stressors, nitrogen losses are part of the body's response and cannot be reversed until after the hypermetabolic period peaks. If urinary losses are known, intake may be planned to replace them. When losses are not known, protein intake is based on energy needs. (See Chapter 16.)

An individual in good nutritional status with a mild infection can meet all needs with additional protein and calorie-rich foods added to the regular diet. An individual in poor nutritional status may require supplements.

Obviously, the primary goal is to identify and eliminate the agent causing the infection. Many antibiotics used in controlling infections have gastrointestinal disturbances as a side effect. (See Chapter 18.)

FEVERS

Fever is a classic sign of infection, although not all fevers are caused by infection. Nor are all elevations in body temperature fever. In heat stroke, for example, the body produces more heat than it can eliminate.

In infections phagocytes are activated and release a protein called endogenous pyrogen (EP) or interleukin-1 (Il-1). Endogenous pyrogen mediates the formation and release of prostaglandins, which raise the body's set point. Heat-producing and heat-sparing mechanisms elevate body temperature. With a rise in body temperature, there is vasoconstriction and the patient has chills and complains of feeling cold.

Fever: increase in set point that allows body temperature to be regulated at a higher level

Set Point (temperature): temperature around which the body temperature is regulated

Thermoregulation: hypothalamus-mediated processes whereby heat production and loss are regulated and body temperature maintained

Hyperthermia:
  temperature above set point

With the removal of the stimulus, the set point declines. Heat loss mechanisms including sweating, increased respiration, and vasodilation restore normal body temperature. Fevers may be acute and of short duration as in colds, chronic as in tuberculosis, or intermittent as in malaria.

Metabolic Effects

Elevated body temperature leads to numerous metabolic changes. The severity and duration determine the nutritional consequences.

Nutritional Considerations

METABOLIC RESULTS OF ELEVATED BODY TEMPERATURE

↑ BMR: 7%/°F; 13%/°C
↑ Breakdown of body proteins
↑ Loss of body water, sodium, and potassium
↓ Glycogen and adipose tissue stores

In an acute fever, adequate intake of fluids and electrolytes is of primary importance. Small frequent feedings of liquid and soft foods help maintain intake when appetite is poor. For a prolonged fever, a high-protein, high-calorie diet is prescribed. (See Table 9–3 and Figure 20–3.)

Medications

Antipyretic agents such as *acetometaphen, ibuprofen, indomethacin,* salicylates, and steroids block synthesis of prostaglandins that are necessary to increase the set point. Most of these medications are also analgesics that help relieve pain and increase comfort. They should be taken with food to decrease gastric irritation.

TUBERCULOSIS

RISK FACTORS FOR ACTIVE TUBERCULOSIS

HIV infection
Homelessness
Illicit drug use
Alcohol abuse
Immigration from area with high incidence of tuberculosis
Nursing home residence

Worldwide, tuberculosis remains a leading cause of illness and death. In the United States, improved housing, nutrition, sanitation, medications, and screening programs as well as education had contributed to a decline in the incidence, but in recent years there has been an increase in cases, some of them drug resistant.

When a person with active tuberculosis coughs or sneezes, droplets containing *Myobacterium tuberculosis* are sprayed into the air. Most people who breathe in the bacteria do not become infected. In those who do, the immune system responds by activating macrophages through T cell mediation, leaving the organism dormant. About one-tenth of those so infected develop active tuberculosis at some time in their lives. Approximately 90 percent of active cases occur through reactivation of dormant infections. People with reduced immune response are at increased risk of developing active tuberculosis.

Tuberculosis most frequently affects the lungs, although other organs may be involved. The bacillus sets up an inflammatory process and healing occurs with calcification of the tuberculous lesions. Cough, expectoration, fever, and tissue wasting accompany the disease. The acute form resembles pneumonia with high fever. The chronic form involves a low grade fever.

Dietary Considerations

The diet for chronic tuberculosis includes:

- Adequate calories (2,500–3,000)
- Liberal protein (75–110 g)
- Iron supplement and increased dietary sources of vitamin C if there has been lung hemorrhage
- Preformed vitamin A in diet and possibly vitamin A supplements. Carotenes are poorly converted to vitamin A activity

Isoniazid is a vitamin B_6 antagonist and B_6 supplements are essential to prevent peripheral **neuritis,** that is, inflammation of the nerves, often degenerative. *Rifampin* can cause nausea and vomiting, *pyrazinamide* can cause gastrointestinal irritation, and both can be toxic to the liver. Failure to comply with or complete medication regimens is a common cause of setbacks, recurrences, and development of medication-resistant strains of the bacteria.

PROTEIN DEFICIENCY

Protein deficiency in the healthy segment of the United States population is rare, but there are some groups that are at risk. On a global basis, protein-energy lack is the main nutritional problem. (See Chapter 7.)

   Fatigue, decreased resistance to infection, and weight loss are present in protein deficiency. Since these symptoms are not specific to protein deficiency, they are of limited diagnostic value. Anthropometric and laboratory methods of assessing protein and energy nutriture are discussed in Chapters 3, 9, and 16.

High-Protein Diet

As in all therapeutic diets, the planning for high protein intake begins with the normal diet. The Food Guide Pyramid (see Figure 1-5) furnishes between 70 and 130 g protein. To achieve a level of 130 g protein means that concentrated sources of protein would be useful so that there is no excessive bulk to be consumed. Adequate calories are necessary so protein can be used for anabolism. If the patient likes meat, fish, or poultry, the modest amounts in the basic diet may be increased by 2 or 3 oz. (See Figure 20-3.) See also Pattern C of the Basic Diet, Table 2-4.

   Additional milk is usually included. One way to increase the protein in the diet at low cost and with little increase in volume of the diet is to use nonfat dry milk. Depending on the brand used, 4 to 5 tablespoons added to 8 oz lowfat milk will double the amount of protein. Nonfat dry milk also may be added to mashed potatoes, cream sauces, cream soups, baked custard, and other foods. High-protein supplements* are useful for some patients.

   Many patients who require a high-protein diet have poor appetites. It is better to start with the patient's present food intake and to gradually increase the protein and calorie intake. The portions served should be of such size that the patient is able to eat all the foods offered. An evening snack is a good way to increase protein and calorie intakes. Two examples of such snacks are

Sandwich
   2 slices bread
   2 teaspoons margarine, mayonnaise
   1½ oz roast beef
   1 oz cheese
Lowfat milk, 1 cup

Protein, 29 g; kcal, 535

High-protein milk shake
   1 cup lowfat milk
   4-5 tablespoons nonfat dry milk
   2 tablespoons chocolate syrup

Protein, 16 g; kcal, 295

RISK OF PROTEIN DEFICIENCY

Inadequate Intake

Some older people living
   alone
Some weight loss regimens
Drug use, alcohol abuse
Chronic anorexia, nausea,
   vomiting, diarrhea

Increased Protein Loss

Acute phase stress reaction
Hemorrhage
Immobilization
Malabsorption
Some kidney and liver
   diseases

---

*Casec, Mead Johnson & Company, Evansville, IN; ProMod™, Ross Laboratories, Columbus, OH.

4 cups lowfat milk
1 egg
7–9 oz lean meat, poultry, fish, lowfat cheese
3–4 servings vegetables
   1 green leafy or deep yellow
   1 raw
   1 potato or substitute
3 servings fruit, including 1 citrus fruit
6–11 servings cereals, breads
Desserts such as lowfat puddings, custard, ice milk, frozen yogurt
Margarine, salad dressings
Sugar, jelly, jam

The following menu illustrates the kinds of foods used for a diet providing about 120 g protein:

BREAKFAST

Orange juice
Wheat flakes *with*
Lowfat milk, 1 cup
Sugar
Toast, enriched, 2 slices
Margarine
Jelly
Coffee with milk, sugar

DINNER

Swiss steak, 4 oz
Parsley noodles
French green beans with slivered almonds
Grapefruit avocado salad on lettuce
French dressing
Dinner roll with margarine
Chocolate pudding
Lowfat milk, 1 cup
Tea or coffee, if desired

LUNCH

Cold sliced turkey, cheese, 3 oz
Potato salad
Lettuce, sliced tomato
Mayonnaise
Rye roll
Margarine
Frozen lowfat yogurt
Peanut butter cookie
Lowfat milk, 1 cup
Tea, if desired

EVENING SNACK

Egg salad sandwich
Lowfat milk, 1 cup

FIGURE 20–3
High-protein diet (120 g).

PUTTING IT ALL TOGETHER

Various stressors, infections, and fevers can interfere with the maintenance of good nutrition by increasing body needs and reducing body reserves. Diets high in protein and calories are often essential for the rehabilitation of such individuals. Of especial concern is the resurgence of the incidence of tuberculosis, which involves both infection and fever.

1. What is the best way to keep a cold or other infection from spreading?

Wash your hands before eating or handling foods, after going to the bathroom, after sneezing, coughing, or blowing your nose. Dispose of used tissues. Don't share eating utensils or other objects.

2. Does chicken soup really help to cure a cold? When I was a child my mother insisted it did. It really smelled and tasted good, but did it help the cold?

There is no evidence that chicken soup will "cure" a cold. The liquid will help keep you hydrated. The soup may also help keep nasal passages open and mucous membranes moist. Soup is easy to eat and the pleasant taste (and memories) may help you forget the symptoms.

3. I have a cold and want an antibiotic and my doctor won't prescribe one. I'm thinking of changing doctors.

The doctor *may* believe you do not need an antibiotic. Overuse and inappropriate use has resulted in drug-resistant strains of microorganisms.

4. I would like to have a manicure at a nail salon, but wonder if there is any danger of getting an infection from the scissors and instruments that have been used on other people before me.

Before you have the manicure, learn if the salon and nail technicians are licensed. Determine how equipment is sterilized. Heat is preferable, but germicidal solutions are acceptable if implements are soaked for at least 10 minutes between customers. Artificial nails can be a source of infections if they separate from the natural nail or are left in place too long, allowing moisture and dirt to accumulate.

✔ CASE STUDY: A Homeless Woman Develops Tuberculosis

Marie M., age 35, left school in the 10th grade. An early marriage ended in divorce. She has held several minimum-wage jobs, barely managing to get along. Last year she left an abusive relationship and moved in with her sister. When this didn't work out she became homeless, sometimes living on the street and sometimes staying in shelters.

Recently she developed a persistent cough. An outreach team observed this and arranged for her to visit a clinic of the local health department for evaluation. A sputum culture proved positive for tuberculosis. A six-month course of *isoniazid (INH)* and *rifampin* along with *pyrazinamide* for the first two months is ordered.

The outreach team meets to evaluate and make a care plan. Obtaining housing for Marie is a top priority.

Marie's meals are somewhat chaotic, especially when she is living on the street where food is provided by passers-by, outreach workers, or a mission where Marie sometimes stops. Regular meals are provided at the shelters, although their composition is sometimes determined by what food is donated. Marie has few food preparation skills.

Questions

1. State four realistic health-related goals you think Marie should accomplish in the next year.

2. Assuming that housing with cooking facilities is located for Marie, state four reasonable food-related objectives that Marie might accomplish in the next six months.

3. Make suggestions for improving food intake while Marie is homeless and after she moves into housing.

4. What risks were present in Marie's lifestyle that may have favored developing tuberculosis?

5. Failure to take medications is a problem with some patients with tuberculosis. How might Marie be encouraged to take her medications?

6. Discuss side effects of Marie's medications and how these effects might be managed.

7. Suggest community agencies to which Marie might be referred and services to be provided.

FOR ADDITIONAL READING

Boutotte, J: "T.B. the Second Time Around . . . and How You Can Help to Control It," *Nurs 93,* **23:**42–50, May 1993.

Cook, DJ, *et al:* "Risk Factors for Gastrointestinal Bleeding in Critically Ill Patients," *New Engl J Med,* **330:**377–81, 1994.

Kurtzweil, P: "Fingernails: Looking Good While Playing Safe," *FDA Consumer,* **29:**20–24, December 1995.

Lewis, R: "The Rise of Antibiotic-Resistant Infections," *FDA Consumer,* **29:**11–15, September 1995.

Reiss, PJ: "Battling the Super Bug!" *RN,* **59:**36–41, March 1996.

Rodman, MJ: "OTC Interactions: Cough, Cold, and Allergy Preparations," *RN,* **56:**38–42, February 1993.

Saper, CB, and Breder, CD: "The Neurologic Basis of Fever," *New Engl J Med,* **330:**1880–86, 1994.

Sheldon, JE: "18 Tips for Infection Control at Home," *Nurs 95,* **25:**32PP–32QQ, August 1995.

Simon, HB: "Hyperthermia," *New Engl J Med,* **329:**483–87, 1993.

"Teaching Your Patients About O.T.C. Cold Products," *Nurs 94,* **24:**32G–32H, September 1994.

Thomas, DO: "Fever in Children: Friend or Foe?" *RN,* **58:**42–48, April 1995.

"Understanding Oral Infections," *Nurs 95,* **25:**32P–32R, August 1995.

Nutritional Considerations in Burns, Surgery, and Lung Disease

Chapter Outline

Reduce by at least 10 percent the incidence of surgical wound infections and nosocomial infections in intensive care patients.*

Slow the rise in deaths from chronic obstructive pulmonary disease to achieve a rate of no more than 25 per 100,000.*

Reduce asthma morbidity, as measured by a reduction in asthma hospitalizations to no more than 160 per 100,000.*

✔ Case Study Preview

In this chapter's case studies you will plan for the care and rehabilitation of a man whose life was abruptly changed by a fire at his place of employment. You will also plan strategies for the management of emphysema and improvement of the quality of life for a man who has been a longtime smoker.

---

*Healthy People 2000: National Health Promotion and Disease Prevention Objectives. Public Health Service, U.S. Department of Health and Human Services, Washington, DC, 1991, Objectives 20.5, 3.3, and 11.1, pp. 122, 95, and 105.

Τhis chapter continues the discussion of the most intense stress response affecting humans, that occasioned by thermal injury or burns. Nutritional implications of the stress of surgery, both general surgery and surgeries specific to the gastrointestinal tract, are presented. Finally, the stress of lung disease is discussed.

BURNS

Burns are the third leading cause of accidental death in the United States. Each year about 70,000 individuals are hospitalized with thermal injuries and more than 9,000 die from these injuries and their complications. Burns can be classified as follows:

Partial thickness, mild: only epidermis injured; treat with pain relief and oral fluid. Formerly called first-degree burns.

Partial thickness, moderate: epidermis and dermis injured, nerve endings exposed, very painful. Frequently heal spontaneously with adequate nutrients, fluids, and oxygen if there is no infection. Formerly called second-degree burns.

Full thickness: dermis destroyed; muscle, tendons, and bone may be damaged. No pain because nerve endings are destroyed. Skin grafts necessary. Nutrition support needed if burns involve more than 10 percent total body surface area. Formerly called third-degree burns.

Clinical Findings

The most intense response to stress in humans is that caused by extensive burns. Severity of the burns is determined by the depth of the cellular damage, the percentage of total body surface area (TBSA) affected, burns to special areas such as the face, feet, hands, and genitals, age, cause of burns, associated trauma, and preexisting disease and illness.

Capillary permeability is increased and plasma proteins, fluids, and electrolytes escape into the burn area and interstitial space, causing edema. In extensive burns, plasma volume may be reduced by 50 percent. **Ileus,** or absence of peristalsis, is common and, until peristalsis returns, nothing can be taken by any enteral route. Urinary nitrogen losses are aggravated by lack of muscular activity, poor nutritional intake, and ambient temperature below 95°F (35°C). Wound infection and **sepsis,** disease-producing organisms in the blood, are common; sepsis is the leading cause of death among burn patients.

Therapy

The patient is evaluated as soon as possible. Depending on the extent of the burns and other factors, transfer may be made to a regional burn center.

Fluid and electrolyte resuscitation is the first priority to prevent **hypovolemia,** diminished blood volume, and shock. Intravenous replacement of 10 or more liters of fluid a day is not uncommon.

Nutritional Needs Various equations have been devised for estimating the energy needs of burn patients. (See inside back cover.) The protein need is increased to cover the nitrogen losses through the urine and surfaces of the burn, plus the needs for wound healing. A calorie to nitrogen ratio of 150:1 to 80–100:1 is suggested, depending on the extent of the burns. (See Chapter 16.)

SKIN FUNCTIONS COMPROMISED BY BURNS

protection against infection, control of body temperature, fluid balance, sense of touch, cosmetic appearance (See Figure 21–1.)

Sepsis: also known as septicemia

In the first days after injury, while the hypermetabolic state is at its height, providing adequate energy and protein may be impossible. When the stress response moderates, providing adequate amounts of these nutrients will be easier. Layered dressings increase ambient temperature and passive exercise helps reduce protein loss. Electrolytes and nutrients lost through **exudate,** or fluid leaking out of capillaries, and urine should obviously be replaced. Up to 1 g of vitamin C daily and a zinc supplement are frequently given to aid wound healing.

Mode of Feeding If ileus occurs, parenteral nutrition (see Chapter 17) is initially required. When gastrointestinal function has returned, transition is made to a tube feeding or oral intake, depending on the site and nature of the burns. Special formulas for hypermetabolic states are available;* protein supplements or modular feedings (see Chapter 17) can be used.

Assessment Nutritional assessment from time to time is essential, not only to monitor the patient's status, but also to predict complications. The individual

FIGURE 21-1
This burn patient, with the assistance of the occupational therapist, practices range-of-motion exercises to assist in performing activities of daily living (ADL) when she returns home from the rehabilitation center. (*Source:* Anthony Magnacca/Merrill.)

---

*Perative®, Ross Laboratories, Columbus, OH; TraumaCal®, Mead Johnson & Company, Evansville, IN.

Fluid replacement: prevents
  hypovolemia
Oxygen therapy: prevents
  hypoxia of gastric mucosa
Nutrition support: nourishes
  gastric mucosa
Antacids and/or ranitidine
  maintains gastric pH greater
  than 5

situation will dictate which assessment techniques are appropriate. Skinfold and midarm circumference measurements may not be feasible because the areas have been traumatized. Edema and heavy dressings can distort body weights. Multiple blood transfusions can render estimation of visceral protein status suspect.

Complications Patients who are at risk for sepsis and wound infection are treated with antibiotics. **Stress ulcers,** multiple acute superficial erosions of the gastric mucosa without evident previous erosion, are a threat because in severe stress, oxygen and nutrient delivery to the gastric mucosa is impaired. Mucus production ceases and the mucosa becomes more permeable. The ability to remove or buffer acid is reduced.

SURGICAL CONDITIONS

Understanding the mechanisms of stress provides insight into the nutritional needs and risks of surgical patients. Good nutritional status before and after surgery is associated with effective wound healing, fewer infections, shorter hospital stays, and decreased morbidity and mortality.

Many surgical patients will lose nutrients through blood loss, wound drainage, possible hemorrhage, and vomiting. Immobilization accelerates protein loss. The stress of surgery increases energy needs (see inside back cover). On the other hand, food intake will probably be prohibited or inadequate for a period of time. Anorexia following surgery is common. For a minor procedure on a person in good nutritional status, a deficient intake for a few days poses no serious problem. However, for the depleted person, especially if the surgery is major, the consequences can be grave and alternate methods of feeding may be indicated. Screening upon admission will identify individuals at nutritional risk. (See Chapter 16.)

Nutritional Considerations

Protein status may be the most critical factor in ensuring a satisfactory outcome. Response to the stress of surgery will increase protein losses, and for several days the patient will probably be in negative nitrogen balance.

Nutrients to Be Carefully Monitored and Their Functions

- *Protein:* promote wound healing, resistance to infection, protect liver from toxicity of anesthesia
- *Energy:* maintain or restore healthy weight and spare protein
- *Vitamin C:* supplements before and after surgery to promote wound healing
- *Iron:* supplements or transfusions to cover blood loss
- *Zinc:* supplements to promote wound healing, cell-mediated immunity

Preoperative Diet

Depleted patients become better surgical risks if their nutritional status can be improved prior to surgery. If the surgery is not an emergency procedure, and if the disease process is such that improved nutritional status is a likely outcome, nutritional rehabilitation by an enteral or parenteral route may be attempted. Weight gain using a high-calorie, high-protein diet reduces the risk imposed by undernutrition. Likewise, the risk status of obese patients is improved by weight loss. Rapid weight loss, however, results in tissue breakdown, which will compromise the nutritional status of the individual. Any postponement of surgery would, of course, be viewed from a risk-benefit standpoint.

Foods and fluids are usually prohibited after midnight of the day before surgery. The stomach will then be empty when anesthesia is given, decreasing the chances of vomiting and aspiration. For gastrointestinal surgery, a low-residue diet (see Chapter 22) is sometimes ordered prior to surgery to reduce fecal residue. Elemental diets (see Chapter 17) sometimes replace low-residue diets.

Postoperative Diet

After a minor procedure, liquids may be tolerated within a few hours and return to a normal diet is rapid. Following major surgery, peristalsis ceases, and food and fluids are usually withheld postoperatively. When the effects of anesthesia abate, patients are progressed through ice chips, sips of water, to a clear-liquid, full-liquid, soft, and then normal diet. (See Chapter 17.) The rate of progression is dictated by the patient's condition and tolerance of food. If the individual is unable to take food orally or intake remains inadequate, tube feeding or parenteral nutrition should be considered.

SPECIFIC SURGICAL CONDITIONS

Several types of surgery involving the gastrointestinal tract require particular dietary adjustment. For multiple tooth extractions the diet is first limited to liquids, followed by soft foods. Radical surgery on the mouth and/or esophagus usually requires full liquids or pureed foods, or, depending on the procedure, gastrostomy tube feeding. (See Chapter 17.) After tonsillectomy, cold bland foods low in fiber are given. The diet is progressed to soft foods, avoiding extremes in temperature, and then to the regular diet.

Gastric Restrictive Surgery

Gastric surgery is sometimes used for morbidly obese individuals who meet certain criteria. The surgery has replaced the jejunoileal bypass for extreme obesity because of the severe side effects of the intestinal procedure. The goal of the gastric surgery is to limit intake without compromising digestion or absorption. The procedures have been modified to achieve these goals and prevent complications such as esophagitis, dilation, obstruction, or ulceration of the stoma, staple line disruption, and excessive vomiting.

Gastric Partitioning In vertical banded gastroplasty, using surgical staples, a small pouch (10–20 ml capacity) is created in the upper stomach with a small reinforced opening or stoma to the remainder of the stomach. The pouch fills rapidly so the patient feels satisfied. Food passes from the pouch into the remaining part of the stomach; thus, none of the digestive tract is bypassed.

Gastric Bypass A small pouch is created by stapling and anastamosed to the jejunum. The remaining part of the stomach becomes nonfunctional. The postoperative diet for both procedures is similar.

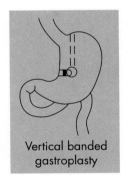

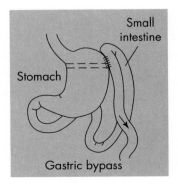

CRITERIA FOR GASTRIC RESTRICTIVE SURGERY

100 lb or more over desirable weight
Documented inability to lose weight and maintain loss
Psychiatrically stable
Motivated to comply with follow-up
Understands procedure
Realistic expectations

Gastroplasty: another name for gastric partitioning

Vertical banded gastroplasty

Small intestine

Stomach

Gastric bypass

**BEHAVIOR MODIFICATION
TECHNIQUES FOR GASTRIC
RESTRICTIVE SURGERY**

Choose nutrient-dense foods
Measure food and beverages
Eat three small meals a day
Avoid snacking
Sip beverages slowly
Take beverages between
 meals
Chew foods until semiliquid
Stop eating at first sign of
 fullness
Do not lie down after a meal

Dietary Considerations Some individuals experience decreased tolerance of red meats and other high-protein foods. Eventually foods of all textures may be used, providing they are well chewed. Vitamin and mineral supplements usually are indicated. Since pills can obstruct the stoma, the supplements or any medication should be in liquid or chewable form or pulverized before ingestion.

Gastrectomy

Gastrectomy, removal of all or part of the stomach, and **vagotomy,** cutting the vagus nerve to relieve pain, are sometimes performed for gastric carcinoma and occasionally for bleeding, perforated, or unmanageable peptic ulcer. Following gastrectomy a number of problems arise. In the absence of the gastric juices the entire digestion of protein must occur in the small intestine. Fat is less well utilized because of inadequate mixing of food with digestive juices. Without gastric juice, iron is less well absorbed and iron-deficiency anemia occurs more often. Since the intrinsic factor is no longer produced from the stomach, the absorption of vitamin B_{12} does not occur. Several years later the patient will have pernicious anemia unless vitamin B_{12} is given by injection. Weight loss is common, and many patients fail to return to desirable weight levels. Any procedure severing the vagus nerve speeds the flow of food through the gastrointestinal tract, with possible resulting diarrhea and steatorrhea.

Immediately following gastrectomy, 60–90 ml of clear fluids is given hourly. Afterward, a full-liquid or very-low-residue diet is usually allowed. The foods are introduced gradually, however, keeping meals very small and at frequent intervals. Eggs, custards, cereals, milk, cream soups, and fruit purées are introduced first; then cottage cheese, tender chicken, and pureed vegetables are added. The emphasis is on foods high in protein, moderate in fat and low in simple carbohydrate. Fluids are better tolerated if they are taken between meals.

Dumping Syndrome Certain patients who have had a gastrectomy complain of nausea, weakness, sweating, and dizziness shortly after meals. Vomiting, diarrhea, and weight loss are common.

The condition is caused by rapid entry of the food material directly into the intestinal tract. The large amount of simple carbohydrate draws water from the blood circulation into the small intestine and thus reduces the circulating blood volume. The sugars are rapidly absorbed into the blood. This causes overproduction of insulin, and in a short time the blood sugar drops to very low levels. Thus, the patient has the symptoms of insulin shock and also the symptoms that accompany reduction of the circulating blood volume. Proteins, complex carbohydrates, and fat are better tolerated.

The diet used for this condition is outlined in Figure 21–2.

Anticholinergics, used to reduce acid secretion, can cause dry mouth; rinsing the mouth with water before meals can help. Also, bulking agents or medications to slow intestinal motility can be taken before meals.

Intestinal Surgery

Malignancy, ileitis, colitis, perforation, trauma, or congenital malformation can necessitate removing a section of the ileum or colon. Depending on the situation, an opening may be made into the abdominal wall and the proximal end of the ileum or colon attached to it for elimination of wastes.

CHARACTERISTICS OF THE DIET

Avoid simple carbohydrates: sugars, syrups, sweet desserts, etc. Include up to 100–120 g complex carbohydrates daily.

Include liberal protein and moderate fat.

Medium-chain triglycerides (MCTs) may be indicated if steatorrhea is present.

Take six small meals on a regular basis.

Take liquids 30–60 minutes after solid foods and limit to $\frac{1}{2}$–1 cup

Avoid extremely hot or cold food.

Pectin may be used.

Rest before and after meals, eat slowly, and chew food well.

Vitamin and mineral supplements are included.

Tailor diet to individual, e.g., milk often is not tolerated.

FOODS ALLOWED DAILY

Meat, fish, poultry: all kinds, baked, broiled, poached, 12 oz.

Eggs: 1, poached, scrambled, soft or hard cooked

Lowfat milk: 1½ cups if tolerated—take 30–60 minutes after solid foods

Bread: enriched, whole grain, zwieback, Melba toast—only one slice per meal

Bread substitutes: saltines, soda crackers, noodles, macaroni, spaghetti, rice, boiled or mashed potato, sweet potato, grits, lima beans, sweet corn, cooked dried peas and beans

Cereals, unsweetened: one serving daily

Vegetables: all kinds, only one serving per meal

Fruits: fresh, canned, frozen without sugar, drained. Not more than one serving per meal.

Fruit juices: without sugar—take 30–60 minutes after solid food

Fats: butter, margarine, bacon, cream cheese, cream, oil, salad dressing

Nuts: as tolerated

FIGURE 21–2
High-protein, moderate-fat, low-carbohydrate diet (postgastrectomy).

Ileostomy

Ileostomy is an opening of the ileum onto the abdominal wall. Since the colon is no longer utilized, its ability to absorb water, as well as voluntary muscle function are lost, and discharge of waste material is continuous. An appliance to contain the discharge must be worn. At first the discharge is fluid. Later the remaining small intestine may take over some absorbing function and the fecal material may become semisolid. Losses of fluids, electrolytes, and other nutrients occur. Fat absorption is reduced and vitamin B_{12} absorption is reduced or absent.

Clear fluids are given following surgery, and then a very low-residue diet is introduced. (See Chapter 22.) Gradually foods with more fiber are added one at a time. Tolerance is an individual matter. Foods high in fiber or with seeds or hard-to-digest kernels such as celery, corn on the cob, nuts, pineapple, and popcorn can cause blockage. Thorough chewing of food is important. Some foods can cause gas and odor. Experimentation with different foods will reveal which, if any, create problems. A diet that is adequate nutritionally should be emphasized. See the Food Guide Pyramid (Figure 1–5) and the Basic Diet (Table 2–4.)

Weight loss is common and a high-protein, high-caloric diet (see Table 9–3 and Figure 20–3) is indicated. Periodic injections of vitamin B_{12} are required. Good sources of sodium and potassium should be emphasized since electrolyte losses can occur. Emotional support is essential. Many communities have ostomy support groups that can provide assistance.

Colostomy

Colostomy is an opening of the colon onto the abdominal wall. Since the colon is still functional, the ability to absorb water remains and the feces are more formed than in the ileostomy. Initially these patients are given a clear-liquid and then a very-low-residue diet. Foods higher in residue are added gradually, one at a time. Most persons are able to eat an essentially normal diet, eliminating any food for which there is an individual intolerance.

Short Bowel Syndrome

Massive resection of the small intestine results in malabsorption, which has severe and potentially life-threatening consequences. The normal small bowel is estimated to vary from 300 to 650 cm in length. The minimum length necessary to maintain optimum nutritional status is unknown, as many other factors affect digestion and absorption and individual adaptation varies.

Clinical findings reveal many digestive and absorptive problems. Ileal resections are associated with the most serious consequences since the ileum is best able to control transit time and is the site of bile salt and vitamin B_{12} absorption. The remaining jejunum cannot assume these functions and has limited ability to absorb water. Bile salts and fatty acids pass into the colon, increasing osmolality and producing diarrhea. Unabsorbed fatty acids combine with calcium and magnesium to form insoluble soaps that are excreted in the feces. Dietary oxalate, normally bound in calcium, is now absorbed in large amounts with a resulting potential for oxalate kidney stones. Calcium malabsorption often causes decreased bone mineral content. Fat malabsorption favors deficiencies of fat-soluble vitamins.

Pancreatic function is diminished due to the loss of stimulation from secretin and cholecystokinin, which were secreted by the bowel that has been resected. Disease processes in the remaining gastrointestinal tract may further reduce digestive and absorptive capacity. (See Chapter 22.)

The status of the **ileocecal valve** between the ileum and colon is critical because it controls the flow from the small to the large intestine and prevents bacterial migration from the colon to the ileum.

Therapy Following Intestinal Resection

After resection intravenous feedings are given until bowel sounds return. The type and route of feedings are determined by the patient's condition and the extent of the procedure. Total parenteral nutrition is frequently necessary initially, after which oral feedings are gradually introduced. For a few patients, total parenteral nutrition may be permanently indicated.

Oral feedings are preferable when tolerated. A diet moderate to low in fat with medium-chain triglycerides to supply calories is frequently prescribed. Since weight loss is often considerable, a high-protein, high-calorie diet with vitamin and mineral supplementation is indicated. Restriction of dietary oxalate and lactose may be necessary. For some patients, tube feeding or some combination of oral intake and tube feeding is used. Antibiotics are given to control bacterial proliferation. Bile acid sequestrants decrease bile acid diarrhea; antidiarrheals may also be given. Histamine$_2$ receptor antagonists reduce gastric hypersecretion.

NUTRITIONAL CONSEQUENCES OF RESECTION

% resected	
50–75	Malabsorption problems
75–90	Intense nutritional management
90–100	Total parenteral nutrition necessary

FACTORS INFLUENCING DIGESTION AND ABSORPTION AFTER RESECTION

Site and extent of resection
Status of ileocecal valve and terminal ileum
Adaptation of remaining small bowel
Condition of remaining gastrointestinal tract

TABLE 21-1

Characteristics of Chronic Obstructive Pulmonary Disease

Disease	Symptoms and Pathology	Predisposing Factors
Asthma	↑ responsiveness of trachea and bronchi to stimuli, narrowing of airways, airway wall edema, excess mucus, bronchospasm, dyspnea, wheezing, coughing	Usually allergy
Chronic bronchitis	Excess mucus, coughing, cyanosis, inflamed airways, hyperactive bronchi, impeded air outflow	Usually smoking
Emphysema	↓ surface area in lung, destruction of alveolar walls, destruction of air spaces, wheezing, chronic cough, usually thin, rosy complexion, barrel-shaped chest due to overwork and overinflation of chest muscles	Male gender, genetic, smoking

CHRONIC OBSTRUCTIVE PULMONARY DISEASE (COPD)

Chronic obstructive pulmonary disease (COPD) is a group of diseases with common characteristics of chronic air-flow limitation or obstruction. Although the pathophysiology of COPD receives much attention, nutrition implications frequently are overlooked. The lungs have considerable ability to adjust to stressors. Injury to as much as 60 percent of the 300 million **alveoli** (air sacs in the lungs) is necessary before symptoms appear. The characteristics and possible causes of COPD appear in Table 21–1. In the United States, COPD affects more than 14 million people and is the fifth leading cause of death. Eighty to 90 percent of the more than 50,000 annual deaths from emphysema and chronic bronchitis are attributable to smoking.

Clinical Findings

Compromised nutritional status and weight loss are common. **Dyspnea,** labored or difficult breathing, interferes with eating as do breathing problems due to diaphragmatic restriction caused by a full stomach, taste interference from chronic sputum production, and gastric irritation due to bronchodilators and steroids. Reduced oxygen supply to the gastric mucosa decreases peristalsis and slows down digestion with resulting anorexia. The decreased oxygen supply and side effects of medications increase the likelihood of gastric ulceration. Between 20 and 25 percent of patients with COPD have peptic ulcers. Nutritional depletion is usually greater in individuals with emphysema than chronic bronchitis.

Therapy

Patients who smoke should cease doing so. Prevention or correction of malnutrition is a primary goal. Assessment will frequently show basal energy needs considerably above those predicted for a reference person of comparable size and age. Energy expended for breathing may be up to 10 times greater than in the reference individual.

Energy Adequate energy must be provided, but excess intake can raise the demand for oxygen and the production of carbon dioxide beyond the capacity of the patient with reduced respiratory function. These individuals have limited

Pulmonary: relating to the lungs

Chronic Obstructive Pulmonary Disease: also known as Chronic Obstructive Lung Disease

GOALS FOR DIETARY TREATMENT OF COPD

Prevent or correct malnutrition
Select foods of high nutrient density
Prevent or correct infection and acidosis
Alleviate difficulty in chewing and swallowing

ENERGY FOR BREATHING

Normal—36–72 kcal/day
COPD—430–720 kcal/day

347

ability to excrete carbon dioxide, so any dietary intake that reduces production of carbon dioxide will decrease the potential for its retention and possible resulting acidosis. The **respiratory quotient (RQ),** ratio of CO_2 produced to O_2 consumed, for fat is lower than that of carbohydrate. Nonprotein calories are sometimes provided by a high fat to carbohydrate ratio to decrease carbon dioxide production. Formulas with such proportions for use as supplements or in tube feeding are available.*

RQ: carbohydrate = 1.0; protein = 0.8; fat = 0.7

Strategies for Mealtimes Patients will eat better if they rest before meals. Bloating is a common problem in COPD. **Aerophagia,** or swallowing air, accompanying dyspnea is a contributing factor. Slow, deep breathing and relaxation techniques assist in managing dyspnea. Avoidance of gas-producing foods and prevention of reflux (see Chapter 22) are helpful. Medications containing simethicone decrease intestinal gas. Omitting stringy and tough foods will ease difficulties in chewing and swallowing that accompany shortness of breath. Anorexia can be reduced with small frequent feedings that decrease stomach fullness, allow freer movement of the diaphragm, and decrease fatigue. (See Figure 21–3.)

Medications Bronchodilators taken orally or by inhalation expand the **bronchi,** the two main branches leading from the trachea to the lungs, and

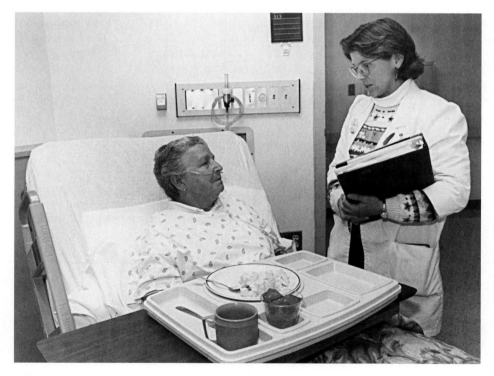

FIGURE 21–3
The respiratory therapist gives hints about conserving energy while eating to a patient with emphysema who is receiving oxygen. (*Source:* Tom Wilcox/Merrill.)

---

*Pulmocare®, Ross Laboratories, Columbus, OH; Respalor®, Mead Johnson & Company, Evansville, IN.

liquefy secretions. *Theophylline* may produce caffeine-like side-effects, so caffeine should be avoided by patients taking this medication. Anticholinergics may also be used as bronchodilators. If symptoms do not respond to bronchodilators, steroids are used to depress mucus formation and relieve inflammation. Steroids, taken orally or by aerosol spray, promote gastric irritation and fluid retention and depress immune response. They should be taken with food; dietary sodium restriction may be necessary. Antibiotics are used for infections.

PUTTING IT ALL TOGETHER

Burns produce the most intense response of any stressor in humans. Surgery can also be a severe stressor and many gastrointestinal surgeries have food intake implications. Both burns and surgery can increase nutrient needs for recovery and rehabilitation. Lung diseases accompanied by difficulty in breathing can render obtaining adequate food a challenge.

? QUESTIONS CLIENTS ASK

1. *My cousin smokes and none of us can persuade her to stop. She just had surgery and is eager to recover as rapidly as possible. I wonder if smoking has any effect on her recovery. If it does, we might be able to motivate her to stop smoking.*

Smoking does restrict the flow of blood to the area of the incision and can therefore lengthen the healing time. This might serve as a short-term motivation to stop smoking. Despite lack of success in the past, you can also encourage her to enter a program to end smoking as a long-term goal.

2. *After my father had abdominal surgery he was ravenous and wanted to eat, but at first they wouldn't give him anything. He begged me to bring him a pizza, but I didn't think I should. They started him on sips of water and ice chips. Then he had clear liquids. It took awhile*
until he was on a regular diet. I'd think he'd need a good food intake to heal his incision.

With surgery, peristalsis, the waves that propel food through the digestive tract, stops. Food is not permitted postoperatively until "bowel sounds" return. The intake progresses from water to clear liquids and through the postoperative progression with monitoring to make certain he can tolerate the food. The health care team does want all surgical patients to advance to a regular diet as soon as safely possible, since it is more likely to be adequate and support healing. The purpose of the dietary progression should have been explained to your father.

3. *When I was a child, if anybody had a burn, my mother put butter on it. Now I hear this is not good. What should you use on a minor burn?*

It is best to cover it with a cool, clean, wet cloth. Butter often contains salt, which may irritate the burn.

✔ CASE STUDY: A Middle-Aged Man with Burns

Richard K., age 49, a warehouse worker, was severely burned when there was a fire at his place of employment. He was first taken to a nearby hospital and then transferred to the regional burn center. Full thickness burns covered 40 percent of his body. Neck burns traumatized his esophagus, making swallowing impossible.

Evaluation of Mr. K.'s nutritional needs led to an order of total parenteral nutrition with 4,000 kcal and 150 g protein daily. After a week his condition stabilized, and a gastrostomy tube was inserted. A balanced complete tube feeding was ordered.

Mr. K. is very depressed. He grieves for his buddies who were killed in the fire and feels he should have done more to try to save them. His income was always modest, and he has little in the way of savings. He faces months of hospitalization. The burn center is located many miles from his home, so his family and friends seldom can visit. He is concerned that scars will leave him repulsive to others. He wonders if the warehouse will reopen and worries that he might not have a job when he is discharged from the burn center.

Questions

1. Mr. K.'s weight is 154 lb (70 kg). What was his daily protein allowance before the burns? Contrast it to his present allowance.

2. Why are his protein and calorie needs now elevated?

3. Why was TPN chosen at first for Mr. K.? (See Chapter 17.)

4. State some advantages and disadvantages of TPN for Mr. K.

5. Why was Mr. K.'s diet changed to tube feeding? (See Chapter 17.)

6. The tube feeding was first given at half strength and gradually increased to full strength. Why was this done?

7. Describe a balanced complete formula and its advantages for Mr. K.

8. How can the nurse and dietitian help Mr. K. with his problems and concerns? Suggest how other members of the health care team might assist him and his family.

9. *Ranitidine* is prescribed for Mr. K. Explain the rationale for the use of this medication.

10. Would you expect Mr. K. to feel pain from his burns? Explain. How might this affect his nourishment?

11. Explain the difference between partial thickness and full thickness burns. How does this relate to Mr. K.'s nutritional care?

✔ CASE STUDY: A Man with Emphysema

Harry R., age 59, began smoking cigarettes as a teenager and has had a two-pack-a-day habit for over 40 years. For 20 years he has had a cough. Increasingly frequent coughing spells, wheezing, and dyspnea have forced him to give up his job as a construction worker and go on disability. His wife and children have tried without success to get him to stop smoking.

Mr. R. is 68 in. (170 cm) tall and weights 120 lb (54.5 kg). He once weighed 145 lb and has lost 8 lb in the past year. Coughing spells and shortness of breath interfere with eating and he usually feels bloated before he has eaten much food. He takes *theophylline*.

The nurse at the union clinic where the family receives their health care discusses ways to reverse Mr. R.'s weight loss with him and his wife.

Questions

1. What would be an appropriate weight for Mr. R., assuming a medium frame? (See Appendix Table D–1.)

2. Why would a rapid weight gain be undesirable?

3. Would you expect Mr. R. to have a normal, elevated, or reduced energy requirement? Explain.

4. What advantage would a high fat diet have for Mr. R. in addition to increasing caloric intake?

5. Mr. R. has always consumed a lot of coffee. He wonders if he may continue. What would you suggest?

6. Suggest some mealtime strategies for increasing comfort and food intake.

7. What is the function of *theophylline*? What are its nutritional side effects? When should it be taken in relation to food intake?

8. Suggest some community resources that might be available to assist and support Mr. R. in an attempt to stop smoking.

Bove, LA: "How Fluids and Electrolytes Shift after Surgery," *Nurs 94,* **24:**34–40, August 1994.

Calistro, A: "Burn Care Basics and Beyond," *RN,* **56:**26–32, March 1993.

Chronic Obstructive Pulmonary Disease. NIH Pub. No. 95-2020, Public Health Service, Bethesda, MD, 1995.

Ferguson, GT, and Cherniack, RM: "Management of Chronic Obstructive Pulmonary Disease," *New Engl J Med,* **328:**1017–22, 1993.

Grant, JP, *et al:* "Malabsorption Associated with Surgical Procedures and Its Treatment," *Nutr Clin Prac,* **11:**43–52, 1996.

Metzler, DJ, and Fromm, CG: "Laying Out a Care Plan for the Elderly Postoperative Patient," *Nursing 93,* **23:**66–74, April 1993.

Peterson, KJ, and Solie, CJ: "Interpreting Lab Values in Chronic Obstructive Pulmonary Disease," *Am J Nurs,* **94:**56A–56F, August 1994.

Rodman, MJ: "OTC Interactions: Asthma Medications," *RN,* **56:**40–46, April 1993.

Soper, NJ, *et al:* "Laparoscopic General Surgery," *New Engl J Med,* **330:**409–19, 1994.

Weinberger, M, and Hendels, L: "Theophylline in Asthma, "*New Engl J Med,* **334:**1380–88, 1996.

Weinberger, SE: "Recent Advances in Pulmonary Medicine," *New Engl J Med,* Part I. **328:**1389–97, 1993; Part II, **328:**1462–70, 1993.

CHAPTER 22

Nutrition in Diseases of the Gastrointestinal Tract

Chapter Outline

General Dietary Considerations
Constipation
Diarrhea
Gastroesophageal Reflux
Hiatus Hernia
Peptic Ulcer
Lactose Intolerance
Gluten Enteropathy
Regional Enteritis

Ulcerative Colitis
Irritable Bowel Syndrome
Diverticulitis
Putting It All Together
Questions Clients Ask
Case Study: A Carpenter with Peptic Ulcer
Case Study: A Young Woman with
 Ulcerative Colitis

The dietary factors most frequently associated with gastrointestinal illnesses are alcohol (liver disease and cancer); inadequate fiber (constipation, hemorrhoids, diverticular disease, and possibly some types of cancer); fat (gallbladder disease and possibly some types of cancer); and substances such as gluten in wheat (celiac disease in genetically predisposed individuals).*

✔ CASE STUDY PREVIEW

Do you think that people with peptic ulcers are counseled to eat lots of soft, bland foods? You will find that therapy has changed as you help Roy, a carpenter, plan lifestyle changes. You will also plan dietary and lifestyle changes for Shirley, who has ulcerative colitis, a condition not uncommon in younger people.

**The Surgeon General's Report on Nutrition and Health*. U.S. Department of Health and Human Services, 1988, p. 414.

D ietary modifications for disease of the gastrointestinal tract are discussed in two chapters: the present chapter covers general factors followed by conditions affecting the esophagus, stomach, small and large intestine, roughly in that order. Diseases of the liver, gallbladder and pancreas are covered in Chapter 23. Since any diseases of the gastrointestinal tract are closely associated with its function, you should first review the normal digestive processes. (See Chapter 4.)

GENERAL DIETARY CONSIDERATIONS

In some diseases there is a physiological basis for dietary modification, for example, lactose intolerance, gluten-induced enteropathy, cirrhosis of the liver, and pancreatic insufficiency. For others there is no sound rationale for diet therapy. The diets used in such conditions are often based on tradition and are sometimes unnecessarily restricted. Clinical research has indicated that for conditions such as peptic ulcer, essentially normal diets are just as effective as those more limited in food choice.

Dietary Fiber and Residue

The terms *fiber* and *residue* are often confused and misused. **Residue,** the bulk remaining in the lower part of the intestinal tract, includes dietary fiber, cells sloughed off from the intestinal mucosa, intestinal bacteria, and their remains. Milk, for example, contains no dietary fiber, but lactose favors the growth of certain types of bacteria in the intestine. As a result, milk leaves considerable residue in the colon.

Reducing Fiber Content of Diets The diet may be progressively reduced in fiber in the following ways.

Selecting only young, tender vegetables.

Omitting those foods that have seeds, tough skins, or much structural fiber, for example, berries, celery, corn, cabbage.

Peeling fruits and vegetables such as apples, potatoes, and broccoli stalks.

Cooking foods to soften the fiber.

Using refined cereals, pastas, and white bread in place of whole-grain products.

Blending foods in a blender or food processor or pressing them through a sieve.

Omitting fruits and vegetables entirely; using only strained juices.

Strained fruits and vegetables and ground meats lack appearance, texture, and flavor appeal and are often unpopular with patients. Such foods are rarely used in therapy for peptic ulcer or colitis, and only for the early stages of treating diverticulitis. On the other hand, such foods are useful for the therapy of bleeding esophageal varices, some types of dysphagia, and for blenderized tube feedings.

Food and Gastric Acidity

The sight, smell, and taste of food stimulate the secretion of gastric juice. Foods have a pH higher than that of gastric acid (see Chapter 4). Thus, no foods, including citrus and other acid-tasting fruits, have a sufficiently low pH to lower the acidity of the stomach contents.

CONSIDERATIONS IN DIETARY PLANNING

Nutritional status of the individual
Adequacy of proposed diet
Secretion of enzymes, gastric juice, and bile
Motility of the gastrointestinal tract
Integrity of absorptive surfaces of intestines
Stress factors
Individual beliefs and tolerances regarding food

Protein-rich foods neutralize gastric acid. Hence, the use of milk was emphasized for many years. The neutralizing effect lasts for only ½ to 2 hours, and diet alone cannot be depended on to fully neutralize the acids that are formed. Calcium is thought by some to increase gastric secretion; therefore milk in large quantities is no longer recommended to control acidity.

Fats reduce acid production and also decrease motility. The long-used regimen of milk and cream feedings for peptic ulcer therefore was intended to neutralize and to reduce acid production. Because saturated fats may increase the risk of atherosclerosis, cream is now seldom used. Gastric acid production is increased by caffeine-containing beverages, alcohol, tobacco, salicylates, *indomethacin,* and *phenylbutazone.*

Food Tolerance

Many people believe that certain foods cause discomfort such as heartburn, abdominal distention, and flatulence. Among foods commonly cited are strongly flavored vegetables that contain sulfur compounds, among them broccoli, brussels sprouts, cabbage, cauliflower, cucumbers, leeks, onions, radishes, and turnips; melons; and dry beans. Tolerance to these foods is highly individual, and one person may eat onions with no subsequent problems, while another experiences discomfort after eating even a small quantity. Therefore, these foods should not be arbitrarily excluded for all patients, but should be allowed according to individual tolerance.

Patients with gastrointestinal disturbances are often nervous, anxious, overly concerned about their work, and tense. Their emotions strongly influence the digestion of foods. They also may have preconceived ideas about foods. Perhaps in no disease condition is individualization of diet more important than it is for diseases of the gastrointestinal tract. On the other hand, some firmness is often necessary to emphasize the need for nutritional adequacy.

Rapid eating, incomplete mastication, and failure to rest are frequently noted. Behavioral modification techniques such as eating at regular times, relaxing before and after meals, decreasing eating speed, and reducing stress may prove helpful.

The role of the bacillus *Helicobacter pylori* in the etiology of gastrointestinal diseases has received much attention. Found only in the gastric epithelium, its presence increases with age. The organisms can cause gastritis and have been linked to stomach cancer. Almost all individuals with duodenal ulcer and 80 percent of those with gastric ulcer have *H. pylori* infections.

CONSTIPATION

Constipation is the most common digestive complaint in the United States. Atonic constipation occurs most frequently in older persons, those who have little or no exercise, and/or a low intake of insoluble fiber. Their food selection may be limited to low-fiber items. Patients often limit fluids to prevent **nocturia,** that is, urination, especially excessive, during the night. Some medications cause or aggravate constipation. Individuals also become dependent on laxatives, yet have irregular elimination habits.

Unless otherwise indicated, the patient should be encouraged to drink at least 1,200–1,500 ml fluid daily, to exercise regularly, and to increase intake of insoluble fiber. (see Chapter 5 and Table 2–3), with the goal of developing regular habits of elimination.

DIARRHEA

Clinical Findings

Diarrhea is the passage of liquid or semisolid stools of greater than normal volume and frequency; abdominal pain and cramps occur often. Diarrhea is not a disease, but a symptom. In diarrhea, food passes through the gastrointestinal tract so rapidly that digestion and absorption are reduced. Fecal matter leaves the colon so quickly that water cannot be reabsorbed.

Most people have experienced diarrhea that begins abruptly and usually ends within hours to a few days, so nutritional deficiencies are not a problem. Prolonged diarrhea leads to serious losses of fluids, electrolytes and other minerals, vitamins, proteins, fats, carbohydrates, and body weight.

Therapy

The first step is to identify the cause and remove it or adapt therapy to the condition. If diarrhea is prolonged and severe, fluids and electrolytes may at first be given intravenously to allow the gastrointestinal tract to rest. Food intake may at first be restricted to fluids with gradual progression to a very-low-residue diet, soft fiber-restricted diet, and normal diet taken in small frequent feedings.

Very-Low-Residue Diet This diet is designed to furnish a minimum of fiber and also lead to a minimum of residue in the intestinal tract. The diet allows tender meats, poultry, fish, eggs, white bread, macaroni, noodles, simple desserts, clear soups, tea, and coffee. It omits all fruits, vegetables, and usually milk, since milk, while free of fiber, leaves considerable residue. Such a diet is obviously lacking in calcium, iron, and vitamins and should be used for only a few days. A typical menu follows:

Breakfast	*Lunch*	*Dinner*
Strained citrus juice	Tomato bouillon	Strained fruit juice
Cream of wheat	Crackers	Broiled fish
Lowfat milk for cereal	Roast chicken	Baked potato without
Sugar	Rice	skin, margarine
Soft-cooked egg	White bread or roll	Roll with margarine
White toast	Margarine	Whipped raspberry
Margarine	White cake with icing	gelatin
Coffee	Tea with lemon and	Plain sugar cookies
	sugar	Tea with lemon and sugar

For chronic diarrhea, oral rehydration, mineral and vitamin supplements, and a high-calorie, high-protein diet (see Chapter 20) will replenish losses. Antidiarrheals may be prescribed to slow intestinal motility. Behavior modification to reduce stress will help with anxiety-related diarrheas.

GASTROESOPHAGEAL REFLUX

Clinical Findings

Most people have experienced the **heartburn,** the burning sensation caused by reflux of gastric contents into the esophagus, of gastroesophageal reflux. If the symptoms become frequent or chronic, there is danger of esophagitis. Normally, pressure of the **lower esophageal sphincter (LES),** a band of muscle between

POSSIBLE CAUSES OF DIARRHEA

Acute
Dietary excesses
Anxiety
Food poisoning

Chronic
Malabsorptive diseases
Laxative and enema abuse
Continuing stress
Lactose intolerance
Gastrointestinal infection
Hyperosmolar tube feedings
Intestinal cancer
Radiation therapy
Certain intestinal surgeries
HIV-AIDS-related infections

the esophagus and stomach, prevents stomach contents from entering the esophagus. Any factor that decreases LES pressure or increases gastric pressure may favor reflux. It is more frequent during pregnancy and in the obese.

Therapy

Small meals eaten slowly, avoidance of tight garments and belts, and weight reduction where indicated will reduce gastric pressure. Postural precautions such as not lying down, bending over, or exercising after meals and sleeping with the head of the bed elevated will help to prevent symptoms.

Foods known to reduce lower esophageal sphincter pressure causing reflux are chocolate, coffee, tea, and other caffeine-containing foods, alcohol, peppermint and spearmint extracts, and meals high in fat. Protein, which increases LES pressure, should be emphasized. Cigarette smoking, oral contraceptives, and anticholinergic medications also decrease LES pressure.

With esophagitis, citrus fruits may be limited to prevent irritation. Antacids and histamine$_2$ receptor antagonists are sometimes employed to reduce gastric acidity.

Hiatus Hernia: also known as hiatal hernia

Normal Stomach

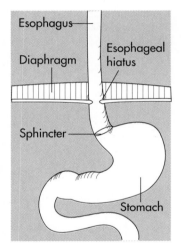

HIATUS HERNIA

Hiatus hernia is an abnormal gap in the diaphragm caused by weakened muscles that permits part of the stomach or parts of the stomach and esophagus to protrude into the chest cavity. It is more common in people over 45 years of age and in those who are obese. Often there are no symptoms, but some patients complain of heartburn, belching, and hiccoughing, especially after meals or when lying down. Although distinct from gastroesophageal reflux, hiatus hernia presents some of the same symptoms and similar therapy may prove effective.

Usually a normal diet is tolerated, but limiting foods that favor heartburn, avoiding tight clothing, and employing postural modifications (see above) are often helpful in preventing symptoms. Weight reduction is essential for the overweight. Antacids and histamine$_2$ receptor antagonists may be prescribed.

Hiatus Hernia

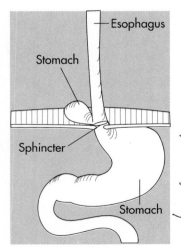

PEPTIC ULCER

Clinical Findings

Peptic ulcer, an erosion of the gastrointestinal mucosa, may occur in the esophagus, stomach, or jejunum, but in the United States, about 80 percent are duodenal. (See Table 22-1.) Ten percent of the population of the United States may be affected at some time in their lives. The normal resistance of the lining of the gastrointestinal tract to digestive juices breaks down. Hypermotility of the gastrointestinal tract is common. There is some tendency for ulcer development to be familial. Stress is known to play a role in the etiology of peptic ulcers; however, stress is difficult to measure, and coping mechanisms may modify the individual's susceptibility to developing ulcers. Aspirin, steroids, and nonsteroidal antiinflammatory agents have been linked to gastric ulcer, possibly by altering the gastric mucosa. Cigarette smoking doubles the risk of developing an ulcer. *H. pylori* may be a causative factor.

Typically, the individual will feel a dull, gnawing, burning epigastric pain a few hours after eating. The pain occurs when hydrochloric acid comes in contact

Factor	Gastric Ulcer	Duodenal Ulcer
HCl secretion	normal	increased
Age of onset	40s+	20s–30s
Gender risk	none	male
Incidence	20% of cases	80% of cases
Risk factors	stress, smoking, alcohol abuse, ulcerogenic medications, gastritis	stress, smoking, alcohol abuse, chronic pancreatitis, chronic renal failure, COPD, cirrhosis
	more likely to be malignant	

with the exposed nerve fibers in the eroded area. Some people discover that taking antacids and milk will provide temporary relief and therefore postpone seeking medical attention. Limiting food intake to items not associated with discomfort can result in low serum proteins, anemia, and weight loss.

Therapy

Most patients require some guidance in modifying their lifestyles; emphasis is placed on stress management. The physician prescribes one or more of these medications: a **histamine₂ receptor antagonist** such as *cimetidine, famotidine, nizatidine,* or *ranitidine,* which inhibits gastric acid and pepsin secretions; an antacid; an anticholinergic; an **antispasmodic,** which reduces motility and delays gastric emptying; and a **cytoprotective agent** such as *sucralfate,* which forms a protective coating over the ulcer. *Sucralfate* should be taken on an empty stomach. (See Chapter 18.) *H. pylori* infections are treated with antibiotics.

The objectives of dietary management are to restore and maintain good nutrition, to supply nutrients needed to heal the ulcer, especially protein and vitamin C, and to provide a diet consistent with the individual's preferences and lifestyle. Modifications are similar regardless of the location of the ulcer.

**OBJECTIVES OF THERAPY
FOR PEPTIC ULCER**

Relieve pain
Promote healing of the ulcer
Prevent recurrence

Normal Diet without Stimulants

The restricted diets of the past have been largely abandoned in favor of an essentially normal diet without stimulants. However, if the ulcer is located in the esophagus or is bleeding, a more conservative approach may be used.

The normal diet without stimulants follows the normal diet (see Chapter 17) with emphasis on these points:

Individualize the diet to the patient's habits and preferences. Explain fully the reasons for any modifications.

Regulate energy intake according to weight status.

Include recommended amounts of food from the Food Guide Pyramid and/or Basic Diet to ensure nutritive adequacy.

Use moderation in food selection; avoid excessive use of any one food group, for example, milk and high-fiber foods.

Allow all foods that the patient tolerates, except for items listed below.

FIGURE 22-1
What practical, realistic suggestions might you give to this worker with peptic ulcers who has been advised to relax before meals and eat in a pleasant setting? He carries a packed lunch. (*Source:* Anthony Magnacca/Merrill.)

Omit alcohol, tobacco, meat extractives, pepper, chili powder, mustard, and nutmeg. Small amounts of coffee or tea with milk may be permitted, although some clinicians feel that tea and coffee (including decaffeinated) increase acid production.

Regular mealtimes are important. Physicians who favor a liberal diet usually recommend three meals a day since more frequent meals are believed to stimulate acid secretion. Others recommend four to six feedings daily.

Rest before and after meals. Eat slowly in a pleasant, relaxed setting. Avoid mealtime confrontations or situations that are upsetting. (See Figure 22-1.)

LACTOSE INTOLERANCE

When lactose intake exceeds lactase activity, excess lactose is not broken down to glucose and galactose, but accumulates in the intestines where it is fermented by bacteria, producing flatulence, cramps, diarrhea, and abdominal pain. Clinical diagnosis is made from these symptoms and through a breath test that measures hydrogen produced by lactose or examination of the stool for pH and reducing substances. Self-diagnosis is common and fostered by some advertising. The World Health Organization (WHO) has recommended consistent terminology to describe low levels or absence of lactase.

Congenital Lactase Deficiency

Upon receiving milk the infant develops flatulence, diarrhea, and fails to gain weight. A formula free of lactose, such as soybean, or amino acid hydrolysate is

WHO TERMINOLOGY

Congenital Lactase Deficiency: a rare condition in which the infant is born without lactase activity

Lactase Nonpersistence: age-related decreased lactase activity, also known as acquired or primary lactase deficiency

Lactase Deficiency: low lactase activity secondary to disease or medication, also known as secondary lactase deficiency

given.* As foods are added to the diet, great care must be taken to exclude all sources of lactose.

Lactase Nonpersistence

Most individuals have adequate lactase activity during infancy and early child-hood, but levels often drop with age. This condition is common in Asian, African, and Middle Eastern populations, but is rare in Caucasians. Since the decrease is normal for a majority of the world's population, the World Health Organization does not consider it a true deficiency, and favors the term lactase nonpersistence over acquired lactase deficiency or primary lactase deficiency.

Lactase Deficiency

In some individuals, lactase deficiency develops secondary to ulcerative colitis, enteritis, gluten enteropathy, cystic fibrosis, AIDS, following gastrectomy or extensive intestinal surgery, in protein-energy malnutrition, and with *neomycin* or *colchicine* therapy.

Lactose-Restricted Diet

Reducing lactose intake to correspond to lactase activity causes the symptoms to disappear. Since individuals vary widely in their level of lactase activity, the diet must be tailored to the individual. Few children or adults have total absence of lactase, and eliminating all foods containing lactose can lead to dietary deficiencies.

When lactase is totally absent, a rare occurrence, the diet must be planned to eliminate all sources of lactose. Nonmilk dietary sources of calcium should be emphasized. Calcium and possibly vitamin D supplements will be necessary. Many prepared foods, some supplements, and medicines contain lactose, so careful label reading is essential. (See Figure 22–2.)

GLUTEN ENTEROPATHY

Clinical Findings

Gluten enteropathy involves an intolerance to **gluten,** a component of wheat, oats, rye, and barley. **Gliadin** is a protein fraction of gluten. The cause is not understood; it may be an immune factor involving T cells (see Chapter 19) or an inherited mucosal defect. When gluten-containing foods are eaten, changes occur in the epithelial cells of the jejunum, with resulting decreased absorption of sugars, fats, and amino acids. The stools are foul smelling and foamy because of fermentation of undigested carbohydrate. Steatorrhea is also present and fat-soluble vitamins are poorly absorbed.

Untreated patients show many signs of malnutrition such as weight loss, stunting of growth in children, muscle wasting, protruding abdomen, sore mouth, bone pain, increased risk of fractures, peripheral neuritis, and prolonged bleeding time. Lactase deficiency due to damage of the mucosal cells is common. Diagnosis is made by intestinal biopsy.

SOME STRATEGIES FOR CONSUMING LACTOSE-CONTAINING FOODS

Space intake throughout day
Take with meals to delay emptying of stomach
Whole milk better tolerated than skim
Aged cheese and yogurt with active culture ↓ lactose
Add lactase† to milk to break down lactose
Milk, ice cream, cottage cheese treated with lactase† available in some areas
Take lactase† tablets with food

Celiac Disease: gluten enteropathy in children

Nontropical Sprue: gluten enteropathy in adults

---

*Isomil®, Ross Laboratories, Columbus, OH; Nutramigen® and ProSobee®, Mead Johnson & Company, Evansville, IN; Nursoy®, Wyeth Laboratories, Philadelphia, PA.

†Lactaid®, McNeil Consumer Products, Fort Washington, PA; Dairy Ease®, Bayer Corporation, Myerstown, PA.

Milk in all forms—fresh, evaporated, condensed, dry, fermented, malted, whole, low-fat, nonfat
Beverages containing milk or milk powder—Cocomalt, Ovaltine, cocoa, chocolate
Cheeses—all types
Yogurt—all types
Breads and rolls—made with milk, sweet rolls, bread mixes, pancakes, waffles, zwiebach, French toast
Cereals—any containing nonfat milk powder. *Read labels*
Desserts—any made with milk including cakes, cookies, custard, ice cream, pies made with milk, cheesecake, puddings, sherbets
Fats—butter, margarines containing milk solids, cream, cream substitutes, sour cream
Meat—frankfurters and luncheon meats with milk powder. *Read labels*
Sauces and soups—cream, dried soup mixes, chowders, white or cheese sauce
Sweets—caramel or milk chocolate candy; artificial sweeteners with lactose; items containing Simplesse®
Vegetables—seasoned with butter or margarine, with cream or cheese sauces, mashed potatoes
Pareve or *Parve* on label indicates that no milk is present.
Key label words—butter, casein, caseinate, cheese, curds, dried milk solids, milk, whey

FIGURE 22–2
Lactose-restricted diet—Some examples of foods that may contain lactase.

SO GOOD MARGARINE

Ingredients: partially hydrogenated soybean oil, water, salt, sweet dairy whey, colored with beta carotene, vitamin D added

Therapy

The elimination of all sources of gluten in the diet allows regeneration of the affected cells and usually brings remarkable improvement. All products containing wheat, rye, oats, or barley must be omitted from the diet, no matter how minute the amount may be. Corn, rice, arrowroot, soy, and potato flours and starches replace the other grains. This diet must be followed indefinitely. (See Figure 22-3.) Corticosteroids are often prescribed.

The diet may be progressed gradually. Medium-chain triglycerides (see Chapter 17) may be used at first to reduce steatorrhea while increasing the caloric intake. Small amounts of unsaturated fats may then be used, progressing to harder fats later. Fiber may be reduced initially by using only cooked fruits and vegetables. Strongly flavored vegetables may be poorly tolerated at first. Lactase deficiency may necessitate lactose restriction. Aqueous multivitamins and iron supplements are usually prescribed.

The patient and family require in-depth counseling and support regarding the foods allowed, how to prepare them, how to interpret labels, how to choose foods when eating out, and so on.

Crohn's Disease: another name for regional enteritis

Inflammatory Bowel Disease: umbrella term including regional enteritis and ulcerative colitis

REGIONAL ENTERITIS

Clinical Findings

Regional enteritis most commonly affects the ileum, but may appear in any part of the intestine. Involvement may be confined to one area, but usually appears in several sites with unaffected segments in between—hence the name regional enteritis. Of unknown cause and occurring mainly in young adults, it is seen with

Beverages—ale, beer, instant coffee containing cereal, malted milk, Postum, products containing cereal

Breads—all containing wheat, rye, oats, or barley; bread crumbs, muffins, pancakes, rolls, rusks, waffles, zwieback; all commercial yeast and quick bread mixes; all crackers, pretzels, Ry-Krisp

Cereals—cooked or ready-to-eat breakfast cereals containing wheat, rye, oats, or barley; macaroni, noodles, pasta, spaghetti; wheat germ

Desserts—cake, cookies, doughnuts, pastries, pie, commercial ice cream, ice cream cones; prepared mixes containing wheat, rye, oats, or barley; puddings thickened with wheat flour

Fats—salad dressing with flour thickening

Flour—barley, oat, rye, and many kinds of wheat flours including bread, cake, entire wheat, graham, self-rising, whole-wheat

Meat—breaded, creamed, croquettes, luncheon meats, frankfurters and sausage unless all meat, meat loaf, stuffings with bread, scrapple, thickened stew, tuna canned with hydrolyzed protein

Soups—thickened with flour, containing noodles, barley, etc.

Sweets—candies containing wheat products

Vegetables—creamed if thickened with wheat, oat, rye, or barley products

Miscellaneous—gravies and sauces thickened with flours not permitted

Key label words—wheat, oats, rye, barley, malt, bran (unless corn), farina, seminola

FIGURE 22–3
Diet without gluten—Some examples of foods that may contain gluten.

SPICY SOY SAUCE

Ingredients: water, soybeans, wheat, salt

increasing frequency. Inflammation of the intestinal wall results in abdominal pain, diarrhea, steatorrhea, weight loss, fever, and weakness. Decreased absorption, increased excretion, and reduced intake can lead to vitamin and mineral deficiencies, anemia, and protein loss.

Therapy

A diet high in calories and protein along with vitamin and mineral supplements will help overcome deficiencies. (See Table 9–3.) In acute stages, when rest of the affected areas is desired to promote healing, a very-low-residue diet is first tried. Total parenteral nutrition (see Chapter 17) may be indicated to rest the involved areas or to reverse growth retardation in teenagers with Crohn's disease. Medium-chain triglycerides (see Chapter 17) are used instead of usual dietary fats to reduce steatorrhea.

A gradual return to a regular diet is then made, eliminating foods known to aggravate the condition but making sure the diet is nutritionally adequate. Antiinflammatory medications are usually prescribed.

ULCERATIVE COLITIS

Clinical Findings

Ulcerative colitis, a chronic inflammation of the mucosa of the colon with periodic flare-ups and periods of remission, occurs most frequently in young adults. Its cause is unknown. Many patients are nervous and apprehensive. During acute attacks there is considerable loss of electrolytes, protein, and blood in the numerous watery stools. Abdominal discomfort, weight loss, dehydration, hemorrhage, anemia, fever, and generalized weakness are common.

Therapy

Because tissue wasting is great, a diet supplying 2,500–3,500 kcal and 100–150 g protein is required. See Basic Diet, Pattern C (Table 2–4), Table 9–3 and Figure 20–3. Supplements of iron and vitamins are usually indicated. Lactose intolerance is common. Small frequent feedings are often preferable to large meals.

Initially a very-low-residue diet is prescribed with progression to a soft fiber-restricted diet. Sometimes a semisynthetic, fiber-free diet is used, and occasionally total parenteral nutrition is employed to rest the colon and promote healing. Apprehension, questions, and complaints are common, and visits by the nurse or dietitian, especially at mealtime, will reassure the patient of the importance of food in recovery.

Antiinflammatory agents are commonly prescribed. Behavior modification to alleviate and manage stress is helpful. Sometimes surgery is necessary if the condition does not respond to other therapies. Patients with ulcerative colitis are at increased risk of developing colon cancer.

IRRITABLE BOWEL SYNDROME

Clinical Findings

Irritable Bowel Syndrome: also known as spastic colon, mucous colitis, irritable colon syndrome

Irritable bowel syndrome is a common condition of irregular bowel motility without apparent metabolic or anatomic cause. Abdominal pain due to spasmlike contractions of the colon is a frequent symptom. Diarrhea, constipation, or alternating episodes of both are common. Bloating, heartburn, headache, or mucus in the stools sometimes occur.

Stress and emotional upsets may precipitate an attack. Some individuals experience the symptoms after eating certain foods. Overactive intestinal nerves, laxative abuse, inadequate fluid intake, excessive caffeine, and irregular sleep, rest, and bowel habits have all been linked to the syndrome. Some cases are **idiopathic,** or without known cause.

Therapy

The first step is to find the cause and treat or eliminate it. Helping patients to cope with stress is often useful, as is eliminating foods that seem to trigger attacks, taking care to have an adequate diet. Since exercise affects intestinal motility and lowers stress, an exercise program should be implemented. Establishment of regular habits of sleep, rest, fluid, and food intake is essential. Laxatives are prohibited.

For patients with constipation, increased dietary fiber is recommended. Sometimes adding bran to the diet will help. For frequent diarrhea, a reduction in fiber may be indicated. Occasionally, antispasmodics, analgesics, or antidepressants are prescribed.

DIVERTICULITIS

Clinical Findings

Diverticula

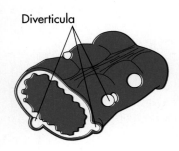

Diverticula are tiny sacs or pouches in the intestinal wall. Although they can occur in any part of the intestinal tract, the colon is the most frequent site. More

common in older people, **diverticulosis** is the presence of diverticula without inflammation or symptoms.

The sacs are caused by pressure in the intestine that brings about herniation of the mucosa through the intestinal wall. A diet low in fiber increases the pressure needed to move the fecal mass through the colon, while food that leaves large amounts of residue in the colon decreases pressure. Therefore, a diet high in insoluble fiber is recommended both to prevent diverticulosis and to reduce the chances of existing diverticula filling with fecal matter with resulting bacterial infection.

If the sacs become inflamed, the patient may experience severe lower abdominal pain and distention. Perforation of the inflamed diverticula and bleeding are possible complications. This condition is known as **diverticulitis.**

Therapy

During acute diverticulitis, bed rest, antibiotics, and a clear-liquid diet followed by a very-low-residue diet are prescribed. Chemically defined formulas may be offered to improve adequacy of intake, but are often poorly accepted by patients. (See Chapter 17.) Surgery is sometimes indicated.

The second stage in dietary management is a gradual increase in insoluble fiber content until a high-fiber intake is achieved. Patients are likely to be fearful of increasing their fiber intake. They need to understand that a high-fiber intake results in lowering the pressure that builds up in the colon, that fiber increases the bulk of the stool, that there is a shorter time for food wastes and bacterial growth to accumulate, and that elimination is improved. Ample fluids are essential. Some persons experience flatulence at first, but this is not cause for discontinuing the diet. The adjustment is more readily made if fiber intake is increased gradually.

PUTTING IT ALL TOGETHER

A variety of conditions involving the gastrointestinal tract affect many people. Determining and then treating or eliminating the cause is essential. Numerous diet modifications are employed to help manage or cure these problems.

[?] QUESTIONS CLIENTS ASK

1. My father had ulcers years ago and his diet was milk, cream, and baby foods. Now I have an ulcer and I can eat almost anything. Why the difference?

Research showed that ulcers heal as quickly with a more liberal diet. Also, milk in large amounts may increase the secretion of acid in the stomach. There are medicines available now, the histamine$_2$ receptor antagonists, that block production of stomach acid more effectively than medicines previously available.

2. I hear now that peptic ulcers are caused by bacteria. Could I "catch" an ulcer from somebody?

A vast majority, but not all, of patients with peptic ulcer have *H. pylori* infections. However, most people with *H. pylori* infections do not develop ulcers. Eradication of *H. pylori* with antibiotics is recommended for patients with gastric or duodenal ulcers who have the infection. *H. pylori* infections are not easily transmitted when there is good sanitation and a safe water supply.

3. My mother believes it is very important to have a daily bowel movement and has been taking mineral oil daily for years to "keep regular." I've heard mineral oil isn't good for you. Is that true?

Mineral oil can interfere with the absorption of fat-soluble vitamins in the body. Increasing dietary fiber with plenty of fluids and increasing activity is a better approach. If this simply does not work, a stool softener such as Metamucil is a possibility.

4. My aunt refuses to drink citrus juices because she claims they give her "acid stomach." Is this true?

Citrus juices have a pH, which indicates the degree of acidity or alkalinity, of 3–3.5. The lower the numbers, the more acid. The pH of the stomach is 1–3. Since the stomach is more acid than the citrus juice, the juices could not make it more acid.

✔ CASE STUDY: A Carpenter with Peptic Ulcer

Roy H., age 34, is a perfectionist, meticulous about his grooming and the appearance of his home and garden, and at his work as a carpenter. Recently he has been worried that he may be laid off because his company failed to get several big contracts on which it had bid. Roy's wife works as a teacher, but they would have difficulty getting along on her salary alone, and jobs in his field are scarce.

Roy is a chain smoker. His usual breakfast is orange juice, toast, and several cups of black coffee. He sometimes takes a sandwich and thermos of coffee along to work. Often he becomes so preoccupied in his work that he does not bother to stop for lunch, although he will drink the coffee throughout the day. Dinner is eaten at home and usually includes meat, fish, or poultry, a starch, a cooked vegetable, bread, and a salad. He has little interest in food and rushes through meals so he can get on to some project in his home workshop or garden. He will eat most foods, but has asked his wife to leave cucumbers, radishes, and onions out of the dinner salad because they give him gas.

Roy has been having occasional abdominal pains that seem to come on especially stressful days. He has found that drinking milk or taking an antacid will relieve the pain.

One day everything seems to be going wrong. His wife is home sick, the appearance of the work he is doing does not satisfy him, his supervisor is trying to hurry him to finish the job, and his coworkers are talking about expected layoffs. In the midst of this, Roy receives a call that his eight-year-old son has fallen off his bicycle and broken his arm. Roy leaves work to take his son to the hospital emergency room.

That night Roy awakes with an excruciating pain in his abdomen. Although it is relieved with an antacid, he decides, at his wife's urging, to see a physician.

An upper GI series reveals a duodenal ulcer. An *H. pylori* infection is found. Occult blood is found in his stool, and his hemoglobin is 12 g/dl. *Ranitidine, amoxicillin,* an iron supplement, and a normal diet as tolerated are prescribed.

Questions

1. Although Roy's prescribed diet is basically normal, list some principles to keep in mind when planning it.
2. What foods and beverages should Roy avoid?
3. Why is regularity in meals important?
4. Suggest some lunches Roy might pack to take to work. Usually no refrigerator is available at the worksite.
5. Suggest some lunches Roy might purchase from street vendors or at fast food restaurants located near the worksite.
6. Roy and his family occasionally have dinner in a restaurant. Give some suggestions to help Roy in ordering food there.
7. Can *ranitidine* and/or *amoxicillin* affect nutrition in any way? If so, how?
8. Explain the role of *H. pylori* in the etiology of peptic ulcer.
9. What factors in Roy's history, lifestyle, and personality might make him at risk to develop an ulcer?
10. Rest before and after meals is advised. How might Roy do this when at work?
11. Give some practical suggestions for modifying factors other than diet that may have contributed to the development of the ulcer.

Shirley T., age 25, is from a low-income family with a mother who is very ambitious for her only daughter. Shirley tries hard to excel and was an "A" student in college, which she attended with the aid of scholarships and part-time jobs. A hectic schedule and her desire to do well in both school and work put great pressure on her. After graduation she went to work for an interior decorator and was very successful, although demanding clients and an erratic schedule gave her a high level of stress.

She recently has begun her own business and has many worries about obtaining and pleasing customers, making a reputation, and finances. Before leaving home in the morning, she tries to have fruit or juice, cereal with milk, toast and coffee, because her other meals are unpredictable. Often she skips lunch because of work pressures. Sometimes she goes out to lunch with a client or business associate. Frequently when she gets home at night, she is so tired she grabs anything handy from the refrigerator and sometimes skips dinner completely. Her business requires much night work.

For some time she has been losing weight. She is always tired and her joints ache. The appearance of a bloody diarrhea prompts her to see her physician.

On examination she appears pale, nervous, and somewhat dehydrated. Her height is 65 in. (165 cm) and her weight, 110 lb. (50 kg); hemoglobin, 11 g/dl; hematocrit, 35%. She is hospitalized for further evaluation. The physician arrives at a diagnosis of ulcerative colitis and anemia, while tests also reveal a lactose intolerance.

A diet high in protein and calories, low in fiber, and restricted in lactose is ordered, plus a vitamin and mineral supplement. Corticosteroids are prescribed.

Questions

1. What factors in Shirley's background and lifestyle may have contributed to the ulcerative colitis?

2. Shirley is very apprehensive and has many questions and complaints. At mealtime she is afraid to eat for fear that her symptoms will be aggravated. How might health care team members assist and support Shirley, whom some consider a "difficult" patient?

3. What practical suggestions might you make to help Shirley reduce the level of stress in her life?

4. Shirley likes milk and dairy products and wonders if she can have them. How might you answer?

5. She knows that milk is a good source of protein and calcium and is concerned about getting enough of these nutrients in her lactose-restricted diet. What might you suggest?

6. She has always been weight conscious and although she has lost weight, she is concerned about gaining excessive weight on a high-calorie diet. What might you tell her?

7. Write a day's menu for Shirley following the sample menu for a high-calorie diet (Table 9–3), incorporating fiber restriction and lactose restriction.

8. Shirley has decided that when she returns to work it will be most practical to pack a lunch. Following the pattern in question 7 above, plan two lunches that might be packed and kept in the office refrigerator. A microwave oven is available.

9. Shirley will sometimes have to go to restaurants for business lunches. What suggestions might you give for ordering?

10. What are possible nutritional side effects of corticosteroids?

FOR ADDITIONAL READING

Chase, SL: "OTC Interactions: GI Remedies," *RN,* **56:**30–35, October 1993.

Donowitz, M, *et al:* "Evaluation of Patients with Chronic Diarrhea," *New Engl J Med,* **332:**725–29, 1995.

Doughty, DB: "What You Need to Know About Inflammatory Bowel Disease," *Am J Nurs,* **94:**24–31, July 1994.

Halsted, CH: "The Many Faces of Celiac Disease," *New Engl J Med,* **334:**1190–91, 1996.

"Helicobacter Pylori in Peptic Ulcer Disease," *NIH Consens Statement,* **12:**1–22, February 1994.

Hertzler, SR, *et al:* "How Much Lactose Is Low Lactose?" *J Am Diet Assoc,* **96:**243-46, 1996.

Laine, L, and Peterson, WL: "Bleeding Peptic Ulcer," *New Engl J Med,* **331:**717-27, 1994.

Lynn, RB, and Friedman, LS: "Irritable Bowel Syndrome," *New Engl J Med,* **329:**1940-45, 1993.

Marchiondo, K: "When the Dx is Diverticular Disease," *RN,* **57:**42-47, February 1994.

Meissner, JE: "Caring for Patients with Ulcerative Colitis," *Nurs 94,* **24:**54-55, July 1994.

Pope, CE II: "Acid-Reflux Disorders," *New Engl J Med,* **331:**656-60, 1994.

Schneeman, B: "Nutrition and Gastrointestinal Function," *Nutr Today,* **28:**20-24, January/February 1993.

Soll, AH: "Consensus Statement: Medical Treatment of Peptic Ulcer Disease, Practice Guidelines," *JAMA,* **275:** 622-29, 1996.

Suarez, FL, *et al:* "A Comparison of Symptoms after the Consumption of Milk or Lactose-Hydrolyzed Milk by People with Self-Reported Severe Lactose Intolerance," *New Engl J Med,* **333:**1-4, 1995.

Tolbert, CG, and Pratt, JC: "Bleeding Gastric Ulcer," *Am J Nurs,* **96:**48, February 1996.

Walsh, JH, and Peterson, WL: "The Treatment of *Helicobacter Pylori* Infection in the Management of Peptic Ulcer Disease," *New Engl J Med,* **333:**984-91, 1995.

Nutrition in Diseases of the Liver, Gallbladder, and Pancreas

Chapter Outline

Reduce hepatitis B infections among occupationally exposed workers to an incidence of no more than 1,250 cases.*

Increase hepatitis B immunization levels to 90 percent among occupationally exposed workers.*

Reduce cirrhosis deaths to no more than 6 per 100,000 people.*

✔ CASE STUDY PREVIEW

The liver performs so many functions that any illness or damage to the liver will have a profound effect on many body systems. Preventing further deterioration is essential. You will plan for the care, stabilization, and rehabilitation of a man with cirrhosis caused by alcohol abuse and malnutrition.

---

*Healthy People 2000: National Health Promotion and Disease Prevention Objectives. Public Health Service, U.S. Department of Health and Human Services, Washington, DC, 1991, Objectives 10.5, 10.9, and 4.2, pp. 105 and 97.

Hepatobiliary disease includes a heterogeneous group of diseases of the liver and biliary system caused by viral, bacterial, and parasitic infections, neoplasia, toxic chemicals, alcohol consumption, poor nutrition, metabolic disorders, and cardiac failure.*

DISEASES OF THE LIVER

Functions of the Liver

Hepatic: having to do with the liver

The role of the liver in nutrient metabolism is summarized in Table 23-1. Since the liver is involved in so many nutrition-related functions, it is understandable that liver disease can have wide ranging nutritional consequences.

Jaundice, yellow color of the skin and tissues due to elevated blood levels of bile pigments, is a frequent symptom of diseases of the liver and biliary tract. It may be caused by obstruction of bile flow, destruction of blood cells, drugs, poisons, or viral infections.

Goals of Therapy

In any liver disease, primary goals are to protect the organ from further stress and enable it to function as normally as possible. An adequate diet is central to achieving these objectives; poor nutritional status can result in permanent damage. Liver tissue is capable of regenerating after injury.

Except in hepatic failure, generous allowances of good quality protein are indicated for tissue regeneration and prevention of fatty infiltration. High carbohydrate intake will spare protein, and provide adequate glycogen. Some individuals can tolerate a normal fat intake; others will need a moderately restricted fat allowance. Vitamin supplements may be necessary. If edema and ascites are present, sodium restriction will be necessary.

CLASSIFICATION OF HEPATITIS AND CAUSES

Viral (more common)

Type A: contaminated water, food, or sewage; formerly called infectious hepatitis

Type B: unsterile needles; saliva, blood, semen; formerly called serum hepatitis

Type C: formerly called posttransfusion non-A, non-B hepatitis

Type D: co-infection with Type B; also known as Delta Hepatitis

Type E: contaminated water; also known as enteric non-A, non-B hepatitis

Drug-induced

Alcohol, heroin, marijuana, reaction to medicines, hepatotoxic drugs, poisons such as carbon tetrachloride

Ischemia from trauma or injury

HEPATITIS

Etiology

Hepatitis is an infectious disease that may be viral or drug induced. Type B is more severe than Type A. Individuals can become chronic carriers of the Type B virus. There are estimated 200 to 300 million carriers worldwide. Some 300,000 new cases are reported annually in the United States, but it remains an underreported disease; 4,000 to 5,000 people die each year of hepatitis B–related cirrhosis. Because it occurs so frequently in certain parts of the world, hepatitis B is considered by some to be the most important viral hepatitis. Hepatitis B carriers are at increased risk of later developing liver cancer. Immunization by hepatitis B vaccine is recommended for high-risk groups—homosexual men, intravenous drug users, children of immigrants from areas where hepatitis B is common, health care workers, and infants of hepatitis B–carrier mothers.

Clinical Findings

Hepatitis involves inflammation and degeneration of the liver. Symptoms are similar regardless of the type and include anorexia, nausea, vomiting, fever, abdominal pain, diarrhea, and weight loss, followed by jaundice.

Diet and Health: Implications for Reducing Chronic Disease Risk. National Research Council, National Academy Press, Washington, DC, 1989, p. 633.

TABLE 23–1
Functions of the Liver

Nutrient	Functions
Protein	Amino acid deamination, plasma protein and urea synthesis
Carbohydrate	Glycogen synthesis, storage, and release; heparin synthesis
Lipids	Bile, cholesterol, lipoprotein, and phospholipid synthesis; fatty acid oxidation
Minerals	Iron, copper, and other mineral storage
Vitamins	Vitamin A and D storage; conversion of carotene to vitamin A; hydroxylation of vitamin D to calcidiol
Other	Drug, poison, and waste product detoxification; alcohol metabolism; formation of prothrombin; fluid and electrolyte balance

Therapy

Bed rest and diet are the principal treatments. The goals are to regenerate affected tissue and prevent further damage. Because of nausea and vomiting in the early stages, it may be necessary to utilize parenteral fluids or tube feedings. Preliminary use of interferon therapy shows promise. (See Chapter 25.)

The appetite usually is poor, so it is especially important that meals be attractive and appealing to the taste. Each meal should include only the amounts of food that the patient can be expected to eat. Four to six evenly spaced feedings are often preferable to three meals. The diet is based on the following considerations:

As soon as the patient is able to eat, six small feedings of a full-liquid diet (see Chapter 17) may be given. This is followed by a soft diet (see Chapter 17) and then a normal diet.

If weight loss is considerable and there are other signs of malnutrition, the caloric intake should be increased to 3,000 kcal or more, and the protein to 100 g or more. Proteins of high biologic value should be emphasized. (See Chapter 9 for the High-Calorie Diet and Table 2–4, Pattern C.)

Most patients tolerate a normal fat intake, although a low-fat diet is necessary when there is obstruction of the biliary tract.

A liberal carbohydrate intake enhances the caloric level, ensures a continuous synthesis of glycogen, and spares protein for repair of liver cells.

Vitamin and mineral supplements are prescribed. Fat-soluble vitamins are in an aqueous form.

Some patients experience an intolerance to strongly flavored vegetables, rich desserts, and highly seasoned foods. Since this varies from person to person, any restriction of such foods should be an individual matter.

CIRRHOSIS

Cirrhosis is a chronic disease with loss of liver cells, fatty infiltration, and fibrosis. It is sometimes the outcome of inadequately treated hepatitis, metabolic problems such as hemochromatosis, glycogen storage disease, or Wilson's Disease, toxins, malignancies, or extended **biliary statis,** stagnation of bile flow. More often it is associated with chronic alcohol abuse accompanied by malnutrition. The most common type of cirrhosis in the United States is Laennec's cirrhosis.

Laennec's Cirrhosis: also known as alcoholic cirrhosis or portal cirrhosis

Clinical Findings

Chronic liver injury in cirrhosis causes regenerating nodules to form. These nodules cause changes in the structure and function of the liver.

Blood flow from the portal vein is inhibited by the fibrous tissue and forced back through the blood vessels of the spleen, intestines, stomach, and esophagus, with resulting high blood pressure known as **portal hypertension.** The portal hypertension forces fluid out of the blood vessels into the abdominal cavity causing **ascites,** massive fluid accumulation in the abdominal cavity. The affected blood vessels become engorged with blood (**varices**) and are prone to rupture. The enlarged spleen causes low red blood cell and white blood cell counts.

Cirrhosis interferes with the liver's ability to excrete **bilirubin,** a product of red blood cell breakdown, both by impairing conversion of bilirubin to a water-soluble form that can be excreted in bile and by obstructing bile flow from the liver. Bilirubin is deposited in elastin, causing jaundice.

Impairment of the liver's ability to produce clotting factor results in prolonged clotting time, pinpoint hemorrhages, and bruising. Low serum albumin is caused by diminished protein metabolism and utilization. Anorexia, nausea, and vomiting may be present.

Therapy

A normal diet is satisfactory if there are no complications. When a patient is poorly nourished, a high-calorie diet is indicated. The protein intake should be high enough to maintain nitrogen balance, yet low enough to prevent hepatic encephalopathy, approximately 60–70 g/day. Vitamin and mineral supplements are sometimes prescribed. As in hepatitis, many patients with cirrhosis have poor appetites, and the members of the health care team must be prepared to improve food intake by counseling the patient concerning the importance of diet, by adjusting menus to the patient's preferences, and sometimes by using supplementary foods rich in nutrients.

When ascites is present, the diet must be restricted in sodium to about 1,000 mg daily. (See Chapter 24.) Fluid restriction may also be indicated. Because of the distention of the abdomen, patients are unable to eat large amounts of food at one time. Thus, six meals daily are indicated.

Esophageal varices may be irritated by coarse, fibrous foods, by swallowing a large bolus of food, or by coffee, tea, tobacco, pepper, and chili seasoning, with attendant risk of rupture followed by severe hemorrhage. Thus, small frequent feedings of a soft diet (see Chapter 17) are used for patients with esophageal varices. If further restriction seems indicated, a full-liquid or very-low-fiber diet (see Chapters 17 and 22) is ordered.

Hepatic Failure: also known as end-stage liver disease and hepatic coma

BLOOD PROFILE IN HEPATIC FAILURE

Elevated ammonia
Elevated aromatic amino acids
Lowered branched-chain amino acids

HEPATIC FAILURE

Hepatic failure results from a decreased number of functioning liver cells and diminished delivery of nutrients due to shunting of the portal blood circulation with a progressive deterioration in function. It may be caused by viral hepatitis or drugs or toxins that injure the liver. Ascites, edema, jaundice, central nervous system dysfunction, coagulation changes, infections, and cachexia are present. Immune response may be compromised because of decreased phagocyte function in the liver.

Clinical Findings

The liver loses the ability to convert ammonia, produced by bacteria in the intestines, to urea. The ammonia is toxic to the central nervous system. There is a decreased breakdown of aromatic amino acids (phenylalanine, tyrosine, tryptophan) that are metabolized in the liver, and they accumulate in the blood. The branched-chain amino acids (leucine, isoleucine, valine) are broken down in the peripheral muscle for energy and their blood level decreases.

The neurological and neuromuscular abnormalities in patients with severe hepatic dysfunction are known as **hepatic encephalopathy.** The symptoms may be mild at first, but will progress to coma if untreated. Death will occur without therapeutic intervention.

Therapy

The basic dietary principle is to decrease protein to minimize ammonia production.

A protein-free to low-protein diet, about 20–30 g, is followed at first. (See Chapter 28.) With improvement, the diet is cautiously advanced by 10 g protein every few days until a normal diet is achieved. Patients may remain on a 40–50 g protein intake for long periods of time and still have an adequate diet if caloric intake is sufficient to maintain weight. Branched-chain amino acids taken orally or enterally* improve the amino acid profile and enteropathy.

Include about 1500–2000 kcal from carbohydrate and fat to prevent tissue breakdown.

If the patient is comatose or otherwise unable to be adequately nourished through oral intake, tube feedings or parenteral nutrition are indicated.

Antibiotics such as *neomycin* and *ampicillin* will reduce bacterial growth and hence ammonia production. Diuretics may be prescribed to relieve ascites. **Lactulose,** a synthetic disaccharide of lactose and fructose, acidifies the colon through bacterial action, causing the ammonia to remain there to be excreted in the feces instead of entering the bloodstream. *Levodopa* is used to treat psychiatric symptoms.

Liver Transplant If dietary and medication therapies fail, liver transplant may be a viable alternative. Postoperatively, the diet is individualized to the patient's status. Protein and calories must be adequate for wound healing and positive nitrogen balance. Fluid restriction may be necessary if ascites is present. If ileus persists, total parenteral nutrition may be indicated. Some combination of enteral and parenteral nutrition may be used until adequate oral intake is possible.

Long-term nutritional care involves calorie control to prevent excessive weight gain. Concentrated carbohydrates and fats are restricted to prevent hyperglycemia and hyperlipidemia.

Prednisone, commonly used as an immunosuppressive, favors negative nitrogen balance, increased appetite favoring weight gain, hyperglycemia, fluid retention, peptic ulcers, and bone loss. The diet should feature increased protein, complex carbohydrates instead of sugars, and mild sodium restriction. *Cyclosporine* is an immunosuppressive that protects against transplant rejection. Common side effects are hyperlipidemia, hyperkalemia, and hypertension, which call for

---

*NutriHep™, Clintec Nutrition, Deerfield, IL.

SYMPTOMS OF HEPATIC ENCEPHALOPATHY

Fetor hepaticus—fecal odor to breath
Asterixis—flapping tremor of arms and legs
Disorientation
Delirium
Coma

reduction of dietary fat, potassium, and sodium. *Azathioprine,* which fosters bone marrow suppression, may cause stomatitis and gastrointestinal upsets that can interfere with the patient's food intake.

DISEASES OF THE GALLBLADDER

Gallbladder disease affects an estimated 20 million Americans, the majority of whom are asymptomatic. It is the fifth leading cause of hospitalization, and **cholecystectomy,** removal of the gallbladder, is the most common elective abdominal operation in Western countries.

Functions of the Gallbladder

The gallbladder concentrates and stores bile formed in the liver. When dietary fat enters the duodenum, secretion of cholecystokinin is stimulated. This hormone is carried by the bloodstream to the gallbladder and forces its contraction so that bile is released into the common duct and on into the duodenum. Bile emulsifies fats so they can be digested by the fat-splitting enzymes, the lipases. If there is interference with the flow of bile, fat digestion is impaired. Inflammation of the gallbladder (**cholecystitis**) or **cholelithiasis,** stones blocking the gallbladder or common duct, prevent the flow of bile and result in pain when the gallbladder contracts. This pain is accompanied by abdominal distention, nausea, and vomiting.

Therapy

Treatment may involve diet, medicine, and/or surgery. Since gallbladder disease is much more common among obese than normal weight individuals, those at risk should achieve and maintain normal weight. (See Chapter 9.) Fat intake may be limited to 50–60 g/day since dietary fat causes the gallbladder to contract. During acute attacks of cholecystitis, the patient receives no food at first and then progresses to clear fluids, followed by a soft fiber-restricted diet, limiting fat to 20–30 g daily. Studies have failed to show a relationship between dietary cholesterol and gallstone formation.

Chenodeoxycholic acid and ursodeoxycholic acid, secondary bile acids, may be taken over a period of time to dissolve small stones in selected patients. **Lithotripsy** has been used to shatter stones with ultrasonic shock waves.

Recently, surgery to remove the gallbladder has been modified from a major abdominal procedure to one far less invasive, laparoscopic laser cholecystectomy. This operation involves an overnight stay or may be done on an outpatient basis.

Following cholecystectomy, most individuals can return to a normal diet. The common duct, connecting the liver and small intestine, takes over the function of storing bile. Since surgery does not remove the underlying cause of gallstone formation, continued maintenance of appropriate body weight is helpful.

DISEASES OF THE PANCREAS

Functions of the Pancreas

The pancreas secretes enzymes involved in the digestion of carbohydrate, fat, and protein. The secretions, known collectively as pancreatic juice, are carried to the small intestine via the pancreatic and common bile ducts.

Pancreatic Insufficiency

Pancreatic insufficiency involves inadequate production of pancreatic enzymes; therefore, the digestion of fats, protein, and starch is reduced. Undigested fat, protein, and starch are present in the stools in increased amounts. Fat-soluble vitamins are poorly absorbed. Such losses, if not corrected, lead to generalized malnutrition. Pancreatic insufficiency can result from cancer of the pancreas, pancreatic resection, pancreatitis, and cystic fibrosis.

Acute Pancreatitis

Blockage of the flow of pancreatic juice coupled with continued activation and release of pancreatic enzymes leads to inflammation and edema of the pancreas and ultimately to its autodigestion. Alcohol abuse is the most common cause, while gallstones and other biliary tract diseases and trauma can also be responsible. Cases vary from mild to severe; the latter with extensive **necrosis** or death of tissue and risk of mortality from failure of the cardiovascular, renal, and/or pulmonary systems.

The dietary rationale initially is to halt pancreatic secretions. To this end, the patient is given nothing by mouth (NPO) and receives total parenteral nutrition. Histamine$_2$ receptor antagonists may be prescribed to decrease hydrochloric acid production, which in turn will reduce stimulation of the pancreas.

When the patient has stabilized, oral intake progresses gradually from clear liquids to a soft diet as tolerated. For maintenance the diet should be high in protein, moderate to high in carbohydrate, and low in fat. If pancreatic function has been severely damaged, pancreatic enzymes are taken with food to improve digestion and absorption; insulin may also be necessary.

NPO (*nil per os*): nothing by mouth

GOALS OF DIET THERAPY FOR PANCREATITIS

Prevent or reverse malnutrition
Improve digestion and
  absorption
Manage pain

Chronic Pancreatitis

Inflammation that fails to subside or recurs at intervals characterizes chronic pancreatitis. Large amounts of fibrous tissue lead to decreased enzyme production resulting in impaired digestion and the appearance of undigested protein, fat, and sometimes carbohydrate in the stool. Fat-soluble vitamin deficiencies develop. In extensive damage or blockage, large amounts of enzymes build up within the pancreas and autodigestion can occur.

Diet during attacks is identical to that for acute pancreatitis. At other times a high-protein, high-calorie, low-fat diet divided into six meals is used. Water-soluble preparations of fat-soluble vitamins are frequently necessary. Medium-chain triglycerides (see Chapter 17) are sometimes used to increase caloric levels and palatability. Pancreatic enzymes, antacids, and histamine$_2$ receptor antagonists may be indicated.

CYSTIC FIBROSIS

Cystic fibrosis, which occurs in approximately one in every 2,000 live births, is the most common life-threatening genetic disease in children. Abnormally thick exocrine gland secretions may obstruct the pancreatic and bile ducts, intestines, and bronchi. The gene that causes cystic fibrosis has been identified and gene therapy trials have begun. These are in an experimental stage.

Clinical Findings

Cystic fibrosis is a multisystem disorder in which the symptoms are extremely variable. For example, about 15 percent of patients do not have pancreatic involvement. The major finding is an abnormally high concentration of electrolytes in sweat, which provides the standard diagnostic tool.

Thick mucus in the lungs obstructs the bronchi and coughing, dyspnea, and recurrent respiratory infections are common. With pancreatic duct blockage, enzymes cannot reach the duodenum and malabsorption is often severe. Stools are frequent and foul, containing much undigested fat, protein, and starch. Tissue wasting is evident. Disorders of the liver, intestines, heart, and kidneys are secondary results. Diabetes mellitus may be a complication. Reproductive system disorders may become evident by the teenage or young adult years.

Therapy

An interdisciplinary approach involving the physician, nurse, dietitian, physical therapist, and social worker along with the patient and the family is necessary.

Pancreatic enzyme microspheres coated with a pH-sensitive substance that will not dissolve until they reach the small intestine are prescribed. These **enteric-coated** enzyme preparations, taken with meals and snacks, improve, but do not normalize, digestion and absorption of protein, fat, and carbohydrate.

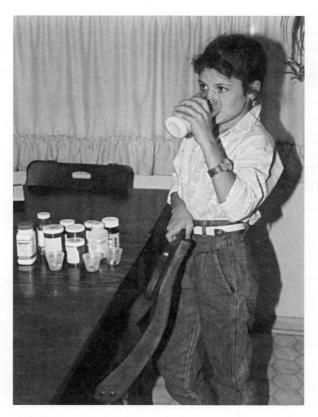

FIGURE 23–1
A child with cystic fibrosis needs an array of medications, digestive enzymes, and vitamin and mineral supplements to maintain growth. (*Source:* Courtesy of the Cystic Fibrosis Foundation.)

Dietary intake will vary with the needs of the individual. For the infant, child, and adolescent, the goal is to provide for normal growth and development. Infants are given proprietary formulas or breast-feeding supplemented with milk-based formula along with pancreatic enzymes. When other foods are added, those with high caloric density should be chosen. Foods of low nutrient density are to be avoided. One-quarter to one-half teaspoon of salt is added to the diet daily. For older children, adolescents, and adults, diet will depend on the medical and nutritional status, age, medications, and individual needs. Protein needs may be 50 percent above the Recommended Dietary Allowances and energy requirements up to twice the normal needs. Fat should supply 30–50 percent of the caloric requirement, depending on tolerance and energy needs. Vitamin supplements in water-soluble form and mineral supplements are taken. (See Figure 23–1.) Extra salt is needed in hot weather and during extensive physical activity.

Women with cystic fibrosis have delivered healthy babies and some mothers have breast-fed their infants. Nutritional guidance before conception and during pregnancy and lactation is essential.

PUTTING IT ALL TOGETHER

The pancreas and gallbladder are essential to the digestion of food. The liver has many functions that involve metabolism of nutrients. Disease of any of these organs can have profound effects on the utilization of food. Various dietary strategies can help preserve function and promote rehabilitation of these organs.

? QUESTIONS CLIENTS ASK

1. *My husband has been diagnosed with cirrhosis of the liver. He never drank alcoholic beverages. I thought only alcoholics got cirrhosis.*

Cirrhosis can be caused by hepatitis that was not treated or not treated adequately, drugs or poisons that were toxic to the liver, or prolonged blockage of the bile flow, as well as by alcoholism. The causes may have occurred years ago.

2. *I've heard that gene replacement is being used to treat cystic fibrosis. Is this true, and if so, how does it work?*

This therapy is still in an experimental stage. Researchers alter a virus by replacing one of its genes with the gene missing in cystic fibrosis. The modified virus is delivered to the patient's lungs where it will produce the missing protein, which will help to keep the airways free of mucus.

3. *My mother had gallbladder surgery 30 years ago and I remember it was a major operation. She was in the hospital for some time and then had a long convalescence at home. Recently my sister had her gallbladder removed and it was done on an outpatient basis. This is hard to believe.*

Your sister probably had laparoscopic laser cholecystectomy. It involves making four small punctures instead of a large open incision. The laparoscope with a small video camera attached is inserted into one of the punctures. The camera transmits an image of the abdominal organs and tissues to a monitor, which the surgeon can watch while operating.

✔ CASE STUDY: A Man with Cirrhosis of the Liver

Ernest M., age 43, has been an alcohol abuser for many years. His drinking has separated him from his family and caused him to lose his job. He now lives in a room in a rundown hotel and does odd jobs. He drinks a fifth of whisky a day, and his food intake is unpredictable. He keeps a few nonperishable food items around his room and sometimes buys food from street vendors, vending machines, fast-food restaurants, or luncheonettes. Some days he does not bother to eat, for food does not interest him.

Swollen ankles, ascites, pain when swallowing food and liquor, plus the appearance of jaundice led him to visit a health clinic in his neighborhood. A diagnosis of cirrhosis was made, and Mr. M. was admitted to a hospital. Tests revealed hemoglobin, 11 mg/dl; serum albumin, 2.5 g/dl; and prothrombin time, 25 sec (normal, 10–20). A soft diet of 3,000 kcal, 60–70 g protein, 1,000 mg sodium, with six meals a day, plus vitamin and mineral supplements, was ordered. *Neomycin, lactulose,* and *levodopa* were prescribed.

1. Why is Mr. M.'s diet restricted in sodium?

2. Why is the diet soft?

3. High-calorie diets, such as that in Chapter 9, are usually high in protein. Why was a lower protein allowance ordered?

4. How might the high-calorie diet in Table 9–3 be modified to comply with the 60–70 g protein allowance? The 1,000 mg sodium intake?

5. Prepare a meal plan incorporating the various modifications based on the high-calorie diet mentioned in question 4.

6. Write a day's menus based on this meal plan.

7. Mr. M. has a very poor appetite. What might be done to improve his food intake?

8. After several days in the hospital, he becomes lethargic, irritable, uncoordinated, and shows signs of hepatic failure. The diet is changed to soft, fiber-restricted 1,500 kcal, and 20 g protein. How might the 1,500-kcal diet in Table 9–2 be modified to conform to the lowered protein intake?

9. Prepare a meal plan based on the diet prescription in question 8.

10. Write a day's menus based on this meal plan.

11. What are the functions and possible side effects and medication-nutrient interactions of *neomycin, lactulose,* and *levodopa?*

12. What arrangements and support do you think will be required when Mr. M. is discharged? What community resources might be available to assist him?

FOR ADDITIONAL READING

Ambrose, MS, and Dreher, HM: "Pancreatitis: Managing A Flare-up," *Nurs 96:* **26:**33–40, April 1996.

Butler, RW: "Managing the Complications of Cirrhosis," *Am J Nurs,* **94:**46–49, March 1994.

Dowsett, J: "Nutrition in the Management of Cystic Fibrosis," *Nutr Rev,* **54:**31–33, 1996.

Jackson, MM, and Rymer, TE: "Viral Hepatitis: Anatomy of a Diagnosis," *Am J Nurs,* **94:**43–48, January 1994.

Johnston, DE, and Kaplan, MM: "Pathogenesis and Treatment of Gallstones," *New Engl J Med,* **328:**412–21, 1993.

Lee, WM: "Acute Liver Failure," *New Engl J Med,* **329:**1862–72, 1993.

Lieber, CS: "Herman Award Lecture, 1993: A Personal Perspective on Alcohol, Nutrition, and the Liver," *Am J Clin Nutr,* **58:**430–42, 1993.

Lowenfels, AB, *et al:* "Pancreatitis and the Risk of Pancreatic Cancer," *New Engl J Med,* **328:**1433–37, 1993.

Marx, JF: "Viral Hepatitis: Unscrambling the Alphabet," *Nurs 93,* **23:**34–42, January 1993.

O'Hanlon-Nichols, T: "Portal Hypertension," *Am J Nurs,* **95:**38–39, November 1995.

Ondrusek, RS: "Cholecystectomy: An Update," *RN,* **56:**28–33, January 1993.

Ramsey, BW, *et al:* "Nutritional Assessment and Management in Cystic Fibrosis: A Consensus Report," *Am J Clin Nutr,* **55:**108–16, 1992.

Steinberg, W, and Tenner, S: "Acute Pancreatitis," *New Engl J Med,* **330:**1198–1210, 1994.

Nutrition in Cardiovascular Diseases

Chapter Outline

Increase to at least 60 percent the proportion of adults with high blood cholesterol who are aware of their condition and are taking action to reduce their blood cholesterol to recommended levels*

Increase to at least 90 percent the proportion of people with high blood pressure who are taking action to help control their blood pressure*

✔ CASE STUDY PREVIEW

Hypertension affects almost 50 million people in this country, making it one of the most common health problems. You will plan for the management of a man whose elevated blood pressure was diagnosed only when he went to see the doctor about another matter, a common occurrence. You will also follow an older man with heart failure, which affects 2 million Americans, from the acute stage until his discharge. You will help him and his wife plan for maintenance at home to prevent reoccurrence.

---

*Healthy People 2000: National Health Promotion and Disease Prevention Objectives. Public Health Service, U.S. Department of Health and Human Services, Washington, DC, 1991, Objectives 15.8 and 15.5, p. 112.

Although the death rate from cardiovascular diseases has decreased since 1960 because of lifestyle changes, risk modification, improved medications, and new technologies, these diseases continue to kill almost as many Americans as all other diseases combined. They are also a major cause of disability.

Coronary heart disease (CHD) is the leading cause of illness and death of both men and women in the United States. About 12 million Americans have CHD. It results in about 1.25 million heart attacks and half a million deaths annually. The economic cost of CHD is estimated between $50 and $100 billion annually.

Development of Coronary and Blood Vessel Disease

STAGES OF DEVELOPMENT

Plaque (in blood vessels): also known as atheroma

It has been proposed that cardiovascular diseases develop in three stages. First there is *initiation* where arterial damage is caused, probably by hypertension, smoking, and/or lipid oxidation products. In *progression*, plaque composed of cholesterol and triglyceride is deposited in connective tissue in the arterial wall, causing progressive thickening and rigidity. Narrowing of the arterial lumen makes delivery of oxygen and nutrients increasingly difficult.

Atherosclerosis affects most people in industrialized countries. Many do not develop overt disease. In some, one or more of the arteries becomes seriously thickened and a blood clot forms at the narrowed area, blocking the artery. Local tissue dies because of lack of oxygen through inadequate blood supply.

Termination is a myocardial infarction caused by thrombosis or spasm. *Occlusion* or blockage of coronary arteries results in a heart attack, while shutting off of cerebral arteries causes a cerebrovascular accident or stroke. Blockage of a vessel in the leg can lead to gangrene.

Myocardial Infarction (MI): also known as heart attack, coronary occlusion, or coronary thrombosis

Coronary Heart Disease: also known as coronary artery disease or ischemic heart disease

RISK FACTORS

Many interrelated factors increase the risk of atherosclerosis and coronary heart disease. The four major risk factors in coronary heart disease are elevated serum cholesterol, hypertension, cigarette smoking, and lack of activity. The presence of one of these doubles the risk of coronary heart disease, and if all four exist simultaneously, the risk is more than 10 times greater. Dietary components, especially saturated fat and cholesterol, are associated with elevated blood lipids, while a high sodium intake may be involved in hypertension.

CORONARY HEART DISEASE RISK FACTORS

Major: elevated serum cholesterol
hypertension
cigarette smoking
lack of activity

Other: family history of premature CHD
low HDL cholesterol
diabetes mellitus
history of cerebrovascular or occlusive peripheral vascular disease
severe obesity
high fat intake

BLOOD LIPIDS

The measurement of blood lipids is useful for diagnosis and also to evaluate therapies. Hyperlipidemia and its subsets hypercholesterolemia and hypertriglyceridemia are risk factors in developing coronary artery disease.

The National Cholesterol Education Program (NCEP) recommends that all individuals 20 years and older have their total and HDL cholesterol measured. If values are normal, the tests are to be repeated in five years. If total cholesterol alone is slightly elevated, in one year. In other cases, a total lipid profile to determine LDL cholesterol is recommended.

Cholesterol

Many clinicians and researchers believe that elevated serum cholesterol is the single best indicator of risk of atherosclerosis. For adults, above 200 mg/dl is frequently pinpointed as the turning point where risk of coronary heart disease escalates. The average level among adults in the United States is 210–215 mg/dl. For blood cholesterol levels representing borderline and high risk for different age groups, see Appendix Table D-2. Since most plasma cholesterol is transported in low-density lipoproteins (LDL), elevated levels of this fraction are associated with increased risk. (See Figure 24-1.) High-density lipoprotein (HDL) protects against the formation of atheroma.

PREVENTION OF ATHEROSCLEROSIS

Dietary Factors

Since the atherosclerotic process begins in childhood, the National Cholesterol Education Program (NCEP) recommends that all healthy persons over two years of age consume a diet with the following characteristics.

Energy Because obesity is an important risk factor in cardiovascular disease and weight loss is associated with lowering of blood lipids, a caloric intake that will lead to attaining and maintaining a healthy weight is suggested. (See Chapter 9.) Adequate energy for growth in children and adolescents is essential, however.

Lipids Modification of dietary fat is the cornerstone of prevention and control of coronary heart disease. Currently the typical American diet contains about 34 percent of its calories as fat with approximately 13 percent as saturated fat.

A prudent diet with fat intake of 30 percent or less of total energy intake is recommended with saturated and polyunsaturated fat each making up one-third of this total. Some authorities suggest a fat intake of 20 percent of calories. Others believe that while this may be desirable, it is unrealistically low for most people. A cholesterol intake of 300 mg or less is suggested. For the cholesterol and saturated fat content of some foods, see Appendix Table A-1.

Carbohydrates A carbohydrate intake of 55–60 percent of total calories is recommended. Since sucrose may cause hypertriglyceridemia, complex carbohydrate should be emphasized.

Fiber Water-soluble fibers such as gums, pectins, and oat bran lower serum cholesterol. The more fibrous components such as cellulose and hemicellulose have little effect on serum cholesterol although they are of value in normal bowel function. A daily recommendation is 25–50 g (see Chapters 5 and 6 and Table 2-3).

Antioxidant Nutrients There is some evidence that vitamins A, C, E, and beta carotene levels in the diet vary inversely with incidence of cardiovascular disease. There is no justification for taking supplements, but consuming a total of five servings from the fruit and vegetable groups daily is prudent.

TOTAL SERUM CHOLESTEROL VALUES

< 200mg/dl	desirable
200–239 mg/dl	borderline high
≥ 240 mg/dl	high

LDL CHOLESTEROL VALUES

< 130 mg/dl	desirable
130–159 mg/dl	borderline high
≥ 160 mg/dl	high

HDL CHOLESTEROL VALUES

≤ 35 mg/dl	increased risk
≥ 35 mg/dl	acceptable

Progression of Atherosclerosis

Normal Vessel

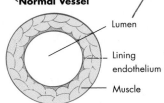

Early Injury

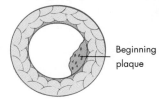

Advanced Injury

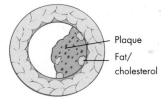

TRIGLYCERIDE VALUES

< 250 mg/dl	normal
250–500 mg/dl	borderline hypertrigly-ceridemia
> 500 mg/dl	frank hypertrigly-ceridemia

FIGURE 24–1
A diet low in total fat, saturated fat, and cholesterol together with aerobic exercise have helped this woman to lower her total blood cholesterol and increase her HDL cholesterol. (*Source:* Dunwoody Village, Newtown Square, PA, and Peter Zinner, photographer.)

Alcohol Alcohol raises plasma triglycerides in some persons while in others a moderate intake increases the level of protective high-density lipoprotein cholesterol. (See Chapter 6.)

EFFECTS OF DIETARY CHANGE

Lower Dietary:	= Lower Serum:
Cholesterol	LDL, IDL, VLDL*
Total fat	LDL
Saturated fat	LDL
Sugar, alcohol	VLDL*
Calories	LDL, IDL, VLDL*

*VLDL are primarily triglycerides.

FAT-MODIFIED DIETS

Two-Step Diet

The National Cholesterol Education Program advocates a stepwise approach to lowering serum lipids.

Step 1 is identical to the diet recommended for the general population. Individuals with borderline high serum cholesterol follow this diet and have their serum cholesterol rechecked within a year. If serum cholesterol does not respond despite dietary compliance, a Step 2 diet is begun. Some persons with serum cholesterol of 240 mg/dl or above also experience significant reduction on this diet.

Step 2 is planned for persons with a total blood cholesterol in excess of 240 mg/dl, who should also have a lipoprotein profile determination. This diet is also

TABLE 24–1

Two-Step Diets (National Cholesterol Education Program)[a]

	Step 1	Step 2
Fat, % kcal	<30	<30
Saturated, %	8–10	<7
Polyunsaturated, %	Up to 10	Up to 10
Monounsaturated, %	Up to 15	Up to 15
Cholesterol, mg	<300	<200
Carbohydrate, % kcal	55 or more	55 or more
Protein, % kcal	Approximately 15	Approximately 15

Source: Adapted from *Second Report of the Expert Panel on Detection, Evaluation, and Treatment of High Blood Cholesterol in Adults,* Bethesda, MD: National Heart, Lung, and Blood Institute, 1993.

[a] Caloric intake to attain and maintain healthy weight and provide for growth and development in children.

appropriate for clients with moderate elevations of blood cholesterol who have accompanying risk factors and for those who do not respond to Step 1 diets.

For step diet characteristics and a sample menu, see Tables 24–1 and 24–2. If the Step 2 diet fails to lower blood cholesterol in 6 months, the diet is continued and medication therapy added. (See Table 24–3.)

MEAL PLANNING AND FOOD LISTS

The *Exchange Lists for Meal Planning* (Appendix Table A–2) may be used for making food choices. Only 1 percent fat milk or skim milk and lean or very lean low-fat meat exchanges are used. Low-cholesterol egg substitutes may replace a meat exchange. Fats are selected from those with a high polyunsaturate content. Reading labels on packaged foods is essential to determine total fat, saturated fat, and cholesterol content, as well as fiber content. The label may also contain nutrient claims such as "Fat free" or "Good source of fiber." (See Chapter 1.)

Sugars, Sweets, and Low-Fat Desserts

In limited amounts, one starch exchange may be replaced by 1 tablespoon sugar, honey, jelly, or jam, ½ oz jelly beans, hard candy, or mints (not chocolate), 6 oz sweetened carbonated beverage, 3 oz fruit-flavored drink, ¼ cup sherbet or water ice, ½ cup pudding made with skim milk and alternative sweetener, 1½-inch cube angel food cake, or ½ cup gelatin.

Alcohol

In excessive amounts alcohol increases blood triglyceride levels. With the physician's approval, it may be used with discretion as a substitute for up to two starch exchanges. One starch exchange is equivalent to 1 oz gin, rum, vodka, or whiskey, 1½ oz sweet or dessert wine, 2½ oz dry wine, or 5 oz beer.

MEAL PREPARATION STRATEGIES

A diet pattern and sample menu for a 1,600 kcal diet that would be appropriate for a Step 1 or Step 2 diet appears in Table 24–2. (See also the 1,600 kcal Basic Diet, Table 2–4.) The 1,800-kcal diet with 80 g protein using exchange lists (see

TABLE 24–2
1,600-kcal Fat-Modified Diet[a,b]

Groups	Number of Servings	DIET PATTERN C = 185 g P = 75 g F = 55 g			
		50% kcal C (g)	20% kcal P (g)	30% kcal F (g)	Cholesterol (mg)
Bread	6	90	18	–	–
Other carbohydrate	1	15	–	–	–
Meat	5	–	35	25	Variable
Vegetable	3	15	6	–	–
Fruit	2	30	–	–	–
Milk, skim	2	24	16	–	–
Fat	6	–	–	30	–
Added sugar	(6 teaspoons)	25	–	–	–
Total		185	75	55	

SAMPLE MENU

Breakfast
Honeydew melon—1 slice
Dry or cooked cereal—¾ cup
Whole-wheat toast—1 slice
Soft margarine—1 teaspoon
Jelly—1 tablespoon
Skim milk—1 cup
Coffee, if desired
Sugar—1 teaspoon

Lunch
Sandwich
  Rye bread—2 slices
  Sliced turkey—2 oz
  Mayonnaise—1 teaspoon
Carrot sticks and pepper strips
Banana—1 small

Dinner
Broiled flounder—3 oz
Parslied potato—1 small
Broccoli—1 cup
Oil—2 teaspoons (for fish and vegetables)
Tossed green salad
French dressing—1 tablespoon
Dinner roll—1
Soft margarine—1 teaspoon
Angel food cake—1 small piece
Tea if desired
Sugar—1 teaspoon

Snack
Fat-free yogurt—1 cup

[a] Suitable for Step 1 and Step 2 diets.
[b] Based on Basic Diet, Pattern A, Table 2–4.

Table 26–4) would also be appropriate for use in planning these fat-modified diets.

Booklets for patients are available from the National Heart, Lung, and Blood Institute and the American Heart Association. They provide detailed information on the purposes of the diet, the lists of foods that may be used and those to be avoided, typical meal patterns, how to shop for allowed food, how to prepare food, and what to do when eating meals away from home. The *Exchange Lists for Meal Planning* may be used in place of these booklets (see Appendix Table A–2).

Preparation of Meats

Only lean meats are used. All visible fat must be removed. If meats are roasted or broiled, they should be placed on a rack so that the drippings are removed. If the meat is stewed, it may be cooked a day ahead, cooled in the refrigerator, and the fat skimmed off the top of the liquid. Meats, fish, and poultry may be basted or

TABLE 24-3
Medications That Modifiy Blood Lipids

Medication	Total Cholesterol	LDL	HDL	Triglycerides
Bile acid resins				
Cholestyramine	↓	↓	↑a	↑a
Colestipol	↓	↓	↑a	↑a
Fibric acids				
Gemfibrizol	↓a	↓a	↑a	↓
Clofibrate	↓a	↓a	↑a	↓
Probucol	↓	↓	↓	—
Nicotinic acid	↓	↓	↑a	↓
HMG-CoA reductase inhibitors				
Lovastatin	↓	↓	↑a	↓
Pravastatin	↓	↓	↑a	↓
Simvastatin	↓	↓	↑a	↓
Fluvastatin	↓	↓	↑a	↓

a Modest or minimal change.

marinated with tomato juice, lemon juice, wine, or bouillon, and baked in aluminum foil.

Seasoning Vegetables

Lemon juice, vinegar, and herbs lend variety to vegetable flavors. See suggestions for flavoring vegetables later in this chapter. Sauces may be prepared using nonfat milk and fat allowance.

Using Oils and Margarine

Choose fats from the unsaturated list (see fat list, Appendix Table A-2). Soft margarine should be used. Corn, soybean, sunflower, and safflower oils contain a higher percentage of polyunsaturated fatty acids; olive, peanut, and canola oils are rich in monounsaturates.

Behavior Modification

To bring about dietary change and promote the long-term adherence that is required, behavioral modifications are essential. Increased knowledge of food composition, heightened awareness of eating habits, automatic label reading while shopping, and appropriate choices in restaurants and social situations will facilitate compliance. A program of increased activity will benefit most persons. Many of the behavioral techniques helpful in weight control or diabetes mellitus can be applied to or adapted for regimens designed to lower blood lipids. (See Chapters 9 and 26.)

MEDICATIONS

Numerous medications are used to lower blood lipids. (See Table 24-3.) Some interfere with absorption of fat-soluble vitamins, and vitamin supplements may be necessary. The bile acid sequestrants, fibric acids, and nicotinic acid may cause a variety of gastrointestinal side effects and should be taken with food. Nicotinic

USES FOR OILS

Marinating meat
Pan frying meat, fish, poultry
In biscuits, muffins, pancakes, waffles
With herbs to flavor vegetables
In mashed potatoes with skim milk
In salad dressings
In sauces with skim milk

acid (niacin) in pharmacological doses causes flushing, itching, intracranial pressure, and rashes. Alcohol may intensify this effect.

Some clinicians are concerned that reliance on lipid-lowering medications will lead people to abandon dietary and other lifestyle changes that have led to a reduction in the average serum cholesterol in the past two decades. The possibility exists that the medications may cause long-term negative effects not currently known. Anti-hypertensive medications can reduce the lipid-lowering effects of dietary changes. Since hypertension and elevated serum lipids are both risk factors for coronary heart disease, the benefit of using the antihypertensives must be weighed against the risk of higher blood lipid levels.

The Food and Drug Administration (FDA) has approved a machine that filters LDL cholesterol from the blood of patients with very high levels that do not respond to diet or medications.

Diseases of the Heart and Blood Vessels

Several common diseases of the heart and blood vessels have nutrition implications in their therapy.

ANGINA PECTORIS

Angina pectoris is caused by the narrowing of the arterial lumen and resulting inadequate blood supply to the heart muscle. It is marked by tight chest pain, often radiating to the shoulder, arm, and hand. Attacks are brought on by physical exertion, exposure to cold wind, excitement, or the digestion of a large meal.

If the individual is overweight, weight loss is helpful, as are small, easily digested meals. *Nitroglycerine* is used to relax smooth muscles in the walls of the blood vessels and relieve angina episodes. *Nitroglycerine* may cause dry mouth, nausea, and vomiting. Beta blockers and calcium channel blockers may be prescribed.

MYOCARDIAL INFARCTION (MI)

Each year approximately 1.25 million people in the United States have a heart attack. About 250,000 die within an hour, many before they receive medical help.

Acute Phase

The care of the patient who has had a heart attack must be tailored to the individual's condition. The patient is continuously monitored through an electrocardiogram. Rest and avoidance of straining during coughing, defecating, or repositioning are of primary importance, and minimizing the work of the heart muscle in supplying oxygen for digestion is central to therapy. *Nigroglycerine, morphine,* and beta blockers as well as oxygen may be given.

LOW-FAT LIQUID DIET

Broth
Fruit juice
Skim milk
Decaffeinated coffee
Weak tea

Diet Often the patient is given nothing by mouth or is permitted sips of cool water supplemented by parenteral dextrose solutions at first. Then a low-fat liquid diet supplying 500–800 kcal and 1,000–1,500 ml fluid is given in small feedings. Some clinicians believe that foods should be at room temperature because extremes of temperature may cause arrhythmias. Because of its stimulating effect, caffeine is sometimes excluded.

Then the patient usually progresses to a soft diet with the following characteristics:

1,000–1,200 kcal to limit circulation required for the digestive and absorptive processes and to begin weight loss if obese.

Five or six small, easily digested meals, especially if the patient is dyspneic or has angina.

Less than 30 percent of calories as total fat; saturated fat 10 percent of calories or less. Cholesterol restricted to 300 mg or less.

Mild sodium restriction (2,000 mg) to prevent excess fluid retention. Greater sodium restriction if there is congestive heart failure.

Avoidance of foods the patient finds to be gas producing.

Fluid restriction if edema is present.

Rehabilitative Phase

Diet is based on weight status and blood lipid levels. Gradual weight loss is indicated if the patient is overweight. A diet low in saturated fat and cholesterol may be useful in reducing the likelihood of a recurrence of the heart attack. (See Table 24–1.) A program of regular exercise is helpful.

CONGESTIVE HEART FAILURE (CHF)

Approximately 400,000 patients are diagnosed with CHF annually. Congestive heart failure leads to nearly a million hopitalizations each year. The Agency for Health Care Policy and Research (AHCPR) has recommended that it be called heart failure since congestion is only one aspect of the condition.

Clinical Findings

Decompensation, or inability of the heart to maintain adequate circulation to the tissues, usually results in congestive heart failure. Reduced pumping ability results in congestion of the pulmonary and systemic circulation. Pulmonary edema and dyspnea occur. With the reduced circulation, excretion of sodium by the kidneys is greatly reduced. Sodium accumulates in the extracellular fluid, retaining water with it. Edema develops first in the extremities and then spreads to other areas of the body such as the chest and abdomen. Diminished blood supply to the gastrointestinal tract slows digestion, resulting in anorexia, distention, and sometimes vomiting.

Therapy

Bed rest is essential, usually with the patient in a semisitting position to permit greater lung expansion. Oxygen may be needed. Vasodilators, diuretics, and angiotensin converting enzyme (ACE) inhibitors are first-line medications.

Diet The aim is to decrease the workload of the heart. The dietary progression is much like that used in myocardial infarction. Sodium may be restricted and fluid restriction may also be necessary. Caffeine and alcohol are often limited because they may cause increased heart rate and arrhythmias. Small, frequent meals eaten slowly decrease the workload of the heart.

SOME CAUSES OF CONGESTIVE HEART FAILURE

Untreated or poorly controlled hypertension
Coronary heart disease
Secondary to myocardial infarction
Heart damage from rheumatic fever
Advanced emphysema
Bacterial or viral infection of the heart

HEART SURGERY

Bypass Surgery

When severe atherosclerosis blocks coronary arteries, coronary artery bypass graft surgery is sometimes performed. This operation has become one of the most frequently used in the United States. New sources of blood supply to the heart are fashioned using pieces of vessel taken from elsewhere in the body.

After the surgery the diet progresses from clear liquids to full liquids to solids as tolerated. For long-term maintenance the patient follows a diet restricted in total fat, saturated fat, and cholesterol to prevent recurrence of the coronary artery disease. The Step 2 diet is usually recommended for secondary prevention. A caloric intake to attain and maintain healthy weight and a planned program of exercise are essential parts of the postsurgery regimen. Diuretics, *digitalis,* and *quinidine* are commonly prescribed. (See Chapter 18.)

Percutaneous transluminal coronary angioplasty (PTCA) involves inserting a deflated balloon into the blocked artery. The balloon is inflated at the blockage site where it then compresses plaque against the artery wall. Laser beams have been used to vaporize plaque in leg arteries.

Heart Transplant

In end-stage heart disease that no longer responds to other therapies, cardiac transplant may be performed. Nutritional considerations relate to postoperative care, long-term care, and management of the side effects of immunosuppresive therapy.

Postoperative The diet is high in calories and protein to allow wound healing and tissue repair. Moderate sodium restriction is frequently prescribed. Small frequent feedings of high-calorie foods, rest between meals, and nutritional supplements help ensure adequate intake.

Immunosuppressants necessary to prevent rejection have numerous nutrition-related side effects. (See Chapter 23.)

Long-Term Care Diabetes mellitus sometimes occurs several years after the transplant as *prednisone* favors glucose intolerance. Triglyceride levels are often elevated in connection with altered carbohydrate metabolism. Increased appetite fosters excessive weight gain. Atherosclerosis, possibly due to immunologic damage to the arteries, is common. A lipid-lowering diet including soluble fiber is prudent for long-term maintenance. Moderate sodium restriction is common. Attempts are frequently made to taper off *prednisone* therapy.

HYPERTENSION

Hypertension, elevated blood pressure, is the most common circulatory problem in humans and is one of the most widespread health problems in the United States today. It is a major risk factor for congestive heart failure, coronary heart disease, renal disease, and stroke. Almost one American in three has hypertension; 70–75 percent of the cases are mild.

Blood pressure is measured by the force required to drive a column of mercury up a tube and is expressed in mm mercury (Hg). Normal blood pressure is less than 130 mm Hg **systolic** (when heart is contracting) and less than 85 mm **diastolic** (when heart is relaxing).

Etiology

Hypertension is a condition with multiple causative factors. Essential hypertension of unknown cause accounts for about 90 percent of cases. Atherosclerosis and hypertension interact in a vicious cycle. Atherosclerosis causes resistance to blood flow, so the heart must pump harder, driving blood pressure up. In turn the elevated blood pressure further injures arterial walls and worsens the atherosclerosis.

Dietary Factors Obesity is often linked to hypertension, since excess weight increases the work of the heart and there is more tissue to be supplied with blood. About 20 to 30 percent of hypertension in the United States can be linked to excess weight. For many people, weight loss will bring about a reduction in blood pressure and in mild cases may be sufficient therapy. However, not all obese people have hypertension, nor are all normal weight or underweight people normotensive.

About 20 percent of people are sensitive to sodium. When sodium is taken in excess, more water is drawn into the circulation, increasing the volume of blood to be pumped. Other people appear to be relatively sodium resistant, and sodium intake has little effect on blood pressure. Currently there is no way to determine which patients will respond to sodium restriction. Salt intake should be limited to no more than 6 g (2,400 mg sodium) daily. Research has linked increased potassium intake to lowered blood pressure. Some studies have shown that calcium appears to have a protective effect. Alcohol intake in excess of 2 oz daily has a hypertensive effect that increases with the amount consumed.

Prevention

The National High Blood Pressure Education Program has developed an initiative for primary prevention of hypertension. This program focuses on weight control, increased physical activity, moderation of sodium intake, and limited intake for those who consume alcoholic beverages.

Therapy

Blood pressure control is one of the most effective means of decreasing mortality in adults.

Lifestyle Modifications The Joint National Committee on Detection, Evaluation, and Treatment of High Blood Pressure (JNC) recommends lifestyle changes including weight management, reduction of dietary fat and cholesterol, moderate sodium restriction, adequate calcium, magnesium and potassium intake, regular exercise tailored to the individual, stress management, cessation of smoking and, if alcohol is consumed, moderate intake, for all patients. A program combining all these factors, phased in gradually, may prove most effective. For Stage 1 patients, this may be adequate to normalize blood pressure.

Pharmacologic Therapy If blood pressure remains above 140/90 mm Hg after three to six months, despite lifestyle changes, medication therapy is initiated. Pharmacologic management involves use of the fewest medications with the lowest doses, while the patient maintains lifestyle changes. In the United States, agents to control blood pressure are among the most commonly prescribed medications. (See Table 18–1.)

Essential Hypertension: also known as primary hypertension or idiopathic hypertension

Lifestyle Modifications: formerly known as nonpharmacologic therapy

JNC CLASSIFICATIONS OF BLOOD PRESSURE

Stage 1: systolic 140 to 159 mm Hg, diastolic 90 to 99 mm Hg

Stage 2: systolic 160 to 179 mm Hg, diastolic 100 to 109 mm Hg

Stage 3: systolic 180 to 209 mm Hg, diastolic 110 to 119 mm Hg

Stage 4: systolic $\geq$ 210 mm Hg; diastolic $\geq$ 120 mm Hg

Unless the systolic pressure is 210 mm Hg or higher and/or the diastolic is 120 mm Hg or more, hypertension is not diagnosed on the basis of a single blood pressure reading.

Medication choice is determined by the patient's condition, age, coexisting diseases, and so on. If one medication is not effective, the dose may be adjusted or another medication added or substituted for the first.

For initial therapy, the JNC recommends **diuretics,** which reduce blood volume, or **beta blockers,** which lessen the heart rate, lower the heart's workload, and reduce release of renin in the kidneys. Only if these are contraindicated or ineffective should other medications be used. These include **angiotensin converting enzyme (ACE) inhibitors,** which prevent the conversion of angiotension I to angiotension II (see Chapter 28); **calcium channel blockers,** which relax the walls of the arteries; **alpha$_1$-receptor blockers,** which relax the smooth muscle in blood vessels; **alpha$_2$-agonists,** which lower peripheral resistance; and an **alpha-beta blocker,** which acts as both an alpha and beta blocker. **Peripheral adrenergic antagonists,** which prevent vasoconstriction, and vasodilators are also sometimes employed.

Side Effects Patients often fail to comply with medication therapies because of gastrointestinal irritation, diarrhea, constipation, exercise intolerance, altered taste sensations, sexual dysfunction, and/or general interference with the quality of life. Possible nutritional side effects of some diuretics, ACE inhibitors, and beta blockers and dietary strategies for avoiding or modifying these symptoms appear in Table 18–1. Diuretics and beta blockers may elevate blood lipids, at least initially, and the effect is increased when they are used together. A diet modified in saturated fat and cholesterol is followed in long-term therapy. Hyperuricemia can be caused by thiazide and loop diuretics and may trigger attacks of gout in susceptible individuals. ACE inhibitors may cause nausea, vomiting, or diarrhea; calcium channel blockers may produce nausea, gastrointestinal upsets, or diarrhea. *Methyldopa,* which blocks the nerve impulse that signals the arteries to constrict, may cause sodium and water retention.

Some medications such as diet pills and decongestants containing *phenylpropanolamine* elevate blood pressure, as do nonsteroidal antiinflammatory drugs and oral contraceptives. Antacids high in sodium and steroidal antiinflammatory agents that favor fluid retention can interfere with therapy for hypertension. Interactions with medications the patient is taking for other conditions may produce additional side effects.

SODIUM-RESTRICTED DIETS

Nomenclature

The usual daily sodium intake in the United States is 3–7 g (3,000-7,000 mg) and can be much higher. A sodium-restricted diet contains a specified amount of sodium that ranges from a mild to an extreme restriction. Terms such as "salt free," "salt poor," or "low salt" are so vague that the patient might well receive much more sodium than indicated or be unnecessarily restricted. For a discussion of the functions of sodium, see Chapter 11.

Sodium-restricted diets should involve the least amount of restriction necessary to produce the desired results. Sodium modifications may be utilized in the therapy for hypertension, congestive heart failure, kidney disease, impaired liver function, or any condition that involves fluid retention. Reduced sodium intake can assist the action of diuretics. Some medications such as steroidal antiinflammatories foster fluid retention, and sodium restriction is necessary. Levels of sodium restriction commonly prescribed are

- 500–700 mg (22–30 mEq*): severe restriction. No salt in cooking or at the table. No canned or processed foods containing salt. Low sodium breads. Some vegetables naturally high in sodium are omitted. Meat limited to 6 oz and milk to 8 oz daily. Use only for short periods of time. Practical only in institutional setting.

- 1,000–1,500 mg (43–65 mEq): moderate restriction. No salt in cooking or at the table. No canned or processed foods containing salt. Four servings of regular bread per day are allowed.

- 2,000–3,000 mg (87–130 mEq): mild restriction. Small measured amount of salt in cooking; no salt at the table; no salty foods. No high-sodium processed foods.

SOURCES OF SODIUM

Naturally Occurring Sodium

All living things, plants as well as animals, require some sodium. Hence, one would expect to find sodium in foods as they naturally occur before they are processed by the manufacturer or cooked in the home. Animal foods are relatively high in sodium, and plant foods, with few exceptions, are low. Meat, fish, and poultry are naturally high in sodium, so their amounts must be controlled on all levels except the mild restriction. Eggs are especially high in sodium, but most of this is in the white and not in the yolk. In planning diets it is important to include adequate amounts of these foods, even though they are high in naturally occurring sodium.

Most vegetables are low in sodium, but several, such as beets, spinach, chard, and kale, contain too much sodium to be permitted on diets restricted to less than 1,000 mg.

Fruits, unsalted cereals, unsalted bread, and unsalted butter and margarine as well as oils and sugar contain small amounts of sodium, or none at all, and may be used without restriction as far as sodium is concerned.

If all foods in the basic diet were processed and prepared, and eaten without adding salt or any sodium compound, the sodium content would be about 1,000 mg. (See Table 24–4.)

Salt

The principal source of sodium in the diet is salt (sodium chloride) used in numerous ways in food processing, in baking and cooking of foods, and at the table.

Salt is about 40 percent sodium. Thus, a teaspoon of salt that weighs 6 g would provide 2.4 g (2,400 mg) sodium. If a recipe calls for 1 teaspoon of salt and serves six people, one serving of that food would provide 400 mg sodium from the addition of the salt.

Sodium-Containing Compounds

Numerous compounds containing sodium are used in home preparation or by the manufacturer of food products. It is essential to form the habit of looking for the words sodium, salt, and soda in the list of ingredients on any label. The sodium content in milligrams per serving must be included on nutrition labels. This can be of great assistance in determining the sodium content of processed foods.

COMMON SODIUM-CONTAINING COMPOUNDS

Baking powder
Baking soda
Brine
Monosodium glutamate (MSG)
Sodium acetate
Sodium alginate
Sodium benzoate
Sodium citrate
Sodium chloride (table salt)
Sodium propionate
Sodium sulfite

---

*1 mEq sodium = 23 mg, thus, 500 ÷ 23 ≈ 22 mEq. See Table 4–8.

TABLE 24-4

1,000-mg Sodium Diet at Three Caloric Levels[a] (All Food Prepared without Added Salt)

Food List	1,200 kcal Servings	1,600 kcal[c] Servings	2,200 kcal Servings	Sodium in 1,600 kcal (mg)
Bread[b]	5	6	9	540
Meat, (3-4 eggs/week)	6	6	6	200
Vegetables	3	3	4	135
Fruit	3	3	3	12
Milk, skim	2	2-3	2-3	250
Fat	2	5	9	-
Sugars and sweets	0	6 teaspoons	12 teaspoons	-
Total				1,137

[a] 2,000-3,000 mg sodium: food may be lightly salted in cooking. Use regular bread and butter. Omit salt at the table. Omit salty foods, such as potato chips, pretzels, pickles, relishes, meat sauces, salty meats, and fish. Unrestricted calories: provide additional calories from fruits, low-sodium breads and cereals, low-sodium fats, sugars and sweets.

[b] Up to 4 slices regular bread may be used. Additional amounts must be sodium-free.

[c] Based on Basic Diet, Patterns A and B, Table 2-4.

Some drinking waters are high in sodium, especially if water softeners are used. Local health officials should be able to provide information about the municipal water supply. Many over-the-counter products such as alkalizers, laxatives, toothpastes, and sedatives contain sodium, and label reading is essential. The client should be cautioned against self-medication with baking soda and antacids.

FOOD SELECTION

In diets restricted to 1,000 mg sodium or less, no salt is used in preparation or processing. Labels must be checked to determine the sodium content of a serving. Numerous foods canned or prepared without added salt are available commercially.

The *Exchange Lists for Meal Planning* (Appendix Table A-2) can be modified for sodium-restricted diets. The exchange lists indicate all foods containing 400 mg or more of sodium per serving, and these should be omitted at most levels of sodium restriction. The level of sodium restriction will dictate if additional foods from the exchange lists must be omitted.

Some other foods that may be too high in sodium are as follows:

TOMATO SAUCE

Ingredients: tomatoes, salt, monosodium glutamate, spices

- **Starch/Other Carbohydrates:** products containing salt, baking powder, or baking soda; many regular breads, muffins, rolls, dry breakfast cereals, bread, cake, and quick bread mixes, crackers, self-rising flour, salty snacks, starchy vegetables canned with salt.

- **Fruits:** dried fruit if treated with sodium sulfite, maraschino cherries, glazed fruit

- **Milk:** at levels less than 1,000 mg sodium, buttermilk, milk shakes, ice cream, ice milk, sherbet

- **Vegetables:** canned vegetables unless canned without salt. Rinsing canned vegetables will lower salt content. At levels less than 1,000 mg sodium, avoid beet greens, beets, carrots, celery, chard, collards, dandelion greens, kale, mustard greens, spinach, white turnips.

- ■ **Meat:** dried salted meat or fish, caviar, kosher meat, some peanut butters
- ■ **Fats:** salted butter and margarine, salted nuts, bacon

Dietary Plans

The selection of foods for three calorie levels of a 1,000-mg sodium diet is shown in Table 24–4. The calculations for sodium in the 1,600-kcal diet are based on the average values assigned to each food list. The 1,800-kcal diets using exchange lists (see Table 26–4) might also be used as a pattern for the 1,800-kcal, 1,000-mg sodium diet. The 1,600-kcal modified fat diet (see Table 24–2) could be used as a basis for diets with 1,000 mg or more sodium.

Two sample menus for the 1,000-mg sodium diet appear in Table 24–5. The 1,000-kcal soft diet illustrates the food choices that might be permitted after the third to fifth day following a myocardial infarction or congestive heart failure, while the 1,600-kcal regular diet might be appropriate for maintenance. When the patient returns home, the sodium level might be increased to 2,000 mg or more with calories adjusted to attain and maintain a healthy body weight.

Preparation of Food

Patients who are accustomed to using a great deal of salt may complain about the flat taste of the food. However, after about three months on a sodium-restricted diet, the taste preference for salt declines. Therefore, if a patient can be encouraged and supported over such a period of time, compliance will be improved. The salt shaker should be removed from the table so the client does not salt food without thinking. Salt substitutes may be useful. Because some are potassium or ammonia compounds, they may not be appropriate for individuals with kidney or liver disease or those using potassium-sparing diuretics.

Many flavoring extracts, spices, and herbs may be used to lend interest to the diet. Usually a dash of spices or a small pinch of herbs is sufficient for most family-size recipes. The flavor should be delicate and subtle rather than strong and overpowering. Meats may be marinated in wine, vinegar, low-sodium French dressing, or sprinkled with lemon juice before cooking.

Lemon juice, onion, garlic, pepper, bay leaves, sage, green pepper, thyme, and rosemary may be used with meat, fish, and poultry. A dash of sugar while cooking vegetables brings out the flavor. Lemon juice, onion, chives, green pepper, and chopped unsalted nuts also enhance vegetables.

Homemade quick breads, biscuits, and muffins may be made by using low-sodium baking powder instead of regular baking powder. For each teaspoon of regular baking powder, it is necessary to use 1½ teaspoons low-sodium baking powder. The salt specified in the recipe should be omitted.

Homemade bread, waffles, and rolls may be made by using yeast and omitting the salt from the recipe. The yeast dough may be rolled out, spread with unsalted butter or margarine, and sprinkled with sugar and cinnamon for delicious cinnamon rolls.

PUTTING IT ALL TOGETHER

Cardiovascular diseases are the leading cause of death in this country. Choosing the proper diet is important to both prevent and postpone such conditions as well as help to manage them. Modifications of dietary fat and sodium intake are major strategies that can be implemented.

TABLE 24–5

Two Sample Menus for the 1,000-mg Sodium Diet

1,000-kcal Soft Diet[a] (No Salt Used in Cooking)	1,600-kcal Regular Diet[c] (No Salt Used in Cooking)
Breakfast	**Breakfast**
Orange sections	Orange sections
Puffed rice	Shredded wheat
Skim milk—½ cup	Milk, skim—1 cup
No sugar	Sugar—2 teaspoons
Toast—1 slice[b]	Toast—1 slice[b]
No butter or margarine	Margarine,[d] unsalted—1 teaspoon
	Marmalade—2 teaspoons
Lunch	**Lunch**
Sliced tender chicken (ground, if necessary)—2 oz	Salad bowl:
Asparagus tips with lemon wedge	Lettuce, endive, escarole, raw cauliflower, green pepper, tomato wedges
Roll, soft—1	Sliced chicken strips—2 oz
No butter or margarine	French dressing, low-sodium—1 tablespoon
Peaches, unsweetened, canned—2 halves	Roll—1
Milk, skim—1 cup	Margarine, unsalted—1 teaspoon
	Milk, skim—1 cup
	Peaches, fresh, sliced
Dinner	**Dinner**
Tender roast beef (ground, if necessary)—3 oz	Roast beef—4 oz with currant jelly—1 tablespoon
Baked potato without skin—1 small	Potato, baked—1 medium with unsalted chive margarine—2 teaspoons
Peas, canned, without salt	Fresh peas with mushrooms
Milk, skim—½ cup	Roll—1
Banana—½	Margarine, unsalted—1 teaspoon
	Orange sherbet—½ cup

[a] Give fruit and milk between meals during early stages of recovery.
[b] Allow up to 4 slices regular bread or rolls. Additional amounts must be sodium-free.
[c] Based on Basic Diet, Pattern A, Table 2–4.
[d] Use margarine or oils high in polyunsaturated fat.

? QUESTIONS CLIENTS ASK

1. The doctor wants me to follow a sodium-restricted diet for hypertension, but I don't want to have to do short-order cooking for myself while making different food for the rest of my family.

Your menus should be as much like those of the rest of the family as possible. Seasonings other than salt that are low in sodium can be used. Since hypertension tends to run in families, it is wise to have your children practice moderation in sodium intake. Also, healthy eating guides suggest that everybody reduce sodium intake.

2. I've heard that soft water prevents heart disease because hard water leaves deposits in blood vessels just as it does in water pipes.

Studies in areas where the water supply is naturally soft or hard show that there is a somewhat lower mortality from cardiovascular disease in areas where the water is hard. This *may* be caused by calcium and other minerals in the hard water, which appear to protect against hypertension. Heart diseases are caused by many factors, and the softness or hardness of one's water supply does not appear to be an important factor.

3. *A business associate of mine refused bypass surgery and is getting chelation therapy instead. What is this and why don't we hear more about it?*

Chelation therapy involves taking an amino acid, EDTA, along with minerals and vitamins. This is supposed to break down plaque in the arteries and clean them out, preventing various types of cardiovascular problems. There is no scientific evidence that chelation therapy works and it may have harmful side effects.

✔ CASE STUDY: An Older Man with Congestive Heart Failure

Michael V., 74, a retired barber, and his wife live simply but comfortably on Social Security and income from savings. Their children are grown, but they, along with other relatives, live nearby. There are always large family gatherings at holidays.

Mrs. V. enjoys cooking and prepares most foods from "scratch." They eat little meat, but are fond of pastas, cheeses, sausage, along with fruits, vegetables, and bread from a local bakery.

Lately, Mr. V. has complained of feeling tired and short of breath when he takes his daily walk. He has been sleeping propped up so he can breathe more easily. He refuses to see a doctor in spite of his wife's urging. One day while on a walk, he has extreme difficulty in breathing and collapses. Neighbors call an ambulance, and he is admitted to the acute coronary care unit of a local hospital.

Symptoms include shortness of breath, ascites, and edema of the ankles, lower legs, and chest cavity. Height is 70 in. (178 cm), weight, 155 lb (70 kg), and blood pressure, 180/100.

Bed rest, oxygen, a diuretic, and a 1,000-mg sodium liquid diet are ordered. After two days he is advanced to a 1,000-mg sodium, 1,000-kcal soft diet. His diet and activity are progressed during his hospital stay. He is discharged with a 2,000 to 3,000-mg sodium diet, digitalis, and *captopril*.

Questions

1. Write a day's menu for the 1,000-mg sodium liquid diet.

2. Write a day's menu for the 1,000-mg sodium, 1,000-kcal, soft diet.

3. Since Mr. V. is not overweight, why was he placed on a 1,000-kcal diet?

4. Mrs. V. is concerned because her husband is very set in his eating habits and she is afraid he will not follow his diet at home. What suggestions might you make?

5. Write a day's menus for the 2,000- to 3,000-mg sodium diet, keeping Mr. V.'s preferences in mind.

6. What are some features of Mr. V.'s lifestyle that may favor compliance with his diet?

7. Mr. and Mrs. V. are counseled to avoid foods high in potassium. Why? What foods should be avoided? Suggest substitutes.

8. What possible effects on nutrition might digitalis have? *Captopril*?

9. Mrs. V. wonders about spices and herbs in the diet since her husband likes highly seasoned foods. What might you tell her?

10. Mrs. V. is famous in her neighborhood for her homemade lasagna. What changes might she make in preparing it so her husband can eat it?

✔ CASE STUDY: A Construction Worker with Hypertension

Walter S., 52, is a construction worker. He is 69 in (175 cm) tall and weighs 159 lb (72 kg). During a recent visit to his physician for a minor work-related injury, his blood pressure was 190/110. This reading was confirmed on a subsequent measurement. There were no other significant findings. His physician prescribed *furosemide* and told Mr. S. that he should reduce his intake of sodium to 2,400 mg daily. He was also counseled to emphasize foods that were good sources of potassium.

Questions

1. Obesity is often associated with hypertension. Is this a problem for Mr. S.? Explain.

2. Mr. S. likes pretzels and beer while watching television. Is this an appropriate snack for him? Explain. If not, suggest snacks that might be appropriate.

3. Make a list of at least 10 foods that Mr. S. should use sparingly. Suggest appropriate substitutions for each.

4. Suggest flavoring aids to prepare the following foods without the addition of salt: broiled steak, roast chicken, green beans, mashed potatoes, lettuce and tomato salad with Italian dressing.

5. Sometimes Mr. S. and his coworkers go to a fast-food restaurant for lunch. Suggest some foods he might choose there.

6. Examine the "Nutrition Facts" on the labels of 10 food items for the sodium content per serving. Which could be recommended for Mr. S.? For any that are not appropriate, suggest substitutes in the same food group.

7. Why did the physician recommend that Mr. S. select foods high in potassium?

8. Make a list of 10 foods that are good sources of potassium.

9. Are there any nutritional side effects of *furosemide?* If so, suggest ways to minimize them.

FOR ADDITIONAL READING

Adams, SO, *et al:* "Consumer Acceptance of Foods Lower in Sodium," *J Am Diet Assoc,* **95:**447–53, 1995

Allred, JB: "Too Much of a Good Thing?" *J Am Diet Assoc,* **95:**417–18, 1995.

Barnett, HJM, *et al:* "Drugs and Surgery in the Prevention of Ischemic Stroke," *New Engl J Med,* **332:**238–48, 1995.

Burrows, ER, *et al:* "Nutritional Applications of a Clinical Low Fat Dietary Intervention to Public Health Change," *J Nutr Educ,* **25:**167–75, 1993.

Cochran, I, *et al:* "Stroke Care; Piecing Together the Long-Term Picture," *Nurs 94,* **24:**34–42, June 1994.

Corti, M-C, *et al:* "HDL Cholesterol Predicts Coronary Heart Disease Mortality in Older Persons," *JAMA,* **274:**539–44, 1995.

Cuddy, RP: "Hypertension: Keeping Dangerous Blood Pressure Down," *Nurs 95,* **25:**34–43, August 1995.

Dracup, K, *et al:* "Rethinking Heart Failure," *Am J Nurs,* **95:**22–28, July 1995.

Gilboy, MB: "Compliance-Enhancing Counseling Strategies for Cholesterol Management," *J Nutr Educ,* **26:**228–32, 1994.

Janowski, MJ: "Managing Heart Failure," *RN,* **59:**34–39, February 1996.

Johannsen, JM: "Update: Guidelines for Treating Hypertension," *Am J Nurs,* **93:**42–53, March 1993.

Kurtzweil, P: "The New Food Label: Help in Preventing Heart Disease," *FDA Consumer,* **28:**19–24, December 1994.

Larkin, M: "Lowering Cholesterol," *FDA Consumer,* **28:**27–31, March 1994.

Neaton, JD, *et al:* "Treatment of Mild Hypertension Study: Final Results," *JAMA,* **270:**713–24, 1993.

Oster, G, and Thompson, D: "Estimated Effects of Reducing Dietary Saturated Fat Intake on the Incidence and Costs of Coronary Heart Disease in the United States," *J Am Diet Assoc,* **96:**127–31, 1996.

Reusser, ME, and McCarron, DA: "Micronutrient Effects on Blood Pressure Regulation," *Nutr Rev,* **52:**367–75, 1994.

Riegel, B, *et al:* "Coronary Precautions: Fact or Fiction?" *Nurs 95,* **25:**52–53, October 1995.

Solomon, J: "Hypertension: New Drug Therapies," *RN,* **57:**26–33, January 1994.

Valantine, HA, and Schroeder, JS: "Recent Advances in Cardiac Transplantation," *New Engl J Med,* **333:**660–61, 1995.

Verschuren, WMM, *et al:* "Serum Total Cholesterol and Long-Term Coronary Heart Mortality in Different Cultures," *JAMA,* **274:**131–36, 1995.

White, E: "Managing Hyperlipidemia: New Approaches to an Old Problem," *Nurs 94,* **24:**66–69, August 1994.

Yacone-Morton, LA: "Cardiovascular Drugs: First-Line Therapy for CHF," *RN,* **58:**38–44, February 1995.

Nutrition in Cancer

Chapter Outline

Increase to at least 75 percent the proportion of primary care providers who routinely counsel patients about tobacco use cessation, diet modification, and cancer screening recommendations.*

 CASE STUDY PREVIEW

Cancer of the large bowel afflicts more than 150,000 Americans annually and is second to lung cancer as a cause of death from cancer. You will assist an independent older woman who has received a colostomy and chemotherapy for colorectal carcinoma.

**Healthy People 2000: National Health Promotion and Disease Prevention Objectives.* Public Health Service, U.S. Department of Health and Human Services, Washington, DC, 1991, Objective 16.10, p. 115.

ore than half a million people die from cancer each year, making it the second leading cause of death in the United States. About one million new cases are diagnosed annually.

CANCER

Cancer cells are derived from normal cells. The development of cancer occurs in two stages. First, there is *initiation* when an irreversible genetic mutation occurs in a cell, turning it into a latent cancer cell. Although the alteration cannot be reversed, the affected cell may be dormant for many years.

Promotion, the stimulation of the "initiated" cell by a promoting agent, is required for a progression to malignancy to occur. A promoter will not act as an initiator. There are some agents known as complete **carcinogens** or cancer-causing agents, however, that both initiate and promote.

Some nutrients can enhance or inhibit promotion. For example, dietary fats are thought to enhance promotion. Vitamin A and carotene inhibit promotion. Since most initiators and promoters are environmental in origin, the majority of cancers, at least theoretically, can be prevented.

Clinical Findings

Changes in metabolism may lead to alterations in taste perceptions such as aversions to bitter taste and heightened sensitivity to salt and glucose. Stress hormones depress insulin production; since insulin increases appetite, anorexia results. Stress hormones also increase production of lactic acid in the blood that can lead to nausea. Stress may also stimulate the emetic center in the brain. Certain foods that are associated with nausea and vomiting may be refused. Protein-energy malnutrition is common.

Cachexia is a syndrome whose pathophysiology is little understood and that often responds poorly to nutritional therapy. Cachexia may be primary, resulting from a host/tumor interaction, or secondary to side effects of therapies resulting in inadequate food intake.

Therapy

Well-nourished patients tolerate cancer therapy better than those in poor nutritional status. Screening to determine those patients at nutritional risk, nutritional assessment, planning nutritional care, and monitoring nutritional status are essential. (See Chapters 3 and 16.)

Little can be said with certainty about nutritional requirements in cancer. A protein intake of 1.5 – 2 g/kg body weight allows for metabolic alterations. Energy needs are also increased, possibly by 20 percent. Vitamin and mineral deficiencies are common, and supplements are usually indicated.

Meal patterns and food choices should be adjusted to promote optimum intake and enjoyment. Meals taken when the patient is least likely to be tired and suffering the side effects of therapy, for example breakfast, are most likely to be consumed. The patient should be encouraged to select foods that are preferred. In a hospital setting, the patient should make menu choices when feeling best.

SYMPTOMS OF CACHEXIA

Alterations in metabolism of carbohydrate, protein, and fat

Negative nitrogen balance in spite of adequate intake

Progressive weight loss

Emaciation

Decreased immune response

Decreased tolerance to therapy

Deterioration of functional ability

GOALS OF NUTRITIONAL THERAPY IN CANCER

Achieve and maintain healthy weight

Prevent or correct nutritional deficiencies

Maximize benefits of other therapies

Enhance quality of life

CANCER THERAPIES AND THEIR NUTRITIONAL SIDE EFFECTS

Radiation Therapy

Although symptoms are specific to the area irradiated, this therapy can have numerous effects on the gastrointestinal system, varying with the site and dosage. Radiation to the mouth and esophagus results in decreased saliva production, dryness of the mouth, and dental decay caused by the lost buffering capacity of saliva. Anorexia, dysgeusia, dysphagia, stomatitis, esophagitis, or esophageal stricture are other possible side effects. Nausea, vomiting, and diarrhea as well as malabsorption, lactose intolerance, and radiation enteritis can follow radiation to the abdominal area.

Chemotherapy

Chemotherapy disrupts the cell cycle, affecting both cancer cells and normal cells. Some patients stop the chemotherapy because of the nausea and vomiting that almost invariably accompany it. Chemotherapeutic agents such as *fluorouracil, methotrexate,* and *procarbazine* have gastrointestinal side effects leading to decreased food intake and weight loss. Dry mouth, dysgeusia, stomatitis, constipation, and diarrhea are common. All antineoplastics stimulate the emetic center in the brain, resulting in nausea and vomiting. Fluid retention and bloating may occur.

Immunotherapy

Biological response modifiers may restore or strengthen the immune response or affect the tumor directly. An example is interferon, which has antiviral, antitumor, and immune response modifying properties. A flulike syndrome with fever, chills, fatigue, and diarrhea is experienced by 90 percent of patients after the first treatment. *Acetaminophen* is used as an antipyretic. Anorexia, dysgeusia, and early satiety are common as therapy continues.

Surgery

The nature of the surgery will dictate the nutritional consequences. Head and neck surgery may interfere with the ability to chew and swallow. Gastric surgery may lead to the dumping syndrome (see Chapter 21) or absence of the intrinsic factor, affecting vitamin B_{12} absorption. Intestinal resection may decrease absorption of nutrients. Complications such as fistulas, strictures, and ulcers further compromise nutritional status.

Combined Therapy

The above therapies are not mutually exclusive, but are frequently used together. Radiation, for example, is employed in about half of patients, often in combination with surgery or chemotherapy. Any of the therapies can cause dehydration because of inadequate intake due to weakness, nausea, vomiting, or diarrhea. Hypokalemia can result from vomiting or diarrhea.

Control of Side Effects

Antiemetics such as *haloperidol* are frequently given. *Metoclopramide* has antiemetic properties and increases gastric motility to speed peristalsis and gastric

TABLE 25-1
Some Nutrition-Related Side Effects of Cancer and Its Therapies

Symptom	Nutritional Strategies[a]
Anorexia (Lack of appetite)	Small frequent feedings. Foods of high nutrient density. Feedings at time when appetite is best. Fluids between meals. Cater to personal tastes. Maintain calorie counts.
Nausea and Vomiting	Antiemetic medicines before treatments and before meals. Dry bland foods at mealtime. Fluids between meals. Small frequent feedings. Simply prepared foods. Avoid greasy, spicy, overly sweet foods. Avoid foods with strong odors. Use air freshener. Avoid favorite foods when nauseated to prevent aversion.
Stomatitis (Sore mouth)	Modify texture to soft or liquid. Avoid very hot or very cold foods. Avoid hard, fibrous, acidic foods. Use a straw. Good oral care.
Xerostomia (Dry mouth)	Emphasize soft, moist foods. Add sauces, gravy, juice. Liquids with meals. Avoid dry, salty, spicy, very fibrous, acidic foods and alcohol. Sugarless gum or hard candy to stimulate saliva. Suck on ice chips. Take plenty of cool fluids. Vaporizer to keep oral tissues moist. Good oral hygiene. Artificial saliva.
Esophageal lesions	Modify texture to soft or liquid. Avoid very hot or very cold foods. Avoid fibrous, acidic, or spicy foods. Plenty of liquids.
Early satiety	Small frequent feedings of calorie-dense foods. Liquids between meals. Cold foods leave stomach more rapidly. Avoid high fat foods, carbonated beverages, gas-producing foods. Rest after meals with head elevated.
Dysphagia (Difficulty with swallowing)	Small frequent feedings. Sit up to eat. Soft moist foods. Add butter, gravy, sauces. Avoid hard, fibrous foods. Extremes in temperature stimulate sensation.
Dysgeusia (Alteration of taste)	Food at room temperature. Adequate fluids. Good oral hygiene. Rinse mouth before eating.
Hypogeusia (Diminished taste sensation)	Strongly flavored foods. Extremes in temperature of food. Vary texture of foods.
Fatigue	Largest meal in morning. Nutrient-dense foods. Foods with minimal chewing. Rest before meals. Home delivered meals.

[a] These strategies are useful to alleviate the symptoms regardless of their cause.

emptying, counteracting the reverse motion of emesis. Distractions such as music or television can help.

Some dietary and behavioral strategies for controlling side effects of cancer and its therapies appear in Table 25-1. (See Figure 25-1.)

Alternative Therapies

Alternative medicine is described as a practice that lacks sufficient documentation about its effectiveness and safety, is not customarily taught in United States medical schools, and usually is not reimbursable by insurance carriers.

About 30 percent of cancer patients try an unorthodox form of therapy such as herbs or megadoses of vitamins. Some dietary regimens eliminate certain foods or aim for balance between yin and yang, such as the macrobiotic diet.

The National Institutes of Health has established an Office of Alternative Medicine to study unorthodox therapies under valid controlled scientific conditions. Currently, clinical trials that are regulated by FDA are being run on several alternative therapies. It is suggested that anyone interested in an alternative treatment try to join a clinical trial.

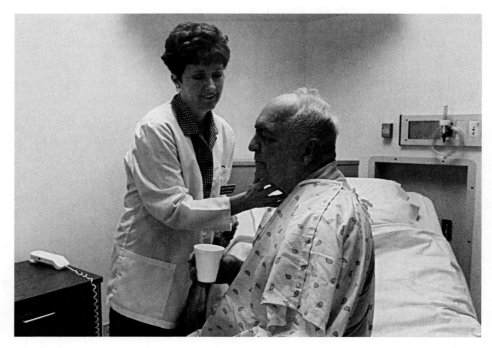

FIGURE 25–1
A speech language pathologist evaluates the swallowing abilities of a patient with
dysphagia to assist in providing foods of appropriate consistency and texture.
(*Source:* Tom Wilcox/Merrill.)

If clinical trials are not available for the particular therapy, the Office of
Alternative Medicine suggests obtaining objective information about the therapy
and the background of the practitioner, and consider costs, since alternative
therapies may not be covered by health insurance.

The alternative treatment should be discussed with the health care provider.
Since some alternative nutritional therapies are inadequate nutritionally, they may
cause debilitation. Any patient embarking on such a therapy should be encouraged to consume an adequate diet. See the Food Guide Pyramid (Figure 1–5)
and/or Basic Diet (Table 2–4).

PUTTING IT ALL TOGETHER

Since certain foods, nutrients, and methods of preparation are involved in the
promotion of cancer, nutrition can play a major role in its prevention. Cancer and
its therapies can have profound nutritional consequences. Dietary modifications
and support can play an important role in the recovery and rehabilitation of
persons with cancer.

? QUESTIONS CLIENTS ASK

1. *When I read ingredient labels on foods, I often*
see the terms BHA or BHT. A friend tells me
they can cause cancer. Is this true?

BHA (butylated hydroxyanisole) and BHT (butylated
hydroxytoluene) are antioxidants that prevent spoilage by oxidation in foods. They are especially used in

items high in fat to prevent rancidity. BHA and BHT have been the subject of much controversy. Scientific studies show that at the levels consumed in the average diet they do not cause cancer and may have a protective effect against it.

2. I hear that the rate of stomach cancer in the United States has decreased a great deal in the last 60 years or so and is the lowest in the world. Does this have anything to do with our diet?

It may be that with widespread refrigeration, we rely less on smoked and salted foods, which have been linked to stomach cancer. Refrigeration and better food handling have reduced the incidence of food-borne illness poisoning. We also have fresh fruits and vegetables available year round, which provide anti-oxidant nutrients. Any or all of these factors may play a role in the decrease in stomach cancer.

3. My friend is trying an alternative therapy for cancer. It sounds appealing, but I am concerned that it may harm her.

Your friend should try to obtain objective information about the treatment, check out the background of the person giving it, learn about costs, and discuss it with her health care provider. She might also participate in a clinical trial for the therapy if one is available.

✔ **CASE STUDY: An Older Woman with Colorectal Cancer**

Rita W., age 71 and widowed, lives alone in a small apartment. Her son, his wife, and their children live nearby and want Mrs. W. to make her home with them, but she values her independence and prefers to live alone.

About six months ago, she began developing stomach cramps and diarrhea, and occasionally passed bloody stools. Her appetite, once very good, decreased, and she began vomiting from time to time. Over a period of six months she lost 15 lb. She first thought it was a touch of indigestion and tried to hide her symptoms from her family. They became alarmed at her lack of appetite and loss of weight and insisted that she see a physician. Tests revealed a tumor of the colon, which proved to be malignant. A colostomy was performed, and Mrs. W. also received intermittent chemotherapy (*fluorouracil* and *levamisole*).

She complains of lack of appetite, nausea, occasional vomiting, stomatitis, difficulty and pain on swallowing, and periodic bouts of diarrhea. All foods have a "strange" taste, and she continues to lose weight. A low-residue diet as tolerated is ordered. (See Chapter 22.)

Questions

1. What dietary modifications are commonly made for a patient with a colostomy?

2. Suggest additional dietary adjustments that could be made to improve Mrs. W.'s nutritional status.

3. Mrs. W. wonders whether the colostomy will ever permit her to eat a regular diet. What might you tell her?

4. She asks what she can do to control the diarrhea. What might you suggest?

5. Mrs. W. looks forward to returning to her own apartment. Since she is weak, she is concerned about food preparation there. What might you suggest?

6. What community resources might be available to assist Mrs. W. when she returns home?

7. She asks if the spicy foods she enjoyed all her life might have caused the cancer. How might you answer?

8. Mrs. W. is concerned that her children and grandchildren may develop cancer and wonders if they should change their diets in any way to prevent this from happening. What might you tell her?

9. What are possible nutritional side effects of *fluorouracil? Levamisole?* Suggest mealtime strategies for minimizing such effects.

FOR ADDITIONAL READING

Ausman, LM: "Fiber and Colon Cancer: Does the Current Evidence Justify a Preventive Policy?" *Nutr Rev,* **51:**57–63, 1993.

Eisenberg, DM, *et al:* "Unconventional Medicine in the United States," *New Engl J Med,* **328:**246–52, 1993.

Greenberg, ER, *et al:* "A Clinical Trial of Antioxidant Vitamins to Prevent Colorectal Adenoma," *New Engl J Med,* **331:**141–47, 1994.

Greenberg, ER, and Sporn, MB: "Antioxidant Vitamins, Cancer, and Cardiovascular Disease," *New Engl J Med,* **334:**1189–90, 1996.

Greifzu, S: "Chemo Quick Guide: Antimetabolites," *RN:* **59:**32–33, March 1996.

Held, JL: "Cancer Care: Correcting Fluid and Electrolyte Imbalances," *Nurs 95,* **25:**71, April 1995.

Hunter, DJ, *et al:* "Cohort Studies of Fat Intake and the Risk of Breast Cancer—A Pooled Analysis," *New Engl J Med,* **334:**356–61, 1996.

Hwang, H, *et al:* "Diet, *Helicobacter pylori* Infection, Food Preservation and Gastric Cancer Risk: Are There New Roles for Preventative Factors?" *Nutr Rev,* **52:**75–83, 1994.

Latkany, L, *et al:* "Development of Adult and Pediatric Oncology Nutrition Screening Tools," *Top Clin Nutr.* **10:**85–89, September 1995.

Mehler, EL: "Colorectal Cancer: Early Detection is Your Priority," *Am J Nurs,* **94:**16A–16D, August 1994.

Moertel, CG: "Chemotherapy for Colorectal Cancer," *New Engl J Med,* **330:**1136–42, 1994.

"Questionable Methods of Cancer Management: 'Nutritional' Therapies," *CA—A Cancer Journal for Clinicians,* **43:**309–19, 1993.

Stehlin, IB: "An FDA Guide to Choosing Medical Treatments," *FDA Consumer,* **29:**10–14, June 1995.

Weber, MS: "Chemotherapy-Induced Nausea and Vomiting," *Am J Nurs,* **95:**34–35, April 1995.

Nutrition in Diabetes Mellitus

Chapter Outline

Types
Diagnosis
Therapy
Acute Complications
Chronic Complications

Putting It All Together
Questions Clients Ask
Case Study: A Teenager with
  Insulin-Dependent Diabetes Mellitus

Reduce diabetes-related deaths to no more than 34 per 100,000 people. *

*Reduce the most severe complications of diabetes as follows: end-stage renal
disease, blindness, lower extremity amputation, perinatal mortality, major
congenital malformations.* *

✔ CASE STUDY PREVIEW

Alex, a teenager, has had diabetes mellitus for five years. You will answer
some of his questions and help him develop a pattern of eating and activity appropriate for his teenage lifestyle, but also planned to prevent long-
term complications.

Healthy People 2000: National Health Promotion and Disease Prevention Objectives. Public Health
Service, U.S. Department of Health and Human Services, Washington, DC, 1991, Objectives 17.9 and
17.10, p. 117.

Diabetes mellitus is a chronic disease of the endocrine system characterized by changes in carbohydrate, protein, and fat metabolism. It has been estimated that more than 12 million people in the United States have diabetes; half of them are unaware of the fact. Diabetes increases the risk of heart disease, stroke, and kidney disease, and shortens the life span.

The beta cells of the pancreas secrete insulin that regulates carbohydrate, fat, and protein metabolism. The alpha cells of the pancreas produce glucagon that raises blood glucose levels. A balance between insulin and glucagon activity maintains blood glucose in a normal range. In diabetes mellitus there is either insufficient production of insulin or resistance to insulin by the body's cells.

TYPES

There are three mutually exclusive types of diabetes mellitus plus other abnormalities in glucose tolerance.

Type I, Insulin-Dependent Diabetes Mellitus (IDDM)

IDDM: formerly called juvenile-onset, brittle, or ketosis-prone diabetes

About 10–20 percent of known cases of diabetes are Type I. Beta cells are destroyed or reduced in number, leading to absence of insulin or inadequate insulin to regulate blood glucose. Although the cause of the cell destruction is poorly understood, it may be due to an autoimmune reaction, viral infection, genetic factors, and/or stress.

Typically, the individuals affected are of normal weight or underweight and may recently have lost weight. While IDDM may occur at any age, many people are young. Nearly all diabetes diagnosed before age 20 is of this type. Onset is abrupt with classic symptoms of **polydipsia** or excessive thirst, **polyphagia** or excessive hunger, **polyuria** or excessive urination, weight loss, fatigue, and possibly ketoacidosis resulting from buildup of ketone bodies. Exogenous insulin is necessary to prevent ketoacidosis and sustain life.

Type II, Non-Insulin-Dependent Diabetes Mellitus (NIDDM)

NIDDM: formerly called maturity-onset or adult-onset diabetes

Type II diabetes accounts for 80–90 percent of known cases. There are probably an equal number of undiagnosed cases. African-Americans, Hispanics, and Native Americans have a higher rate than others. Insulin production may be slightly decreased, normal, or increased, but insulin receptor response is decreased. The cause is unknown, but heredity and overweight are risk factors. Most patients are obese. Although NIDDM can occur at any adult age, it is usually diagnosed after age 40. Onset is gradual. Individuals are not prone to ketoacidosis and do not usually require exogenous insulin except possibly during periods of stress. The classic symptoms of diabetes mellitus are not present.

The diagnosis is often made when an abnormal blood glucose level is found during a routine medical examination. Sometimes the patient complains of itching of the genital and rectal area or has noticed that cuts do not heal readily. In others, the disease may have progressed over a period of time so that complications such as changing vision or numbness and tingling of the extremities are present. Characteristics of IDDM and NIDDM appear in Table 26–1.

Other Categories

A third type of diabetes is associated with or secondary to other conditions such as chronic pancreatitis or medications such as some antihypertensives, glucocor-

TABLE 26–1

Characteristics of Insulin-Dependent (IDDM) and Non-Insulin-Dependent Diabetes Mellitus (NIDDM)

	IDDM	NIDDM
Percentage of cases	10–20	80–90
Age of onset	Often under 20	Usually over 40
Symptoms	Classic	Often asymptomatic
Onset of symptoms	Sudden	Gradual
Insulin dependent	Yes	No
Oral hypoglycemics	No	Sometimes
Weight	Normal or underweight	Usually overweight
Genetic	Rarely	Frequently
Beta cell function	Little or none	Erratic
Insulin receptors	Normal	Decreased or defective

ticoids and immunosuppressives. Impaired glucose tolerance features plasma glucose levels higher than normal but lower than established diagnostic standards for diabetes. **Gestational diabetes** is glucose intolerance first occurring during pregnancy. The blood glucose usually returns to normal after delivery, but many of these women later develop diabetes mellitus.

DIAGNOSIS

Blood Measurements

Because blood glucose cannot be utilized in the cells, **hyperglycemia** or elevated blood glucose results. A fasting blood glucose level of 140 mg/dl or higher on two occasions is diagnostic of diabetes, as is a random level of 200 mg/dl or greater coupled with the classic symptoms. An oral glucose tolerance test may be ordered to confirm the diagnosis. After determining the fasting blood glucose, a known amount of glucose is consumed and blood glucose levels are measured at intervals. Blood levels of 200 mg/dl or above after the glucose is consumed are indicative of diabetes. The individual shows a curve that begins at a higher level and stays higher than the normal curve. Normally the curve comes down sharply, but in diabetes, it returns slowly. (See Figure 26–1.)

The **glycosylated hemoglobin (Hb A_{1c}),** hemoglobin to which glucose is attached, is elevated when blood glucose is increased, and the level falls only when it is replaced by new red blood cells. It is an index of blood glucose levels over a three- to four-month period, the life span of the red blood cell.

Hb A_{1c}: normal, 6.05%; poorly controlled diabetes, 10%

Urine Measurements

When the blood sugar exceeds the renal threshold of approximately 180 mg/dl, sugar is spilled into the urine. Since glycosuria occurs in other conditions, it is not by itself an adequate diagnostic test for diabetes. Ketonuria may also be present.

Ketonuria: also known as acetonuria

THERAPY

Since diabetes mellitus is a chronic condition and permanent lifestyle changes are necessary, it is especially important that treatment plans be consistent with individual preferences, resources, and values.

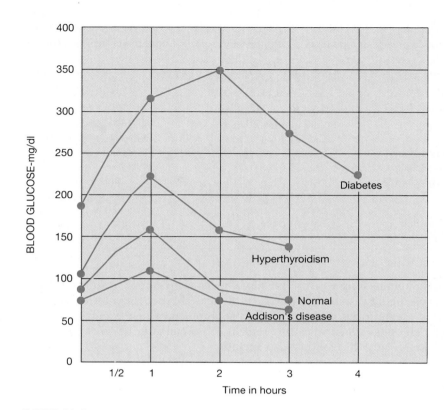

FIGURE 26–1
The glucose tolerance curve shows the varying responses to a standard dose of glucose given after fasting. Note the differences in the fasting blood sugar as well as the time at which maximum response occurred.

GOALS OF THERAPY

Restore and maintain near
 normal blood glucose and
 optimal lipid levels
Improve overall health
Blood pressure management
Healthy weight maintenance
Meal plan appropriate to
 individual's lifestyle
Consistent meal timing
Normal growth rate for
 children and adolescents
Adequate nutrition for
 pregnant and lactating
 women

ENERGY NEEDS BASED ON DESIRABLE WEIGHTS

	kcal/kg
Weight loss needed	20
Bed patient	25
Light activity	30
Medium activity	35
Heavy activity	40

Nutritional management is the cornerstone of therapy. Insulin is necessary in Type I diabetes mellitus; oral hypoglycemic agents are sometimes used in Type II. An individualized program of physical activity and good personal hygiene is essential.

Dietary Recommendations

The American Diabetes Association has published nutrition recommendations for individuals with diabetes mellitus. These guidelines, which are similar to those of the American Heart Association, American Cancer Society, and the 1995 U.S. Dietary Guidelines and Nutrition Recommendations for Canadians include the following recommendations.

Energy Energy needs are the same as for other individuals, and a primary objective is to attain and to maintain a healthy weight. Weight loss by obese individuals often leads to improved or normal glucose tolerance.

Protein Intake should be similar to that of the general population, 10–20 percent of energy intake, derived from both animal and vegetable sources. Growth periods, pregnancy, lactation and catabolic states may require increased intake. At signs of nephropathy (see Chapter 28), intake should be similar to the RDA, 0.8 g/kg body weight for adults.

Total Fat Lipid intake should feature less than 10 percent of calories from saturated fats and up to 10 percent from polyunsaturated, along with less than 300 mg dietary cholesterol. For individuals of reasonable weight with normal serum lipid levels and for children and adolescents, total fat can contribute up to 30 percent of energy intake. For individuals who are obese or have elevated LDL or triglyceride levels, total fat and its subsets may be adjusted. (See Chapter 24.)

Carbohydrates and Sweeteners Carbohydrate intake will vary with the individual's eating habits and glucose and lipid goals. Emphasis is on total carbohydrate rather than specific sources since research has shown that some complex carbohydrates have a glycemic response (See Chapter 5) similar to sucrose. Sucrose, fructose, and other nutritive sweeteners may be substituted for other carbohydrates in moderate amounts, not merely added. Alternative sweeteners approved by the FDA are also safe to use.

Fiber Recommendations are the same as for the general public— 20 – 35 g of dietary fiber from a wide variety of food sources. Water-soluble fiber helps in regulating blood glucose. (See Table 5 – 2.)

Vitamins and Minerals When the diet is adequate, there is no need for vitamin and mineral supplements.

Alcohol For individuals with well-controlled diabetes using insulin, two or fewer alcoholic beverages can be taken with and in addition to the usual meal plan. Since alcohol may increase the risk of hypoglycemia in people taking insulin or sulfonylureas, it should be consumed with meals. Pregnant women should avoid alcohol. When alcohol is calculated as part of total calories, one alcoholic beverage should substitute for two fat exchanges. Alcohol taken with the oral hypoglycemic agent *chlorpropamide* sometimes causes an antabuse-like effect. (See Chapter 18.)

Dietary Planning

Meal plans should be realistic, flexible, and attractive. Economic status, time and place of meals, food preparation facilities, and cultural and religious factors must be considered when planning daily meals. Each client must have an individualized educational program that should involve family members and significant others. Meals should be as much like those of other family members as possible. (See Figure 26 – 2.)

Meal Distribution When insulin is used, calories—especially carbohydrate—should be distributed to coincide with the action timing of the insulin. (See Table 26 – 2.) Since there are many types of insulin, it is usually possible to devise an insulin regimen consistent with the individual's preferred eating patterns. Large carbohydrate loads at any one meal should be avoided. Timing and consistency of food intake are especially critical in IDDM because insulin dosages are arranged to coincide with anticipated food intake. (See Table 26 – 3.) Some patients follow intensive insulin therapy in which they monitor their blood glucose several times a day. Insulin is given three or more times daily by injection or by pump with dosages adjusted to the blood glucose. For some clients with NIDDM, a reduced calorie, reduced fat diet based on customary eating habits may be most realistic. The Food Guide Pyramid, Figure 1 – 5, and/or Basic Diet, Table 2 – 4, are good guides.

FIGURE 26–2
Which member of this family has diabetes mellitus? Since the diet for individuals with diabetes is similar to that recommended for other family members, there is no way of knowing. (*Source:* Anthony Magnacca/Merrill.)

TABLE 26–2
Action Timing of Insulins

Type	Onset (Hours)	Peak (Hours)	Duration (Hours)
Regular (Rapid-acting)	¼–1	2–4	5–8
Semilente (Rapid-acting)	1–2	4–8	12–16
Lente (Intermediate-acting)	1–4	6–16	18–24
NPH (Intermediate-acting)	1–4	6–16	18–24
Protamine zinc (PZI) (Long-acting)	6–8	14–24	24–36+
Ultralente (Long-acting)	4–8	14–24	24–36+

TABLE 26–3
Distribution of Calories and Carbohydrate

Type of Insulin	Breakfast	Noon	Afternoon	Evening	Bedtime
None: *Pattern 1* or	⅓	⅓		⅓	Optional
Pattern 2	⅕	⅖		⅖	Optional
Short-acting (before breakfast and dinner)	⅖	⅕		⅖	Optional
Intermediate-acting NPH	⅐	2/7	⅐	2/7	⅐
Long-acting	⅕	⅖		⅖	20–40 g carbohydrate
Long-acting with regular insulin at breakfast	⅓	⅓		⅓	20–40 g carbohydrate

A dietary assessment (see Chapter 3) will reveal the patient's dietary pattern. All strengths of the current eating habits should be incorporated into the plan. Planning should be done with those concerned and if the patient is not the person primarily responsible for meal planning and preparation, the individual who does this must be involved.

Include basic foods to ensure adequate levels of minerals and vitamins; two cups milk (three or more for children and pregnant or lactating women); three servings vegetables; two servings fruit, including a good source of vitamin C; five meat exchanges; whole-grain or enriched bread and cereal. Exchanges and a sample menu for an 1,800 kcal diet appear in Table 26-4.

Food Preparation and Service The exchange lists (see Appendix Table A-2) are the most widely used system for planning diabetic diets. All foods are measured according to the amounts in the exchange lists. (see Table 26-4).

TABLE 26–4
1,800-kcal Diet Using Exchange Lists[a] with Sample Menu

| | | DIET PATTERN | | |
| | | C = 235 g; P = 80 g; F = 60 g | | |
Exchange List	Exchanges	C (g)	P (g)	F (g)
Milk, skim	2	24	16	—
Vegetables	3	15	6	—
Fruit	5	75	—	—
Starch/bread	8	120	24	—
Meat, medium-fat	5	—	35	25
Fat	7	—	—	35
Total		234	81	60
% of Calories		53	18	29

SAMPLE MENU

Breakfast

Orange juice—½ cup
Oatmeal—½ cup with raisins—4 tablespoons
Whole-wheat toast—1 slice
Margarine—1 teaspoon
Skim milk—1 cup
Coffee

Lunch

Hamburger
   Lean ground meat—2 oz cooked
   Hamburger bun—1
   Onion
Tossed salad with peppers and tomatoes = 1 cup
French dressing—2 tablespoons
Strawberries—1¼ cups
Iced tea, alternative sweetener

Dinner

Broiled chicken—3 oz
Pasta with oil and garlic—1 cup
   (Use 3 teaspoons oil for chicken and pasta)
Rye roll—1 small
Margarine—1 teaspoon
Asparagus spears—½ cup
Carrot sticks—½ cup
Banana (1 small) in sugar-free gelatin
Tea with lemon

Snack

Lowfat yogurt—¾ cup
Graham crackers—3

[a] See Appendix Table A–2.

When purchasing meat, for 3 oz cooked meat allow: 4 oz raw, lean meat, fish, or poultry if there is no waste; 5 oz meat, fish, or poultry if there is a small amount of bone or fat; and 6 oz raw meat, fish, or poultry if there is much waste. Meat, fish, and poultry may be broiled, baked, roasted, or stewed. If it is fried, some of the fat allowance must be used.

Foods are prepared using only those ingredients and amounts indicated in the meal plan. Many recipes are available from cookbooks for diabetics and can be adapted to the individual's diet plan. Some fast-food restaurants and food manufacturers have translated their products into diabetic exchange equivalents.

Fruits canned without sugar are available in most food markets and may be used according to the exchange lists. It is important to read labels carefully. (See Figure 26–3.)

Snacks are permitted if they are calculated in the diet plan. They are necessary with long-acting insulins. Some foods such as coffee, tea, fat-free broth, sugar-free gelatin, and sugar-free diet soft drinks are so low in calories they need not be calculated. (See "Free Foods" listed in Appendix Table A–2.)

In the hospital setting, each patient's tray is a teaching aid. The patient should be instructed to become visually accustomed to portion sizes, learning to

FIGURE 26–3
Reading the nutrition information on food labels when shopping for and planning meals is essential for people with diabetes mellitus as well as all individuals whether on regular or modified diets. (*Source:* Courtesy of FDA Consumer.)

FIGURE 26–4
Patients with diabetes mellitus require individualized counseling regarding (1) techniques
of insulin administration or use of oral compounds, (2) diet, (3) general hygiene,
(4) exercise, and (5) complications of the disease. (*Source:* Metropolitan Medical
Center, Minneapolis, and Jeffrey Grosscup, photo-journalist.)

relate the specific foods on the tray to the exchange lists. The patient's tray
should be checked after each meal to determine food intake and acceptance.

Insulin

Insulin must be taken by injection, because, since it is a hormone, it would be
digested like any other protein if taken orally. (See Figure 26–4.) The dosage
depends on individual needs and may be reduced by exercise and increased by
stressors such as infections. Most individuals take a single dose in the morning or
two injections a day. Insulins with various onsets, peaks, and durations of activity
are available. (See Table 26–2.)

Two methods of insulin delivery attempt more closely to follow normal
insulin secretion. For some selected clients, an intensive insulin therapy is used.
They take an intermediate or long-acting insulin in the morning to approximate
basal insulin. Then, using commercially available kits, they test their blood
glucose level, usually before meals and at bedtime, by obtaining a drop of
capillary blood through a finger prick. (see Figure 26–5.) They give themselves

FIGURE 26–5
Before eating lunch this man with IDDM will determine his blood sugar and give himself insulin if indicated by the results. (*Source:* Courtesy of FDA Consumer.)

ORAL HYPOGLYCEMIC AGENTS

Sulfonylureas
First generation:
Acetohexamide
Chlorpropamide
Tolazamide
Tolbutamide
Second generation:
Glipizide
Glyburide
Biguanide:
  Metformin
Oligosaccharide:
  Acarbose

regular insulin, the amount determined by the blood test. Some individuals use continuous subcutaneous insulin infusion delivered by a programmable pump. The pump infuses a steady amount of insulin and also provides a bolus dose at mealtimes.

Oral Hypoglycemic Agents

Oral agents are sometimes used in NIDDM when diet and exercise alone are not effective. Since they require the ability of the pancreas to secrete insulin, oral agents are not appropriate in IDDM.

Sulfonylurea compounds initially stimulate beta cells to increase insulin production. Later insulin production returns to pretreatment levels, but the control of glucose levels remains. This effect is not clearly understood; it is thought that oral agents may increase the number of insulin receptor sites. Six oral agents currently are in use; four of the so-called first generation and two of the more potent second generation, which also have fewer side effects.

Metformin, a biguanide, has been approved by the FDA. It suppresses glucose production in the liver and increases tissue sensitivity to insulin, but does not stimulate insulin secretion. *Acarbose,* an oligosaccharide, acts to slow carbohydrate digestion.

Exercise

Planned daily activity tailored to the individual's capacity and needs is an integral part of any treatment plan. Advantages include weight maintenance, blood lipid control, cardiovascular conditioning, and stress management. Before and during exercise, individuals with IDDM may have to decrease their insulin dosage or take carbohydrate-containing snacks to prevent exercise-induced hypoglycemia.

Behavior Modification

Many of the behavior modification techniques suggested for weight control (see Chapter 9) are useful for persons with diabetes mellitus. Eating in restaurants and refusing offers of inappropriate foods from others present special challenges to compliance. Responses such as requesting alternative menu items in restaurants and learning refusal skills that are socially acceptable may be practiced in anticipation of such problems.

Illness

Individuals with diabetes should report illness to their health care provider. Insulin or an oral hypoglycemia agent should be continued if they are used. Food should be taken as usual, adjusting texture to soft or liquid if necessary. Adequate fluids should be consumed. A formula low in carbohydrate and high in fiber for patients with abnormal glucose tolerance can be used for tube feeding or as supplementation.*

Experimental Treatments

Whole-pancreas and beta-cell transplants have been performed, usually in conjunction with kidney transplants where immunosuppressants have already been given. Researchers who believe that IDDM is an autoimmune condition suggest immunosuppression for prevention or therapy. Currently, identifying at-risk individuals is difficult. Both approaches carry the drawbacks of long-term immunosuppression.

Monitoring Control

In the office or case setting, the clinican may review blood glucose and/or glycosylated hemoglobin, thereby giving both a day-to-day and long-term index of control. Many individuals perform self-monitoring of blood glucose. A drop of blood obtained by finger prick is placed on a reagent strip. Either by comparing the strip visually to a standard or by using a portable meter, the person can determine the blood glucose level. Some meters are small enough to fit in a purse or pocket and can therefore be taken and used anywhere.

Urine glucose can also be monitored using tablets or paper strips. Substances such as megadoses of vitamin C, salicylates, and *levodopa* can produce false-positive or false-negative results. This test cannot reliably detect hypoglycemia or

---

*Glucerna®, Ross Laboratories, Columbus, OH.

CONTRAINDICATIONS FOR ORAL HYPOGLYCEMICS

Allergy to sulfa drugs (sulfonylureas)
Impaired liver function (metabolized in liver)
Impaired renal function (enter kidney in active form)
Pregnancy (may harm fetus)
Lactation (may cause hypoglycemia or jaundice in infant)
Surgery
Chronic intestinal disease (*acarbose*)

the degree of hyperglycemia and its main value is to confirm elevated blood glucose and monitor ketonuria.

ACUTE COMPLICATIONS

Hypoglycemia

Hypoglycemia occurs when blood glucose falls below about 50 mg/dl. It may be caused by an overdose of insulin or oral hypoglycemic agent, by a decreased supply of glucose because of delay or omission of eating, by vomiting, diarrhea, or an increase in activity without food or insulin adjustment. Since the brain requires glucose for energy, prolonged glucose lack can impair brain function. Extended hypoglycemia can cause permanent damage to the central nervous system.

If the patient is conscious, taking fruit juice, sugar, hard candy, syrup, or sugar-containing carbonated beverages brings rapid relief of symptoms. Glucagon may be used. For unconscious persons, intravenous glucose is indicated. (See Table 26–5.)

The client should carry sugar, glucose tablets, or hard candy and wear medical alert jewelry identifying the presence of diabetes. Medications, both prescription and over-the-counter, can have a hypoglycemic or hyperglycemic effect.

Acidosis and Coma

Diabetic ketoacidosis is caused by severe insulin deficiency, usually coupled with stress, fever, infection, dehydration, acute myocardial infarction, and/or excess of catabolic hormones. Individuals with IDDM diabetes are at greatest risk. Prevention involves avoiding or correcting the associated factors. If early symptoms of hyperglycemia (see Table 26–5) are recognized and treated promptly, progression to coma can be prevented. Small repeated doses of insulin with small carbohydrate feedings are given.

Diabetic coma is a medical emergency and should be treated in a hospital. Insulin, saline, and potassium therapy are indicated.

In older patients with NIDDM diabetes, nonketotic, hyperglycemic, hyperosmolar coma sometimes occurs. Extreme hyperglycemia in excess of 600 mg/dl, severe dehydration, and plasma hyperosmolality create a life-threatening situation. Infections, chronic diseases, extensive burns, dialysis, total parenteral nutrition, surgery, and some medications can be involved. The cause must be determined and corrected, along with insulin, fluid, and potassium therapy. Preventing dehydration is especially important.

CHRONIC COMPLICATIONS

Macrovascular Disease

Atherosclerosis of major arteries occurs earlier than in nondiabetic persons, with greatest involvement of the carotid, cerebral, and coronary arteries. The risk of cardiovascular disease is greatly increased. Individuals with diabetes mellitus are twice as likely to die from coronary artery disease.

Microvascular Disease

Thickening of capillaries with lesions in the retina and glomerulus leads to retinopathy and nephropathy. Most individuals with long-standing diabetes develop retinopathy; it is the leading cause of new cases of blindness.

TABLE 26-5
Symptoms of Hypoglycemia and Hyperglycemia

	Hypoglycemia	Hyperglycemia
Onset	sudden	gradual
Skin	pale, moist, cold	flushed, dry, warm
Behavior	excited, irritable	drowsy
Breathing	normal to rapid	labored
Breath odor	normal	fruity, acetone
Basic needs	hungry	thirsty

Renal failure is the leading cause of death among patients with IDDM. Renal disease is complicated by hypertension, although some antihypertensive medications also create renal complications. No specific therapy exists, although dialysis or kidney transplants may be feasible. (See Chapter 28.)

Neuropathy

The most common chronic complication, affecting about two-thirds of patients, is **neuropathy,** functional or pathological disturbances in the peripheral nervous system. Neuropathy produces sensory loss and patients complain of burning and numbness in the feet. Neuropathy plus an inadequate blood supply resulting from peripheral vascular disease leaves the foot susceptible to injury, ulceration, infection, gangrene, and ultimately amputation. Foot examination and hygiene are extremely important.

About 20–30 percent of individuals with IDDM develop **gastroparesis,** delayed gastric emptying with abdominal distress, early satiety, heartburn, nausea, and vomiting. The cause is unknown, although it has been linked to neuropathy of the vagus nerve with loss of peristalsis. Decreased gastric acid production may be related to the neuropathy and possibly explains the fact that duodenal ulcers are less common in persons with diabetes. *Metoclopramide* has been used to increase gastric contractions and relax the pyloric sphincter.

The Diabetes Control and Complications Trial (DCCT)

DCCT was a large-scale, eight-year clinical trial to compare intensive and conventional therapy for IDDM to determine if blood sugar level was related to the development of chronic complications. It has changed the course of therapy for IDDM.

Conventional therapy included one or two daily injections of insulin, daily self-monitoring of urine or blood glucose, and education about diet and exercise. Intensive therapy included self-monitoring of blood glucose at least four times a day and insulin administered three or more times daily by injection or pump with dosage adjusted to blood glucose, dietary intake, and exercise. The goal was to maintain near-normal blood glucose and Hb A_{1c} levels. Retinopathy, nephropathy, and neuropathy were significantly lower in patients on intensive therapy. Intensive therapy group members did have more incidents of hypoglycemia and did gain more weight than those in the conventional therapy group.

It would appear that with intensive therapy, maintaining individuals with IDDM in good control will reduce long-term complications. It is unclear if this approach is applicable to individuals with NIDDM.

Diet alone or in conjunction with insulin or medications is central to the management of diabetes. Recently, dietary recommendations have been made more flexible. Since diabetes mellitus is a chronic condition, lifelong diet and lifestyle changes are necessary and must be geared to the preferences of the individual.

? QUESTIONS CLIENTS ASK

1. I have heard that eating sugar causes diabetes. If I eliminate sweets from my diet, will this reduce my chances of developing diabetes?

There is no evidence that dietary sugar causes diabetes. This idea probably got its start because people with diabetes can have a high level of sugar in the blood. Sugar does, of course, add to total calories, and overweight is a risk factor for Type II (NIDDM) diabetes. Sugar is also of low nutrient density and a factor in tooth decay. Diets for individuals with diabetes mellitus have been liberalized recently, and sugar in moderate amounts, no more than 10 percent of total carbohydrate calories, is acceptable as part of the total carbohydrate intake.

2. I have diabetes and I think it is unfair that I can't eat cake at a birthday party or pie at a family gathering.

Under the present guidelines both cake and pie can be consumed by individuals with diabetes. They must be calculated into your menu patterns, not taken in addition. Consult your registered dietitian.

3. Why must insulin be injected? I don't like needles and it would be a lot easier to take it by mouth.

Insulin is protein in nature and if taken orally, it would be digested like other proteins. Then it would not be able to perform its function of controlling blood glucose.

✔ CASE STUDY: A Teenager with Insulin-Dependent Diabetes Mellitus

Five years ago, when he was nine years old, Alex L. began losing weight, although his appetite increased and he ate more than previously. He was diagnosed as having IDDM diabetes mellitus.

For several years the diabetes was well controlled by diet and insulin, 25 units NPH plus 5 units regular before breakfast; then monitoring blood glucose during his waking hours and taking insulin as needed. Recently, he has become rebellious and refuses to adhere to his diet. He sometimes skips breakfast, spends his school lunch money for sweets and soft drinks, neglects to monitor his blood sugar, and sometimes does not take insulin. While attending his eighth grade graduation party he collapsed and was admitted to the hospital emergency room in diabetic coma.

Questions

1. What symptoms did Alex possibly exhibit while in diabetic coma?

2. What treatment would Alex receive for diabetic coma?

3. Before Alex was diagnosed as having diabetes, what symptoms other than weight loss might he have had?

4. What tests on Alex would establish the diagnosis of diabetes?

5. Although Alex's diet is essentially a normal one, how might it differ from those of his friends?

6. A 2,700-kcal diet is ordered for Alex. Does this seem appropriate? Why? (See Chapter 13.)

7. Twenty percent of his calories are to come from protein and 55 percent from carbohydrate. Calculate the grams of protein, carbohydrate, and fat.

8. Why would a bedtime snack be included in Alex's menu plan?

9. Alex's bedtime snack includes the following exchanges: 1 fruit, 1 starch/bread, and 1 non-

fat milk. Plan four snacks using these exchanges. (See Appendix Table A-2.)

10. Alex asks whether he could take pills instead of insulin. How would you answer?

11. Alex likes pizza. How is this food divided into exchanges? What exchanges would be included in this food? (See Appendix Table A-2.)

12. A friend asks Alex what would happen if he took too much insulin. What symptoms might Alex describe? What might he do to prevent insulin shock?

13. Alex is interested in sports and asks if he should change his insulin and/or diet on days when he is very active. How might you answer?

14. Alex's mother wonders if she should have his younger sister and brother tested for diabetes. They have no symptoms. How might you answer?

FOR ADDITIONAL READING

Atkinson, MA, and MacLaren, NK: "The Pathogenesis of Insulin-Dependent Diabetes Mellitus," *New Engl J Med,* **331**:1428–36, 1994.

Cirone, N: "Diabetes in the Elderly: Unmasking a Hidden Disorder," *Nurs 96:***26**:34–39, March 1996.

Cirone, N, and Schwartz, N: "Diabetes in the Elderly: Finding the Balance for Drug Therapy," *Nurs 96:***26**:40–45, March 1996.

Clark, CM, and Lee, DA: "Prevention and Treatment of the Complications of Diabetes Mellitus," *New Engl J Med,* **332**:1210–17, 1995.

Daly, A: "Diabesity: The Deadly Pentad Disease," *The Diabetes Educator,* **20**:156–62, 1994.

Diabetes Control and Complications Trial Research Group: "The Effect of Intensive Treatment of Diabetes on the Development and Progression of Long-Term Complications in Insulin-Dependent Diabetes Mellitus," *New Engl J Med,* **329**:977–86, 1993.

Fagen, C, *et al:* "Nutrition Management in Women with Gestational Diabetes Mellitus: A Review of ADA's Diabetes Care and Education Dietetic Practice Group," *J Am Diet Assoc,* **95**:460–67, 1995.

Hoyson, PM: "Diabetes 2000: Oral Medications," *RN,* **58**:34–40, May 1995.

Kestel, F: "Using Blood Glucose Meters: What You and Your Patient Need to Know," *Nurs 93,* **23**:34–42, March 1993.

Kurtzweil, P: "The New Food Label: Coping with Diabetes," *FDA Consumer,* **28**:20–25, November 1994.

Norton, RA: "Diabetes 2000: The Right Mix of Diet and Exercise," *RN,* **58**:20–25, April 1995.

"Nutrition Recommendations and Principles for People with Diabetes Mellitus," *Diabetes Care,* **18**(Suppl 1):16–19, 1995.

"Office Guide to Diagnosis and Classification of Diabetes Mellitus and Other Categories of Glucose Intolerance," *Diabetes Care,* **18**(Suppl 1):4, 1995.

O'Hanlon-Nichols, T: "Hyperglycemic Hyperosmolar Nonketotic Syndrome: How to Recognize and Manage This Diabetic Emergency," *Am J Nurs,* **96**:38–39, March 1996.

Peragallo-Dittko, V: "Diabetes 2000: Acute Complications," *RN,* **58**:36–42, August 1995.

Reising, DL: "Acute Hyperglycemia: Putting a Lid on the Crisis," *Nurs 95,* **25**:33–41, February 1995.

Reising, DL: "Acute Hypoglycemia: Keeping the Bottom from Falling Out," *Nurs 95,* **25**:41–49, February 1995.

Robertson, C: "Diabetes 2000: Chronic Complications," *RN:***58**:34–41, September 1995.

Schlundt, DG, *et al:* "Situational Obstacles to Dietary Adherence for Adults with Diabetes," *J Am Diet Assoc,* **94**:874–79, 1994.

Tinker, LF, *et al:* "Commentary and Translation: 1994 Nutrition Recommendations for Diabetes," *J Am Diet Assoc,* **94**:507–11, 1994.

Nutrition in Endocrine and Metabolic Disorders

Increase to at least 95 percent the proportion of newborns screened by State-sponsored programs for genetic disorders and other disabling conditions and to 90 percent the proportion of newborns testing positive for disease who receive appropriate treatment. *

 CASE STUDY PREVIEW

Sandra's osteoarthritis has affected her lifestyle, leaving her almost immobile. The pain is aggravated by overweight. You will help her plan to lose weight. Her energy needs are already low and exercise is difficult.

Healthy People 2000: National Health Promotion and Disease Prevention Objectives. Public Health Service, U.S. Department of Health and Human Services, Washington, DC, 1991, Objective 14.15, p. 111.

The Human Genome Project, an effort to map all genetic material in the human body, has located the site of the genetic defect of some hereditary diseases with nutritional implications, including cystic fibrosis and some forms of osteoarthritis. Genes associated with obesity, hypertension, insulin sensitivity, some cancers, and vitamin D receptors have been identified. As this vast project proceeds, insights into the cause, treatment, and cure of genetic diseases, possibly through genetic engineering, will result.

Endocrine Disorders

HYPOTHYROIDISM

Clinical Findings

Decreased output by the thyroid gland results in a lowering of the metabolic rate. Mild hypothyroidism is not uncommon, affecting 6 to 7 million Americans, and is 5 to 8 times more common in women than in men. Because of decreased metabolism, overweight is common. All states require screening of newborns for hypothyroidism so therapy can begin promptly. The incidence is approximately one per 4,000 live births.

Myxedema, severe hypothyroidism in adults, features greatly reduced metabolic rate, weight gain, muscle flabbiness, intolerance to cold, lethargy, constipation, and elevated blood lipids. Cretinism, severe hypothyroidism originating in fetal life, is discussed in Chapter 11.

Therapy

The objective is to return the patient to normal thyroid status. Since most patients are overweight, a reduced-calorie diet is indicated. Increased fiber will help prevent constipation. The missing thyroid hormone is replaced with a synthetic hormone. *Levothyroxine* is among the 20 most frequently prescribed medications in the United States. (See Table 18–1).

HYPERTHYROIDISM

Clinical Findings

The most common cause is the production of thyroid-stimulating antibodies. Excessive secretion by the thyroid gland increases the metabolic rate by as much as 50 percent. Some of the symptoms are weight loss, increased appetite, hyperactivity, nervousness, rapid heartbeat, double vision, prominent eyes, and enlarged thyroid gland. The increased metabolism leads to rapid loss of liver glycogen and some tissue wasting, and in severe cases to signs of cardiac failure. A glucose tolerance test shows a peak concentration much higher than a normal individual's, followed by a rapid decline. (See Figure 26–1.) Calcium and phosphorus excretion are often increased and osteoporosis may result. It is believed to be an autoimmune disorder that leads the body to produce excessive thyroid hormone.

Hyperthyroidism: also known as exophthalmic goiter, thyrotoxicosis, Graves' disease, Basedow's disease

Therapy

To reverse weight loss and tissue wastage, a diet supplying 4,000–5,000 kcal and 100–125 g protein (see Chapter 20) is needed until nutritional rehabilitation is completed. Between-meal snacks will help satisfy hunger. Mineral and vitamin supplements are often prescribed. Caffeine-containing foods are usually eliminated because of their stimulating effect.

Most patients are treated with antithyroid medications or radioactive iodine to reduce the metabolic rate to normal. *Propranolol, atenolol,* or *nadolol* and tranquilizers control cardiac effects and hyperactivity. A subtotal thyroidectomy is sometimes performed.

ADRENOCORTICAL INSUFFICIENCY

Clinical Findings

Andrenocortical Insufficiency: also known as Addison's disease

This is a rare disease featuring impaired production of the hormones of the adrenal cortex. There is reduced function of the adrenal glands or failure of the pituitary to produce adrenocorticotropic hormone (ACTH), which stimulates the adrenal glands. Although the causes are multiple, the primary one is autoimmune destruction of the adrenal glands.

Hydrocortisone: pharmacological name for cortisol

Lack of aldosterone, a mineral corticoid which regulates electrolyte balance, leads to excessive loss of sodium and water with potassium retention, resulting in decreased blood volume, hypotension, and dehydration.

Insufficient cortisol, a glucocorticoid concerned with carbohydrate, protein, and fat metabolism, causes rapid depletion of liver glycogen and hypoglycemia a few hours after meals. If no food has been eaten for 10–12 hours, hypoglycemia is severe. A glucose tolerance test shows a lower peak and a more rapid return to pretest levels than normally. (See Figure 26–1.)

SYMPTOMS OF ADRENOCORTICAL INSUFFICIENCY

Salt craving
Thirst
Weakness
Anorexia
Hypotension
Vomiting
Diarrhea
Heart rhythm changes
Rapid weight loss
Hypoglycemia

Therapy

Mild insufficiency can sometimes be controlled by taking five to six meals daily and increasing salt intake. A diet high in protein and moderate in carbohydrate, divided into frequent feedings all containing protein, will control hypoglycemia by reducing insulin stimulation. Simple sugars are avoided and a liberal salt and fluid intake is stressed. Cortisone is usually prescribed to replace the missing hormone. Patients should wear medical alert jewelry.

Enzyme Deficiencies

An **inborn error of metabolism** is a genetic enzyme deficiency causing metabolic dysfunction. More than a hundred enzyme deficiencies causing metabolic disorders have been identified. Occurring at conception, many are evident shortly after birth; others appear later. Some deficiencies that respond to dietary modification appear in Table 27–1. Deficiency of intestinal enzymes leads to malabsorption, which is discussed in Chapter 22.

PHENYLKETONURIA

Clinical Findings

Phenylketonuria (PKU) occurs in about 1 of every 15,000 births. The liver enzyme phenylalanine hydroxylase, necessary to convert phenylalanine to tyrosine, is absent or inactive. Phenylalanine, an amino acid, accumulates in the blood.

TABLE 27–1

421

NUTRITION IN ENDOCRINE AND
METABOLIC DISORDERS

Some Inborn Errors of Metabolism

Disorder	Dietary Modification
Phenylketonuria	Low phenylalanine
Maple syrup urine disease	Low leucine, isoleucine, valine
Homocystinuria	Low methionine with added cystine; pyridoxine supplements
Galactosemia	Eliminate galactose and lactose
Fructosemia	Eliminate fructose, sucrose, sorbitol
Glycogen storage diseases	Small frequent feedings around the clock

Phenylketones, by-products of abnormal phenylalanine metabolism, are excreted in the urine—hence the name phenylketonuria. High blood levels of phenylalanine cause severe mental retardation, hyperactivity, irritability, and sometimes eczema and seizures. Because tyrosine is involved in pigment formation, children with untreated PKU usually have lighter hair and skin color than their siblings. Phenylketones cause a musty or gamy odor of the skin and urine.

Early diagnosis and treatment are essential to prevent mental retardation, which is not reversible. Most states require that newborn infants be screened and those with elevated blood phenylalanine levels be further tested.

Serum Phenylalanine in Infants (mg/dl): normal, 2; phenylketonuria, 20+

Therapy

Dietary modification is the only treatment for PKU. The objective is to control blood phenylalanine levels while supplying nutrients to promote normal physical and mental development.

Food proteins contain about 5 percent phenylalanine. However, since phenylalanine is an essential amino acid, some must be provided to meet growth needs.

Special formulas are used for these infants.* With the addition of measured amounts of breast milk or regular infant formula, they supply just enough phenylalanine for growth. As the infant grows, ordinary foods that are low in phenylalanine are added. For older infants and children, other products are available.† They provide all essential nutrients except phenylalanine, but have fewer calories. This permits greater variety in the diet with the added foods providing all the phenylalanine and a greater proportion of the calories. A team approach is essential to provide care, counseling, and support for the child and family. (See Figure 27–1.)

Successful treatment requires frequent testing of blood phenylalanine levels and evaluation of growth, development, and food intake. The diet is adjusted to provide for growth needs. Caregivers receive detailed lists of fruits, vegetables, breads, and cereals from which to choose. Even low-protein foods such as fruits contain some phenylalanine and are taken in measured amounts. Careful instructions for measuring are needed. The alternative sweetener aspartame is prohibited because phenylalanine is a basic component. Packages of aspartame and

*Lofenalac®, Mead Johnson & Company, Evansville, IN.

†Phenyl-Free™, Mead Johnson & Company, Evansville, IN.; Phenex®, Ross Laboratories, Columbus, OH.

FIGURE 27–1
A dietitian and nurse, as members of the health care team, provide detailed dietary counseling to the parents of a child with an inborn error of metabolism. Follow-up is provided periodically. (*Source:* Drexel University and The Children's Hospital of Philadelphia, Peter Groesbeck, photographer.)

aspartame-sweetened foods and beverages contain a warning for those with phenylketonuria.

The low-phenylalanine diet is continued throughout childhood and into adulthood. Women with phenylketonuria should be monitored closely before conception and during pregnancy to reduce the possibility of birth defects in their infants.

GALACTOSEMIA

Clinical Findings

Galactosemia occurs in about 1 of every 60,000 to 80,000 live births. The liver enzyme galactose-1-phosphate-uridyltransferase, necessary to convert galactose to glucose, is absent, and galactose accumulates in the blood at toxic levels. Shortly after birth the infant develops vomiting, diarrhea, edema, drowsiness, weight loss, and liver failure. Mental retardation and cataracts occur in untreated infants that survive.

Therapy

To prevent brain damage, a diet free of galactose and lactose must be started in the first days of life. Symptoms disappear dramatically, but if any mental retardation has occurred it cannot be reversed.

Since all milk from mammals contains lactose, soy base formulas are used. (See Chapter 22.) All milk-containing foods must be rigidly excluded. (See Lactose-Restricted Diet, Figure 22–2.) Organ meats also contain galactose and are eliminated.

ARTHRITIS

Some 20 million Americans are affected by arthritis. For 4.4 million there is interference with daily activity; 1.5 million have partial disability; and an equal number, complete disability. The chronic, painful, and often disabling nature of the disease leads patients to spend nearly $1 billion each year seeking relief through unproven diets, drugs, and devices. The Arthritis Foundation advises that warning signs of a questionable therapy include claims that it is all natural, inexpensive, without side effects, works immediately and permanently, and makes visiting a doctor unnecessary. There may be a claim that it can cure a wide variety of disorders, but even if it is promoted for only one condition, this does not mean that it is effective or safe. No diet will cure arthritis. Since individuals with arthritis are easy prey of food faddists (see Chapter 2), a balanced diet from the Food Guide Pyramid (Figure 1–5) and/or Basic Diet (Table 2–4) should be emphasized. The aims of therapy are to relieve pain, strengthen neighboring muscles to maintain or restore function, and protect affected joints.

OSTEOARTHRITIS

Clinical Findings

Osteoarthritis involves deterioration of the cartilage and growth of bone spurs, especially at weight-bearing joints. It is the most common joint disease, affecting almost 16 million adults in the United States. Once considered a "wear-and-tear" degenerative disease, it is now thought to involve metabolically active remodeling, and ways to influence this process are being explored. Osteoarthritis is more common among older people, often those who are overweight, due partly to inactivity.

Therapy

For overweight individuals, weight reduction will ease the burden on weight-bearing joints. An individualized program of range-of-motion, aerobic, and muscle-strengthening exercises will maintain muscle strength, which helps joints function, and will also moderately increase energy expenditure.

Aspirin and nonsteroidal anti-inflammatory drugs (NSAIDs) are the cornerstone of medication therapy. (See Chapter 18). They should be enteric coated and/or taken with meals to reduce gastric irritation. In advanced cases that can no longer be managed with other therapies, joint replacement may be performed.

RHEUMATOID ARTHRITIS

Clinical Findings

This condition involves inflammatory changes in joints and related structures resulting in crippling deformities. Affecting more than 2 million people in the United States, it most commonly begins in the third or fourth decade, although it can occur at any time. Autoimmunity is a factor in the progression of the disease, although there is less certainty that it is a cause. Periods of joint inflammation may

Osteoarthritis: also known as degenerative joint disease

Normal Joint

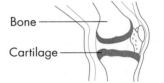

Bone

Cartilage

Osteoarthritis

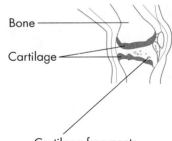

Bone

Cartilage

Cartilage fragments

be followed by periods of remission. Medication side effects, opportunistic infections, pain, fatigue, and restricted mobility may reduce food intake, resulting in underweight which is often severe.

Therapy

Increased caloric intake will reverse underweight. (See Chapter 9.) For patients with limited function, occupational and physical therapists can suggest methods and adaptive equipment to assist in preparing and consuming foods. Increased dietary sources of omega-3 fatty acids may provide pain relief by reducing inflammation. Adequate rest is essential. Non-weight-bearing exercises will maintain muscle tone.

Aspirin and NSAIDs are widely used. Low doses of cortisone to reduce inflammation or *methotrexate,* an anticancer medication that interferes with rapid cell division, may be utilized. (See Chapter 25.) Immunosuppressives are sometimes employed to retard the autoimmune progression.

GOUT

Clinical Findings

Purines are nitrogen-containing compounds that are broken down to uric acid in the body. **Gout** is a condition of abnormal purine metabolism in which the individual has reduced ability to excrete uric acid and may also produce it in excessive amounts. Uric acid builds up in the blood and precipitates out in needle-like crystals which eventually form clumps called **tophi,** especially in the joints. The uric acid crystals irritate the skin covering the affected area. Risk of uric acid kidney stones is increased in individuals with gout. Gout usually occurs in periodic acute attacks of severe pain and inflammation of the metatarsal, knee, and toe joints. About a million Americans are affected.

Therapy

Since the body can synthesize purines, many question the value of a purine-restricted diet, and gout is managed with medications. Some clinicians suggest consuming a diet restricted in purines (see Table 27–2) and alcohol for a week or more, followed by evaluation of serum and urinary uric acid levels. These lifestyle changes may cause a substantial reduction in uric acid values. Many individuals will still require medications.

Since proteins provide the basic materials for purine synthesis, a moderate protein intake is prudent. Fats are believed to reduce uric acid excretion, so moderate intake is also advised. Alcohol increases uric acid production, so it should be omitted or used in moderation.

Many patients with gout are obese. Weight loss should be gradual, as rapid weight loss, especially fasting, elevates blood uric acid levels and can trigger an attack. A reduced-calorie diet should not be initiated during an acute attack because this could further aggravate the symptoms.

During an acute attack, some clinicians recommend the omission of foods both moderate and high in purines. When the attack has subsided, small amounts of meat, fish, and poultry are gradually introduced.

NSAIDS are used to relieve the inflammation and pain of an acute attack and as a preventive during times when there are no symptoms. They should be taken with food to reduce gastric irritation. Steroids may be used if NSAIDs fail to

---

Serum Uric Acid Levels (mg/dl): normal, 2–7; hyperuricemia; 7–20

RISK FACTORS FOR GOUT

Genetic
Male gender
Middle age and beyond
Overweight

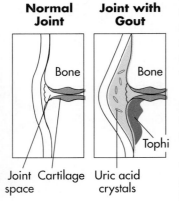

Normal Joint	**Joint with Gout**
Bone	Bone
	Tophi
Joint Cartilage space	Uric acid crystals

Low	Moderate	High
Milk	Meat[a]	Organ meats
Cheese	Fish and shellfish[a]	Wild game
Egg	Poultry	Asparagus
Bread, cereal, pasta[a]	Legumes	Mushrooms
Fruit	Oatmeal	Anchovies
Vegetables[a]	Spinach	Herring
		Sardines
		Meat extracts
		Meat broths
		Meat gravy
		Sweetbreads
		Liver
		Kidneys
		Brains
		Mackerel
		Scallops

[a] Unless listed under moderate or high.

control the attack. *Allopurinol* decreases uric acid production; it should be accompanied by increased fluids to reduce the risk of kidney stones. *Probenecid* or *sulfinpyrazone* is used to increase uric acid excretion and dissolve uric acid crystals. Thiazide diuretics used to control hypertension tend to elevate uric acid levels, so a person with elevated blood pressure and gout might require another medication.

PUTTING IT ALL TOGETHER

Many and varied conditions fall into these categories: endocrine disorders, enzyme deficiencies, different types of arthritis. In some, such as inborn errors of metabolism, diet is the only therapy. In others, for example arthritis, no diet will cure or control the condition, but an adequate diet will improve well-being and quality of life.

❓ QUESTIONS CLIENTS ASK

1. *My mother has arthritis, which causes great discomfort. Her neighbor, who also has arthritis, is taking megadoses of vitamins somebody is selling her, and she says they give her relief. My mother is thinking of trying them. I am dubious, but I know how much pain my mother has and I want her to be more comfortable.*

There is no scientific evidence that megadoses of vitamins help arthritis, and large doses of fat-soluble vitamins over a period of time can be toxic. Symptoms of arthritis do go into remission periodically,

and this may be what your mother's neighbor is experiencing.

2. *Another friend has advised my mother to try acupuncture for relief from arthritis pain. What do you think about this?*

Currently there is a clinical trial utilizing acupuncture to treat osteoarthritis. Your mother can check with her health care provider about it. If she cannot join the trial, she should check about the results when available.

3. *My husband has been diagnosed with gout. I though it only affected wealthy people who drank a lot of alcohol and ate a lot of rich food. He's overweight, but we certainly aren't rich and he's a moderate social drinker. His uncle did have gout, I remember.*

Heredity, overweight, male gender, and the middle years of life are all risk factors for developing gout.

Although gout is historically associated with affluence and gluttony, it can affect people of more modest means. The only modifiable risk factor your husband has is overweight. Weight loss to a healthy weight should be an objective of his, although it should not be attempted during an acute episode.

✔ **CASE STUDY: An Obese Client with Osteoarthritis**

Mrs. Sandra E., age 60, is 63 in. (160 cm) tall and weights 180 lb (82 kg). She has osteoarthritis that especially affects her knees, making it painful for her to move around. Doing household chores has become increasingly difficult, and lack of mobility has forced her to give up volunteer work that she enjoyed. Her physician recommended a 1,000-kcal diet to reduce the burden on her knees.

Questions

1. Assume that Mrs. E.'s caloric needs are 1,250 kcal a day. On a 1,000-kcal diet, how much weight should she lose in a week?

2. Mrs. E. and her husband like to relax over an evening snack. How might she continue to do this on a 1,000-kcal diet?

3. Mrs. E. and her husband occasionally go out to dinner in a local restaurant. What are some suggestions you might give her for ordering?

4. Mrs. E. and her husband sometimes take a day trip by car for which she packs a lunch. What suggestions might you give for such a meal?

5. On Sundays, Mrs. E. and her husband usually visit the homes of relatives for dinner. How might she continue these visits while on the diet?

6. What are some behavior modification techniques you might suggest to Mrs. E. that would be helpful in weight management?

7. Plan three days' menus for Mrs. E., using the food allowances in Table 9–2. Include an evening snack and on at least one day a lunch that might be packed.

8. Suggest some ways in which Mrs. E. might increase her physical activity, keeping in mind the osteoarthritis and decreased mobility.

9. Mrs. E. has been taking *ibuprofen*, an anti-inflammatory. What nutritional side effects may result? What dietary strategies might she use to decrease side effects?

FOR ADDITIONAL READING

Acosta, PB, and Yannicelli, S: "Nutrition Support of Inherited Disorders of Amino Acid Metabolism," *Top Clin Nutr,* Part I. **9:**65–82, December 1993; Part II, **10:**48–72, March, 1995.

Angelucci, PA: "Caring for Patients with Hypothyroidism," *Nurs 95,* **25:**60–61, May 1995.

Bowers, DF, and Allred, JB: "Advances in Molecular Biology: Implications for the Future of Clinical Nutrition Practice," *J Am Diet Assoc,* **95:**53–59, 1995.

Cash, JM, and Klippel, JH: "Second-Line Drug Therapy for Rheumatoid Arthritis," *New Engl J Med,* **330:**1368–75, 1994.

Emmerson, BT: "The Management of Gout," *New Engl J Med,* **334:**445–51, 1996.

Flieger, K: "Getting to Know Gout," *FDA Consumer,* **29:**19–22, March 1995.

Franklyn, JA: "The Management of Hyperthyroidism," *New Engl J Med,* **330:**1731–38, 1994.

Leiden, JM: "Gene Therapy—Promise, Pitfalls, and Prognosis," *New Engl J Med,* **333:**871–73, 1995.

"The Maternal Phenylketonuria Collaborative Study: A Status Report," *Nutr Rev,* **52:**390–93, 1994.

McCain, J, and O'Hanlon-Nichols, T: "Commonly Asked Questions About Rheumatoid Arthritis," *Am J Nurs,* **96:**16-A–16-D, April 1996.

Napier, K: "Unproven Medical Treatments Lure Elderly," *FDA Consumer,* **28:**32–37, March 1994.

Orth, DN: "Cushing's Syndrome," *New Engl J Med,* **332:**791–803, 1995.

Pelcovitz, D, and Goldberg, T: "Enhancing Nutrition Compliance in Children: Inborn Errors of Metabolism as a Paradigm," *Top Clin Nutr,* **10:**73–81, March 1995.

Service, FJ: "Hypoglycemic Disorders," *New Engl J Med,* **332:**1144–52, 1995.

Strange, CJ: "Coping with Arthritis in Its Many Forms," *FDA Consumer,* **30:**17–21, March 1996.

Toft, AD: "Thyroxine Therapy," *New Engl J Med,* **331:**174–80, 1994.

Vance, ML: "Hypopituitarism," *New Engl J Med,* **330:**1651–62, 1994.

CHAPTER 28

Nutrition in Diseases of the Kidney

Chapter Outline

Structural Units of the Kidney
Functions of the Kidney
Glomerulonephritis
Nephrotic Syndrome
Acute Renal Failure
Chronic Renal Failure
Dialysis

Kidney Transplant
Urinary Calculi
Putting It All Together
Questions Clients Ask
Case Study: A Man with Chronic Renal
  Failure

Reverse the increase in end-stage renal disease (requiring maintenance dialysis or transplantation) to attain an incidence of no more than 13 per 100,000.*

✔ CASE STUDY PREVIEW

The kidneys perform so many functions that when their function is compromised many body systems are affected. You will help a man with kidney failure and his wife plan food intake appropriate for declining renal function.

---

*Healthy People 2000: National Health Promotion and Disease Prevention Objectives. Public Health Service, U.S. Department of Health and Human Services, Washington, DC, 1991, Objective 15.3 p. 112.

The goals of nutritional therapy for both acute and chronic renal failure are to maintain optimal nutritional status, to minimize the toxic effects of excess urea in the blood, to prevent loss of lean body mass, to promote patient well-being, to retard the progression of renal failure, and to postpone initiation of dialysis. In children, an additional goal is to maintain growth rates as close to normal as possible.*

STRUCTURAL UNITS OF THE KIDNEY

The structural and functional unit of the kidney is the nephron consisting of a glomerulus, a tuft of capillaries attached to a long winding tubule that empties into collecting ducts. Each nephron functions independently to produce urine. The glomerulus filters blood that circulates through it. Large molecules such as blood proteins are held back in the circulation. Water together with glucose, amino acids, urea, sodium chloride, and other small molecules filter into the proximal tubules. This is called the glomerular filtrate. The glomerular filtration rate (GFR) is the amount of plasma filtered in 1 minute; normally 125 ml/minute. Selective reabsorption occurs in the winding tubules and the filtrate proceeds through the distal tubules into collecting ducts and on to the ureter. (See Figure 28–1.)

Renal: having to do with the kidney

Each kidney contains over a million nephrons that provide tremendous reserve functional capacity. Loss of half of the nephrons, as when a healthy adult donates a kidney for transplant or loses one in an accident, has little, if any effect on function. Kidney function gradually diminishes with age, and by age 80 the filtration rate may be one-half to two-thirds of what it was at age 30. Nevertheless, function is still adequate unless kidney disease occurs. By adaptation the kidneys can function adequately when up to two-thirds of the nephrons have been damaged or destroyed.

Each glomerulus filters only a tiny drop of fluid a day, but the volume of plasma filtered by the two million glomeruli amounts to 125 ml/minute or 180 liters in 24 hours. The amounts of glucose, sodium chloride, and other substances filtered are equally large; for example, the sodium chloride filtered is over 1 kg (2.2 lb), which is roughly 100 times the daily intake of salt!

Normally, the urine volume ranges from 1,000 to 2,000 ml, which means that over 99 percent of the filtered water has been returned to the circulation. Likewise, all of the glucose and vitamin C, and almost all of the amino acids, sodium, and other substances have been returned to the blood. If you eat foods containing more salt than your body needs, the renal excretion of water and sodium will be increased. On the other hand, if you greatly reduce your salt intake or if the body sodium is depleted, the excretion in the urine will be very small.

FUNCTIONS OF THE KIDNEY

The overall role of the kidneys is to maintain normal composition and volume of the blood and other body fluids. This is accomplished through numerous interrelated controls.

Excrete metabolic wastes: Urea, uric acid, creatinine, ammonia, and other products of metabolism; medications, drugs, and toxic substances

Regulate fluid balance: **Antidiuretic hormone (ADH),** secreted by the pituitary gland, and aldosterone, a hormone from the adrenal gland,

---

The Surgeon General's Report on Nutrition and Health. U.S. Department of Health and Human Services, 1988, pp. 387–88.

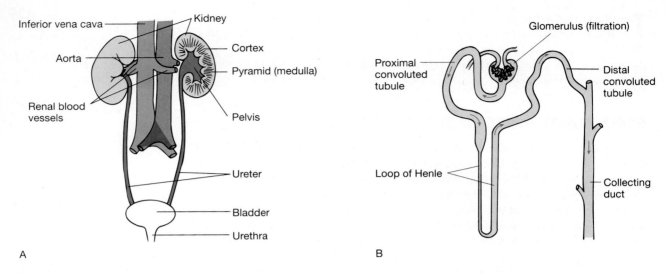

FIGURE 28–1

(A) The components of the urinary system. (B) The structure of an individual nephron. (*Source:* Reprinted with the permission of Macmillan Publishing Company from *Clinical Nutrition and Dietetics,* Second Edition by Frances J. Zeman. Copyright © 1990 Macmillan Publishing Company, Inc.)

influence the amount of water and sodium the kidney reabsorbs or excretes and thereby control fluid balance. **Renin,** an enzyme produced and stored in the kidneys, influences the production of aldosterone.

Regulate electrolyte balance: Sodium and potassium

Help regulate acid-base balance: Along with the lungs, the kidneys are the main regulators of blood pH. The kidneys mediate the balance between plasma bicarbonate, a base, and acids. They also synthesize ammonia. (See Chapter 4.)

Help regulate blood pressure: Blood pressure is controlled by the excretion and reabsorption of sodium and water by the renin-angiotens in system. Renin acts in the conversion of angiotensin I to **angiotensin II,** a vasopressor, which influences sodium reabsorption and thereby increases blood pressure.

Influence production of red blood cells: **Erythropoietin,** a hormone produced by the kidney, stimulates the bone marrow to produce red blood cells and also influences their maturation and life span.

Calcitriol: also known as 1,25(OH)$_2$D$_3$

Convert calcidiol to calcitriol (active vitamin D$_3$ hormone): In its active form, vitamin D$_3$ hormone mediates the absorption of calcium and phosphorus from the gastrointestinal tract and regulates calcium and phosphorus levels in the blood. (See Chapter 10.)

It is easy to see the far-reaching effects that kidney dysfunction can produce. Disease may affect the glomerulus or tubule or both and may be acute or chronic.

GLOMERULONEPHRITIS

As the name suggests, this is an inflammation of the nephron confined largely to the glomerulus.

Acute Glomerulonephritis

This condition frequently follows a streptococcal infection of the respiratory tract or scarlet fever. It is seen most often in children and young adults. Sometimes the infection is so mild the person is unaware of it, and permanent damage may be done that will only be discovered many years later. Others have nausea, vomiting, fever, hypertension, hematuria, and **oliguria,** decreased output of urine, usually less than 400 ml/day and insufficient to excrete wastes.

Dietary Considerations During the acute stage when there are nausea and vomiting, nonprotein liquids such as sweetened fruit juices, sweetened tea, ginger ale, fruit ices, and high-carbohydrate, low-electrolyte supplements* are given to help reduce tissue breakdown. Fluid intake usually is restricted in proportion to urine output.

As soon as the patient is able to eat, a diet adequate in calories to maintain weight should be given. Protein is not restricted unless oliguria or renal failure exists, in which case it is limited to 40 g protein of high biological value. (See Chapter 7.) See food allowances for 40 g protein diet in Table 28–1, later in chapter. If oliguria is present, fluids are limited. Sodium is restricted to 1,000 mg. Most of these patients recover completely.

Chronic Glomerulonephritis

This condition can result from an immunological cause of unknown origin or, less frequently, from untreated acute glomerulonephritis. In the early stages the only indications may be abnormal urinalysis results: protein and red and white blood cells in the urine. As the disease progresses, fatigue, proteinuria, hematuria, edema, hypertension, and blurring of vision frequently occur. Since the kidneys are unable to concentrate urine, there is frequency of urination and nocturia. In some patients, the condition advances through the nephrotic syndrome to chronic renal failure.

Dietary Considerations The diet is tailored to the functional capacity of the individual's kidneys. When the kidneys are able to excrete wastes, normal intake of protein is allowed. If proteinuria exists, the protein intake is increased to cover urinary losses. Proteins of high biological value are emphasized. (See Chapter 7.) As the disease progresses and the **blood urea nitrogen (BUN)**, waste product of protein metabolism, rises, the protein allowance is usually decreased to 40 g/day or less. Sufficient calories from carbohydrate and fat must be taken to prevent tissue breakdown.

Sodium is restricted only in the presence of edema. Since the kidneys can no longer concentrate urine, considerable sodium may be lost and a sodium-restricted diet could lead to weakness and shock. Anemia is common and iron supplements are often prescribed.

Medications Steroid therapy, immunosuppressants, and anticoagulants may be prescribed. See Chapter 23 for a discussion of side effects of steroids and immunosuppressants.

NEPHROTIC SYNDROME

Nephrotic syndrome is a protein-wasting disorder that may be a phase of the disease progression of chronic glomerulonephritis, diabetic nephropathy, or hypertension, or occur secondary to lupus erythematosus, hepatitis B, cancer, or heavy-metal poisoning.

Acute Glomerulonephritis: also known as hemorrhagic nephritis

Blood Urea Nitrogen (BUN): normal 4–22 mg/dl

Nephrotic Syndrome: also known as nephrosis

---

*Polycose®, Ross Laboratories, Columbus, OH; Moducal®, Mead Johnson & Company, Evansville, IN.

Clinical Findings

Injury to the glomerulus is involved and changes in the permeability of the glomerular capillaries permit the filtration of albumin into the urine. Massive albuminuria is a common characteristic of all cases of the nephrotic syndrome, followed by hypoalbuminemia and edema. Loss of plasma proteins leads to malnutrition, tissue wasting, anemia, fatty liver, and increased susceptibility to infection. Hyperlipidemia is usually present although the cause is unclear—possibly increased production or decreased removal of circulating low density lipoproteins (LDL), decreased synthesis of high density lipoproteins (HDL), and possibly secondary to long-term use of antihypertensives or corticosteroids. Oliguria is present.

Dietary Considerations

The diet should furnish sufficient calories for tissue repletion and maintenance of weight in adults and normal growth in children. Traditionally, diets have been high in protein to correct the lowered serum protein and compensate for the urinary losses, up to 120 g a day for adults and 3–4 g per kilogram for young children. (See high protein diet, Figure 20–3.) Research findings that high protein intake increases urinary losses without replenishing serum levels, coupled with concern that high protein intake may hasten the progression of various renal diseases, have led some to recommend 0.6 g to 0.8 g per kilogram plus 1 g of high biologic protein for each gram of urinary protein lost daily.

If severe edema is present, sodium may be restricted to 1,000–2,000 mg/day. Dietary fat and cholesterol are frequently limited to control the hyperlipidemias. (See Tables 24–1 and 24–2.) A vegan soy-protein regimen showed promising results although patient compliance and possible resulting nutritional deficiencies are problematic.

Medications

Diuretics are used to control the edema. Corticosteroids and immunosuppressants may be used.

ACUTE RENAL FAILURE

Sudden, often reversible, kidney shutdown in an individual who previously had adequate renal function characterizes acute renal failure.

Clinical Findings

The glomerular filtration rate drops rapidly and when it falls below 20 ml/minute, serum urea and creatinine rise rapidly. At less than 10 ml/minute, symptoms of **uremia,** an excess of urea and other nitrogenous waste in the blood, appear. Oliguria or **anuria,** lack of urine or urine output of less than 100 ml/day, is present. Patients may be drowsy, weak, tired, or have headache, itching, or blurred vision.

A quarter of patients with acute renal failure die. The rate rises to half when there is associated trauma or the patient is over age 75. Most common causes of death are complications of fluid and electrolyte imbalance such as pulmonary edema, cardiac arrest, and respiratory paralysis. Acute renal failure can be divided into three distinct stages.

Most, but not all, patients go through the *oliguric phase* that usually lasts 1–2 weeks. Dialysis is usually instituted until kidney function returns.

Uremia: also known as uremic syndrome or azotemia

POSSIBLE CAUSES OF ACUTE RENAL FAILURE

Obstruction from calculi, tumors, or prostatic disease
Hemorrhage
Septicemia
Myocardial infarction
Renal ischemia
Extensive burns
Severe crushing injuries
Shock from surgery or other causes
Ingestion or inhalation of poison
Ingestion of nephrotoxic medicines
Severe acute glomerulonephritis
Mismatched blood transfusions
Iodinated radiocontrast media with preexisting renal insufficiency
Hemolytic uremic syndrome

Food and fluid by mouth are restricted for the first 24–48 hours. Initially, anorexia, nausea, and vomiting limit oral intake, and intravenous dextrose is given. Tube feeding or total parenteral nutrition is sometimes employed. Because of the volume of fluid required for TPN, dialysis is essential when total parenteral feeding is employed. Sufficient carbohydrate and fat should be furnished to prevent protein catabolism. Sometimes continuous arteriovenous hemofiltration (CAVH), which filters wastes from the patient's blood and infuses a plasma-like solution, is utilized. The volume of fluid removed is greater than that replaced thereby avoiding fluid overload, so TPN can be used.

Before a patient goes on dialysis, a protein-free diet may be utilized. Intravenous glucose with essential amino acids is another approach. Formulas restricted in protein, fluids, and electrolytes are available.*

Fluid is restricted to urine output plus approximately 500 ml to allow for insensible water losses. An extra allowance is made if vomiting, diarrhea, draining wounds, or fever are present. Sodium and potassium allowances are based on serum levels and whether or not the patient is on dialysis.

During the *diuretic phase,* urine output gradually increases and the glomerular filtration rate rises. Polyuria can occur. Energy intake should be sufficient to prevent catabolism. A protein allowance of 20–40 g is permitted with gradual increases as kidney function improves until normal levels of intake are reached. Monitoring of serum electrolytes with appropriate dietary adjustments is essential as excessive losses can occur through polyuria. Likewise, urine output is measured and fluid intake is based on it with allowances for insensible water losses.

As the *recovery phase* progresses, diet is modified as the patient gradually regains normal kidney function. If tissue wasting has occurred, a high-protein high-calorie diet is required. Some patients have varying amounts of permanent loss of kidney function.

CHRONIC RENAL FAILURE

Changes in Function of the Kidney

Chronic renal failure often begins insidiously with a gradual loss of kidney function, but as the glomerular filtration rate begins to decline, the renal insufficiency runs a relentless course toward end-stage renal disease (ESRD), the later stages of chronic renal failure. Because of their vast reserve capacity, the kidneys can support life comfortably through much of this progression. When the glomerular filtration rate drops below 30 ml/minute, dietary intervention is instituted. Some clinicians favor earlier modification of protein and phosphorus intake to slow the course of the disease and postpone dialysis. A national study, Modification of Diet in Renal Disease (MDRD), researched the effect of protein and phosphorus restriction and blood pressure control on the progression of chronic renal disease. The research failed to show significant differences at various levels of protein intake, but did point up benefits of blood pressure control. When the GFR reaches around 3 ml/minute, dietary control is no longer sufficient and dialysis or kidney transplant is required to preserve life.

In end-stage renal disease, symptoms resulting from the failure of many functions of the kidney are apparent.

*Suplena®, Ross Laboratories, Columbus, OH: Renalcal™, Clintec Nutrition, Deerfield, IL.

CHRONIC RENAL FAILURE SECONDARY TO

Diabetic nephropathy
Recurrent acute or chronic glomerulonephritis
Acute kidney failure where normal function failed to return
Nephrosclerosis (hardening of renal arteries)
Chronic pyelonephritis (inflammation caused by bacterial infection)
Cardiac failure
Extensive atherosclerosis
Malignant hypertension (severe, life-threatening hypertension)
HIV-related nephropathy

Excrete metabolic wastes. Nitrogenous products accumulate in the blood. The blood urea nitrogen is elevated and nausea and vomiting are common. While the retained wastes are largely nitrogenous, the kidneys also fail to excrete poisons and medication. Medication doses must be adjusted to the decreased and delayed excretory capacity.

Regulate fluid balance. The kidneys can no longer concentrate urine and the patient is oliguric or anuric. Fluids and sodium are retained and can cause congestive heart failure.

Regulate electrolyte balance. The kidneys lose the ability to excrete excess potassium leading to hyperkalemia and irregular heart rhythm. The kidneys usually retain sodium and this, together with fluid retention, results in pitting edema. Some patients are "sodium wasters," losing large amounts of sodium and water in the urine.

Help regulate acid-base balance. The kidneys cease to excrete acid ions, so free hydrogen and ammonia are retained. Bicarbonate synthesis and reabsorption are also reduced. This leads to metabolic acidosis resulting in anorexia, nausea, fatigue, disorientation, mental deterioration, and bone demineralization.

Help regulate blood pressure. Changes in sodium balance affect blood pressure. In addition, renin levels rise, producing more angiotensin II, which elevates the blood pressure. Most patients with chronic renal failure have hypertension. (See Figure 28–2.)

FIGURE 28–2
A nurse makes a home visit to a patient. She checks blood pressure, signs of edema, medications, and answers questions about the sodium-restricted diet. (*Source:* Dunwoody Village, Newtown Square, PA, and Peter Zinner, photographer.)

Influence production of red blood cells. Erythropoietin output is reduced, decreasing the production of red blood cells, their maturation, and life span. Ammonia, urea, and excessive parathyroid hormone in the blood provide a hostile environment for the red blood cells to survive. The hemoglobin level is inversely related to the blood urea nitrogen. With anemia comes fatigue and weakness.

Convert calcidiol to calcitriol. The kidneys cease to produce the active vitamin D hormone. Serum phosphorus levels rise because the kidney no longer can excrete phosphorus. Hyperphosphatemia lowers serum calcium levels. Also, the individual cannot absorb calcium from the gastrointestinal tract because calcitriol is not present, further aggravating the hypocalcemia.

Parathyroid hormone output rises in response to the low blood calcium levels and a secondary hyperparathyroidism results. The excess parathyroid hormone increases calcium resorption from the bone, leading to renal osteodystrophy, osteomalacia and deposit of calcium in the soft tissues. (See Chapter 11.)

As the disease progresses, elevation of serum triglycerides, hyperglycemia, and impaired glucose tolerance are commonly seen. Progressive weakness, anorexia, jaundice, ammoniacal breath odor, mouth ulceration, hiccups, blurred vision, and disorientation frequently occur.

Therapy

The stage of the disease, biochemical and functional profile, etiology of the end-stage renal disease, coexisting diseases, and whether or not the patient is receiving dialysis influence the planning.

Dietary Considerations

The diet takes into account the symptoms, the blood levels of urea and electrolytes, and the nutritional status. Laboratory studies from time to time determine whether the diet needs to be adjusted. The diet may be controlled for some or all of these factors: protein, potassium, sodium, phosphorus, and fluids. For patients who are awaiting dialysis or who are unable to receive dialysis, the diet is more severely restricted than for patients who are on dialysis.

Energy Adequate caloric intake will prevent tissue breakdown with its resulting release of nitrogen and potassium into the circulation. Carbohydrates and fats are the principal sources of energy. Their metabolic end products, small amounts of water and carbon dioxide, are excreted through the lungs, sweat glands, and bowel and create no problems for individuals with impaired kidney function. High carbohydrate, low protein, low electrolyte supplements and beverages and desserts made from nondairy creamer, oil, sugar, and flavoring can be used to increase total calories and spare protein. Low-protein, low-electrolyte formulas are available.*

Protein The end products of protein metabolism—urea, creatinine, uric acid, sulfate, and organic acids—are excreted in the urine, and problems arise when kidney function is impaired. Protein intake is decreased as the glomerular filtration rate declines. On diets restricted in protein, 2/3 to 3/4 of the protein should be of high biologic value, distributed evenly throughout the day. Alternatively, a greater

_____
*Suplena®, Ross Laboratories, Columbus, OH; Renalcal™, Clintec Nutrition, Deerfield, IL.

OBJECTIVES OF NUTRITIONAL MANAGEMENT OF ESRD

Maintain optimal nutritional status
Provide normal growth and development in children
Prevent protein catabolism
Prevent uremia
Prevent renal osteodystrophy
For pre-dialysis patients slow progression of renal failure and postpone dialysis

GLOMERULAR FILTRATION RATE

GFR ml/min	g protein/ kg/day
15–20	1.0
10–15	0.7
4–10	0.55–0.6 not less than 35–40 g/day

variety of protein sources is permitted and essential amino acids are provided by supplements or keto analogs. The body can synthesize nonessential amino acids from the excess nitrogenous constituents of the blood. Possible daily meal patterns at various levels of protein, based on a modification of the *Exchange Lists for Meal Planning* (Appendix Table A-2), appear in Table 28-1.

Potassium When the urine volume is still adequate, potassium is excreted without much difficulty and restriction is not necessary. As renal insufficiency increases, hyperkalemia becomes common, and a 2,000-3,000 mg potassium-restricted diet is often prescribed. If potassium-losing diuretics are used, additional potassium may be needed. For potassium content of foods, see Chapter 11 and Appendix Table A-1.

Sodium The sodium intake depends on blood and urine levels. Restrictions ranging from 2,000 to 3,000 mg are necessary if edema and hypertension are present. For sodium content of foods, see Chapters 11 and 24 and Appendix Table A-1.

Phosphorus Since serum phosphorus rises as kidney function declines, and elevated phosphorus levels are involved in renal osteodystrophy and metabolic

TABLE 28-1

Suggested Daily Meal Pattern for Controlled Protein Diet[a]

		Protein			
	Measure	20 g	40 g	60 g	80 g
Breakfast					
Fruit	1 serving	1	1	1	1
Egg	1	—	1	1	1
Cereal	1 serving	1	1	1	1
Bread, low-protein	1 slice	2	—	—	—
Bread, enriched	1 slice	—	1	1	1
Milk	1 cup	¼	¼	¾	¾
Lunch					
Egg	1	1	—	—	—
Meat or equivalent	1 oz	—	1	2	3
Starch	1 serving	—	1	2	2
Low-protein bread	1 slice	2	—	—	—
Vegetable, free	1 serving	—	1	1	1
Milk	1 cup	—	½	¼	¼
Dessert, low-protein	1 serving	1	1	1	1
Fruit	1 serving	1	1	1	1
Dinner					
Meat or equivalent	1 oz	—	1	2	3
Starch	1 serving	1	1	2	2
Bread, low-protein	1 slice	2	—	—	—
Vegetable	1 serving	1	1	2	2
Fruit	1 serving	—	1	2	2
Milk	1 cup	¼	—	—	1
Nondairy milk substitute	½ cup	1	1	1	—
Dessert, low protein	1 serving	1	—	—	—

[a] Other low-protein foods may be added to increase calorie level.

acidosis, the dietary phosphorus is often restricted to 600 – 1,200 mg/day. For phosphorus content of foods, see Chapter 11 and Appendix Table A-1.

Other Minerals and Vitamins   The restricted diets do not provide recommended allowances of calcium, iron, and vitamin B complex. Supplements of these nutrients should be provided. Calcitriol may be provided in supplement form since the kidneys are unable to produce the vitamin D hormone. Vitamin C should not be supplemented because it favors formation of oxalates. Neither should vitamin A supplements be given because of possible toxicity.

Fluid   When there is oliguria, the fluid is restricted to the daily urine volume plus about 500 ml for insensible losses. For example, a urinary excretion of 250 ml daily would permit a fluid intake of 750 ml. This includes the water present in foods as well as beverages. A 100 g portion of fruits and vegetables supplies 80 – 90 ml water, and 100 ml milk is equal to 87 ml water. See Table 4 – 7 and Appendix Table A-1 for water content of foods.

National Renal Diet   Multiple restrictions can lead to very limited diets and difficulties with patient compliance. Hospitals and medical centers throughout the country have developed their own diet plans. A National Renal Diet to provide consistency from one institution to another and uniform teaching tools have been developed by the Renal Dietitian's Practice Group of the American Dietetic Association and the Council on Renal Nutrition under the National Kidney Foundation. A professional guide and client education booklets about kidney disease, hemodialysis, and peritoneal dialysis, for clients with and without diabetes, are available.

Medications

The decreased ability of the kidney to excrete medications must be considered in prescribing any medication. Patients should be cautioned about over-the-counter medications that often contain restricted components such as sodium, interact with prescription medications, or interfere with therapy.

In early phases of renal failure thiazide diuretics may be prescribed to control edema, but they become ineffective as kidney function deteriorates and then *furosemide* is used. Increasingly large doses may be necessary to produce diuresis. Beta blockers often cause hyperkalemia as do nonsteroidal anti-inflammatories.

For severe hyperkalemia, an exchange resin such as sodium polystyrene sulfonate can be given by mouth or enema to bind potassium. Since the potassium is exchanged for sodium, sodium retention can occur, aggravating hypertension or edema. Sorbitol is usually given to produce osmotic diarrhea, thereby assisting in the excretion of the resin.

Bicarbonate therapy used to treat acidosis increases sodium intake that can cause edema and elevate blood pressure. To control these symptoms, increased doses of diuretics may be necessary.

Calcium-containing binders sequester phosphorus in the gastrointestinal tract and decrease its absorption. These binders can also provide supplemental calcium. These compounds should be taken with meals to maximize the sequestering of the phosphorus. Aluminum- and magnesium-based binders should be avoided because of toxicity.

Supplements of iron or transfusions have been used to correct anemia. The availability of genetically engineered erythropoietin has provided a therapeutic tool to increase red blood cells. Calcium supplements help normalize serum

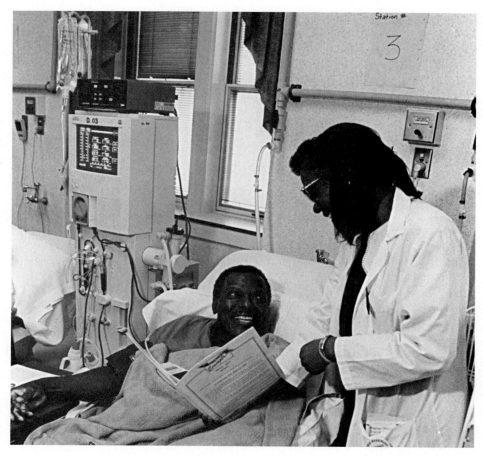

FIGURE 28–3
The registered dietitian answers questions about favorite foods for a patient receiving hemodialysis. (*Source:* Courtesy of FDA Consumer.)

levels. Calcium as calcium carbonate additionally buffers the metabolic acidosis. If serum calcium levels do not respond, vitamin D as calcitriol is given. Diseases associated with pyelonephritis should be treated with antibiotics.

DIALYSIS

Dialysis is frequently used in acute and chronic renal failure on a short-term or long-term basis. Dialysis does not correct metabolic problems of kidney failure and dietary modification and supplementation continue to be necessary. (See Figure 28–3.) A moderate-protein formula restricted in fluids and electrolytes is available.* Most patients with end-stage renal disease lose remaining kidney function after beginning dialysis. More than 150,000 patients are receiving dialysis in the United States.

---

*Nepro®, Ross Laboratories, Columbus, OH.

Hemodialysis

The patient's blood circulates outside the body through semipermeable membranes bathed in dialyzing fluid to remove nitrogenous wastes. The patient is attached to the dialyzer for 4 hours three times a week, usually at a dialysis center. Urea kinetic modeling, which uses mathematical equations to compare urea production to urea removal, enables dialysis treatments to be individualized to the needs of the patient. This approach can replace traditional nitrogen balance determinations in patients whose kidney function is impaired or absent.

Between dialyses nitrogenous waste products, potassium, sodium, and fluids accumulate; a fluid gain of no more than 2 lb/day is suggested. Some serum amino acids and water-soluble vitamins are lost in the dialysate. Water-soluble vitamin supplementation is necessary.

Peritoneal Dialysis

This form of dialysis uses the **peritoneum,** a sac that lines the abdominal cavity, as the dialyzing membrane. A permanent catheter that penetrates the abdominal wall is surgically implanted and through it a hyperosmolar glucose solution is introduced.

No restriction of fluids, sodium, or potassium is usually necessary. Phosphorus is usually restricted and calcium supplementation may be necessary.

Intermittent Peritoneal Dialysis (IPD) Dialysis is performed four times a week, usually for 10 hours. Patients may have low body weight with decreased muscle mass and serum proteins.

Continuous Ambulatory Peritoneal Dialysis (CAPD) Dialysis solution in a plastic pouch is introduced and drained by gravity, usually five times a day, seven days a week.

Continuous Cyclic Peritoneal Dialysis (CCPD) A machine delivers and drains the solution at night. Some solution is left in the peritoneal cavity during the day. In CAPD and CCPD some of the glucose from the dialysate is absorbed, so weight management may be a problem and calorie restriction coupled with an exercise program may be indicated.

There is evidence that some patients on dialysis develop a deficiency of **carnitine,** a metabolite of lysine found especially in tissues that preferentially use fatty acids for energy, since the small carnitine molecule is easily lost in the dialysate. A deficiency is associated with plasma lipid disorders and in some cases carnitine supplementation has reduced hypertriglyceridemia and increased high-density lipoprotein levels. (See Chapter 6.)

KIDNEY TRANSPLANT

Renal transplant provides the individual with a functioning kidney. Immunosuppressants to prevent rejection are necessary and *azathioprine, prednisone,* and/or *cyclosporine* are used, sometimes together as triple therapy. *Cyclosporine* has reduced the need for large doses of steroids and moderated their side effects. Unfortunately it is nephrotoxic. For side effects of immunosuppressants see Chapter 23.

Dialyzer: also known as artificial kidney

Urea Kinetic Modeling: also known as urea mass balance

DIETARY GOALS DURING DIALYSIS

Sufficient energy intake to spare protein
Emphasize proteins of high biological value
   Hemodialysis: 1.2 g/kg healthy body weight
   Peritoneal dialysis: 1.2–1.5 g/kg healthy body weight
Maintain near normal serum sodium and potassium levels
   Sodium: individual basis
   Potassium: individual basis
Prevent fluid overload or dehydration
   1,000 ml (hemodialysis)
   2,000 ml (peritoneal dialysis)/day + urine output, if any

Postoperatively, a liquid to solid dietary progression is followed according to individual tolerance. A high protein, low carbohydrate, and sodium restricted diet may be followed to counteract the effects of high dose immunosuppressive therapy. When a maintenance dose of steroids is reached, a regular diet with 2,000 to 4,000 mg sodium for blood pressure control and reduced in fat to control hyperlipidemias is used. Exercise is another component of control of hyperlipidemia. Simple sugars are sometimes restricted to prevent hyperglycemia and hyperinsulinemia.

URINARY CALCULI

Etiology

Urinary calculi may be found in the bladder, kidney, ureter, or urethra. They are formed by deposit of crystalline material in an organic matrix and vary greatly in size. As stones form they usually move toward the ureter. Small smooth stones can pass into the ureter, but large rough stones may block the ureter opening causing **renal colic,** extreme pain in the flank, sometimes radiating to the groin. Chills, fever, nausea, and vomiting sometimes accompany the pain. Approximately 90 percent of stones are small enough to pass through the ureter.

About 90 percent of the stones contain calcium, sometimes in combination with ammonia, carbonate, magnesium, oxalate, and/or phosphate. The remaining stones are composed of uric acid or, more rarely, xanthine or cystine. About 1 person in 1,000 in the United States has kidney stones.

Calculi may be secondary to hypervitaminosis D, osteoporosis, excessive calcium intake with alkali, high urinary excretion of calcium, hyperparathyroidism, intestinal resection, or gout. Many cases are idiopathic.

Therapy

If the cause is known, it should be treated, eliminated, or corrected. Determining the predominant component of the stones is helpful in planning therapy.

Liberal fluid intake of 3,000 ml or more per day is recommended regardless of the type of stone to prevent the formation of urine so concentrated that salts precipitate out. No dietary modification can dissolve existing stones, but it may be helpful in preventing the development of new stones.

In the past, depending on the composition of the stone, an acid ash or alkaline ash diet was prescribed. (See Chapter 4.) These diets are seldom used today. Medications are commonly employed.

Calcium Stones If stones are idiopathic, dietary calcium may be limited to 600 mg/day or less, although the value of restricting dietary calcium is controversial. Increased fiber is taken to bind excess calcium. Thiazide diuretics to "flush out" the stones or exchange resins to bind calcium in the intestine and prevent its absorption may be employed. If water supplies are high in calcium, an alternate source of drinking and cooking water may be necessary. There is some evidence that high sodium intake increases urinary calcium.

Oxalate Stones If the stones are caused by hyperoxaluria, a diet low in oxalate is indicated. Vitamin C should be taken only in normal amounts; megadoses can cause stones in predisposed people. If steatorrhea is present, dietary fat is reduced to less than 50 g/day. *Cholestyramine* (see Chapter 24), which binds with oxalate, may be prescribed.

Urinary Calculi: also known as urolithiasis, kidney stones, renal calculi, or nephrolithiasis

RISK FACTORS FOR KIDNEY STONES

Gender—male
Race—Asian and white
Age—30 to 50
Geography—high in Southeast Asia; low in Africa
Sedentary lifestyle
Immobility causing urinary stasis
Persistent urinary tract infections
Genetic—altered cystine and purine metabolism

FOODS HIGH IN OXALATE

asparagus, green and wax beans, beets and beet greens, cashews, chard, chocolate, cocoa, endive, grapes, okra, plums, raspberries, rhubarb, spinach, sweet potatoes, tea

Uric Acid Stones   These stones are related to altered purine metabolism and are sometimes a complication of gout. A diet reduced in purines is sometimes prescribed. (See Chapter 27.) *Allopurinal* and/or *probenecid* may be used in conjunction with or instead of dietary purine restriction.

Other Treatments   If stones are too large to pass through the ureter and are causing symptoms, surgical removal may be necessary to prevent infection, pain, and deterioration of renal function. Another therapy, extracorporeal shock wave lithotripsy (ESWL), shatters stones into particles small enough to pass through the ureter.

PUTTING IT ALL TOGETHER

The kidneys perform so many functions that any disease of these organs is going to have far-reaching effects on body systems. Sometimes multiple dietary modifications are essential to compensate for the decreased functional capacity of the kidneys. Prevention and treatment of kidney stones is also necessary.

? QUESTIONS CLIENTS ASK

1. I've heard that a diet high in sodium causes kidney stones. Is this true?

It does appear that individuals prone to form calcium stones excrete more calcium in the urine as dietary sodium intake increases. High sodium intakes also decrease urinary citrate, which helps prevent crystallization of calcium. Whether limiting sodium will prevent renal calculi formation has not been proven, but it appears prudent for individuals at risk of developing calcium stones to moderate sodium intake. Of course, moderation in dietary sodium has other health benefits.

2. I am on hemodialysis and I find it hard to know what to choose when I go out to eat at a restaurant since I must restrict sodium, potassium, and fluids. Do you have any suggestions?

If possible, you may want to call ahead to ask about the menu. Restaurants with a large variety of foods make choices easier. Surveys show that most chefs and cooks are willing to prepare menu items without sauces or seasonings. Take time to study the menu once you are at the restaurant. Don't be hesitant about asking questions. Many people are health and nutrition conscious nowadays and it is common to inquire. If portions are too large, request that leftovers be wrapped to take home. Be sure to take your phosphate binders.

3. Can the average person change his or her diet to prevent the kidneys from failing?

For a person without any risk factors such as diabetes or evidence of chronic renal insufficiency, there are no dietary modifications necessary. Just eat a variety of foods in moderation for a well-balanced diet. For a person with renal insufficiency, there is evidence that controlling dietary protein and phosphorus can slow the progression of the disease.

✔ CASE STUDY: A Man with Chronic Renal Failure

Ten years ago Earl S., now 59 years old, had a routine physical examination. Urinalysis revealed the presence of protein. The blood urea nitrogen was moderately elevated, and a diagnosis of chronic glomerulonephritis was made. For several years he had no other symptoms, but then began developing hypertension and blurring of vision. As his illness progressed, he was unable to continue at his job as a truck driver and has gone on disability. He has been on a diet of approximately 2,100 kcal and 70 g protein.

Recently he has been losing weight, having dizzy spells, and developing mouth sores, along with a "funny" taste in his mouth. When his wife noticed that his breath had an odor of ammonia, he visited his physician who had him hospitalized for evaluation.

On admission his height was 68 in. (173 cm); weight, 136 lb (62 kg); blood pressure, 200/100; BUN, 85 mg/dl; GFR, 10 ml/minute; hemoglobin, 9.2 g/dl; hematocrit, 19%. A diagnosis of chronic renal failure was made, and a diet of 60 g protein and 2,000 mg sodium with fluids restricted to 1,500 ml daily was ordered.

Questions

1. Why is it important that Mr. S. receive adequate nonprotein calories?

2. What would be a healthy weight for Mr. S.?

3. Calculate the sodium content of the 60 g protein diet in Table 28–1. (See Appendix Table A–1.)

4. Write a day's menu for Mr. S.

5. Why does most of his protein come from meat, eggs, and milk?

6. How can the nonprotein calories in the diet be increased?

7. Mr. S. wonders if he can use a salt substitute. How might you answer?

8. Mr. S. has always eaten a lot of vegetables, and his wife asks if there is any special way she could cook them. How might you answer?

9. Why was Mr. S.'s fluid restricted? Calculate the approximate fluid content of the 60 g protein diet in Table 28–1. (See Tables 4–7 and Appendix Table A–1 for water content of foods.) How much fluid, in addition to diet, can Mr. S. have? What is the approximate urine output expected?

FOR ADDITIONAL READING

Beto, JA: "Which Diet for Which Renal Failure: Making Sense of the Options," *J Am Diet Assoc,* **95**:898–903, 1995.

Byers, JF, and Goshorn, J: "How to Manage Diuretic Therapy," *Am J Nurs,* **95**:38–44, February 1995.

Dobell, E, *et al:* "Food Preferences and Food Habits of Patients with Chronic Renal Failure Undergoing Dialysis," *J Am Diet Assoc,* **93**:1129–35, 1993.

Dunn, SA: "How to Care for the Dialysis Patient," *Am J Nurs,* **93**:26–34, June 1993.

Dwyer, J: "Vegetarian Diets for Treating Nephrotic Syndrome," *Nutr Rev,* **51**:44–46, 1993.

Giddens, JF, *et al:* "Risks and Rewards of Kidney Transplant," *RN,* **56**:56–62, June 1993.

Hruska, KA, and Teitelbaum, SL: "Renal Osteodystrophy," *New Engl J Med,* **333**:166–74, 1995.

Kelly, M: "Chronic Renal Failure: How to Detect This Insidious Disease and Provide Your Patient with the Necessary Support," *Am J Nurs,* **96**:36–37, January 1996.

King, BA: "Detecting Acute Renal Failure," *RN,* **57**:34–40, March 1994.

Klahr, S, *et al:* "The Effects of Dietary Protein Restriction and Blood-Pressure Control on the Progression of Chronic Renal Disease," *New Engl J Med,* **330**:877–84, 1994.

Lemann, J, Jr: "Composition of the Diet and Calcium Kidney Stones," *New Engl J Med,* **328**:880–81, 1993.

Massey, LK, and Whiting, SJ: "Dietary Salt, Urinary Calcium, and Kidney Stone Risk," *Nutr Rev,* **53**:131–34, 1995.

"Morbidity and Mortality of Dialysis," *NIH Consens Statement,* **11**:1–33, November 1993.

Remuzzi, G, and Ruggenenti, P: "Slowing the Progression of Diabetic Nephropathy," *New Engl J Med,* **329**:1496–97, 1993.

Ruth-Sahd, LA: "Renal Calculi," *Am J Nurs,* **95**:50, November 1995.

Sanders, HN, *et al:* "Nutritional Implications of Recombinant Human Erythropoietin Therapy in Renal Disease," *J Am Diet Assoc,* **94**:1023–29, 1994.

Stark, JL: "Interpreting B.U.N/Creatinine Levels: It's Not as Simple as You Think," *Nurs 94,* **24**:58–61. September 1994.

Nutrition in Human Immunodeficiency Virus (HIV) Infection and Acquired Immunodeficiency Syndrome (AIDS)

Chapter Outline

Human Immunodeficiency Virus (HIV)
Acquired Immunodeficiency Syndrome
  (AIDS)

Putting It All Together
Questions Clients Ask
Case Study: A Young Man Is HIV-Positive

*Increase to at least 75 percent the proportion of primary care and mental health care providers who provide age-appropriate counseling on the prevention of HIV and other sexually transmitted diseases.**

✔ **CASE STUDY PREVIEW**

Jeremy, a young man with a past history of drug abuse, has been diagnosed as HIV-positive. You will help plan for his care to slow the course of the infection and improve the quality of his life.

---

*Healthy People 2000: National Health Promotion and Disease Prevention Objectives. Public Health Service. U.S. Department of Health and Human Services, Washington, DC, 1991. Objective 19.14, p. 121.

Since the first case of AIDS was identified in 1982, considerable research has been conducted on its causes and management. Indeed, the human immunodeficiency virus (HIV) may be the most thoroughly studied virus in biomedical research. New therapies are constantly being developed and tried for an infection that now affects one million individuals or one in every 250 people in the United States. HIV/AIDS and their therapies can have devastating nutritional consequences. For a review of the effects of the HIV virus on the immune system, consult Chapter 19.

HUMAN IMMUNODEFICIENCY VIRUS (HIV)

HIV is a **retrovirus,** with reverse coding that allows genetic transcribing from RNA to DNA rather than the customary DNA to RNA. After entering the body the virus joins with host cells, especially CD_4 T lymphocytes.

Latent Period

CD_4 T cells: also known as CD_4 cells, CD_4 lymphocytes, T-4 cells

Although the course of HIV/AIDS varies among individuals, there is a general pattern. Three to six weeks after initial infection, 50 to 70 percent of patients develop flu-like symptoms. Immune responses to the virus develop, followed by a period of latency when the patient is usually without outward symptoms. However, during this time there is gradual deterioration of humoral and cell-mediated immune function, especially depletion of the CD_4 T cells. CD_4 cell counts can decline from a normal 1000/cu mm to 500 to 200 to 50 or less.

Therapy

GOALS FOR NUTRITIONAL MANAGEMENT OF HIV

Preserve lean body mass
Prevent malnutrition
Maintain physical and mental functioning
Control malabsorption
Prevent wasting
Maximize nutrition-related immune function
Improve quality of life
Increase life span

A nutritional assessment should be done when the HIV-positive diagnosis is confirmed. Although the patient may be asymptomatic, adequate diet and food safety (see Chapter 2) should be emphasized in counseling. (See Figure 29–1.) Vitamin and mineral supplements may be indicated.

Preventive *isoniazid* therapy may be instituted because the immunosuppression allows *mycobacterium tuberculosis* to proliferate rapidly, increasing the likelihood of developing active tuberculosis. (See Chapter 20.) Antiretroviral therapy may be instituted in HIV-infected individuals whose CD_4 counts fall below 500. Medications may be used as prophylaxis against various opportunistic infections.

ACQUIRED IMMUNODEFICIENCY SYNDROME (AIDS)

With the severely compromised immune system, symptoms begin to appear, culminating in Acquired Immunodeficiency Syndrome (AIDS). The rate of progression varies from individual to individual. For criteria for a diagnosis of AIDS developed by the Centers for Disease Control and Prevention (CDC), see Figure 29–2.

Clinical Findings

Infections Opportunistic infections (see Chapter 19) are frequent. One of the most common symptoms is difficulty or pain with chewing and swallowing due to candidiasis, a fungal infection of the mouth and/or esophagus caused by *Candida albicans.* Among other common infectious agents is *Cytomegalovirus (CMV),* which causes ulceration of the esophagus, stomach, and the small and

445

NUTRITION IN HUMAN
IMMUNODEFICIENCY VIRUS
(HIV) INFECTION AND
ACQUIRED IMMUNODEFICIENCY
SYNDROME (AIDS)

FIGURE 29–1
The home health nurse reviews nutrient-dense foods and the importance of food
sanitation with this patient who has been diagnosed as HIV-positive. (*Source:* Anthony
Magnacca/Merrill.)

large intestine. It may also be involved with hepatitis and visual problems.
Mycobacterium avium intercellulare Complex (MAC) affects the small and large
intestine, blunting the villi and leading to malabsorption, diarrhea, fever, and
extreme fatigue. **Toxoplasmosis** caused by *Toxoplasma gondi* frequently results
in encephalitis with chills, fever, and mental changes. Herpes simplex affects the
mouth and rectum. The parasite *Cryptosporidium* colonizes in the brush border
of the small intestine, causing mucosal damage, diarrhea, and malabsorption. AIDS
enteropathy with diarrhea and malabsorption but without bacterial or parasitic
infection has also been observed.

A patient with diarrhea may have 10 to 30 bowel movements a day measuring
10 to 15 liters. Because of the rapid transit through the small bowel, the
absorption of all nutrients is markedly reduced. The losses of fluids and electrolytes from the large bowel are often great.

Pneumocystis carinii pneumonia *(PCP)* occurs in many patients with accompanying fever and cough. Tuberculosis from the activation of a latent or new
infection is not uncommon. (See Chapter 20.)

Secondary Malignancies Kaposi's sarcoma frequently presents lesions in
the mouth, esophagus, small intestine, and lungs. Dyspnea and coughing may
result. Lymphomas of the small and large bowel and squamous cell carcinoma of
the tongue and rectum can also develop.

Neurological Problems Over half of AIDS patients have neurological
involvement which may be caused by opportunistic infections, lymphomas, or by
direct infection with HIV affecting the brain and spinal cord. Symptoms include
decreased cognitive and motor function and progressive dementia. The peripheral nervous system can also be affected with weakness, pain, and sometimes
paralysis resulting.

1. CD_4 lymphocyte count <200/cu mm (or CD_4 % <14) in person with confirmed HIV infection
2. A diagnosis of any of the conditions below in a person with confirmed HIV infection:

   Candidiasis of bronchi, trachea, or lungs
   Candidiasis, esophageal
   Cervical cancer, invasive
   Coccidioidomycosis, disseminated or extrapulmonary
   Cryptococcosis, extrapulmonary
   Cryptosporidiosis, chronic intestinal (greater than one month's duration)
   Cytomegalovirus disease (other than liver, spleen, or nodes)
   Cytomegalovirus retinitis (with loss of vision)
   Encephalopathy, HIV-related
   Herpes simplex; chronic ulcer(s) (greater than one month's duration); or bronchitis, pneumonitis, or esophagitis
   Histoplasmosis, disseminated or extrapulmonary
   Isosporiasis, chronic intestinal (greater than one month's duration)
   Kaposi's sarcoma
   Lymphoma, Burkitt's (or equivalent term)
   Lymphoma, immunoblastic (or equivalent term)
   Lymphoma, primary, of brain
   Mycobacterium avium complex or *M. kansasii,* disseminated or extrapulmonary
   Mycobacterium tuberculosis, any site (pulmonary or extrapulmonary)
   Mycobacterium, other species or unidentified species, disseminated or extrapulmonary
   Pneumocystis carinii pneumonia (PCP)
   Pneumonia, recurrent
   Progressive multifocal leukoencephalopathy
   Salmonella septicemia, recurrent
   Toxoplasmosis of brain
   Wasting due to HIV

FIGURE 29–2
Criteria for a diagnosis of AIDS developed by Centers for Disease Control and Prevention (CDC). (*Source:* Adapted from Cohen, P.T., ed. *The AIDS Knowledge Base.* 2nd ed. Boston: Little, Brown and Company, 1994.)

Protein-Energy Malnutrition PEM from multifactorial causes develops in most people with AIDS and can appear at any stage of the disease. Metabolic alterations, malabsorption, increased host requirements, and/or decreased intake are involved. Malnutrition can further compromise the immune system. Cachexia (see Chapter 25) is common.

A wasting syndrome involving involuntary weight loss, chronic diarrhea, and fever frequently occurs. In HIV wasting there is a disproportionate loss of lean body mass and less of body fat. Weight loss is often great; in Africa AIDS is called "slim" disease. Depressed serum albumin is associated with increased morbidity and mortality. In children with AIDS, malnutrition can lead to growth failure.

Therapy

As the disease progresses, dietary intervention is dictated by the systems and organs affected and by the patient's tolerance. For infections or lesions in the

447

NUTRITION IN HUMAN
IMMUNODEFICIENCY VIRUS
(HIV) INFECTION AND
ACQUIRED IMMUNODEFICIENCY
SYNDROME (AIDS)

TABLE 29–1
Possible Adverse Effects of Antiretroviral Medications

Medication	Effects
Zidovudine (ZDV)	Anorexia, nausea, anemia, headache, seizures, confusion, hepatitis, myositis (inflamed muscles)
Dideoxyinosine (ddI)	Nausea, diarrhea, confusion, seizures, peripheral neuropathy, pancreatitis
Dideoxycitidine (ddC)	Esophageal ulcers, ulcers on the mucous membranes, stomatitis, skin eruptions, peripheral neuropathy, pancreatitis

oral cavity or esophagus, diet consistency may be modified to soft or liquid. *Sucalfate* (see Chapter 22) to coat the mouth and esophagus has been suggested for relief of mouth pain while eating. A complete supplement may be used to help provide adequate intake. A supplement containing 1.25 kcal/ml and formulated especially for patients with HIV is available.* If lesions are too painful to permit swallowing, a gastrostomy tube or total parenteral nutrition (see Chapter 17) are possible alternate feeding modes. Vitamin and mineral supplementation is frequently necessary, but there is no indication for megadoses. Food safety and prevention of exposure to environmental sources of infection are emphasized.

Diarrhea and malabsorption are often difficult to control. A diet restricted in lactose (see Chapter 22) and fiber and low in fat may be of value. A lactose-free or chemically defined formula diet taken orally and/or by tube can supplement other intake. Gluten-free diets have been suggested when there is great mucosal damage. Diarrheas that do not respond to any of these interventions may necessitate total parenteral nutrition. However, individuals with compromised immune systems are at increased risk of catheter-related infections. In some cases, home TPN may be feasible.

Fevers are treated with increased caloric intake to meet hypermetabolic needs and with antipyretic medications. (See Chapter 20.) Chemotherapy and radiation therapy are employed for the treatment of cancers. Patients with dementia may need assistance with feeding.

Medications Since AIDS manifestations can involve a great variety of organisms with a host of symptoms affecting different organs and systems, many medications are used to treat AIDS. Furthermore, numerous new ones are being developed. Patients with advanced HIV infection may take 20 to 40 pills a day. Side effects include nausea and vomiting, stomatitis, and taste alterations. Some strategies for addressing these problems appear in Table 25–1.

Antiretroviral medications that block HIV replication, thereby increasing survival time, are commonly used. Currently, *zidovudine (ZDV), dideoxyinosine (ddI),* and *dideoxycitidine (ddC)* are in use. Their side effects, which can be considerable and sometimes toxic, appear in Table 29–1.

Zidovudine: formerly called azidothymidine or AZT

Two medications have been approved for anorexia associated with weight loss from HIV/AIDS. With *megestrol acetate*, a synthetic oral progesterone, the weight gain may be primarily as adipose tissue rather than lean body mass. It can also cause impotence in men and is contraindicated in pregnant women.

---

*Advera ®, Ross Laboratories, Columbus, OH.

Dronabinol, a synthetic extract of marijuana, may produce decreased memory and cognitive function along with the desired increased appetite.

Alternative Therapies Many individuals with AIDS turn to unproven therapies such as megadoses of vitamins, yeast-free diets to prevent opportunistic infections, macrobiotic diets, and dietary manipulations intended to boost immune power. Cleansing regimens involving fasting, starvation, and colonic enemas are followed to starve the infections and rid the body of toxins. For patients following such courses, every effort should be made to ensure dietary adequacy. See the Food Guide Pyramid (Figure 1–5) and/or Basic Diet (Table 2–4).

PUTTING IT ALL TOGETHER

HIV/AIDS can take a tremendous toll on the body. Wasting with great weight loss and loss of lean body tissue are common. Currently HIV/AIDS cannot be cured, but aggressive nutrition support can help to extend life and contribute to its quality.

? QUESTIONS CLIENTS ASK

1. Is there any diet that can prevent HIV infection and AIDS?

There is no diet or dietary modification that can prevent HIV infection. Modification of health-related behaviors is essential. If HIV infection is present, nutrition assessment and aggressive nutrition support can modify the course of the disease and improve quality of life.

2. Is there a diet that can boost my immune response?

Studies, primarily on laboratory animals, have shown that nutrient deficiencies can impair immune response. However, there is no evidence that large doses of supplements can improve it. Indeed, some supplements may have the opposite effect. An adequate, varied, balanced diet is the best course.

3. I've heard that people with HIV infections and AIDS are at high risk for foodborne infections. How can such infections be prevented?

People with AIDS, for example, are at least 20 times more likely to develop a *salmonella* infection and six times more likely to develop a blood infection from it than other people. For healthy people, the cramps, vomiting, and diarrhea are unpleasant, but short-lived. For individuals with AIDS, the symptoms are much more severe and may involve infections that resist treatment or may even be fatal. Patronize stores where cleanliness is apparent. Do not consume any raw or undercooked animal food. Refrigerate foods promptly. Wash hands, utensils, counters, and cutting surfaces between preparing different foods and afterward.

✔ CASE STUDY: A Young Man Is HIV-Positive

Jeremy A., age 27, had a history of intravenous drug use in his late teens. He successfully completed a drug rehabilitation program, returned to college, and "turned his life around." Currently he is working as a salesman for a large company with good employment prospects for the future.

Recently, symptoms of fatigue, an unplanned weight loss of 13 lb (6 kg), diarrhea, and a sore mouth led him to visit his HMO. A test for the HIV virus proved positive as did a repeat test.

A team of HIV specialists including a dietitian, nurse, psychologist, physician, and social worker met with Jeremy. An assessment and care plan were prepared.

Jeremy lives alone in a studio apartment in a complex where many other young professional people live. For breakfast, he usually gets coffee and a pastry on the way to work. For lunch, he generally eats with coworkers at a fast-food restaurant or food court. When he is traveling on business, he stops at whatever restaurant is available or may have lunch with a client. For dinner he often gets take-out food. Sometimes he goes to a restaurant with coworkers or on a date.

Questions

1. Why is it more appropriate to compare Jeremy's present weight to his usual weight than to compare it to a weight chart? What would be the disadvantage of the latter?

2. From the available data, does Jeremy's food intake appear inadequate in any group from the Food Guide Pyramid? If so, which?

3. Would you make any suggestions to Jeremy to modify his food intake? If so, justify the suggestions on the basis of nutritional adequacy and his lifestyle.

4. Should Jeremy take vitamin/mineral supplements? Explain.

5. Jeremy eats out a great deal. Make suggestions for food choices in light of nutritional adequacy and food safety.

6. Make suggestions for controlling diarrhea.

7. Make suggestions for managing sore mouth.

8. The physician plans to begin a course of *zidovudine (ZDV)* therapy. What are possible side effects of this medication? How may they be controlled or minimized?

9. What community agencies and services might be suggested to Jeremy for support and assistance?

FOR ADDITIONAL READING

Anastasi, JK, and Lee VS: "HIV Wasting: How to Stop the Cycle," *Am J Nurs,* 94:18–25, June 1994.

Black, RD: "Transmission of HIV-1 in the Breast-Feeding Process," *J Am Diet Assoc,* **96:**267–76, 1996.

Chlebowski, RT, *et al:* "Dietary Intake and Counseling, Weight Maintenance, and the Course of HIV Infection," *J Am Diet Assoc,* **95:**428–35, 1995.

Gorbach, SL, *et al:* "Interactions Between Nutrition and Infection with Human Immunodeficiency Virus," *Nutr Rev,* **51:**226–34, 1993.

Greely, A: "Concern About AIDS in Minority Communities," *FDA Consumer,* **29:**11–15, December 1995.

Kaplan, JE, *et al:* "Reducing the Impact of Opportunistic Infections in Patients with HIV Infection: New Guidelines," *JAMA,* **274:**347–48, 1995.

Managing Early HIV Infection, Agency for Health Care Policy and Research. Pub. No. 94-0573. Public Health Service, Rockville, MD, 1994.

McKinley, MJ, *et al:* "Improved Body Weight Status as a Result of Nutrition Intervention in Adult, HIV-Positive Outpatients," *J Am Diet Assoc,* **94:**1014–17, 1994.

"New Guidelines for Preventing Opportunistic Infections," *Nurs 95,* **25:**32I–32O, November 1995.

Pantaleo, G, *et al:* "The Immunopathogenesis of Human Immunodeficiency Virus Infection," *New Engl J Med,* **328:**327–35, 1993.

Peckham, C, and Gibb, D: "Mother-to-Child Transmission of the Human Immunodeficiency Virus," *New Engl J Med,* **333:**298–302, 1995.

"Position of the American Dietetic Association and the Canadian Dietetic Association: Nutrition Intervention in the Care of Persons with Human Immunodeficiency Virus Infection," *J Am Diet Assoc,* **94:**1042–45, 1994.

Schmidt, J, and Crespo-Fierro, M: "Who Says There's Nothing We Can Do?" *RN,* **58:**30–36, October 1995.

Timbo, BB, and Tollefson, L: "Nutrition: A Cofactor in HIV Disease," *J Am Diet Assoc,* **94:**1019–22, 1994.

Ungvarski, PJ: "Waging War on HIV Wasting," *RN,* **59:**26–33, February 1996.

Whipple, B, and Scura KW: "The Overlooked Epidemic: HIV in Older Adults." *Am J Nurs,* **96:**22–29, February 1996.

Appendices

TABLE A–1 Nutritive Value of the Edible Part of Food (Tr indicates nutrient present in trace amount)

Foods, approximate measures, units, and weight (weight of edible portion only)		Water	Food energy	Protein	Fat	Fatty Acids Saturated	Mono-unsaturated	Poly-unsaturated
BEVERAGES		Grams	Percent	Calories	Grams	Grams	Grams	Grams
						Grams	Grams	Grams
Alcoholic:								
Beer:								
Regular	12 fl oz	360	92	150	1	0	0.0	0.0	0.0
Light	12 fl oz	355	95	95	1	0	0.0	0.0	0.0
Gin, rum, vodka, whiskey:								
80-proof	1½ fl oz	42	67	95	0	0	0.0	0.0	0.0
86-proof	1½ fl oz	42	64	105	0	0	0.0	0.0	0.0
90-proof	1½ fl oz	42	62	110	0	0	0.0	0.0	0.0
Wines:								
Dessert	3½ fl oz	103	77	140	Tr	0	0.0	0.0	0.0
Table:								
Red	3½ fl oz	102	88	75	Tr	0	0.0	0.0	0.0
White	3½ fl oz	102	87	80	Tr	0	0.0	0.0	0.0
Carbonated:[2]								
Club soda	12 fl oz	355	100	0	0	0	0.0	0.0	0.0
Cola type:								
Regular	12 fl oz	369	89	160	0	0	0.0	0.0	0.0
Diet, artificially sweetened	12 fl oz	355	100	Tr	0	0	0.0	0.0	0.0
Ginger ale	12 fl oz	366	91	125	0	0	0.0	0.0	0.0
Grape	12 fl oz	372	88	180	0	0	0.0	0.0	0.0
Lemon-lime	12 fl oz	372	89	155	0	0	0.0	0.0	0.0
Orange	12 fl oz	372	88	180	0	0	0.0	0.0	0.0
Pepper type	12 fl oz	369	89	160	0	0	0.0	0.0	0.0
Root beer	12 fl oz	370	89	165	0	0	0.0	0.0	0.0
Cocoa and chocolate-flavored beverages. See Dairy Products.								
Coffee:								
Brewed	6 fl oz	180	100	Tr	Tr	Tr	Tr	Tr	Tr
Instant, prepared (2 tsp powder plus 6 fl oz water)	6 fl oz	182	99	Tr	Tr	Tr	Tr	Tr	Tr
Fruit drinks, noncarbonated:								
Canned:								
Fruit punch drink	6 fl oz	190	88	85	Tr	0	0.0	0.0	0.0
Grape drink	6 fl oz	187	86	100	Tr	0	0.0	0.0	0.0
Pineapple-grapefruit juice drink	6 fl oz	187	87	90	Tr	Tr	Tr	Tr	Tr
Frozen:								
Lemonade concentrate:								
Undiluted	6-fl-oz can	219	49	425	Tr	Tr	Tr	Tr	Tr
Diluted with 4⅓ parts water by volume	6 fl oz	185	89	80	Tr	Tr	Tr	Tr	Tr
Limeade concentrate:								
Undiluted	6-fl-oz can	218	50	410	Tr	Tr	Tr	Tr	Tr
Diluted with 4⅓ parts water by volume	6 fl oz	185	89	75	Tr	Tr	Tr	Tr	Tr
Fruit juices. See type under Fruits and Fruit Juices.								
Milk beverages. See Dairy Products.								
Tea:								
Brewed	8 fl oz	240	100	Tr	Tr	Tr	Tr	Tr	Tr
Instant, powder, prepared:								
Unsweetened (1 tsp powder plus 8 fl oz water)	8 fl oz	241	100	Tr	Tr	Tr	Tr	Tr	Tr
Sweetened (3 tsp powder plus 8 fl oz water)	8 fl oz	262	91	85	Tr	Tr	Tr	Tr	Tr

Adapted from Gebhardt, S. E., and Matthews, R. H.: *Nutritive Value of Foods.* Home and Garden Bulletin 72, Agricultural Research Service, U.S. Department of Agriculture, Washington, DC, revised June 1991.

[1]Value not determined.

[2]Mineral content varies depending on water source.

Cho-lesterol	Carbo-hydrate	Calcium	Phos-phorus	Iron	Potas-sium	Sodium	Vitamin A value	Thiamin	Ribo-flavin	Niacin	Ascorbic acid
Milli-grams	Grams	Milli-grams	Milli-grams	Milli-grams	Milli-grams	Milli-grams	Retinol equiv. (RE)	Milli-grams	Milli-grams	Milli-grams	Milli-grams
0	13	14	50	0.1	115	18	0	0.02	0.09	1.8	0
0	5	14	43	0.1	64	11	0	0.03	0.11	1.4	0
0	Tr	Tr	Tr	Tr	1	Tr	0	Tr	Tr	Tr	0
0	Tr	Tr	Tr	Tr	1	Tr	0	Tr	Tr	Tr	0
0	Tr	Tr	Tr	Tr	1	Tr	0	Tr	Tr	Tr	0
0	8	8	9	0.2	95	9	([1])	0.01	0.02	0.2	0
0	3	8	18	0.4	113	5	([1])	0.00	0.03	0.1	0
0	3	9	14	0.3	83	5	([1])	0.00	0.01	0.1	0
0	0	18	0	Tr	0	78	0	0.00	0.00	0.0	0
0	41	11	52	0.2	7	18	0	0.00	0.00	0.0	0
0	Tr	14	39	0.2	7	[3]32	0	0.00	0.00	0.0	0
0	32	11	0	0.1	4	29	0	0.00	0.00	0.0	0
0	46	15	0	0.4	4	48	0	0.00	0.00	0.0	0
0	39	7	0	0.4	4	33	0	0.00	0.00	0.0	0
0	46	15	4	0.3	7	52	0	0.00	0.00	0.0	0
0	41	11	41	0.1	4	37	0	0.00	0.00	0.0	0
0	42	15	0	0.2	4	48	0	0.00	0.00	0.0	0
0	Tr	4	2	Tr	124	2	0	0.00	0.02	0.4	0
0	1	2	6	0.1	71	Tr	0	0.00	0.03	0.6	0
0	22	15	2	0.4	48	15	2	0.03	0.04	Tr	[4]61
0	26	2	2	0.3	9	11	Tr	0.01	0.01	Tr	[4]64
0	23	13	7	0.9	97	24	6	0.06	0.04	0.5	[4]110
0	112	9	13	0.4	153	4	4	0.04	0.07	0.7	66
0	21	2	2	0.1	30	1	1	0.01	0.02	0.2	13
0	108	11	13	0.2	129	Tr	Tr	0.02	0.02	0.2	26
0	20	2	2	Tr	24	Tr	Tr	Tr	Tr	Tr	4
0	Tr	0	2	Tr	36	1	0	0.00	0.03	Tr	0
0	1	1	4	Tr	61	1	0	0.00	0.02	0.1	0
0	22	1	3	Tr	49	Tr	0	0.00	0.04	0.1	0

[3]Blend of aspartame and saccharin; if only sodium saccharin is used, sodium is 75 mg; if only aspartame is used, sodium is 23 mg.
[4]With added ascorbic acid.

Foods, approximate measures, units, and weight (weight of edible portion only)		Water		Food energy	Protein	Fat	FATTY ACIDS		
							Satu-rated	Mono-unsaturated	Poly-unsaturated
DAIRY PRODUCTS		Grams	Percent	Calories	Grams	Grams	Grams	Grams	Grams
Butter. See Fats and Oils.									
Cheese:									
Natural:									
Blue	1 oz	28	42	100	6	8	5.3	2.2	0.2
Camembert, 1⅓ oz wedge	1 wedge	38	52	115	8	9	5.8	2.7	0.3
Cheddar:									
Cut pieces	1 oz	28	37	115	7	9	6.0	2.7	0.3
Shredded	1 cup	113	37	455	28	37	23.8	10.6	1.1
Cottage (curd not pressed down):									
Creamed (cottage cheese, 4% fat):									
Large curd	1 cup	225	79	235	28	10	6.4	2.9	0.3
Small curd	1 cup	210	79	215	26	9	6.0	2.7	0.3
With fruit	1 cup	226	72	280	22	8	4.9	2.2	0.2
Lowfat (2%)	1 cup	226	79	205	31	4	2.8	1.2	0.1
Uncreamed (cottage cheese dry curd, less than ½% fat)	1 cup	145	80	125	25	1	0.4	0.2	Tr
Cream	1 oz	28	54	100	2	10	6.2	2.8	0.4
Feta	1 oz	28	55	75	4	6	4.2	1.3	0.2
Mozzarella, made with:									
Whole milk	1 oz	28	54	80	6	6	3.7	1.9	0.2
Part skim milk (low moisture)	1 oz	28	49	80	8	5	3.1	1.4	0.1
Muenster	1 oz	28	42	105	7	9	5.4	2.5	0.2
Parmesan, grated:									
Cup, not pressed down	1 cup	100	18	455	42	30	19.1	8.7	0.7
Tablespoon	1 tbsp	5	18	25	2	2	1.0	0.4	Tr
Ounce	1 oz	28	18	130	12	9	5.4	2.5	0.2
Provolone	1 oz	28	41	100	7	8	4.8	2.1	0.2
Ricotta, made with:									
Whole milk	1 cup	246	72	430	28	32	20.4	8.9	0.9
Part skim milk	1 cup	246	74	340	28	19	12.1	5.7	0.6
Swiss	1 oz	28	37	105	8	8	5.0	2.1	0.3
Pasteurized process cheese:									
American	1 oz	28	39	105	6	9	5.6	2.5	0.3
Swiss	1 oz	28	42	95	7	7	4.5	2.0	0.2
Pasteurized process cheese food, American	1 oz	28	43	95	6	7	4.4	2.0	0.2
Pasteurized process cheese spread, American	1 oz	28	48	80	5	6	3.8	1.8	0.2
Cream, sweet:									
Half-and-half (cream and milk)	1 cup	242	81	315	7	28	17.3	8.0	1.0
	1 tbsp	15	81	20	Tr	2	1.1	0.5	0.1
Light, coffee, or table	1 cup	240	74	470	6	46	28.8	13.4	1.7
	1 tbsp	15	74	30	Tr	3	1.8	0.8	0.1
Whipping, unwhipped (volume about double when whipped):									
Light	1 cup	239	64	700	5	74	46.2	21.7	2.1
	1 tbsp	15	64	45	Tr	5	2.9	1.4	0.1
Heavy	1 cup	238	58	820	5	88	54.8	25.4	3.3
	1 tbsp	15	58	50	Tr	6	3.5	1.6	0.2
Whipped topping, (pressurized)	1 cup	60	61	155	2	13	8.3	3.9	0.5
	1 tbsp	3	61	10	Tr	1	0.4	0.2	Tr
Cream, sour	1 cup	230	71	495	7	48	30.0	13.9	1.8
	1 tbsp	12	71	25	Tr	3	1.6	0.7	0.1
Cream products, imitation (made with vegetable fat):									
Sweet:									
Creamers:									
Liquid (frozen)	1 tbsp	15	77	20	Tr	1	1.4	Tr	Tr
Powdered	1 tsp	2	2	10	Tr	1	0.7	Tr	Tr

Cho-lesterol	Carbo-hydrate	Calcium	Phos-phorus	Iron	Potas-sium	Sodium	Vitamin A value	Thiamin	Ribo-flavin	Niacin	Ascorbic acid
Milli-grams	Grams	Milli-grams	Milli-grams	Milli-grams	Milli-grams	Milli-grams	Retinol equiv. (RE)	Milli-grams	Milli-grams	Milli-grams	Milli-grams
21	1	150	110	0.1	73	396	65	0.01	0.11	0.3	0
27	Tr	147	132	0.1	71	320	96	0.01	0.19	0.2	0
30	Tr	204	145	0.2	28	176	86	0.01	0.11	Tr	0
119	1	815	579	0.8	111	701	342	0.03	0.42	0.1	0
34	6	135	297	0.3	190	911	108	0.05	0.37	0.3	Tr
31	6	126	277	0.3	177	850	101	0.04	0.34	0.3	Tr
25	30	108	236	0.2	151	915	81	0.04	0.29	0.2	Tr
19	8	155	340	0.4	217	918	45	0.05	0.42	0.3	Tr
10	3	46	151	0.3	47	19	12	0.04	0.21	0.2	0
31	1	23	30	0.3	34	84	124	Tr	0.06	Tr	0
25	1	140	96	0.2	18	316	36	0.04	0.24	0.3	0
22	1	147	105	0.1	19	106	68	Tr	0.07	Tr	0
15	1	207	149	0.1	27	150	54	0.01	0.10	Tr	0
27	Tr	203	133	0.1	38	178	90	Tr	0.09	Tr	0
79	4	1,376	807	1.0	107	1,861	173	0.05	0.39	0.3	0
4	Tr	69	40	Tr	5	93	9	Tr	0.02	Tr	0
22	1	390	229	0.3	30	528	49	0.01	0.11	0.1	0
20	1	214	141	0.1	39	248	75	0.01	0.09	Tr	0
124	7	509	389	0.9	257	207	330	0.03	0.48	0.3	0
76	13	669	449	1.1	307	307	278	0.05	0.46	0.2	0
26	1	272	171	Tr	31	74	72	0.01	0.10	Tr	0
27	Tr	174	211	0.1	46	406	82	0.01	0.10	Tr	0
24	1	219	216	0.2	61	388	65	Tr	0.08	Tr	0
18	2	163	130	0.2	79	337	62	0.01	0.13	Tr	0
16	2	159	202	0.1	69	381	54	0.01	0.12	Tr	0
89	10	254	230	0.2	314	98	259	0.08	0.36	0.2	2
6	1	16	14	Tr	19	6	16	0.01	0.02	Tr	Tr
159	9	231	192	0.1	292	95	437	0.08	0.36	0.1	2
10	1	14	12	Tr	18	6	27	Tr	0.02	Tr	Tr
265	7	166	146	0.1	231	82	705	0.06	0.30	0.1	1
17	Tr	10	9	Tr	15	5	44	Tr	0.02	Tr	Tr
326	7	154	149	0.1	179	89	1,002	0.05	0.26	0.1	1
21	Tr	10	9	Tr	11	6	63	Tr	0.02	Tr	Tr
46	7	61	54	Tr	88	78	124	0.02	0.04	Tr	0
2	Tr	3	3	Tr	4	4	6	Tr	Tr	Tr	0
102	10	268	195	0.1	331	123	448	0.08	0.34	0.2	2
5	1	14	10	Tr	17	6	23	Tr	0.02	Tr	Tr
0	2	1	10	Tr	29	12	[5]1	0.00	0.00	0.0	0
0	1	Tr	8	Tr	16	4	Tr	0.00	Tr	0.0	0

[5]Vitamin A value is largely from beta-carotene used for coloring.

Foods, approximate measures, units, and weight (weight of edible portion only)			Water	Food energy	Protein	Fat	FATTY ACIDS Saturated	Mono-unsaturated	Poly-unsaturated
DAIRY PRODUCTS—Continued		Grams	Percent	Calories	Grams	Grams	Grams	Grams	Grams
Cream products, imitation (made with vegetable fat):									
Whipped topping:									
Frozen	1 cup	75	50	240	1	19	16.3	1.2	0.4
	1 tbsp	4	50	15	Tr	1	0.9	0.1	Tr
Powdered, made with whole milk	1 cup	80	67	150	3	10	8.5	0.7	0.2
	1 tbsp	4	67	10	Tr	Tr	0.4	Tr	Tr
Pressurized	1 cup	70	60	185	1	16	13.2	1.3	0.2
	1 tbsp	4	60	10	Tr	1	0.8	0.1	Tr
Sour dressing (filled cream type product, nonbutterfat)	1 cup	235	75	415	8	39	31.2	4.6	1.1
	1 tbsp	12	75	20	Tr	2	1.6	0.2	0.1
Ice cream. See Milk desserts, frozen.									
Ice milk. See Milk desserts, frozen.									
Milk:									
Fluid:									
Whole (3.3% fat)	1 cup	244	88	150	8	8	5.1	2.4	0.3
Lowfat (2%)									
No milk solids added	1 cup	244	89	120	8	5	2.9	1.4	0.2
Milk solids added, label claim less than 10 g of protein per cup	1 cup	245	89	125	9	5	2.9	1.4	0.2
Lowfat (1%):									
No milk solids added	1 cup	244	90	100	8	3	1.6	0.7	0.1
Milk solids added, label claim less than 10 g of protein per cup	1 cup	245	90	105	9	2	1.5	0.7	0.1
Nonfat (skim):									
No milk solids added	1 cup	245	91	85	8	Tr	0.3	0.1	Tr
Milk solids added, label claim less than 10 g of protein per cup	1 cup	245	90	90	9	1	0.4	0.2	Tr
Buttermilk	1 cup	245	90	100	8	2	1.3	0.6	0.1
Canned:									
Condensed, sweetened	1 cup	306	27	980	24	27	16.8	7.4	1.0
Evaporated:									
Whole milk	1 cup	252	74	340	17	19	11.6	5.9	0.6
Skim milk	1 cup	255	79	200	19	1	0.3	0.2	Tr
Dried:									
Buttermilk	1 cup	120	3	465	41	7	4.3	2.0	0.3
Nonfat, instant:									
Envelope, 3.2 oz, net wt.[6]	1 envelope	91	4	325	32	1	0.4	0.2	Tr
Cup	1 cup	68	4	245	24	Tr	0.3	0.1	Tr
Milk beverages:									
Chocolate milk (commercial):									
Regular	1 cup	250	82	210	8	8	5.3	2.5	0.3
Lowfat (2%)	1 cup	250	84	180	8	5	3.1	1.5	0.2
Lowfat (1%)	1 cup	250	85	160	8	3	1.5	0.8	0.1
Milk Beverages:									
Cocoa and chocolate-flavored beverages:									
Powder containing nonfat dry milk	1 oz	28	1	100	3	1	0.6	0.3	Tr
Prepared (6 oz water plus 1 oz powder)	1 serving	206	86	100	3	1	0.6	0.3	Tr

[6]Yields 1 qt of fluid milk when reconstituted according to package directions.

							NUTRIENTS IN INDICATED QUANTITY				
Cholesterol	Carbohydrate	Calcium	Phosphorus	Iron	Potassium	Sodium	Vitamin A value	Thiamin	Riboflavin	Niacin	Ascorbic acid
Milligrams	Grams	Milligrams	Milligrams	Milligrams	Milligrams	Milligrams	Retinol equiv. (RE)	Milligrams	Milligrams	Milligrams	Milligrams
0	17	5	6	0.1	14	19	[5]65	0.00	0.00	0.0	0
0	1	Tr	Tr	Tr	1	1	[5]3	0.00	0.00	0.0	0
8	13	72	69	Tr	121	53	[5]39	0.02	0.09	Tr	1
Tr	1	4	3	Tr	6	3	[5]2	Tr	Tr	Tr	Tr
0	11	4	13	Tr	13	43	[5]33	0.00	0.00	0.0	0
0	1	Tr	1	Tr	1	2	[5]2	0.00	0.00	0.0	0
13	11	266	205	0.1	380	113	5	0.09	0.38	0.2	2
1	1	14	10	Tr	19	6	Tr	Tr	0.02	Tr	Tr
33	11	291	228	0.1	370	120	76	0.09	0.40	0.2	2
18	12	297	232	0.1	377	122	139	0.10	0.40	0.2	2
18	12	313	245	0.1	397	128	140	0.10	0.42	0.2	2
10	12	300	235	0.1	381	123	144	0.10	0.41	0.2	2
10	12	313	245	0.1	397	128	145	0.10	0.42	0.2	2
4	12	302	247	0.1	406	126	149	0.09	0.34	0.2	2
5	12	316	255	0.1	418	130	149	0.10	0.43	0.2	2
9	12	285	219	0.1	371	257	20	0.08	0.38	0.1	2
104	166	868	775	0.6	1,136	389	248	0.28	1.27	0.6	8
74	25	657	510	0.5	764	267	136	0.12	0.80	0.5	5
9	29	738	497	0.7	845	293	298	0.11	0.79	0.4	3
83	59	1,421	1,119	0.4	1,910	621	65	0.47	1.89	1.1	7
17	47	1,120	896	0.3	1,552	499	[7]646	0.38	1.59	0.8	5
12	35	837	670	0.2	1,160	373	[7]483	0.28	1.19	0.6	4
31	26	280	251	0.6	417	149	73	0.09	0.41	0.3	2
17	26	284	254	0.6	422	151	143	0.09	0.41	0.3	2
7	26	287	256	0.6	425	152	148	0.10	0.42	0.3	2
1	22	90	88	0.3	223	139	Tr	0.03	0.17	0.2	Tr
1	22	90	88	0.3	223	139	Tr	0.03	0.17	0.2	Tr

[7]With added vitamin A.

Foods, approximate measures, units, and weight (weight of edible portion only)		Water	Food energy	Protein	Fat	FATTY ACIDS Satu- rated	FATTY ACIDS Mono- unsaturated	FATTY ACIDS Poly- unsaturated
DAIRY PRODUCTS—Continued		Grams Percent	Calories	Grams	Grams	Grams	Grams	Grams
Milk beverages:								
Cocoa and chocolate-flavored beverages:								
Powder without nonfat dry milk	¾ oz	21 1	75	1	1	0.3	0.2	Tr
Prepared (8 oz whole milk plus ¾ oz powder)	1 serving	265 81	225	9	9	5.4	2.5	0.3
Eggnog (commercial)	1 cup	254 74	340	10	19	11.3	5.7	0.9
Malted milk:								
Chocolate:								
Powder	¾ oz	21 2	85	1	1	0.5	0.3	0.1
Prepared (8 oz whole milk plus ¾ oz powder)	1 serving	265 81	235	9	9	5.5	2.7	0.4
Natural:								
Powder	¾ oz	21 3	85	3	2	0.9	0.5	0.3
Prepared (8 oz whole milk plus ¾ oz powder)	1 serving	265 81	235	11	10	6.0	2.9	0.6
Shakes, thick:								
Chocolate	10-oz	283 72	335	9	8	4.8	2.2	0.3
Vanilla	10-oz	283 74	315	11	9	5.3	2.5	0.3
Milk desserts, frozen:								
Ice cream, vanilla:								
Regular (about 11% fat):								
Hardened	1 cup	133 61	270	5	14	8.9	4.1	0.5
	3 fl oz	50 61	100	2	5	3.4	1.6	0.2
Soft serve (frozen custard)	1 cup	173 60	375	7	23	13.5	6.7	1.0
Rich (about 16% fat), hardened	1 cup	148 59	350	4	24	14.7	6.8	0.9
Ice milk, vanilla:								
Hardened (about 4% fat)	1 cup	131 69	185	5	6	3.5	1.6	0.2
Soft serve (about 3% fat)	1 cup	175 70	225	8	5	2.9	1.3	0.2
Sherbet (about 2% fat)	1 cup	193 66	270	2	4	2.4	1.1	0.1
Yogurt:								
With added milk solids:								
Made with lowfat milk:								
Fruit-flavored[8]	8-oz	227 74	230	10	2	1.6	0.7	0.1
Plain	8-oz	227 85	145	12	4	2.3	1.0	0.1
Made with nonfat milk	8-oz	227 85	125	13	Tr	0.3	0.1	Tr
Without added milk solids:								
Made with whole milk	8-oz	227 88	140	8	7	4.8	2.0	0.2
EGGS								
Eggs, large (24 oz per dozen):								
Raw:								
Whole, without shell	1 egg	50 75	75	6	5	1.6	1.9	0.7
White	1 white	33 88	15	4	0	0.0	0.0	0.0
Yolk	1 yolk	17 49	60	3	5	1.6	1.9	0.7
Cooked:								
Fried in margarine	1 egg	46 69	90	6	7	1.9	2.7	1.3
Hard-cooked, shell removed	1 egg	50 75	75	6	5	1.6	2.0	0.7
Poached	1 egg	50 75	75	6	5	1.5	1.9	0.7
Scrambled (milk added) in margarine	1 egg	61 73	100	7	7	2.2	2.9	1.3

[8]Carbohydrate content varies widely because of amount of sugar added and amount and solids content of added flavoring. Consult the label if more precise values for carbohydrate and calories are needed.

Cho-lesterol	Carbo-hydrate	Calcium	Phos-phorus	Iron	Potas-sium	Sodium	Vitamin A value	Thiamin	Ribo-flavin	Niacin	Ascorbic acid
Milli-grams	Grams	Milli-grams	Milli-grams	Milli-grams	Milli-grams	Milli-grams	Retinol equiv. (RE)	Milli-grams	Milli-grams	Milli-grams	Milli-grams
0	19	7	26	0.7	136	56	Tr	Tr	0.03	0.1	Tr
33	30	298	254	0.9	508	176	76	0.10	0.43	0.3	3
149	34	330	278	0.5	420	138	203	0.09	0.48	0.3	4
1	18	13	37	0.4	130	49	5	0.04	0.04	0.4	0
34	29	304	265	0.5	500	168	80	0.14	0.43	0.7	2
4	15	56	79	0.2	159	96	17	0.11	0.14	1.1	0
37	27	347	307	0.3	529	215	93	0.20	0.54	1.3	2
30	60	374	357	0.9	634	314	59	0.13	0.63	0.4	0
33	50	413	326	0.3	517	270	79	0.08	0.55	0.4	0
59	32	176	134	0.1	257	116	133	0.05	0.33	0.1	1
22	12	66	51	Tr	96	44	50	0.02	0.12	0.1	Tr
153	38	236	199	0.4	338	153	199	0.08	0.45	0.2	1
88	32	151	115	0.1	221	108	219	0.04	0.28	0.1	1
18	29	176	129	0.2	265	105	52	0.08	0.35	0.1	1
13	38	274	202	0.3	412	163	44	0.12	0.54	0.2	1
14	59	103	74	0.3	198	88	39	0.03	0.09	0.1	4
10	43	345	271	0.2	442	133	25	0.08	0.40	0.2	1
14	16	415	326	0.2	531	159	36	0.10	0.49	0.3	2
4	17	452	355	0.2	579	174	5	0.11	0.53	0.3	2
29	11	274	215	0.1	351	105	68	0.07	0.32	0.2	1
213	1	25	89	0.7	60	63	95	0.03	0.25	Tr	0
0	Tr	2	4	Tr	48	55	0	Tr	0.15	Tr	0
213	Tr	23	81	0.6	16	7	97	0.03	0.11	Tr	0
211	1	25	89	0.7	61	162	114	0.03	0.24	Tr	0
213	1	25	86	0.6	63	62	84	0.03	0.26	Tr	0
212	1	25	89	0.7	60	140	95	0.02	0.22	Tr	0
215	1	44	104	0.7	84	171	119	0.03	0.27	Tr	Tr

NUTRIENTS IN INDICATED QUANTITY

Foods, approximate measures, units, and weight (weight of edible portion only)		Water	Food energy	Protein	Fat	FATTY ACIDS		
						Satu- rated	Mono- unsaturated	Poly- unsaturated
FATS AND OILS		Grams	Percent	Calories	Grams	Grams	Grams	Grams	Grams

		Grams	Percent	Calories	Grams	Grams	Grams	Grams
Butter (4 sticks per lb):								
Stick	½ cup	113	16	810	1	92	57.1	26.4	3.4
Tablespoon (⅛ stick)	1 tbsp	14	16	100	Tr	11	7.1	3.3	0.4
Pat (1 in. square, ⅓ in. high; 90 per lb)	1 pat	5	16	35	Tr	4	2.5	1.2	0.2
Fats, cooking (vegetable shortenings)	1 cup	205	0	1,810	0	205	51.3	91.2	53.5
	1 tbsp	13	0	115	0	13	3.3	5.8	3.4
Lard	1 cup	205	0	1,850	0	205	80.4	92.5	23.0
	1 tbsp	13	0	115	0	13	5.1	5.9	1.5
Margarine:								
Imitation (about 40% fat), soft	8-oz	227	58	785	1	88	17.5	35.6	31.3
	1 tbsp	14	58	50	Tr	5	1.1	2.2	1.9
Regular (about 80% fat):								
Hard (4 sticks per lb):								
Stick	½ cup	113	16	810	1	91	17.9	40.5	28.7
Tablespoon (⅛ stick)	1 tbsp	14	16	100	Tr	11	2.2	5.0	3.6
Pat (1 in. square, ⅓ in. high; 90 per lb)	1 pat	5	16	35	Tr	4	0.8	1.8	1.3
Soft	8-oz	227	16	1,625	2	183	31.3	64.7	78.5
	1 tbsp	14	16	100	Tr	11	1.9	4.0	4.8
Spread (about 60% fat):								
Hard (4 sticks per lb)								
Stick	½ cup	113	37	610	1	69	15.9	29.4	20.5
Tablespoon (⅛ stick)	1 tbsp	14	37	75	Tr	9	2.0	3.6	2.5
Pat (1 in. square, ⅓ in. high; 90 per lb)	1 pat	5	37	25	Tr	3	0.7	1.3	0.9
Soft	8-oz	227	37	1,225	1	138	29.1	71.5	31.3
	1 tbsp	14	37	75	Tr	9	1.8	4.4	1.9
Oils, salad or cooking:								
Corn	1 cup	218	0	1,925	0	218	27.7	52.8	128.0
	1 tbsp	14	0	125	0	14	1.8	3.4	8.2
Olive	1 cup	216	0	1,910	0	216	29.2	159.2	18.1
	1 tbsp	14	0	125	0	14	1.9	10.3	1.2
Peanut	1 cup	216	0	1,910	0	216	36.5	99.8	69.1
	1 tbsp	14	0	125	0	14	2.4	6.5	4.5
Safflower	1 cup	218	0	1,925	0	218	19.8	26.4	162.4
	1 tbsp	14	0	125	0	14	1.3	1.7	10.4
Soybean oil, hydrogenated (partially hardened)	1 cup	218	0	1,925	0	218	32.5	93.7	82.0
	1 tbsp	14	0	125	0	14	2.1	6.0	5.3
Soybean-cottonseed oil blend, hydrogenated	1 cup	218	0	1,925	0	218	39.2	64.3	104.9
	1 tbsp	14	0	125	0	14	2.5	4.1	6.7
Sunflower	1 cup	218	0	1,925	0	218	22.5	42.5	143.2
	1 tbsp	14	0	125	0	14	1.4	2.7	9.2
Salad dressings:								
Commercial:								
Blue cheese	1 tbsp	15	32	75	1	8	1.5	1.8	4.2
French:								
Regular	1 tbsp	16	35	85	Tr	9	1.4	4.0	3.5
Low calorie	1 tbsp	16	75	25	Tr	2	0.2	0.3	1.0
Italian:								
Regular	1 tbsp	15	34	80	Tr	9	1.3	3.7	3.2
Low calorie	1 tbsp	15	86	5	Tr	Tr	Tr	Tr	Tr
Mayonnaise:								
Regular	1 tbsp	14	15	100	Tr	11	1.7	3.2	5.8
Imitation	1 tbsp	15	63	35	Tr	3	0.5	0.7	1.6
Mayonnaise type	1 tbsp	15	40	60	Tr	5	0.7	1.4	2.7

[9]For salted butter; unsalted butter contains 12 mg sodium per stick, 2 mg per tbsp, or 1 mg per pat.

[10]Values for vitamin A are year-round average.

Cholesterol	Carbohydrate	Calcium	Phosphorus	Iron	Potassium	Sodium	Vitamin A value	Thiamin	Riboflavin	Niacin	Ascorbic acid
Milligrams	Grams	Milligrams	Milligrams	Milligrams	Milligrams	Milligrams	Retinol equiv. (RE)	Milligrams	Milligrams	Milligrams	Milligrams
247	Tr	27	26	0.2	29	[9]933	[10]852	0.01	0.04	Tr	0
31	Tr	3	3	Tr	4	[9]116	[10]106	Tr	Tr	Tr	0
11	Tr	1	1	Tr	1	[9]41	[10]38	Tr	Tr	Tr	0
0	0	0	0	0.0	0	0	0	0.00	0.00	0.00	0
0	0	0	0	0.0	0	0	0	0.00	0.00	0.00	0
195	0	0	0	0.0	0	0	0	0.00	0.00	0.00	0
12	0	0	0	0.0	0	0	0	0.00	0.00	0.00	0
0	1	40	31	0.0	57	[11]2,178	[12]2,254	0.01	0.05	Tr	Tr
0	Tr	2	2	0.0	4	[11]134	[12]139	Tr	Tr	Tr	Tr
0	1	34	26	0.1	48	[11]1,066	[12]1,122	0.01	0.04	Tr	Tr
0	Tr	4	3	Tr	6	[11]132	[12]139	Tr	0.01	Tr	Tr
0	Tr	1	1	Tr	2	[11]47	[12]50	Tr	Tr	Tr	Tr
0	1	60	46	0.0	86	[11]2,449	[12]2,254	0.02	0.07	Tr	Tr
0	Tr	4	3	0.0	5	[11]151	[12]139	Tr	Tr	Tr	Tr
0	0	24	18	0.0	34	[11]1,123	[12]1,122	0.01	0.03	Tr	Tr
0	0	3	2	0.0	4	[11]139	[12]139	Tr	Tr	Tr	Tr
0	0	1	1	0.0	1	[11]50	[12]50	Tr	Tr	Tr	Tr
0	0	47	37	0.0	68	[11]2,256	[12]2,254	0.02	0.06	Tr	Tr
0	0	3	2	0.0	4	[11]139	[12]139	Tr	Tr	Tr	Tr
0	0	0	0	0.0	0	0	0	0.00	0.00	0.00	0
0	0	0	0	0.0	0	0	0	0.00	0.00	0.00	0
0	0	0	0	0.0	0	0	0	0.00	0.00	0.00	0
0	0	0	0	0.0	0	0	0	0.00	0.00	0.00	0
0	0	0	0	0.0	0	0	0	0.00	0.00	0.00	0
0	0	0	0	0.0	0	0	0	0.00	0.00	0.00	0
0	0	0	0	0.0	0	0	0	0.00	0.00	0.00	0
0	0	0	0	0.0	0	0	0	0.00	0.00	0.00	0
0	0	0	0	0.0	0	0	0	0.00	0.00	0.00	0
0	0	0	0	0.0	0	0	0	0.00	0.00	0.00	0
0	0	0	0	0.0	0	0	0	0.00	0.00	0.00	0
0	0	0	0	0.0	0	0	0	0.00	0.00	0.00	0
0	0	0	0	0.0	0	0	0	0.00	0.00	0.00	0
0	0	0	0	0.0	0	0	0	0.00	0.00	0.00	0
3	1	12	11	Tr	6	164	10	Tr	0.02	Tr	Tr
0	1	2	1	Tr	2	188	Tr	Tr	Tr	Tr	Tr
0	2	6	5	Tr	3	306	Tr	Tr	Tr	Tr	Tr
0	1	1	1	Tr	5	162	3	Tr	Tr	Tr	Tr
0	2	1	1	Tr	4	136	Tr	Tr	Tr	Tr	Tr
8	Tr	3	4	0.1	5	80	12	0.00	0.00	Tr	0
4	2	Tr	Tr	0.0	2	75	0	0.00	0.00	0.0	0
4	4	2	4	Tr	1	107	13	Tr	Tr	Tr	0

[11]For salted margarine.

[12]Based on average vitamin A content of fortified margarine.

Foods, approximate measures, units, and weight (weight of edible portion only)		Water	Food energy	Protein	Fat	FATTY ACIDS		
						Satu-rated	Mono-unsaturated	Poly-unsaturated
FATS AND OILS—Continued		Grams	Percent	Calories	Grams	Grams	Grams	Grams
		Grams	Percent	Calories	Grams	Grams	Grams	Grams
Salad dressings:								
Tartar sauce	1 tbsp	14	34	75	Tr	8	1.2	2.6	3.9
Thousand island:								
Regular	1 tbsp	16	46	60	Tr	6	1.0	1.3	3.2
Low calorie	1 tbsp	15	69	25	Tr	2	0.2	0.4	0.9
Prepared from home recipe:								
Cooked type[13]	1 tbsp	16	69	25	1	2	0.5	0.6	0.3
Vinegar and oil	1 tbsp	16	47	70	0	8	1.5	2.4	3.9
FISH AND SHELLFISH								
Clams:								
Raw, meat only	3 oz	85	82	65	11	1	0.3	0.3	0.3
Canned, drained solids	3 oz	85	77	85	13	2	0.5	0.5	0.4
Crabmeat, canned	1 cup	135	77	135	23	3	0.5	0.8	1.4
Fish sticks, frozen reheated (stick, 4 by 1 by ½ in)	1 fish stick	28	52	70	6	3	0.8	1.4	0.8
Flounder or Sole, baked, with lemon juice:								
With butter	3 oz	85	73	120	16	6	3.2	1.5	0.5
With margarine	3 oz	85	73	120	16	6	1.2	2.3	1.9
Without added fat	3 oz	85	78	80	17	1	0.3	0.2	0.4
Haddock, breaded, fried[14]	3 oz	85	61	175	17	9	2.4	3.9	2.4
Halibut, broiled, with butter and lemon juice	3 oz	85	67	140	20	6	3.3	1.6	0.7
Herring, pickled	3 oz	85	59	190	17	13	4.3	4.6	3.1
Ocean perch, breaded, fried[14]	1 fillet	85	59	185	16	11	2.6	4.6	2.8
Oysters:								
Raw, meat only (13–19 medium Selects)	1 cup	240	85	160	20	4	1.4	0.5	1.4
Breaded, fried[14]	1 oyster	45	65	90	5	5	1.4	2.1	1.4
Salmon:								
Canned (pink), solids and liquid	3 oz	85	71	120	17	5	0.9	1.5	2.1
Baked (red)	3 oz	85	67	140	21	5	1.2	2.4	1.4
Smoked	3 oz	85	59	150	18	8	2.6	3.9	0.7
Sardines, Atlantic, canned in oil, drained solids	3 oz	85	62	175	20	9	2.1	3.7	2.9
Scallops, breaded, frozen, reheated	6 scallops	90	59	195	15	10	2.5	4.1	2.5
Shrimp:								
Canned, drained solids	3 oz	85	70	100	21	1	0.2	0.2	0.4
French fried (7 medium)[16]	3 oz	85	55	200	16	10	2.5	4.1	2.6
Trout, broiled, with butter and lemon juice	3 oz	85	63	175	21	9	4.1	2.9	1.6
Tuna, canned, drained solids:								
Oil pack, chunk light	3 oz	85	61	165	24	7	1.4	1.9	3.1
Water pack, solid white	3 oz	85	63	135	30	1	0.3	0.2	0.3
Tuna salad[17]	1 cup	205	63	375	33	19	3.3	4.9	9.2
FRUITS AND FRUIT JUICES								
Apples:								
Raw:								
Unpeeled, without cores:								
2¾-in diam. (about 3 per lb with cores)	1 apple	138	84	80	Tr	Tr	0.1	Tr	0.1
3¼-in diam. (about 2 per lb with cores)	1 apple	212	84	125	Tr	1	0.1	Tr	0.2
Peeled, sliced	1 cup	110	84	65	Tr	Tr	0.1	Tr	0.1
Dried, sulfured	10 rings	64	32	155	1	Tr	Tr	Tr	0.1

[13]Fatty acid values apply to product made with regular margarine.

[14]Dipped in egg, milk, and breadcrumbs; fried in vegetable shortening.

[15]If bones are discarded, value for calcium will be greatly reduced.

Cho-lesterol	Carbo-hydrate	Calcium	Phos-phorus	Iron	Potas-sium	Sodium	Vitamin A value	Thiamin	Ribo-flavin	Niacin	Ascorbic acid
Milli-grams	Grams	Milli-grams	Milli-grams	Milli-grams	Milli-grams	Milli-grams	Retinol equiv. (RE)	Milli-grams	Milli-grams	Milli-grams	Milli-grams
4	1	3	4	0.1	11	182	9	Tr	Tr	0.0	Tr
4	2	2	3	0.1	18	112	15	Tr	Tr	Tr	0
2	2	2	3	0.1	17	150	14	Tr	Tr	Tr	0
9	2	13	14	0.1	19	117	20	0.01	0.02	Tr	Tr
0	Tr	0	0	0.0	1	Tr	0	0.00	0.00	0.0	0
43	2	59	138	2.6	154	102	26	0.09	0.15	1.1	9
54	2	47	116	3.5	119	102	26	0.01	0.09	0.9	3
135	1	61	246	1.1	149	1,350	14	0.11	0.11	2.6	0
26	4	11	58	0.3	94	53	5	0.03	0.05	0.6	0
68	Tr	13	187	0.3	272	145	54	0.05	0.08	1.6	1
55	Tr	14	187	0.3	273	151	69	0.05	0.08	1.6	1
59	Tr	13	197	0.3	286	101	10	0.05	0.08	1.7	1
75	7	34	183	1.0	270	123	20	0.06	0.10	2.9	0
62	Tr	14	206	0.7	441	103	174	0.06	0.07	7.7	1
85	0	29	128	0.9	85	850	33	0.04	0.18	2.8	0
66	7	31	191	1.2	241	138	20	0.10	0.11	2.0	0
120	8	226	343	15.6	290	175	223	0.34	0.43	6.0	24
35	5	49	73	3.0	64	70	44	0.07	0.10	1.3	4
34	0	[15]167	243	0.7	307	443	18	0.03	0.15	6.8	0
60	0	26	269	0.5	305	55	87	0.18	0.14	5.5	0
51	0	12	208	0.8	327	1,700	77	0.17	0.17	6.8	0
85	0	[15]371	424	2.6	349	425	56	0.03	0.17	4.6	0
70	10	39	203	2.0	369	298	21	0.11	0.11	1.6	0
128	1	98	224	1.4	104	1,955	15	0.01	0.03	1.5	0
168	11	61	154	2.0	189	384	26	0.06	0.09	2.8	0
71	Tr	26	259	1.0	297	122	60	0.07	0.07	2.3	1
55	0	7	199	1.6	298	303	20	0.04	0.09	10.1	0
48	0	17	202	0.6	255	468	32	0.03	0.10	13.4	0
80	19	31	281	2.5	531	877	53	0.06	0.14	13.3	6
0	21	10	10	0.2	159	Tr	7	0.02	0.02	0.1	8
0	32	15	15	0.4	244	Tr	11	0.04	0.03	0.2	12
0	16	4	8	0.1	124	Tr	5	0.02	0.01	0.1	4
0	42	9	24	0.9	288	[18]56	0	0.00	0.10	0.6	2

[16]Dipped in egg, breadcrumbs, and flour; fried in vegetable shortening.

[17]Made with drained chunk light tuna, celery, onion, pickle relish, and mayonnaise-type salad dressing.

[18]Sodium bisulfite used to preserve color; unsulfited product would contain less sodium.

Foods, approximate measures, units, and weight (weight of edible portion only)		Water		Food energy	Protein	Fat	FATTY ACIDS		
							Saturated	Mono-unsaturated	Poly-unsaturated
FRUITS AND FRUIT JUICES—Continued		Grams	Percent	Calories	Grams	Grams	Grams	Grams	Grams
Apple juice, bottled or canned[19]	1 cup	248	88	115	Tr	Tr	Tr	Tr	0.1
Applesauce, canned:									
Sweetened	1 cup	255	80	195	Tr	Tr	0.1	Tr	0.1
Unsweetened	1 cup	244	88	105	Tr	Tr	Tr	Tr	Tr
Apricots:									
Raw, without pits (about 12 per lb with pits)	3 apricots	106	86	50	1	Tr	Tr	0.2	0.1
Canned (fruit and liquid):									
Heavy syrup pack	1 cup	258	78	215	1	Tr	Tr	0.1	Tr
	3 halves	85	78	70	Tr	Tr	Tr	Tr	Tr
Juice pack	1 cup	248	87	120	2	Tr	Tr	Tr	Tr
	3 halves	84	87	40	1	Tr	Tr	Tr	Tr
Dried:									
Uncooked (28 large or 37 medium halves per cup)	1 cup	130	31	310	5	1	Tr	0.3	0.1
Cooked, unsweetened, fruit and liquid	1 cup	250	76	210	3	Tr	Tr	0.2	0.1
Apricot nectar, canned	1 cup	251	85	140	1	Tr	Tr	0.1	Tr
Avocados, raw, whole, without skin and seed:									
California (about 2 per lb with skin and seed)	1 avocado	173	73	305	4	30	4.5	19.4	3.5
Florida (about 1 per lb with skin and seed)	1 avocado	304	80	340	5	27	5.3	14.8	4.5
Bananas, raw, without peel:									
Whole (about 2½ per lb with peel)	1 banana	114	74	105	1	1	0.2	Tr	0.1
Sliced	1 cup	150	74	140	2	1	0.3	0.1	0.1
Blackberries, raw	1 cup	144	86	75	1	1	0.2	0.1	0.1
Blueberries:									
Raw	1 cup	145	85	80	1	1	Tr	0.1	0.3
Frozen, sweetened	1 cup	230	77	185	1	Tr	Tr	Tr	0.1
Cantaloupe. See Melons.									
Cherries:									
Sour, red, pitted, canned, water pack	1 cup	244	90	90	2	Tr	0.1	0.1	0.1
Sweet, raw, without pits and stems	10 cherries	68	81	50	1	1	0.1	0.2	0.2
Cranberry juice cocktail, bottled, sweetened	1 cup	253	85	145	Tr	Tr	Tr	Tr	0.1
Cranberry sauce, sweetened, canned, strained	1 cup	277	61	420	1	Tr	Tr	0.1	0.2
Dates:									
Whole, without pits	10 dates	83	23	230	2	Tr	0.1	0.1	Tr
Chopped	1 cup	178	23	490	4	1	0.3	0.2	Tr
Figs, dried	10 figs	187	28	475	6	2	0.4	0.5	1.0
Fruit cocktail, canned, fruit and liquid:									
Heavy syrup pack	1 cup	255	80	185	1	Tr	Tr	Tr	0.1
Juice pack	1 cup	248	87	115	1	Tr	Tr	Tr	Tr
Grapefruit:									
Raw, without peel, membrane and seeds (3¾-in. diam., 1 lb 1 oz, whole, with refuse)	½ grapefruit	120	91	40	1	Tr	Tr	Tr	Tr
Canned, sections with syrup	1 cup	254	84	150	1	Tr	Tr	Tr	0.1
Grapefruit juice:									
Raw	1 cup	247	90	95	1	Tr	Tr	Tr	0.1
Canned:									
Unsweetened	1 cup	247	90	95	1	Tr	Tr	Tr	0.1
Sweetened	1 cup	250	87	115	1	Tr	Tr	Tr	0.1

[19]Also applies to pasteurized apple cider.

[20]Without added ascorbic acid. For value with added ascorbic acid, refer to label.

Cho-lesterol	Carbo-hydrate	Calcium	Phos-phorus	Iron	Potas-sium	Sodium	Vitamin A value	Thiamin	Ribo-flavin	Niacin	Ascorbic acid
Milli-grams	Grams	Milli-grams	Milli-grams	Milli-grams	Milli-grams	Milli-grams	Retinol equiv. (RE)	Milli-grams	Milli-grams	Milli-grams	Milli-grams
0	29	17	17	0.9	295	7	Tr	0.05	0.04	0.2	[20]2
0	51	10	18	0.9	156	8	3	0.03	0.07	0.5	[20]4
0	28	7	17	0.3	183	5	7	0.03	0.06	0.5	[20]3
0	12	15	20	0.6	314	1	277	0.03	0.04	0.6	11
0	55	23	31	0.8	361	10	317	0.05	0.06	1.0	8
0	18	8	10	0.3	119	3	105	0.02	0.02	0.3	3
0	31	30	50	0.7	409	10	419	0.04	0.05	0.9	12
0	10	10	17	0.3	139	3	142	0.02	0.02	0.3	4
0	80	59	152	6.1	1,791	13	941	0.01	0.20	3.9	3
0	55	40	103	4.2	1,222	8	591	0.02	0.08	2.4	4
0	36	18	23	1.0	286	8	330	0.02	0.04	0.7	[20]2
0	12	19	73	2.0	1,097	21	106	0.19	0.21	3.3	14
0	27	33	119	1.6	1,484	15	186	0.33	0.37	5.8	24
0	27	7	23	0.4	451	1	9	0.05	0.11	0.6	10
0	35	9	30	0.5	594	2	12	0.07	0.15	0.8	14
0	18	46	30	0.8	282	Tr	24	0.04	0.06	0.6	30
0	20	9	15	0.2	129	9	15	0.07	0.07	0.5	19
0	50	14	16	0.9	138	2	10	0.05	0.12	0.6	2
0	22	27	24	3.3	239	17	184	0.04	0.10	0.4	5
0	11	10	13	0.3	152	Tr	15	0.03	0.04	0.3	5
0	38	8	3	0.4	61	10	1	0.01	0.04	0.1	[21]108
0	108	11	17	0.6	72	80	6	0.04	0.06	0.3	6
0	61	27	33	1.0	541	2	4	0.07	0.08	1.8	0
0	131	57	71	2.0	1,161	5	9	0.16	0.18	3.9	0
0	122	269	127	4.2	1,331	21	25	0.13	0.16	1.3	1
0	48	15	28	0.7	224	15	52	0.05	0.05	1.0	5
0	29	20	35	0.5	236	10	76	0.03	0.04	1.0	7
0	10	14	10	0.1	167	Tr	[22]1	0.04	0.02	0.3	41
0	39	36	25	1.0	328	5	Tr	0.10	0.05	0.6	54
0	23	22	37	0.5	400	2	2	0.10	0.05	0.5	94
0	22	17	27	0.5	378	2	2	0.10	0.05	0.6	72
0	28	20	28	0.9	405	5	2	0.10	0.06	0.8	67

[21]With added ascorbic acid.
[22]For white grapefruit; pink grapefruit have about 31 RE.

Foods, approximate measures, units, and weight (weight of edible portion only)		Water		Food energy	Protein	Fat	FATTY ACIDS		
							Satu-rated	Mono-unsaturated	Poly-unsaturated
FRUITS AND FRUIT JUICES—Continued		Grams	Percent	Calories	Grams	Grams	Grams	Grams	Grams
Grapefruit juice:									
Frozen concentrate, unsweetened									
Undiluted	6-fl-oz can	207	62	300	4	1	0.1	0.1	0.2
Diluted with 3 parts water by volume	1 cup	247	89	100	1	Tr	Tr	Tr	0.1
Grapes, European type (adherent skin), raw:									
Thompson Seedless	10 grapes	50	81	35	Tr	Tr	0.1	Tr	0.1
Tokay and Emperor, seeded types	10 grapes	57	81	40	Tr	Tr	0.1	Tr	0.1
Grape juice:									
Canned or bottled	1 cup	253	84	155	1	Tr	0.1	Tr	0.1
Frozen concentrate, sweetened:									
Undiluted	6-fl-oz can	216	54	385	1	1	0.2	Tr	0.2
Diluted with 3 parts water by volume	1 cup	250	87	125	Tr	Tr	0.1	Tr	0.1
Kiwifruit, raw, without skin (about 5 per lb with skin)	1 kiwifruit	76	83	45	1	Tr	Tr	0.1	0.1
Lemons, raw, without peel and seeds (about 4 per lb with peel and seeds)	1 lemon	58	89	15	1	Tr	Tr	Tr	0.1
Lemon juice:									
Raw	1 cup	244	91	60	1	Tr	Tr	Tr	Tr
Canned or bottled, unsweetened	1 cup	244	92	50	1	1	0.1	Tr	0.2
	1 tbsp	15	92	5	Tr	Tr	Tr	Tr	Tr
Frozen, single-strength, unsweet-ened	6-fl-oz can	244	92	55	1	1	0.1	Tr	0.2
Lime juice:									
Raw	1 cup	246	90	65	1	Tr	Tr	Tr	0.1
Canned, unsweetened	1 cup	246	93	50	1	1	0.1	0.1	0.2
Mangos, raw, without skin and seed (about 1½ per lb with skin and seed)	1 mango	207	82	135	1	1	0.1	0.2	0.1
Melons, raw, without rind and cav-ity contents:									
Cantaloupe, orange-fleshed (5-in. diam., 2⅓ lb, whole, with rind and cavity contents)	½ melon	267	90	95	2	1	0.1	0.1	0.3
Honeydew (6½-in. diam., 5¼ lb, whole, with rind and cavity contents)	⅒ melon	129	90	45	1	Tr	Tr	Tr	0.1
Nectarines, raw, without pits (about 3 per lb with pits)	1 nectarine	136	86	65	1	1	0.1	0.2	0.3
Oranges, raw:									
Whole, without peel and seeds (2⅝-in. diam., about 2½ per lb, with peel and seeds)	1 orange	131	87	60	1	Tr	Tr	Tr	Tr
Sections without membranes	1 cup	180	87	85	2	Tr	Tr	Tr	Tr
Orange juice:									
Raw, all varieties	1 cup	248	88	110	2	Tr	0.1	0.1	0.1
Canned, unsweetened	1 cup	249	89	105	1	Tr	Tr	0.1	0.1
Chilled	1 cup	249	88	110	2	1	0.1	0.1	0.2
Frozen concentrate:									
Undiluted	6-fl-oz can	213	58	340	5	Tr	0.1	0.1	0.1
Diluted with 3 parts water by volume	1 cup	249	88	110	2	Tr	Tr	Tr	Tr
Orange and grapefruit juice, canned	1 cup	247	89	105	1	Tr	Tr	Tr	Tr
Papayas, raw, ½-in. cubes	1 cup	140	86	65	1	Tr	0.1	0.1	Tr

							NUTRIENTS IN INDICATED QUANTITY				
Cho-lesterol	Carbo-hydrate	Calcium	Phos-phorus	Iron	Potas-sium	Sodium	Vitamin A value	Thiamin	Ribo-flavin	Niacin	Ascorbic acid
Milli-grams	Grams	Milli-grams	Milli-grams	Milli-grams	Milli-grams	Milli-grams	Retinol equiv. (RE)	Milli-grams	Milli-grams	Milli-grams	Milli-grams
0	72	56	101	1.0	1,002	6	6	0.30	0.16	1.6	248
0	24	20	35	0.3	336	2	2	0.10	0.05	0.5	83
0	9	6	7	0.1	93	1	4	0.05	0.03	0.2	5
0	10	6	7	0.1	105	1	4	0.05	0.03	0.2	6
0	38	23	28	0.6	334	8	2	0.07	0.09	0.7	[20]Tr
0	96	28	32	0.8	160	15	6	0.11	0.20	0.9	[21]179
0	32	10	10	0.3	53	5	2	0.04	0.07	0.3	[21]60
0	11	20	30	0.3	252	4	13	0.02	0.04	0.4	74
0	5	15	9	0.3	80	1	2	0.02	0.01	0.1	31
0	21	17	15	0.1	303	[23]2	5	0.07	0.02	0.2	112
0	16	27	22	0.3	249	[23]51	4	0.10	0.02	0.5	61
0	1	2	1	Tr	15	[23]3	Tr	0.01	Tr	Tr	4
0	16	20	20	0.3	217	2	3	0.14	0.03	0.3	77
0	22	22	17	0.1	268	2	2	0.05	0.02	0.2	72
0	16	30	25	0.6	185	[23]39	4	0.08	0.01	0.4	16
0	35	21	23	0.3	323	4	806	0.12	0.12	1.2	57
0	22	29	45	0.6	825	24	861	0.10	0.06	1.5	113
0	12	8	13	0.1	350	13	5	0.10	0.02	0.8	32
0	16	7	22	0.2	288	Tr	100	0.02	0.06	1.3	7
0	15	52	18	0.1	237	Tr	27	0.11	0.05	0.4	70
0	21	72	25	0.2	326	Tr	37	0.16	0.07	0.5	96
0	26	27	42	0.5	496	2	50	0.22	0.07	1.0	124
0	25	20	35	1.1	436	5	44	0.15	0.07	0.8	86
0	25	25	27	0.4	473	2	19	0.28	0.05	0.7	82
0	81	68	121	0.7	1,436	6	59	0.60	0.14	1.5	294
0	27	22	40	0.2	473	2	19	0.20	0.04	0.5	97
0	25	20	35	1.1	390	7	29	0.14	0.07	0.8	72
0	17	35	12	0.3	247	9	40	0.04	0.04	0.5	92

[23]Sodium benzoate and sodium bisulfite added as preservatives.

Foods, approximate measures, units, and weight (weight of edible portion only)		Water	Food energy	Protein	Fat	FATTY ACIDS Saturated	Monounsaturated	Polyunsaturated
FRUITS AND FRUIT JUICES—Continued		Grams	Percent	Calories	Grams	Grams	Grams	Grams
Peaches:								
Raw:								
Whole, 2½-in. diam., peeled, pitted (about 4 per lb with peels and pits)	1 peach	87	88	35	1	Tr	Tr	Tr	Tr
Sliced	1 cup	170	88	75	1	Tr	Tr	0.1	0.1
Canned, fruit and liquid:								
Heavy syrup pack	1 cup	256	79	190	1	Tr	Tr	0.1	0.1
	1 half	81	79	60	Tr	Tr	Tr	Tr	Tr
Juice pack	1 cup	248	87	110	2	Tr	Tr	Tr	Tr
	1 half	77	87	35	Tr	Tr	Tr	Tr	Tr
Dried:								
Uncooked	1 cup	160	32	380	6	1	0.1	0.4	0.6
Cooked, unsweetened, fruit and liquid	1 cup	258	78	200	3	1	0.1	0.2	0.3
Frozen, sliced, sweetened	1 cup	250	75	235	2	Tr	Tr	0.1	0.2
Pears:								
Raw, with skin, cored:								
Bartlett, 2½-in. diam. (about 2½ per lb with cores and stems)	1 pear	166	84	100	1	1	Tr	0.1	0.2
Bosc, 2½-in. diam. (about 3 per lb with cores and stems)	1 pear	141	84	85	1	1	Tr	0.1	0.1
D'Anjou, 3-in. diam. (about 2 per lb with cores and stems)	1 pear	200	84	120	1	1	Tr	0.2	0.2
Canned, fruit and liquid:								
Heavy syrup pack	1 cup	255	80	190	1	Tr	Tr	0.1	0.1
	1 half	79	80	60	Tr	Tr	Tr	Tr	Tr
Juice pack	1 cup	248	86	125	1	Tr	Tr	Tr	Tr
	1 half	77	86	40	Tr	Tr	Tr	Tr	Tr
Pineapple:								
Raw, diced	1 cup	155	87	75	1	1	Tr	0.1	0.2
Canned, fruit and liquid:								
Heavy syrup pack:								
Crushed, chunks, tidbits	1 cup	255	79	200	1	Tr	Tr	Tr	0.1
Slices	1 slice	58	79	45	Tr	Tr	Tr	Tr	Tr
Juice pack:								
Chunks or tidbits	1 cup	250	84	150	1	Tr	Tr	Tr	0.1
Slices	1 slice	58	84	35	Tr	Tr	Tr	Tr	Tr
Pineapple juice, unsweetened, canned	1 cup	250	86	140	1	Tr	Tr	Tr	0.1
Plantains, without peel:								
Raw	1 plantain	179	65	220	2	1	0.3	0.1	0.1
Cooked, boiled, sliced	1 cup	154	67	180	1	Tr	0.1	Tr	0.1
Plums, without pits:								
Raw:								
2⅛-in. diam. (about 6½ per lb with pits)	1 plum	66	85	35	1	Tr	Tr	0.3	0.1
1½-in. diam. (about 15 per lb with pits)	1 plum	28	85	15	Tr	Tr	Tr	0.1	Tr
Canned, purple, fruit and liquid:								
Heavy syrup pack	1 cup	258	76	230	1	Tr	Tr	0.2	0.1
	3 plums	133	76	120	Tr	Tr	Tr	0.1	Tr
Juice pack	1 cup	252	84	145	1	Tr	Tr	Tr	Tr
	3 plums	95	84	55	Tr	Tr	Tr	Tr	Tr
Prunes, dried:								
Uncooked	4 extra large or 5 large prunes	49	32	115	1	Tr	Tr	0.2	0.1
Cooked, unsweetened, fruit and liquid	1 cup	212	70	225	2	Tr	Tr	0.3	0.1

Cho-lesterol	Carbo-hydrate	Calcium	Phos-phorus	Iron	Potas-sium	Sodium	Vitamin A value	Thiamin	Ribo-flavin	Niacin	Ascorbic acid
Milli-grams	Grams	Milli-grams	Milli-grams	Milli-grams	Milli-grams	Milli-grams	Retinol equiv. (RE)	Milli-grams	Milli-grams	Milli-grams	Milli-grams
0	10	4	10	0.1	171	Tr	47	0.01	0.04	0.9	6
0	19	9	20	0.2	335	Tr	91	0.03	0.07	1.7	11
0	51	8	28	0.7	236	15	85	0.03	0.06	1.6	7
0	16	2	9	0.2	75	5	27	0.01	0.02	0.5	2
0	29	15	42	0.7	317	10	94	0.02	0.04	1.4	9
0	9	5	13	0.2	99	3	29	0.01	0.01	0.4	3
0	98	45	190	6.5	1,594	11	346	Tr	0.34	7.0	8
0	51	23	98	3.4	826	5	51	0.01	0.05	3.9	10
0	60	8	28	0.9	325	15	71	0.03	0.09	1.6	[21]236
0	25	18	18	0.4	208	Tr	3	0.03	0.07	0.2	7
0	21	16	16	0.4	176	Tr	3	0.03	0.06	0.1	6
0	30	22	22	0.5	250	Tr	4	0.04	0.08	0.2	8
0	49	13	18	0.6	166	13	1	0.03	0.06	0.6	3
0	15	4	6	0.2	51	4	Tr	0.01	0.02	0.2	1
0	32	22	30	0.7	238	10	1	0.03	0.03	0.5	4
0	10	7	9	0.2	74	3	Tr	0.01	0.01	0.2	1
0	19	11	11	0.6	175	2	4	0.14	0.06	0.7	24
0	52	36	18	1.0	265	3	4	0.23	0.06	0.7	19
0	12	8	4	0.2	60	1	1	0.05	0.01	0.2	4
0	39	35	15	0.7	305	3	10	0.24	0.05	0.7	24
0	9	8	3	0.2	71	1	2	0.06	0.01	0.2	6
0	34	43	20	0.7	335	3	1	0.14	0.06	0.6	27
0	57	5	61	1.1	893	7	202	0.09	0.10	1.2	33
0	48	3	43	0.9	716	8	140	0.07	0.08	1.2	17
0	9	3	7	0.1	114	Tr	21	0.03	0.06	0.3	6
0	4	1	3	Tr	48	Tr	9	0.01	0.03	0.1	3
0	60	23	34	2.2	235	49	67	0.04	0.10	0.8	1
0	31	12	17	1.1	121	25	34	0.02	0.05	0.4	1
0	38	25	38	0.9	388	3	254	0.06	0.15	1.2	7
0	14	10	14	0.3	146	1	96	0.02	0.06	0.4	3
0	31	25	39	1.2	365	2	97	0.04	0.08	1.0	2
0	60	49	74	2.4	708	4	65	0.05	0.21	1.5	6

Foods, approximate measures, units, and weight (weight of edible portion only)		Water	Food energy	Protein	Fat	FATTY ACIDS Satu- rated	FATTY ACIDS Mono- unsaturated	FATTY ACIDS Poly- unsaturated
		Grams	Percent	Calories	Grams	Grams	Grams	Grams

FRUITS AND FRUIT JUICES—Continued

Food	Measure	Grams	Percent	Calories	Grams	Grams	Grams	Grams	Grams
Prune juice, canned or bottled	1 cup	256	81	180	2	Tr	Tr	0.1	Tr
Raisins, seedless:									
Cup, not pressed down	1 cup	145	15	435	5	1	0.2	Tr	0.2
Packet, ½ oz (1½ tbsp)	1 packet	14	15	40	Tr	Tr	Tr	Tr	Tr
Raspberries:									
Raw	1 cup	123	87	60	1	1	Tr	0.1	0.4
Frozen, sweetened	1 cup	250	73	255	2	Tr	Tr	Tr	0.2
Rhubarb, cooked, added sugar	1 cup	240	68	280	1	Tr	Tr	Tr	0.1
Strawberries:									
Raw, capped, whole	1 cup	149	92	45	1	1	Tr	0.1	0.3
Frozen, sweetened, sliced	1 cup	255	73	245	1	Tr	Tr	Tr	0.2
Tangerines:									
Raw, without peel and seeds (2⅜-in. diam., about 4 per lb, with peel and seeds)	1 tangerine	84	88	35	1	Tr	Tr	Tr	Tr
Canned, light syrup, fruit and liquid	1 cup	252	83	155	1	Tr	Tr	Tr	0.1
Tangerine juice, canned, sweetened	1 cup	249	87	125	1	Tr	Tr	Tr	0.1
Watermelon, raw, without rind and seeds:									
Piece (4 by 8 in. wedge with rind and seeds; ¹⁄₁₆ of 32⅔-lb melon, 10 by 16 in.)	1 piece	482	92	155	3	2	0.3	0.2	1.0
Diced	1 cup	160	92	50	1	1	0.1	0.1	0.3

GRAIN PRODUCTS

Food	Measure	Grams	Percent	Calories	Grams	Grams	Grams	Grams	Grams
Bagels, plain or water, enriched, 3½-in. diam.[24]	1 bagel	68	29	200	7	2	0.3	0.5	0.7
Barley, pearled, light, uncooked	1 cup	200	11	700	16	2	0.3	0.2	0.9
Biscuits, baking powder, 2-in. diam. (enriched flour, vegetable shortening):									
From home recipe	1 biscuit	28	28	100	2	5	1.2	2.0	1.3
From mix	1 biscuit	28	29	95	2	3	0.8	1.4	0.9
From refrigerated dough	1 biscuit	20	30	65	1	2	0.6	0.9	0.6
Breadcrumbs, enriched:									
Dry, grated	1 cup	100	7	390	13	5	1.5	1.6	1.0
Soft. See White bread.									
Breads:									
Boston brown bread, canned, slice, 3¼ in. by ½ in.[25]	1 slice	45	45	95	2	1	0.3	0.1	0.1
Cracked-wheat bread (¾ enriched wheat flour, ¼ cracked wheat flour):[25]									
Slice (18 per loaf)	1 slice	25	35	65	2	1	0.2	0.2	0.3
Toasted	1 slice	21	26	65	2	1	0.2	0.2	0.3
French or vienna bread, enriched:[25]									
Slice:									
French, 5 by 2½ by 1 in.	1 slice	35	34	100	3	1	0.3	0.4	0.5
Vienna, 4¾ by 4 by ½ in.	1 slice	25	34	70	2	1	0.2	0.3	0.3
Italian bread, enriched:									
Slice, 4½ by 3¼ by ¾ in.	1 slice	30	32	85	3	Tr	Tr	Tr	0.1
Mixed grain bread, enriched:[25]									
Slice (18 per loaf)	1 slice	25	37	65	2	1	0.2	0.2	0.4
Toasted	1 slice	23	27	65	2	1	0.2	0.2	0.4

[24]Egg bagels have 44 mg cholesterol and 7 RE vitamin A per bagel.

[25]Made with vegetable shortening.

NUTRIENTS IN INDICATED QUANTITY											
Cholesterol	Carbohydrate	Calcium	Phosphorus	Iron	Potassium	Sodium	Vitamin A value	Thiamin	Riboflavin	Niacin	Ascorbic acid
Milligrams	Grams	Milligrams	Milligrams	Milligrams	Milligrams	Milligrams	Retinol equiv. (RE)	Milligrams	Milligrams	Milligrams	Milligrams
0	45	31	64	3.0	707	10	1	0.04	0.18	2.0	10
0	115	71	141	3.0	1,089	17	1	0.23	0.13	1.2	5
0	11	7	14	0.3	105	2	Tr	0.02	0.01	0.1	Tr
0	14	27	15	0.7	187	Tr	16	0.04	0.11	1.1	31
0	65	38	43	1.6	285	3	15	0.05	0.11	0.6	41
0	75	348	19	0.5	230	2	17	0.04	0.06	0.5	8
0	10	21	28	0.6	247	1	4	0.03	0.10	0.3	84
0	66	28	33	1.5	250	8	6	0.04	0.13	1.0	106
0	9	12	8	0.1	132	1	77	0.09	0.02	0.1	26
0	41	18	25	0.9	197	15	212	0.13	0.11	1.1	50
0	30	45	35	0.5	443	2	105	0.15	0.05	0.2	55
0	35	39	43	0.8	559	10	176	0.39	0.10	1.0	46
0	11	13	14	0.3	186	3	59	0.13	0.03	0.3	15
0	38	29	46	1.8	50	245	0	0.26	0.20	2.4	0
0	158	32	378	4.2	320	6	0	0.24	0.10	6.2	0
Tr	13	47	36	0.7	32	195	3	0.08	0.08	0.8	Tr
Tr	14	58	128	0.7	56	262	4	0.12	0.11	0.8	Tr
1	10	4	79	0.5	18	249	0	0.08	0.05	0.7	0
5	73	122	141	4.1	152	736	0	0.35	0.35	4.8	0
3	21	41	72	0.9	131	113	[26]0	0.06	0.04	0.7	0
0	12	16	32	0.7	34	106	Tr	0.10	0.09	0.8	Tr
0	12	16	32	0.7	34	106	Tr	0.07	0.09	0.8	Tr
0	18	39	30	1.1	32	203	Tr	0.16	0.12	1.4	Tr
0	13	28	21	0.8	23	145	Tr	0.12	0.09	1.0	Tr
0	17	5	23	0.8	22	176	0	0.12	0.07	1.0	0
0	12	27	55	0.8	56	106	Tr	0.10	0.10	1.1	Tr
0	12	27	55	0.8	56	106	Tr	0.08	0.10	1.1	Tr

[26]Made with white cornmeal. If made with yellow cornmeal, value is 3 RE.

Foods, approximate measures, units, and weight (weight of edible portion only)		Water		Food energy	Protein	Fat	FATTY ACIDS		
							Satu-rated	Mono-unsaturated	Poly-unsaturated
GRAIN PRODUCTS—Continued		Grams	Percent	Calories	Grams	Grams	Grams	Grams	Grams
Breads:									
Oatmeal bread, enriched:[25]									
Slice (18 per loaf)	1 slice	25	37	65	2	1	0.2	0.4	0.5
Toasted	1 slice	23	30	65	2	1	0.2	0.4	0.5
Pita bread, enriched, white, 6½-in. diam.	1 pita	60	31	165	6	1	0.1	0.1	0.4
Pumpernickel (⅔ rye flour, ⅓ enriched wheat flour):[25]									
Slice, 5 by 4 by ⅜ in.	1 slice	32	37	80	3	1	0.2	0.3	0.5
Toasted	1 slice	29	28	80	3	1	0.2	0.3	0.5
Raisin Bread, enriched:[25]									
Slice (18 per loaf)	1 slice	25	33	65	2	1	0.2	0.3	0.4
Toasted	1 slice	21	24	65	2	1	0.2	0.3	0.4
Rye bread, light (⅔ enriched wheat flour, ⅓ rye flour):[25]									
Slice, 4¾ by 3¾ by ⁷⁄₁₆ in.	1 slice	25	37	65	2	1	0.2	0.3	0.3
Toasted	1 slice	22	28	65	2	1	0.2	0.3	0.3
Wheat bread, enriched:[25]									
Slice (18 per loaf)	1 slice	25	37	65	2	1	0.2	0.4	0.3
Toasted	1 slice	23	28	65	3	1	0.2	0.4	0.3
White bread, enriched:[25]									
Slice (18 per loaf)	1 slice	25	37	65	2	1	0.3	0.4	0.2
Toasted	1 slice	22	28	65	2	1	0.3	0.4	0.2
Slice (22 per loaf)	1 slice	20	37	55	2	1	0.2	0.3	0.2
Toasted	1 slice	17	28	55	2	1	0.2	0.3	0.2
Cubes	1 cup	30	37	80	2	1	0.4	0.4	0.3
Crumbs, soft	1 cup	45	37	120	4	2	0.6	0.6	0.4
Whole-wheat bread:[25]									
Slice (16 per loaf)	1 slice	28	38	70	3	1	0.4	0.4	0.3
Toasted	1 slice	25	29	70	3	1	0.4	0.4	0.3
Bread stuffing (from enriched bread), prepared from mix:									
Dry type	1 cup	140	33	500	9	31	6.1	13.3	9.6
Moist type	1 cup	203	61	420	9	26	5.3	11.3	8.0
Breakfast cereals:									
Hot type, cooked:									
Corn (hominy) grits:									
Regular and quick, enriched	1 cup	242	85	145	3	Tr	Tr	0.1	0.2
Instant, plain	1 pkt	137	85	80	2	Tr	Tr	Tr	0.1
Cream of Wheat®:									
Regular, quick, instant	1 cup	244	86	140	4	Tr	0.1	Tr	0.2
Mix'n Eat, plain	1 pkt	142	82	100	3	Tr	Tr	Tr	0.1
Malt-O-Meal®	1 cup	240	88	120	4	Tr	Tr	Tr	0.1
Oatmeal or rolled oats:									
Regular, quick, instant, nonfortified	1 cup	234	85	145	6	2	0.4	0.8	1.0
Instant, fortified:									
Plain	1 pkt	177	86	105	4	2	0.3	0.6	0.7
Flavored	1 pkt	164	76	160	5	2	0.3	0.7	0.8
Ready to eat:									
All-Bran® (about ⅓ cup)	1 oz	28	3	70	4	1	0.1	0.1	0.3
Cap'n Crunch® (about ¾ cup)	1 oz	28	3	120	1	3	1.7	0.3	0.4
Cheerios® (about 1¼ cup)	1 oz	28	5	110	4	2	0.3	0.6	0.7
Corn Flakes (about 1¼ cup):									
Kellogg's®	1 oz	28	3	110	2	Tr	Tr	Tr	Tr
Toasties®	1 oz	28	3	110	2	Tr	Tr	Tr	Tr

[27]Nutrient added.

[28]Cooked without salt. If salt is added according to label recommendations, sodium content is 540 mg.

[29]For white corn grits. Cooked yellow grits contain 14 RE.

[30]Value based on label declaration for added nutrients.

Cho-lesterol	Carbo-hydrate	Calcium	Phos-phorus	Iron	Potas-sium	Sodium	Vitamin A value	Thiamin	Ribo-flavin	Niacin	Ascorbic acid
Milli-grams	Grams	Milli-grams	Milli-grams	Milli-grams	Milli-grams	Milli-grams	Retinol equiv. (RE)	Milli-grams	Milli-grams	Milli-grams	Milli-grams
0	12	15	31	0.7	39	124	0	0.12	0.07	0.9	0
0	12	15	31	0.7	39	124	0	0.09	0.07	0.9	0
0	33	49	60	1.4	71	339	0	0.27	0.12	2.2	0
0	16	23	71	0.9	141	177	0	0.11	0.17	1.1	0
0	16	23	71	0.9	141	177	0	0.09	0.17	1.1	0
0	13	25	22	0.8	59	92	Tr	0.08	0.15	1.0	Tr
0	13	25	22	0.8	59	92	Tr	0.06	0.15	1.0	Tr
0	12	20	36	0.7	51	175	0	0.10	0.08	0.8	0
0	12	20	36	0.7	51	175	0	0.08	0.08	0.8	0
0	12	32	47	0.9	35	138	Tr	0.12	0.08	1.2	Tr
0	12	32	47	0.9	35	138	Tr	0.10	0.08	1.2	Tr
0	12	32	27	0.7	28	129	Tr	0.12	0.08	0.9	Tr
0	12	32	27	0.7	28	129	Tr	0.09	0.08	0.9	Tr
0	10	25	21	0.6	22	101	Tr	0.09	0.06	0.7	Tr
0	10	25	21	0.6	22	101	Tr	0.07	0.06	0.7	Tr
0	15	38	32	0.9	34	154	Tr	0.14	0.09	1.1	Tr
0	22	57	49	1.3	50	231	Tr	0.21	0.14	1.7	Tr
0	13	20	74	1.0	50	180	Tr	0.10	0.06	1.1	Tr
0	13	20	74	1.0	50	180	Tr	0.08	0.06	1.1	Tr
0	50	92	136	2.2	126	1,254	273	0.17	0.20	2.5	0
67	40	81	134	2.0	118	1,023	256	0.10	0.18	1.6	0
0	31	0	29	[27]1.5	53	[28]0	[29]0	[27]0.24	[27]0.15	[27]2.0	0
0	18	7	16	[27]1.0	29	343	0	[27]0.18	[27]0.08	[27]1.3	0
0	29	[30]54	[31]43	[30]10.9	46	[31,32]5	0	[30]0.24	[30]0.07	[30]1.5	0
0	21	[30]20	[30]20	[30]8.1	38	241	[30]376	[30]0.43	[30]0.28	[30]5.0	0
0	26	5	[30]24	[30]9.6	31	[33]2	0	[30]0.48	[30]0.24	[30]5.8	0
0	25	19	178	1.6	131	[34]2	4	0.26	0.05	0.3	0
0	18	[27]163	133	[27]6.3	99	[27]285	[27]453	[27]0.53	[27]0.28	[27]5.5	0
0	31	[27]168	148	[27]6.7	137	[27]254	[27]460	[27]0.53	[27]0.38	[27]5.9	Tr
0	21	23	264	[30]4.5	350	320	[30]375	[30]0.37	[30]0.43	[30]5.0	[30]15
0	23	5	36	[27]7.5	37	213	4	[27]0.50	[27]0.55	[27]6.6	0
0	20	48	134	[30]4.5	101	307	[30]375	[30]0.37	[30]0.43	[30]5.0	[30]15
0	24	1	18	[30]1.8	26	351	[30]375	[30]0.37	[30]0.43	[30]5.0	[30]15
0	24	1	12	[27]0.7	33	297	[30]375	[30]0.37	[30]0.43	[30]5.0	0

[31]For regular and instant cereal. For quick cereal, phosphorus is 102 mg and sodium is 142 mg.

[32]Cooked without salt. If salt is added according to label recommendations, sodium content is 390 mg.

[33]Cooked without salt. If salt is added according to label recommendations, sodium content is 324 mg.

[34]Cooked without salt. If salt is added according to label recommendations, sodium content is 374 mg.

473

Foods, approximate measures, units, and weight (weight of edible portion only)		Water		Food energy	Protein	Fat	FATTY ACIDS		
							Satu-rated	Mono-unsaturated	Poly-unsaturated
GRAIN PRODUCTS—Continued		Grams	Percent	Calories	Grams	Grams	Grams	Grams	Grams
Breakfast cereals:									
Ready to eat:									
40% Bran Flakes:									
Kellogg's® (about ¾ cup)	1 oz	28	3	90	4	1	0.1	0.1	0.3
Post® (about ⅔ cup)	1 oz	28	3	90	3	Tr	0.1	0.1	0.2
Froot Loops® (about 1 cup)	1 oz	28	3	110	2	1	0.2	0.1	0.1
Frosted Flakes, Kellogg's® (about ¾ cup)	1 oz	28	3	110	1	Tr	Tr	Tr	Tr
Golden Crisp® (about ⅞ cup)	1 oz	28	2	105	2	Tr	Tr	Tr	0.1
Golden Grahams® (about ¾ cup)	1 oz	28	2	110	2	1	0.7	0.1	0.2
Grape-Nuts® (about ¼ cup)	1 oz	28	3	100	3	Tr	Tr	Tr	0.1
Honey Nut Cheerios® (about ¾ cup)	1 oz	28	3	105	3	1	0.1	0.3	0.3
Lucky Charms® (about 1 cup)	1 oz	28	3	110	3	1	0.2	0.4	0.4
Nature Valley® Granola (about ⅓ cup)	1 oz	28	4	125	3	5	3.3	0.7	0.7
100% Natural Cereal (about ¼ cup)	1 oz	28	2	135	3	6	4.1	1.2	0.5
Product 19® (about ¾ cup)	1 oz	28	3	110	3	Tr	Tr	Tr	0.1
Raisin Bran:									
Kellogg's® (about ¾ cup)	1 oz	28	8	90	3	1	0.1	0.1	0.3
Post® (about ½ cup)	1 oz	28	9	85	3	1	0.1	0.1	0.3
Rice Krispies® (about 1 cup)	1 oz	28	2	110	2	Tr	Tr	Tr	0.1
Shredded Wheat (about ⅔ cup)	1 oz	28	5	100	3	1	0.1	0.1	0.3
Special K® (about 1⅓ cup)	1 oz	28	2	110	6	Tr	Tr	Tr	Tr
Smacks® (about ¾ cup)	1 oz	28	3	105	2	1	0.1	0.1	0.2
Total® (about 1 cup)	1 oz	28	4	100	3	1	0.1	0.1	0.3
Trix® (about 1 cup)	1 oz	28	3	110	2	Tr	0.2	0.1	0.1
Wheaties® (about 1 cup)	1 oz	28	5	100	3	Tr	0.1	Tr	0.2
Buckwheat flour, light, sifted	1 cup	98	12	340	6	1	0.2	0.4	0.4
Bulgur, uncooked	1 cup	170	10	600	19	3	1.2	0.3	1.2
Cakes prepared from cake mixes with enriched flour:[35]									
Angel food:									
Piece, 1/12 of cake	1 piece	53	38	125	3	Tr	Tr	Tr	0.1
Coffeecake, crumb:									
Piece, 1/16 of cake	1 piece	72	30	230	5	7	2.0	2.8	1.6
Devil's food with chocolate frosting:									
Piece, 1/16 of cake	1 piece	69	24	235	3	8	3.5	3.2	1.2
Cupcake, 2½-in. diam.	1 cupcake	35	24	120	2	4	1.8	1.6	0.6
Gingerbread:									
Piece, 1/9 of cake	1 piece	63	37	175	2	4	1.1	1.8	1.2
Cakes prepared from cake mixes with enriched flour:[35]									
Yellow with chocolate frosting:									
Piece, 1/16 of cake	1 piece	69	26	235	3	8	3.0	3.0	1.4
Cakes prepared from home recipes using enriched flour:									
Carrot, with cream cheese frosting:[36]									
Piece, 1/16 of cake	1 piece	96	23	385	4	21	4.1	8.4	6.7
Fruitcake, dark:[36]									
Piece, 1/32 of cake, ⅔ in. arc	1 piece	43	18	165	2	7	1.5	3.6	1.6

[35]Except for angel food cake, cakes were made from mixes containing vegetable shortening and frostings were made with margarine.

[36]Made with vegetable oil.

							NUTRIENTS IN INDICATED QUANTITY				
Cho-lesterol	Carbo-hydrate	Calcium	Phos-phorus	Iron	Potas-sium	Sodium	Vitamin A value	Thiamin	Ribo-flavin	Niacin	Ascorbic acid
Milli-grams	Grams	Milli-grams	Milli-grams	Milli-grams	Milli-grams	Milli-grams	Retinol equiv. (RE)	Milli-grams	Milli-grams	Milli-grams	Milli-grams
0	22	14	139	[30]8.1	180	264	[30]375	[30]0.37	[30]0.43	[30]5.0	0
0	22	12	179	[30]4.5	151	260	[30]375	[30]0.37	[30]0.43	[30]5.0	0
0	25	3	24	[30]4.5	26	145	[30]375	[30]0.37	[30]0.43	[30]5.0	[30]15
0	26	1	21	[30]1.8	18	230	[30]375	[30]0.37	[30]0.43	[30]5.0	[30]15
0	26	6	52	[30]1.8	105	25	[30]375	[30]0.37	[30]0.43	[30]5.0	0
Tr	24	17	41	[30]4.5	63	346	[30]375	[30]0.37	[30]0.43	[30]5.0	[30]15
0	23	11	71	1.2	95	197	[30]375	[30]0.37	[30]0.43	[30]5.0	0
0	23	20	105	[30]4.5	99	257	[30]375	[30]0.37	[30]0.43	[30]5.0	[30]15
0	23	32	79	[30]4.5	59	201	[30]375	[30]0.37	[30]0.43	[30]5.0	[30]15
0	19	18	89	0.9	98	58	2	0.10	0.05	0.2	0
Tr	18	49	104	0.8	140	12	2	0.09	0.15	0.6	0
0	24	3	40	[30]18.0	44	325	[30]1,501	[30]1.50	[30]1.70	[30]20.0	[30]60
0	21	10	105	[30]3.5	147	207	[30]288	[30]0.28	[30]0.34	[30]3.9	0
0	21	13	119	[30]4.5	175	185	[30]375	[30]0.37	[30]0.43	[30]5.0	0
0	25	4	34	[30]1.8	29	340	[30]375	[30]0.37	[30]0.43	[30]5.0	[30]15
0	23	11	100	1.2	102	3	0	0.07	0.08	1.5	0
Tr	21	8	55	[30]4.5	49	265	[30]375	[30]0.37	[30]0.43	[30]5.0	[30]15
0	25	3	31	[30]1.8	42	75	[30]375	[30]0.37	[30]0.43	[30]5.0	[30]15
0	22	48	118	[30]18.0	106	352	[30]1,501	[30]1.50	[30]1.70	[30]20.0	[30]60
0	25	6	19	[30]4.5	27	181	[30]375	[30]0.37	[30]0.43	[30]5.0	[30]15
0	23	43	98	[30]4.5	106	354	[30]375	[30]0.37	[30]0.43	[30]5.0	[30]15
0	78	11	86	1.0	314	2	0	0.08	0.04	0.4	0
0	129	49	575	9.5	389	7	0	0.48	0.24	7.7	0
0	29	44	91	0.2	71	269	0	0.03	0.11	0.1	0
47	38	44	125	1.2	78	310	32	0.14	0.15	1.3	Tr
37	40	41	72	1.4	90	181	31	0.07	0.10	0.6	Tr
19	20	21	37	0.7	46	92	16	0.04	0.05	0.3	Tr
1	32	57	63	1.2	173	192	0	0.09	0.11	0.8	Tr
36	40	63	126	1.0	75	157	29	0.08	0.10	0.7	Tr
74	48	44	62	1.3	108	279	15	0.11	0.12	0.9	1
20	25	41	50	1.2	194	67	13	0.08	0.08	0.5	16

Foods, approximate measures, units, and weight (weight of edible portion only)		Water		Food energy	Protein	Fat	FATTY ACIDS		
							Satu-rated	Mono-unsaturated	Poly-unsaturated
GRAIN PRODUCTS—Continued		Grams	Percent	Calories	Grams	Grams	Grams	Grams	Grams
Cakes prepared from home recipes using enriched flour:[37]									
Plain sheet cake:									
Without frosting:									
Piece, 1/9 of cake	1 piece	86	25	315	4	12	3.3	5.0	2.8
With uncooked white frosting:									
Piece, 1/9 of cake	1 piece	121	21	445	4	14	4.6	5.6	2.9
Pound:[38]									
Slice, 1/17 loaf	1 slice	30	22	120	2	5	1.2	2.4	1.6
Cakes, commercial, made with enriched flour:									
Pound:									
Slice, 1/17 of loaf	1 slice	29	24	110	2	5	3.0	1.7	0.2
Snack cakes:									
Devil's food with creme filling (2 small cakes per pkg)	1 small cake	28	20	105	1	4	1.7	1.5	0.6
Sponge with creme filling (2 small cakes per pkg)	1 small cake	42	19	155	1	5	2.3	2.1	0.5
White with white frosting:									
Piece, 1/16 of cake	1 piece	71	24	260	3	9	2.1	3.8	2.6
Yellow with chocolate frosting:									
Piece, 1/16 of cake	1 piece	69	23	245	2	11	5.7	3.7	0.6
Cheesecake:									
Piece, 1/12 of cake	1 piece	92	46	280	5	18	9.9	5.4	1.2
Cookies made with enriched flour:									
Brownies with nuts:									
Commercial, with frosting, 1½ by 1¾ by ⅞ in.	1 brownie	25	13	100	1	4	1.6	2.0	0.6
From home recipe, 1¾ by 1¾ by ⅞ in.[36]	1 brownie	20	10	95	1	6	1.4	2.8	1.2
Chocolate chip:									
Commercial, 2¼-in. diam., ⅜ in. thick	4 cookies	42	4	180	2	9	2.9	3.1	2.6
Cookies made with enriched flour:									
Chocolate chip:									
From home recipe, 2⅓-in. diam.[25]	4 cookies	40	3	185	2	11	3.9	4.3	2.0
From refrigerated dough, 2¼-in. diam., ⅜ in. thick	4 cookies	48	5	225	2	11	4.0	4.4	2.0
Fig bars, square, 1⅝ by 1⅝ by ⅜ in. or rectangular, 1½ by 1¾ by ½ in.	4 cookies	56	12	210	2	4	1.0	1.5	1.0
Oatmeal with raisins, 2⅝-in. diam., ¼ in. thick	4 cookies	52	4	245	3	10	2.5	4.5	2.8
Peanut butter cookie, from home recipe, 2⅝ in. diam.[25]	4 cookies	48	3	245	4	14	4.0	5.8	2.8
Sandwich type (chocolate or vanilla), 1¾-in. diam., ⅜ in. thick	4 cookies	40	2	195	2	8	2.0	3.6	2.2
Shortbread:									
Commercial	4 small cookies	32	6	155	2	8	2.9	3.0	1.1
From home recipe[38]	2 large cookies	28	3	145	2	8	1.3	2.7	3.4
Sugar cookie, from refrigerated dough, 2½-in. diam., ¼ in. thick	4 cookies	48	4	235	2	12	2.3	5.0	3.6
Vanilla wafers, 1¾-in. diam., ¼ in. thick	10 cookies	40	4	185	2	7	1.8	3.0	1.8

[37]Cake made with vegetable shortening; frosting with margarine.

[38]Made with margarine.

NUTRIENTS IN INDICATED QUANTITY

Cholesterol	Carbohydrate	Calcium	Phosphorus	Iron	Potassium	Sodium	Vitamin A value	Thiamin	Riboflavin	Niacin	Ascorbic acid
Milligrams	Grams	Milligrams	Milligrams	Milligrams	Milligrams	Milligrams	Retinol equiv. (RE)	Milligrams	Milligrams	Milligrams	Milligrams
61	48	55	88	1.3	68	258	41	0.14	0.15	1.1	Tr
70	77	61	91	1.2	74	275	71	0.13	0.16	1.1	Tr
32	15	20	28	0.5	28	96	60	0.05	0.06	0.5	Tr
64	15	8	30	0.5	26	108	41	0.06	0.06	0.5	0
15	17	21	26	1.0	34	105	4	0.06	0.09	0.7	0
7	27	14	44	0.6	37	155	9	0.07	0.06	0.6	0
3	42	33	99	1.0	52	176	12	0.20	0.13	1.7	0
38	39	23	117	1.2	123	192	30	0.05	0.14	0.6	0
170	26	52	81	0.4	90	204	69	0.03	0.12	0.4	5
14	16	13	26	0.6	50	59	18	0.08	0.07	0.3	Tr
18	11	9	26	0.4	35	51	6	0.05	0.05	0.3	Tr
5	28	13	41	0.8	68	140	15	0.10	0.23	1.0	Tr
18	26	13	34	1.0	82	82	5	0.06	0.06	0.6	0
22	32	13	34	1.0	62	173	8	0.06	0.10	0.9	0
27	42	40	34	1.4	162	180	6	0.08	0.07	0.7	Tr
2	36	18	58	1.1	90	148	12	0.09	0.08	1.0	0
22	28	21	60	1.1	110	142	5	0.07	0.07	1.9	0
0	29	12	40	1.4	66	189	0	0.09	0.07	0.8	0
27	20	13	39	0.8	38	123	8	0.10	0.09	0.9	0
0	17	6	31	0.6	18	125	89	0.08	0.06	0.7	Tr
29	31	50	91	0.9	33	261	11	0.09	0.06	1.1	0
25	29	16	36	0.8	50	150	14	0.07	0.10	1.0	0

Foods, approximate measures, units, and weight (weight of edible portion only)		Water		Food energy	Protein	Fat	FATTY ACIDS		
							Satu-rated	Mono-unsaturated	Poly-unsaturated
GRAIN PRODUCTS—Continued		Grams	Percent	Calories	Grams	Grams	Grams	Grams	Grams
Corn chips	1-oz package	28	1	155	2	9	1.4	2.4	3.7
Cornmeal:									
Whole-ground, unbolted, dry form	1 cup	122	12	435	11	5	0.5	1.1	2.5
Bolted (nearly whole-grain), dry form	1 cup	122	12	440	11	4	0.5	0.9	2.2
Degermed, enriched:									
Dry form	1 cup	138	12	500	11	2	0.2	0.4	0.9
Cooked	1 cup	240	88	120	3	Tr	Tr	0.1	0.2
Crackers:[39]									
Cheese:									
Plain, 1 in. square	10 crackers	10	4	50	1	3	0.9	1.2	0.3
Sandwich type (peanut butter)	1 sandwich	8	3	40	1	2	0.4	0.8	0.3
Graham, plain, 2½ in. square	2 crackers	14	5	60	1	1	0.4	0.6	0.4
Melba toast, plain	1 piece	5	4	20	1	Tr	0.1	0.1	0.1
Rye wafers, whole-grain, 1⅞ by 3½ in.	2 wafers	14	5	55	1	1	0.3	0.4	0.3
Saltines[40]	4 crackers	12	4	50	1	1	0.5	0.4	0.2
Snack-type, standard	1 round cracker	3	3	15	Tr	1	0.2	0.4	0.1
Wheat, thin	4 crackers	8	3	35	1	1	0.5	0.5	0.4
Whole-wheat wafers	2 crackers	8	4	35	1	2	0.5	0.6	0.4
Croissants, made with enriched flour, 4½ by 4 by 1¾ in.	1 croissant	57	22	235	5	12	3.5	6.7	1.4
Danish pastry, made with enriched flour:									
Plain without fruit or nuts:									
Round piece, about 4¼-in. diam., 1 in. high	1 pastry	57	27	220	4	12	3.6	4.8	2.6
Fruit, round piece	1 pastry	65	30	235	4	13	3.9	5.2	2.9
Doughnuts, made with enriched flour:									
Cake type, plain, 3¼-in. diam., 1 in. high	1 doughnut	50	21	210	3	12	2.8	5.0	3.0
Yeast-leavened, glazed, 3¾-in. diam., 1¼ in. high	1 doughnut	60	27	235	4	13	5.2	5.5	0.9
English muffins, plain, enriched	1 muffin	57	42	140	5	1	0.3	0.2	0.3
Toasted	1 muffin	50	29	140	5	1	0.3	0.2	0.3
French toast, from home recipe	1 slice	65	53	155	6	7	1.6	2.0	1.6
Macaroni, enriched, cooked (cut lengths, elbows, shells):									
Firm stage (hot)	1 cup	130	64	190	7	1	0.1	0.1	0.3
Tender stage:									
Cold	1 cup	105	72	115	4	Tr	0.1	0.1	0.2
Hot	1 cup	140	72	155	5	1	0.1	0.1	0.2
Muffins made with enriched flour, 2½-in. diam., 1½ in. high:									
From home recipe:									
Blueberry[25]	1 muffin	45	37	135	3	5	1.5	2.1	1.2
Bran[36]	1 muffin	45	35	125	3	6	1.4	1.6	2.3
Corn (enriched, degermed cornmeal and flour)[25]	1 muffin	45	33	145	3	5	1.5	2.2	1.4
From commercial mix (egg and water added):									
Blueberry	1 muffin	45	33	140	3	5	1.4	2.0	1.2
Bran	1 muffin	45	28	140	3	4	1.3	1.6	1.0
Corn	1 muffin	45	30	145	3	6	1.7	2.3	1.4
Noodles (egg noodles), enriched, cooked	1 cup	160	70	200	7	2	0.5	0.6	0.6

[39]Crackers made with enriched flour except for rye wafers and whole-wheat wafers.
[40]Made with lard.

NUTRIENTS IN INDICATED QUANTITY

Cholesterol	Carbohydrate	Calcium	Phosphorus	Iron	Potassium	Sodium	Vitamin A value	Thiamin	Riboflavin	Niacin	Ascorbic acid
Milligrams	Grams	Milligrams	Milligrams	Milligrams	Milligrams	Milligrams	Retinol equiv. (RE)	Milligrams	Milligrams	Milligrams	Milligrams
0	16	35	52	0.5	52	233	11	0.04	0.05	0.4	1
0	90	24	312	2.2	346	1	62	0.46	0.13	2.4	0
0	91	21	272	2.2	303	1	59	0.37	0.10	2.3	0
0	108	8	137	5.9	166	1	61	0.61	0.36	4.8	0
0	26	2	34	1.4	38	0	14	0.14	0.10	1.2	0
6	6	11	17	0.3	17	112	5	0.05	0.04	0.4	0
1	5	7	25	0.3	17	90	Tr	0.04	0.03	0.6	0
0	11	6	20	0.4	36	86	0	0.02	0.03	0.6	0
0	4	6	10	0.1	11	44	0	0.01	0.01	0.1	0
0	10	7	44	0.5	65	115	0	0.06	0.03	0.5	0
4	9	3	12	0.5	17	165	0	0.06	0.05	0.6	0
0	2	3	6	0.1	4	30	Tr	0.01	0.01	0.1	0
0	5	3	15	0.3	17	69	Tr	0.04	0.03	0.4	0
0	5	3	22	0.2	31	59	0	0.02	0.03	0.4	0
13	27	20	64	2.1	68	452	13	0.17	0.13	1.3	0
49	26	60	58	1.1	53	218	17	0.16	0.17	1.4	Tr
56	28	17	80	1.3	57	233	11	0.16	0.14	1.4	Tr
20	24	22	111	1.0	58	192	5	0.12	0.12	1.1	Tr
21	26	17	55	1.4	64	222	Tr	0.28	0.12	1.8	0
0	27	96	67	1.7	331	378	0	0.26	0.19	2.2	0
0	27	96	67	1.7	331	378	0	0.23	0.19	2.2	0
112	17	72	85	1.3	86	257	32	0.12	0.16	1.0	Tr
0	39	14	85	2.1	103	1	0	0.23	0.13	1.8	0
0	24	8	53	1.3	64	1	0	0.15	0.08	1.2	0
0	32	11	70	1.7	85	1	0	0.20	0.11	1.5	0
19	20	54	46	0.9	47	198	9	0.10	0.11	0.9	1
24	19	60	125	1.4	99	189	30	0.11	0.13	1.3	3
23	21	66	59	0.9	57	169	15	0.11	0.11	0.9	Tr
45	22	15	90	0.9	54	225	11	0.10	0.17	1.1	Tr
28	24	27	182	1.7	50	385	14	0.08	0.12	1.9	0
42	22	30	128	1.3	31	291	16	0.09	0.09	0.8	Tr
50	37	16	94	2.6	70	3	34	0.22	0.13	1.9	0

Foods, approximate measures, units, and weight (weight of edible portion only)		Water	Food energy	Protein	Fat	FATTY ACIDS		
						Satu-rated	Mono-unsaturated	Poly-unsaturated
GRAIN PRODUCTS—Continued	Grams	Percent	Calories	Grams	Grams	Grams	Grams	Grams
Noodles, chow mein, canned	1 cup	45	11	220	6	11	2.1	7.3	0.4
Pancakes, 4-in. diam.:									
Buckwheat, from mix (with buck-wheat and enriched flours), egg and milk added	1 pancake	27	58	55	2	2	0.9	0.9	0.5
Plain:									
From home recipe using enriched flour	1 pancake	27	50	60	2	2	0.5	0.8	0.5
From mix (with enriched flour), egg, milk, and oil added	1 pancake	27	54	60	2	2	0.5	0.9	0.5
Piecrust, made with enriched flour and vegetable shortening, baked:									
From home recipe, 9-in. diam.	1 pie shell	180	15	900	11	60	14.8	25.9	15.7
From mix, 9-in. diam.	Piecrust for 2-crust pie	320	19	1,485	20	93	22.7	41.0	25.0
Pies, piecrust made with enriched flour, vegetable shortening, 9-in. diam.:									
Apple:									
Piece, ⅙ of pie	1 piece	158	48	405	3	18	4.6	7.4	4.4
Blueberry:									
Piece, ⅙ of pie	1 piece	158	51	380	4	17	4.3	7.4	4.6
Cherry:									
Piece, ⅙ of pie	1 piece	158	47	410	4	18	4.7	7.7	4.6
Creme:									
Piece, ⅙ of pie	1 piece	152	43	455	3	23	15.0	4.0	1.1
Custard:									
Piece, ⅙ of pie	1 piece	152	58	330	9	17	5.6	6.7	3.2
Lemon meringue:									
Piece, ⅙ of pie	1 piece	140	47	355	5	14	4.3	5.7	2.9
Peach:									
Piece, ⅙ of pie	1 piece	158	48	405	4	17	4.1	7.3	4.4
Pecan:									
Piece, ⅙ of pie	1 piece	138	20	575	7	32	4.7	17.0	7.9
Pumpkin:									
Piece, ⅙ of pie	1 piece	152	59	320	6	17	6.4	6.7	3.0
Pies, fried:									
Apple	1 pie	85	43	255	2	14	5.8	6.6	0.6
Cherry	1 pie	85	42	250	2	14	5.8	6.7	0.6
Popcorn, popped:									
Air-popped, unsalted	1 cup	8	4	30	1	Tr	Tr	0.1	0.2
Popped in vegetable oil, salted	1 cup	11	3	55	1	3	0.5	1.4	1.2
Sugar syrup coated	1 cup	35	4	135	2	1	0.1	0.3	0.6
Pretzels, made with enriched flour:									
Stick, 2¼ in. long	10 pretzels	3	3	10	Tr	Tr	Tr	Tr	Tr
Twisted, dutch, 2¾ by 2⅝ in.	1 pretzel	16	3	65	2	1	0.1	0.2	0.2
Twisted, thin, 3¼ by 2¼ by ¼ in.	10 pretzels	60	3	240	6	2	0.4	0.8	0.6
Rice:									
Brown, cooked, served hot	1 cup	195	70	230	5	1	0.3	0.3	0.4
White, enriched:									
Commercial varieties, all types:									
Raw	1 cup	185	12	670	12	1	0.2	0.2	0.3
Cooked, served hot	1 cup	205	73	225	4	Tr	0.1	0.1	0.1
Instant, ready-to-serve, hot	1 cup	165	73	180	4	0	0.1	0.1	0.1
Parboiled:									
Raw	1 cup	185	10	685	14	1	0.1	0.1	0.2
Cooked, served hot	1 cup	175	73	185	4	Tr	Tr	Tr	0.1

Cho-lesterol	Carbo-hydrate	Calcium	Phos-phorus	Iron	Potas-sium	Sodium	Vitamin A value	Thiamin	Ribo-flavin	Niacin	Ascorbic acid
Milli-grams	Grams	Milli-grams	Milli-grams	Milli-grams	Milli-grams	Milli-grams	Retinol equiv. (RE)	Milli-grams	Milli-grams	Milli-grams	Milli-grams
5	26	14	41	0.4	33	450	0	0.05	0.03	0.6	0
20	6	59	91	0.4	66	125	17	0.04	0.05	0.2	Tr
16	9	27	38	0.5	33	115	10	0.06	0.07	0.5	Tr
16	8	36	71	0.7	43	160	7	0.09	0.12	0.8	Tr
0	79	25	90	4.5	90	1,100	0	0.54	0.40	5.0	0
0	141	131	272	9.3	179	2,602	0	1.06	0.80	9.9	0
0	60	13	35	1.6	126	476	5	0.17	0.13	1.6	2
0	55	17	36	2.1	158	423	14	0.17	0.14	1.7	6
0	61	22	40	1.6	166	480	70	0.19	0.14	1.6	0
8	59	46	154	1.1	133	369	65	0.06	0.15	1.1	0
169	36	146	172	1.5	208	436	96	0.14	0.32	0.9	0
143	53	20	69	1.4	70	395	66	0.10	0.14	0.8	4
0	60	16	46	1.9	235	423	115	0.17	0.16	2.4	5
95	71	65	142	4.6	170	305	54	0.30	0.17	1.1	0
109	37	78	105	1.4	243	325	416	0.14	0.21	1.2	0
14	31	12	34	0.9	42	326	3	0.09	0.06	1.0	1
13	32	11	41	0.7	61	371	19	0.06	0.06	0.6	1
0	6	1	22	0.2	20	Tr	1	0.03	0.01	0.2	0
0	6	3	31	0.3	19	86	2	0.01	0.02	0.1	0
0	30	2	47	0.5	90	Tr	3	0.13	0.02	0.4	0
0	2	1	3	0.1	3	48	0	0.01	0.01	0.1	0
0	13	4	15	0.3	16	258	0	0.05	0.04	0.7	0
0	48	16	55	1.2	61	966	0	0.19	0.15	2.6	0
0	50	23	142	1.0	137	0	0	0.18	0.04	2.7	0
0	149	44	174	5.4	170	9	0	0.81	0.06	6.5	0
0	50	21	57	1.8	57	0	0	0.23	0.02	2.1	0
0	40	5	31	1.3	0	0	0	0.21	0.02	1.7	0
0	150	111	370	5.4	278	17	0	0.81	0.07	6.5	0
0	41	33	100	1.4	75	0	0	0.19	0.02	2.1	0

Foods, approximate measures, units, and weight (weight of edible portion only)		Water	Food energy	Protein	Fat	FATTY ACIDS		
						Saturated	Mono-unsaturated	Poly-unsaturated
		Grams	Percent	Calories	Grams	Grams	Grams	Grams	Grams

GRAIN PRODUCTS—Continued		Grams	Percent	Calories	Grams	Grams	Grams	Grams	Grams
Rolls, enriched:									
Commercial:									
Dinner, 2½-in. diam., 2 in. high	1 roll	28	32	85	2	2	0.5	0.8	0.6
Frankfurter and hamburger (8 per 11½ oz pkg.)	1 roll	40	34	115	3	2	0.5	0.8	0.6
Hard, 3¾-in. diam., 2 in. high	1 roll	50	25	155	5	2	0.4	0.5	0.6
Hoagie or submarine, 11½ by 3 by 2½ in.	1 roll	135	31	400	11	8	1.8	3.0	2.2
From home recipe:									
Dinner, 2½-in. diam., 2 in. high	1 roll	35	26	120	3	3	0.8	1.2	0.9
Spaghetti, enriched, cooked:									
Firm stage, "al dente," served hot	1 cup	130	64	190	7	1	0.1	0.1	0.3
Tender stage, served hot	1 cup	140	73	155	5	1	0.1	0.1	0.2
Toaster pastries	1 pastry	54	13	210	2	6	1.7	3.6	0.4
Tortillas, corn	1 tortilla	30	45	65	2	1	0.1	0.3	0.6
Waffles, made with enriched flour, 7-in.-diam.									
From home recipe	1 waffle	75	37	245	7	13	4.0	4.9	2.6
From mix, egg and milk added	1 waffle	75	42	205	7	8	2.7	2.9	1.5
Wheat flours:									
All-purpose, enriched									
Sifted, spooned	1 cup	115	12	420	12	1	0.2	0.1	0.5
Unsifted, spooned	1 cup	125	12	455	13	1	0.2	0.1	0.5
Cake or pastry flour, enriched, sifted, spooned	1 cup	96	12	350	7	1	0.1	0.1	0.3
Self-rising, enriched, unsifted, spooned	1 cup	125	12	440	12	1	0.2	0.1	0.5
Whole-wheat, from hard wheats, stirred	1 cup	120	12	400	16	2	0.3	0.3	1.1

LEGUMES, NUTS, AND SEEDS									
Almonds, shelled:									
Slivered, packed	1 cup	135	4	795	27	70	6.7	45.8	14.8
Whole	1 oz	28	4	165	6	15	1.4	9.6	3.1
Beans, dry:									
Cooked, drained:									
Black	1 cup	171	66	225	15	1	0.1	0.1	0.5
Great Northern	1 cup	180	69	210	14	1	0.1	0.1	0.6
Lima	1 cup	190	64	260	16	1	0.2	0.1	0.5
Pea (navy)	1 cup	190	69	225	15	1	0.1	0.1	0.7
Pinto	1 cup	180	65	265	15	1	0.1	0.1	0.5
Canned, solids and liquid:									
White with:									
Frankfurters (sliced)	1 cup	255	71	365	19	18	7.4	8.8	0.7
Pork and tomato sauce	1 cup	255	71	310	16	7	2.4	2.7	0.7
Pork and sweet sauce	1 cup	255	66	385	16	12	4.3	4.9	1.2
Red kidney	1 cup	255	76	230	15	1	0.1	0.1	0.6
Black-eyed peas, dry, cooked (with residual cooking liquid)	1 cup	250	80	190	13	1	0.2	Tr	0.3
Brazil nuts, shelled	1 oz	28	3	185	4	19	4.6	6.5	6.8
Carob flour	1 cup	140	3	255	6	Tr	Tr	0.1	0.1
Cashew nuts, salted:									
Dry roasted	1 cup	137	2	785	21	63	12.5	37.4	10.7
	1 oz	28	2	165	4	13	2.6	7.7	2.2
Roasted in oil	1 cup	130	4	750	21	63	12.4	36.9	10.6
	1 oz	28	4	165	5	14	2.7	8.1	2.3
Chestnuts, European (Italian), roasted, shelled	1 cup	143	40	350	5	3	0.6	1.1	1.2
Chickpeas, cooked, drained	1 cup	163	60	270	15	4	0.4	0.9	1.9

Cho-lesterol	Carbo-hydrate	Calcium	Phos-phorus	Iron	Potass-ium	Sodium	Vitamin A value	Thiamin	Ribo-flavin	Niacin	Ascorbic acid
Milli-grams	Grams	Milli-grams	Milli-grams	Milli-grams	Milli-grams	Milli-grams	Retinol equiv. (RE)	Milli-grams	Milli-grams	Milli-grams	Milli-grams
Tr	14	33	44	0.8	36	155	Tr	0.14	0.09	1.1	Tr
Tr	20	54	44	1.2	56	241	Tr	0.20	0.13	1.6	Tr
Tr	30	24	46	1.4	49	313	0	0.20	0.12	1.7	0
Tr	72	100	115	3.8	128	683	0	0.54	0.33	4.5	0
12	20	16	36	1.1	41	98	8	0.12	0.12	1.2	0
0	39	14	85	2.0	103	1	0	0.23	0.13	1.8	0
0	32	11	70	1.7	85	1	0	0.20	0.11	1.5	0
0	38	104	104	2.2	91	248	52	0.17	0.18	2.3	4
0	13	42	55	0.6	43	1	8	0.05	0.03	0.4	0
102	26	154	135	1.5	129	445	39	0.18	0.24	1.5	Tr
59	27	179	257	1.2	146	515	49	0.14	0.23	0.9	Tr
0	88	18	100	5.1	109	2	0	0.73	0.46	6.1	0
0	95	20	109	5.5	119	3	0	0.80	0.50	6.6	0
0	76	16	70	4.2	91	2	0	0.58	0.38	5.1	0
0	93	331	583	5.5	113	1,349	0	0.80	0.50	6.6	0
0	85	49	446	5.2	444	4	0	0.66	0.14	5.2	0
0	28	359	702	4.9	988	15	0	0.28	1.05	4.5	1
0	6	75	147	1.0	208	3	0	0.06	0.22	1.0	Tr
0	41	47	239	2.9	608	1	Tr	0.43	0.05	0.9	0
0	38	90	266	4.9	749	13	0	0.25	0.13	1.3	0
0	49	55	293	5.9	1,163	4	0	0.25	0.11	1.3	0
0	40	95	281	5.1	790	13	0	0.27	0.13	1.3	0
0	49	86	296	5.4	882	3	Tr	0.33	0.16	0.7	0
30	32	94	303	4.8	668	1,374	33	0.18	0.15	3.3	Tr
10	48	138	235	4.6	536	1,181	33	0.20	0.08	1.5	5
10	54	161	291	5.9	536	969	33	0.15	0.10	1.3	5
0	42	74	278	4.6	673	968	1	0.13	0.10	1.5	0
0	35	43	238	3.3	573	20	3	0.40	0.10	1.0	0
0	4	50	170	1.0	170	1	Tr	0.28	0.03	0.5	Tr
0	126	390	102	5.7	1,275	24	Tr	0.07	0.07	2.2	Tr
0	45	62	671	8.2	774	[41]877	0	0.27	0.27	1.9	0
0	9	13	139	1.7	160	[41]181	0	0.06	0.06	0.4	0
0	37	53	554	5.3	689	[42]814	0	0.55	0.23	2.3	0
0	8	12	121	1.2	150	[42]177	0	0.12	0.05	0.5	0
0	76	41	153	1.3	847	3	3	0.35	0.25	1.9	37
0	45	80	273	4.9	475	11	Tr	0.18	0.09	0.9	0

[41]Cashews without salt contain 21 mg sodium per cup or 4 mg per oz.

[42]Cashews without salt contain 22 mg sodium per cup or 5 mg per oz.

Foods, approximate measures, units, and weight (weight of edible portion only)		Water	Food energy	Protein	Fat	FATTY ACIDS Satu- rated	Mono- unsaturated	Poly- unsaturated
		Grams	Percent	Calories	Grams	Grams	Grams	Grams

LEGUMES, NUTS, AND SEEDS

Food	Measure	Water (g)	Percent	Calories	Protein (g)	Fat (g)	Saturated (g)	Mono-unsaturated (g)	Poly-unsaturated (g)
Coconut:									
Raw, shredded or grated	1 cup	80	47	285	3	27	23.8	1.1	0.3
Dried, sweetened, shredded	1 cup	93	13	470	3	33	29.3	1.4	0.4
Filberts (hazelnuts), chopped	1 cup	115	5	725	15	72	5.3	56.5	6.9
	1 oz	28	5	180	4	18	1.3	13.9	1.7
Lentils, dry, cooked	1 cup	200	72	215	16	1	0.1	0.2	0.5
Macadamia nuts, roasted in oil, salted	1 cup	134	2	960	10	103	15.4	80.9	1.8
	1 oz	28	2	205	2	22	3.2	17.1	0.4
Mixed nuts, with peanuts, salted:									
Dry roasted	1 oz	28	2	170	5	15	2.0	8.9	3.1
Roasted in oil	1 oz	28	2	175	5	16	2.5	9.0	3.8
Peanuts, roasted in oil, salted	1 cup	145	2	840	39	71	9.9	35.5	22.6
	1 oz	28	2	165	8	14	1.9	6.9	4.4
Peanut butter	1 tbsp	16	1	95	5	8	1.4	4.0	2.5
Peas, split, dry, cooked	1 cup	200	70	230	16	1	0.1	0.1	0.3
Pecans, halves	1 cup	108	5	720	8	73	5.9	45.5	18.1
	1 oz	28	5	190	2	19	1.5	12.0	4.7
Pine nuts (pinyons), shelled	1 oz	28	6	160	3	17	2.7	6.5	7.3
Pistachio nuts, dried, shelled	1 oz	28	4	165	6	14	1.7	9.3	2.1
Pumpkin and squash kernels, dry, hulled	1 oz	28	7	155	7	13	2.5	4.0	5.9
Refried beans, canned	1 cup	290	72	295	18	3	0.4	0.6	1.4
Sesame seeds, dry, hulled	1 tbsp	8	5	45	2	4	0.6	1.7	1.9
Soybeans, dry, cooked, drained	1 cup	180	71	235	20	10	1.3	1.9	5.3
Soy products:									
Miso	1 cup	276	53	470	29	13	1.8	2.6	7.3
Tofu, piece 2½ by 2¾ by 1 in.	1 piece	120	85	85	9	5	0.7	1.0	2.9
Sunflower seeds, dry, hulled	1 oz	28	5	160	6	14	1.5	2.7	9.3
Tahini	1 tbsp	15	3	90	3	8	1.1	3.0	3.5
Walnuts:									
Black, chopped	1 cup	125	4	760	30	71	4.5	15.9	46.9
	1 oz	28	4	170	7	16	1.0	3.6	10.6
English or Persian, pieces or chips	1 cup	120	4	770	17	74	6.7	17.0	47.0
	1 oz	28	4	180	4	18	1.6	4.0	11.1

MEAT AND MEAT PRODUCTS

Food	Measure	Water (g)	Percent	Calories	Protein (g)	Fat (g)	Saturated (g)	Mono-unsaturated (g)	Poly-unsaturated (g)
Beef, cooked:[46]									
Cuts braised, simmered, or pot roasted:									
Relatively fat such as chuck blade:									
Lean and fat, piece, 2½ by 2½ by ¾ in.	3 oz	85	43	325	22	26	10.8	11.7	0.9
Lean only	2.2 oz	62	53	170	19	9	3.9	4.2	0.3
Relatively lean, such as bottom round:									
Lean and fat, piece, 4⅛ by 2¼ by ½ in.	3 oz	85	54	220	25	13	4.8	5.7	0.5
Lean only	2.8 oz	78	57	175	25	8	2.7	3.4	0.3
Ground beef, broiled, patty, 3 by ⅝ in.:									
Lean	3 oz	85	56	230	21	16	6.2	6.9	0.6
Regular	3 oz	85	54	245	20	18	6.9	7.7	0.7
Heart, lean, braised	3 oz	85	65	150	24	5	1.2	0.8	1.6
Liver, fried, slice, 6½ by 2⅜ by ⅜ in.[47]	3 oz	85	56	185	23	7	2.5	3.6	1.3

[43]Macadamia nuts without salt contain 9 mg sodium per cup or 2 mg per oz.

[44]Mixed nuts without salt contain 3 mg sodium per oz.

[45]Peanuts without salt contain 22 mg sodium per cup or 4 mg per oz.

Cho-lesterol	Carbo-hydrate	Calcium	Phos-phorus	Iron	Potas-sium	Sodium	Vitamin A value	Thiamin	Ribo-flavin	Niacin	Ascorbic acid
Milli-grams	Grams	Milli-grams	Milli-grams	Milli-grams	Milli-grams	Milli-grams	Retinol equiv. (RE)	Milli-grams	Milli-grams	Milli-grams	Milli-grams
0	12	11	90	1.9	285	16	0	0.05	0.02	0.4	3
0	44	14	99	1.8	313	244	0	0.03	0.02	0.4	1
0	18	216	359	3.8	512	3	8	0.58	0.13	1.3	1
0	4	53	88	0.9	126	1	2	0.14	0.03	0.3	Tr
0	38	50	238	4.2	498	26	4	0.14	0.12	1.2	0
0	17	60	268	2.4	441	[43]348	1	0.29	0.15	2.7	0
0	4	13	57	0.5	93	[43]74	Tr	0.06	0.03	0.6	0
0	7	20	123	1.0	169	[44]190	Tr	0.06	0.06	1.3	0
0	6	31	131	0.9	165	[44]185	1	0.14	0.06	1.4	Tr
0	27	125	734	2.8	1,019	[45]626	0	0.42	0.15	21.5	0
0	5	24	143	0.5	199	[45]122	0	0.08	0.03	4.2	0
0	3	5	60	0.3	110	75	0	0.02	0.02	2.2	0
0	42	22	178	3.4	592	26	8	0.30	0.18	1.8	0
0	20	39	314	2.3	423	1	14	0.92	0.14	1.0	2
0	5	10	83	0.6	111	Tr	4	0.24	0.04	0.3	1
0	5	2	10	0.9	178	20	1	0.35	0.06	1.2	1
0	7	38	143	1.9	310	2	7	0.23	0.05	0.3	Tr
0	5	12	333	4.2	229	5	11	0.06	0.09	0.5	Tr
0	51	141	245	5.1	1,141	1,228	0	0.14	0.16	1.4	17
0	1	11	62	0.6	33	3	1	0.06	0.01	0.4	0
0	19	131	322	4.9	972	4	5	0.38	0.16	1.1	0
0	65	188	853	4.7	922	8,142	11	0.17	0.28	0.8	0
0	3	108	151	2.3	50	8	0	0.07	0.04	0.1	0
0	5	33	200	1.9	195	1	1	0.65	0.07	1.3	Tr
0	3	21	119	0.7	69	5	1	0.24	0.02	0.8	1
0	15	73	580	3.8	655	1	37	0.27	0.14	0.9	Tr
0	3	16	132	0.9	149	Tr	8	0.06	0.03	0.2	Tr
0	22	113	380	2.9	602	12	15	0.46	0.18	1.3	4
0	5	27	90	0.7	142	3	4	0.11	0.04	0.3	1
87	0	11	163	2.5	163	53	Tr	0.06	0.19	2.0	0
66	0	8	146	2.3	163	44	Tr	0.05	0.17	1.7	0
81	0	5	217	2.8	248	43	Tr	0.06	0.21	3.3	0
75	0	4	212	2.7	240	40	Tr	0.06	0.20	3.0	0
74	0	9	134	1.8	256	65	Tr	0.04	0.18	4.4	0
76	0	9	144	2.1	248	70	Tr	0.03	0.16	4.9	0
164	0	5	213	6.4	198	54	Tr	0.12	1.31	3.4	5
410	7	9	392	5.3	309	90	[48]9,120	0.18	3.52	12.3	23

[46]Outer layer of fat was removed to within approximately ½ inch of the lean. Deposits of fat within the cut were not removed.

[47]Fried in vegetable shortening.

[48]Value varies widely.

Foods, approximate measures, units, and weight (weight of edible portion only)		Water	Food energy	Protein	Fat	FATTY ACIDS Saturated	Monounsaturated	Polyunsaturated
		Grams	Percent	Calories	Grams	Grams	Grams	Grams
MEAT AND MEAT PRODUCTS—Continued		Grams	Percent	Calories	Grams	Grams	Grams	Grams
Beef, cooked:[46]								
Roast, oven cooked, no liquid added:								
Relatively fat, such as rib:								
Lean and fat, 2 pieces, 4⅛ by 2¼ by ¼ in.	3 oz	85	46	315	19	26	10.8	11.4	0.9
Lean only	2.2 oz	61	57	150	17	9	3.6	3.7	0.3
Relatively lean, such as eye of round:								
Lean and fat, 2 pieces, 2½ by 2½ by ⅜ in.	3 oz	85	57	205	23	12	4.9	5.4	0.5
Lean only	2.6 oz	75	63	135	22	5	1.9	2.1	0.2
Steak:								
Sirloin, broiled:								
Lean and fat, piece, 2½ by 2½ by ¾ in.	3 oz	85	53	240	23	15	6.4	6.9	0.6
Lean only	2.5 oz	72	59	150	22	6	2.6	2.8	0.3
Beef, canned, corned	3 oz	85	59	185	22	10	4.2	4.9	0.4
Beef, dried, chipped	2.5 oz	72	48	145	24	4	1.8	2.0	0.2
Lamb, cooked:								
Chops (3 per lb with bone):								
Arm, braised:								
Lean and fat	2.2 oz	63	44	220	20	15	6.9	6.0	0.9
Lean only	1.7 oz	48	49	135	17	7	2.9	2.6	0.4
Loin, broiled:								
Lean and fat	2.8 oz	80	54	235	22	16	7.3	6.4	1.0
Lean only	2.3 oz	64	61	140	19	6	2.6	2.4	0.4
Leg, roasted:								
Lean and fat, 2 pieces, 4⅛ by 2¼ by ¼ in.	3 oz	85	59	205	22	13	5.6	4.9	0.8
Lean only	2.6 oz	73	64	140	20	6	2.4	2.2	0.4
Rib, roasted:								
Lean and fat, 3 pieces, 2½ by 2½ by ¼ in.	3 oz	85	47	315	18	26	12.1	10.6	1.5
Lean only	2 oz	57	60	130	15	7	3.2	3.0	0.5
Pork, cured, cooked:								
Bacon:								
Regular, medium slices	3	19	13	110	6	9	3.3	4.5	1.1
Canadian-style	2 slices	46	62	85	11	4	1.3	1.9	0.4
Ham, light cure, roasted:								
Lean and fat, 2 pieces, 4⅛ by 2¼ by ¼ in.	3 oz	85	58	205	18	14	5.1	6.7	1.5
Lean only	2.4 oz	68	66	105	17	4	1.3	1.7	0.4
Ham, canned, roasted, 2 pieces, 4⅛ by 2¼ by ¼ in.	3 oz	85	67	140	18	7	2.4	3.5	0.8
Luncheon meat:								
Canned, spiced or unspiced, slice, 3 by 2 by ½ in.	2 slices	42	52	140	5	13	4.5	6.0	1.5
Chopped ham (¾ oz slice)	2 slices	42	64	95	7	7	2.4	3.4	0.9
Cooked ham (1 oz slice)								
Regular	2 slices	57	65	105	10	6	1.9	2.8	0.7
Extra lean	2 slices	57	71	75	11	3	0.9	1.3	0.3
Pork, fresh, cooked:								
Chop, loin (3 per lb with bone):								
Broiled:								
Lean and fat	3.1 oz	87	50	275	24	19	7.0	8.8	2.2
Lean only	2.5 oz	72	57	165	23	8	2.6	3.4	0.9

Cho-lesterol	Carbo-hydrate	Calcium	Phos-phorus	Iron	Potas-sium	Sodium	Vitamin A value	Thiamin	Ribo-flavin	Niacin	Ascorbic acid
Milli-grams	Grams	Milli-grams	Milli-grams	Milli-grams	Milli-grams	Milli-grams	Retinol equiv. (RE)	Milli-grams	Milli-grams	Milli-grams	Milli-grams
72	0	8	145	2.0	246	54	Tr	0.06	0.16	3.1	0
49	0	5	127	1.7	218	45	Tr	0.05	0.13	2.7	0
62	0	5	177	1.6	308	50	Tr	0.07	0.14	3.0	0
52	0	3	170	1.5	297	46	Tr	0.07	0.13	2.8	0
77	0	9	186	2.6	306	53	Tr	0.10	0.23	3.3	0
64	0	8	176	2.4	290	48	Tr	0.09	0.22	3.1	0
80	0	17	90	3.7	51	802	Tr	0.02	0.20	2.9	0
46	0	14	287	2.3	142	3,053	Tr	0.05	0.23	2.7	0
77	0	16	132	1.5	195	46	Tr	0.04	0.16	4.4	0
59	0	12	111	1.3	162	36	Tr	0.03	0.13	3.0	0
78	0	16	162	1.4	272	62	Tr	0.09	0.21	5.5	0
60	0	12	145	1.3	241	54	Tr	0.08	0.18	4.4	0
78	0	8	162	1.7	273	57	Tr	0.09	0.24	5.5	0
65	0	6	150	1.5	247	50	Tr	0.08	0.20	4.6	0
77	0	19	139	1.4	224	60	Tr	0.08	0.18	5.5	0
50	0	12	111	1.0	179	46	Tr	0.05	0.13	3.5	0
16	Tr	2	64	0.3	92	303	0	0.13	0.05	1.4	6
27	1	5	136	0.4	179	711	0	0.38	0.09	3.2	10
53	0	6	182	0.7	243	1,009	0	0.51	0.19	3.8	0
37	0	5	154	0.6	215	902	0	0.46	0.17	3.4	0
35	Tr	6	188	0.9	298	908	0	0.82	0.21	4.3	[49]19
26	1	3	34	0.3	90	541	0	0.15	0.08	1.3	Tr
21	0	3	65	0.3	134	576	0	0.27	0.09	1.6	[49]8
32	2	4	141	0.6	189	751	0	0.49	0.14	3.0	[49]16
27	1	4	124	0.4	200	815	0	0.53	0.13	2.8	[49]15
84	0	3	184	0.7	312	61	3	0.87	0.24	4.3	Tr
71	0	4	176	0.7	302	56	1	0.83	0.22	4.0	Tr

NUTRIENTS IN INDICATED QUANTITY

[49]Contains added sodium ascorbate. If sodium ascorbate is not added, ascorbic acid content is negligible.

Foods, approximate measures, units, and weight (weight of edible portion only)		Water	Food energy	Protein	Fat	Satu-rated	Mono-unsaturated	Poly-unsaturated
							FATTY ACIDS	
		Grams	Percent	Calories	Grams	Grams	Grams	Grams	Grams

MEAT AND MEAT PRODUCTS—Continued

		Grams	Percent	Calories	Grams	Grams	Grams	Grams	Grams
Pork, fresh, cooked:									
Chop, loin (3 per lb with bone):									
Pan fried:									
Lean and fat	3.1 oz	89	45	335	21	27	9.8	12.5	3.1
Lean only	2.4 oz	67	54	180	19	11	3.7	4.8	1.3
Ham (leg), roasted:									
Lean and fat, piece, 2½ by 2½ by ¾ in.	3 oz	85	53	250	21	18	6.4	8.1	2.0
Lean only	2.5 oz	72	60	160	20	8	2.7	3.6	1.0
Rib, roasted:									
Lean and fat, piece, 2½ by ¾ in.	3 oz	85	51	270	21	20	7.2	9.2	2.3
Lean only	2.5 oz	71	57	175	20	10	3.4	4.4	1.2
Shoulder cut, braised:									
Lean and fat, 3 pieces, 2½ by 2½ by ¼ in.	3 oz	85	47	295	23	22	7.9	10.0	2.4
Lean only	2.4 oz	67	54	165	22	8	2.8	3.7	1.0
Sausages (See also Luncheon meats.)									
Bologna, 1 oz slice	2 slices	57	54	180	7	16	6.1	7.6	1.4
Braunschweiger, 1 oz slice	2 slices	57	48	205	8	18	6.2	8.5	2.1
Brown and serve (10–11 per 8-oz pkg), browned	1 link	13	45	50	2	5	1.7	2.2	0.5
Frankfurter (10 per 1-lb pkg), cooked (reheated)	1 frankfurter	45	54	145	5	13	4.8	6.2	1.2
Pork link (16 per 1-lb pkg), cooked[50]	1 link	13	45	50	3	4	1.4	1.8	0.5
Salami:									
Cooked type, 1 oz slice	2 slices	57	60	145	8	11	4.6	5.2	1.2
Dry type, slice (12 per 4-oz pkg)	2 slices	20	35	85	5	7	2.4	3.4	0.6
Sandwich spread (pork, beef)	1 tbsp	15	60	35	1	3	0.9	1.1	0.4
Vienna sausage (7 per 4-oz can)	1 sausage	16	60	45	2	4	1.5	2.0	0.3
Veal, medium fat, cooked, bone removed:									
Cutlet, 4⅛ by 2¼ by ½ in., braised or broiled	3 oz	85	60	185	23	9	4.1	4.1	0.6
Rib, 2 pieces, 4⅛ by 2¼ by ¼ in., roasted	3 oz	85	55	230	23	14	6.0	6.0	1.0

MIXED DISHES AND FAST FOODS

		Grams	Percent	Calories	Grams	Grams	Grams	Grams	Grams
Mixed dishes:									
Beef and vegetable stew, from home recipe	1 cup	245	82	220	16	11	4.4	4.5	0.5
Beef potpie, from home recipe, baked, piece, ⅓ of 9-in. diam. pie[51]	1 piece	210	55	515	21	30	7.9	12.9	7.4
Chicken à la king, cooked, from home recipe	1 cup	245	68	470	27	34	12.9	13.4	6.2
Chicken and noodles, cooked from home recipe	1 cup	240	71	365	22	18	5.1	7.1	3.9
Chicken chow mein:									
Canned	1 cup	250	89	95	7	Tr	0.1	0.1	0.8
From home recipe	1 cup	250	78	255	31	10	4.1	4.9	3.5
Chicken potpie, from home recipe, baked, piece ⅓ of 9-in. diam. pie[51]	1 piece	232	57	545	23	31	10.3	15.5	6.6
Chili con carne with beans, canned	1 cup	255	72	340	19	16	5.8	7.2	1.0
Chop suey with beef and pork, from home recipe	1 cup	250	75	300	26	17	4.3	7.4	4.2

[50]One patty (8 per pound) of bulk sausage is equivalent to 2 links.

						NUTRIENTS IN INDICATED QUANTITY					
Cholesterol	Carbohydrate	Calcium	Phosphorus	Iron	Potassium	Sodium	Vitamin A value	Thiamin	Riboflavin	Niacin	Ascorbic acid
Milligrams	Grams	Milligrams	Milligrams	Milligrams	Milligrams	Milligrams	Retinol equiv. (RE)	Milligrams	Milligrams	Milligrams	Milligrams
92	0	4	190	0.7	323	64	3	0.91	0.24	4.6	Tr
72	0	3	178	0.7	305	57	1	0.84	0.22	4.0	Tr
79	0	5	210	0.9	280	50	2	0.54	0.27	3.9	Tr
68	0	5	202	0.8	269	46	1	0.50	0.25	3.6	Tr
69	0	9	190	0.8	313	37	3	0.50	0.24	4.2	Tr
56	0	8	182	0.7	300	33	2	0.45	0.22	3.8	Tr
93	0	6	162	1.4	286	75	3	0.46	0.26	4.4	Tr
76	0	5	151	1.3	271	68	1	0.40	0.24	4.0	Tr
31	2	7	52	0.9	103	581	0	0.10	0.08	1.5	[49]12
89	2	5	96	5.3	113	652	2,405	0.14	0.87	4.8	[49]6
9	Tr	1	14	0.1	25	105	0	0.05	0.02	0.4	0
23	1	5	39	0.5	75	504	0	0.09	0.05	1.2	[49]12
11	Tr	4	24	0.2	47	168	0	0.10	0.03	0.6	Tr
37	1	7	66	1.5	113	607	0	0.14	0.21	2.0	[49]7
16	1	2	28	0.3	76	372	0	0.12	0.06	1.0	[49]5
6	2	2	9	0.1	17	152	1	0.03	0.02	0.3	0
8	Tr	2	8	0.1	16	152	0	0.01	0.02	0.3	0
109	0	9	196	0.8	258	56	Tr	0.06	0.21	4.6	0
109	0	10	211	0.7	259	57	Tr	0.11	0.26	6.6	0
71	15	29	184	2.9	613	292	568	0.15	0.17	4.7	17
42	39	29	149	3.8	334	596	517	0.29	0.29	4.8	6
221	12	127	358	2.5	404	760	272	0.10	0.42	5.4	12
103	26	26	247	2.2	149	600	130	0.05	0.17	4.3	Tr
8	18	45	85	1.3	418	725	28	0.05	0.10	1.0	13
75	10	58	293	2.5	473	718	50	0.08	0.23	4.3	10
56	42	70	232	3.0	343	594	735	0.32	0.32	4.9	5
28	31	82	321	4.3	594	1,354	15	0.08	0.18	3.3	8
68	13	60	248	4.8	425	1,053	60	0.28	0.38	5.0	33

[51]Crust made with vegetable shortening and enriched flour.

Foods, approximate measures, units, and weight (weight of edible portion only)		Water		Food energy	Protein	Fat	FATTY ACIDS		
							Satu-rated	Mono-unsaturated	Poly-unsaturated
MIXED DISHES AND FAST FOODS—Continued		Grams	Percent	Calories	Grams	Grams	Grams	Grams	Grams
Mixed dishes:									
Macaroni (enriched) and cheese:									
Canned:[52]	1 cup	240	80	230	9	10	4.7	2.9	1.3
From home recipe[38]	1 cup	200	58	430	17	22	9.8	7.4	3.6
Quiche Lorraine, 1/8 of 8-in. diam. quiche[51]	1 slice	176	47	600	13	48	23.2	17.8	4.1
Spaghetti (enriched) in tomato sauce with cheese:									
Canned	1 cup	250	80	190	6	2	0.4	0.4	0.5
From home recipe	1 cup	250	77	260	9	9	3.0	3.6	1.2
Spaghetti (enriched) with meatballs and tomato sauce:									
Canned	1 cup	250	78	260	12	10	2.4	3.9	3.1
From home recipe	1 cup	248	70	330	19	12	3.9	4.4	2.2
Fast food entrees:									
Cheeseburger:									
Regular	1 sand-wich	112	46	300	15	15	7.3	5.6	1.0
4 oz patty	1 sand-wich	194	46	525	30	31	15.1	12.2	1.4
Chicken, fried. See Poultry and Poultry Products.									
Enchilada	1 enchi-lada	230	72	235	20	16	7.7	6.7	0.6
English muffin, egg, cheese, and bacon	1 sand-wich	138	49	360	18	18	8.0	8.0	0.7
Fish sandwich:									
Regular, with cheese	1	140	43	420	16	23	6.3	6.9	7.7
Large, without cheese	1	170	48	470	18	27	6.3	8.7	9.5
Hamburger:									
Regular	1 sand-wich	98	46	245	12	11	4.4	5.3	0.5
4 oz patty	1 sand-wich	174	50	445	25	21	7.1	11.7	0.6
Pizza, cheese, 1/8 of 15-in. diam. pizza[51]	1 slice	120	46	290	15	9	4.1	2.6	1.3
Roast beef sandwich	1	150	52	345	22	13	3.5	6.9	1.8
Taco	1 taco	81	55	195	9	11	4.1	5.5	0.8
POULTRY AND POULTRY PRODUCTS									
Chicken:									
Fried, flesh, with skin:[53]									
Batter dipped:									
Breast, 1/2 breast (5.6 oz with bones)	4.9 oz	140	52	365	35	18	4.9	7.6	4.3
Drumstick (3.4 oz with bones)	2.5 oz	72	53	195	16	11	3.0	4.6	2.7
Flour coated:									
Breast, 1/2 breast (4.2 oz with bones)	3.5 oz	98	57	220	31	9	2.4	3.4	1.9
Drumstick (2.6 oz with bones)	1.7 oz	49	57	120	13	7	1.8	2.7	1.6
Roasted, flesh only:									
Breast, 1/2 breast (4.2 oz with bones and skin)	3 oz	86	65	140	27	3	0.9	1.1	0.7
Drumstick, (2.9 oz with bones and skin)	1.6 oz	44	67	75	12	2	0.7	0.8	0.6

[52]Made with corn oil.

[53]Fried in vegetable shortening.

Cho-lesterol	Carbo-hydrate	Calcium	Phos-phorus	Iron	Potas-sium	Sodium	Vitamin A value	Thiamin	Ribo-flavin	Niacin	Ascorbic acid
Milli-grams	Grams	Milli-grams	Milli-grams	Milli-grams	Milli-grams	Milli-grams	Retinol equiv. (RE)	Milli-grams	Milli-grams	Milli-grams	Milli-grams
24	26	199	182	1.0	139	730	72	0.12	0.24	1.0	Tr
44	40	362	322	1.8	240	1,086	232	0.20	0.40	1.8	1
285	29	211	276	1.0	283	653	454	0.11	0.32	Tr	Tr
3	39	40	88	2.8	303	955	120	0.35	0.28	4.5	10
8	37	80	135	2.3	408	955	140	0.25	0.18	2.3	13
23	29	53	113	3.3	245	1,220	100	0.15	0.18	2.3	5
89	39	124	236	3.7	665	1,009	159	0.25	0.30	4.0	22
44	28	135	174	2.3	219	672	65	0.26	0.24	3.7	1
104	40	236	320	4.5	407	1,224	128	0.33	0.48	7.4	3
19	24	97	198	3.3	653	1,332	352	0.18	0.26	Tr	Tr
213	31	197	290	3.1	201	832	160	0.46	0.50	3.7	1
56	39	132	223	1.8	274	667	25	0.32	0.26	3.3	2
91	41	61	246	2.2	375	621	15	0.35	0.23	3.5	1
32	28	56	107	2.2	202	463	14	0.23	0.24	3.8	1
71	38	75	225	4.8	404	763	28	0.38	0.38	7.8	1
56	39	220	216	1.6	230	699	106	0.34	0.29	4.2	2
55	34	60	222	4.0	338	757	32	0.40	0.33	6.0	2
21	15	109	134	1.2	263	456	57	0.09	0.07	1.4	1
119	13	28	259	1.8	281	385	28	0.16	0.20	14.7	0
62	6	12	106	1.0	134	194	19	0.08	0.15	3.7	0
87	2	16	228	1.2	254	74	15	0.08	0.13	13.5	0
44	1	6	86	0.7	112	44	12	0.04	0.11	3.0	0
73	0	13	196	0.9	220	64	5	0.06	0.10	11.8	0
41	0	5	81	0.6	108	42	8	0.03	0.10	2.7	0

Foods, approximate measures, units, and weight (weight of edible portion only)		Water	Food energy	Protein	Fat	FATTY ACIDS Satu-rated	Mono-unsaturated	Poly-unsaturated
		Grams	Percent	Calories	Grams	Grams	Grams	Grams

POULTRY AND POULTRY PRODUCTS—Continued

Chicken:								
Stewed, flesh only, light and dark meat, chopped or diced	1 cup	140	67	250	38	9	2.6	3.3	2.2
Chicken liver, cooked	1 liver	20	68	30	5	1	0.4	0.3	0.2
Duck, roasted, flesh only	½ duck	221	64	445	52	25	9.2	8.2	3.2
Turkey, roasted, flesh only:								
Dark meat, piece, 2½ by 1⅝ by ¼ in.	4 pieces	85	63	160	24	6	2.1	1.4	1.8
Light meat, piece, 4 by 2 by ¼ in.	2 pieces	85	66	135	25	3	0.9	0.5	0.7
Light and dark meat:								
Chopped or diced	1 cup	140	65	240	41	7	2.3	1.4	2.0
Pieces (1 slice white meat, 4 by 2 by ¼ in. and 2 slices dark meat, 2½ by 1⅝ by ¼ in.	3 pieces	85	65	145	25	4	1.4	0.9	1.2
Poultry food products:								
Chicken:								
Canned, boneless	5 oz	142	69	235	31	11	3.1	4.5	2.5
Frankfurter (10 per 1-lb pkg)	1 frankfurter	45	58	115	6	9	2.5	3.8	1.8
Roll, light, 1 oz slice	2 slices	57	69	90	11	4	1.1	1.7	0.9
Turkey:								
Gravy and turkey, frozen	5-oz package	142	85	95	8	4	1.2	1.4	0.7
Ham, cured turkey thigh meat, 1 oz slice	2 slices	57	71	75	11	3	1.0	0.7	0.9
Loaf, breast meat, ¾ oz slice	2 slices	42	72	45	10	1	0.2	0.2	0.1
Patties, breaded, battered, fried (2.25 oz)	1 patty	64	50	180	9	12	3.0	4.8	3.0
Roast, boneless, frozen, seasoned, light and dark meat, cooked	3 oz	85	68	130	18	5	1.6	1.0	1.4

SOUPS, SAUCES, AND GRAVIES

Soups:								
Canned, condensed:								
Prepared with equal volume of milk:								
Clam chowder, New England	1 cup	248	85	165	9	7	3.0	2.3	1.1
Cream of chicken	1 cup	248	85	190	7	11	4.6	4.5	1.6
Cream of mushroom	1 cup	248	85	205	6	14	5.1	3.0	4.6
Tomato	1 cup	248	85	160	6	6	2.9	1.6	1.1
Soups:								
Canned, condensed:								
Prepared with equal volume of water								
Bean with bacon	1 cup	253	84	170	8	6	1.5	2.2	1.8
Beef broth, bouillon, consommé	1 cup	240	98	15	3	1	0.3	0.2	Tr
Beef noodle	1 cup	244	92	85	5	3	1.1	1.2	0.5
Chicken noodle	1 cup	241	92	75	4	2	0.7	1.1	0.6
Chicken rice	1 cup	241	94	60	4	2	0.5	0.9	0.4
Clam chowder, Manhattan	1 cup	244	90	80	4	2	0.4	0.4	1.3
Cream of chicken	1 cup	244	91	115	3	7	2.1	3.3	1.5
Cream of mushroom	1 cup	244	90	130	2	9	2.4	1.7	4.2
Minestrone	1 cup	241	91	80	4	3	0.6	0.7	1.1
Pea, green	1 cup	250	83	165	9	3	1.4	1.0	0.4
Tomato	1 cup	244	90	85	2	2	0.4	0.4	1.0
Vegetable beef	1 cup	244	92	80	6	2	0.9	0.8	0.1
Vegetarian	1 cup	241	92	70	2	2	0.3	0.8	0.7
Dehydrated:								
Unprepared:								
Bouillon	1 pkt	6	3	15	1	1	0.3	0.2	Tr
Onion	1 pkt	7	4	20	1	Tr	0.1	0.2	Tr

Cholesterol	Carbohydrate	Calcium	Phosphorus	Iron	Potassium	Sodium	Vitamin A value	Thiamin	Riboflavin	Niacin	Ascorbic acid
Milligrams	Grams	Milligrams	Milligrams	Milligrams	Milligrams	Milligrams	Retinol equiv. (RE)	Milligrams	Milligrams	Milligrams	Milligrams
116	0	20	210	1.6	252	98	21	0.07	0.23	8.6	0
126	Tr	3	62	1.7	28	10	983	0.03	0.35	0.9	3
197	0	27	449	6.0	557	144	51	0.57	1.04	11.3	0
72	0	27	173	2.0	246	67	0	0.05	0.21	3.1	0
59	0	16	186	1.1	259	54	0	0.05	0.11	5.8	0
106	0	35	298	2.5	417	98	0	0.09	0.25	7.6	0
65	0	21	181	1.5	253	60	0	0.05	0.15	4.6	0
88	0	20	158	2.2	196	714	48	0.02	0.18	9.0	3
45	3	43	48	0.9	38	616	17	0.03	0.05	1.4	0
28	1	24	89	0.6	129	331	14	0.04	0.07	3.0	0
26	7	20	115	1.3	87	787	18	0.03	0.18	2.6	0
32	Tr	6	108	1.6	184	565	0	0.03	0.14	2.0	[54]0
17	0	3	97	0.2	118	608	0	0.02	0.05	3.5	0
40	10	9	173	1.4	176	512	7	0.06	0.12	1.5	0
45	3	4	207	1.4	253	578	0	0.04	0.14	5.3	0
22	17	186	156	1.5	300	992	40	0.07	0.24	1.0	3
27	15	181	151	0.7	273	1,047	94	0.07	0.26	0.9	1
20	15	179	156	0.6	270	1,076	37	0.08	0.28	0.9	2
17	22	159	149	1.8	449	932	109	0.13	0.25	1.5	68
3	23	81	132	2.0	402	951	89	0.09	0.03	0.6	2
Tr	Tr	14	31	0.4	130	782	0	Tr	0.05	1.9	0
5	9	15	46	1.1	100	952	63	0.07	0.06	1.1	Tr
7	9	17	36	0.8	55	1,106	71	0.05	0.06	1.4	Tr
7	7	17	22	0.7	101	815	66	0.02	0.02	1.1	Tr
2	12	34	59	1.9	261	1,808	92	0.06	0.05	1.3	3
10	9	34	37	0.6	88	986	56	0.03	0.06	0.8	Tr
2	9	46	49	0.5	100	1,032	0	0.05	0.09	0.7	1
2	11	34	55	0.9	313	911	234	0.05	0.04	0.9	1
0	27	28	125	2.0	190	988	20	0.11	0.07	1.2	2
0	17	12	34	1.8	264	871	69	0.09	0.05	1.4	66
5	10	17	41	1.1	173	956	189	0.04	0.05	1.0	2
0	12	22	34	1.1	210	822	301	0.05	0.05	0.9	1
1	1	4	19	0.1	27	1,019	Tr	Tr	0.01	0.3	0
Tr	4	10	23	0.1	47	627	Tr	0.02	0.04	0.4	Tr

[54]If sodium ascorbate is added, product contains 11 mg ascorbic acid.

Foods, approximate measures, units, and weight (weight of edible portion only)		Water		Food energy	Protein	Fat	FATTY ACIDS		
							Saturated	Mono-unsaturated	Poly-unsaturated
SOUPS, SAUCES, AND GRAVIES—Continued		Grams	Percent	Calories	Grams	Grams	Grams	Grams	Grams
Soups:									
Prepared with water:									
Chicken noodle	1 pkt (6-fl-oz)	188	94	40	2	1	0.2	0.4	0.3
Onion	1 pkt (6-fl-oz)	184	96	20	1	Tr	0.1	0.2	0.1
Tomato vegetable	1 pkt (6-fl-oz)	189	94	40	1	1	0.3	0.2	0.1
Sauces:									
From dry mix:									
Cheese, prepared with milk	1 cup	279	77	305	16	17	9.3	5.3	1.6
Hollandaise, prepared with water	1 cup	259	84	240	5	20	11.6	5.9	0.9
White sauce, prepared with milk	1 cup	264	81	240	10	13	6.4	4.7	1.7
From home recipe:									
White sauce, medium[55]	1 cup	250	73	395	10	30	9.1	11.9	7.2
Ready to serve:									
Barbecue	1 tbsp	16	81	10	Tr	Tr	Tr	0.1	0.1
Soy	1 tbsp	18	68	10	2	0	0.0	0.0	0.0
Gravies:									
Canned:									
Beef	1 cup	233	87	125	9	5	2.7	2.3	0.2
Chicken	1 cup	238	85	190	5	14	3.4	6.1	3.6
Mushroom	1 cup	238	89	120	3	6	1.0	2.8	2.4
From dry mix:									
Brown	1 cup	261	91	80	3	2	0.9	0.8	0.1
Chicken	1 cup	260	91	85	3	2	0.5	0.9	0.4
SUGARS AND SWEETS									
Candy:									
Caramels, plain or chocolate	1 oz	28	8	115	1	3	2.2	0.3	0.1
Chocolate									
Milk, plain	1 oz	28	1	145	2	9	5.4	3.0	0.3
Milk, with almonds	1 oz	28	2	150	3	10	4.8	4.1	0.7
Milk, with peanuts	1 oz	28	1	155	4	11	4.2	3.5	1.5
Milk, with rice cereal	1 oz	28	2	140	2	7	4.4	2.5	0.2
Semisweet, small pieces (60 per oz)	1 cup or 6 oz	170	1	860	7	61	36.2	19.9	1.9
Sweet (dark)	1 oz	28	1	150	1	10	5.9	3.3	0.3
Fondant, uncoated (mints, candy corn, other)	1 oz	28	3	105	Tr	0	0.0	0.0	0.0
Fudge, chocolate, plain	1 oz	28	8	115	1	3	2.1	1.0	0.1
Gum drops	1 oz	28	12	100	Tr	Tr	Tr	Tr	0.1
Candy:									
Hard	1 oz	28	1	110	0	0	0.0	0.0	0.0
Jelly beans	1 oz	28	6	105	Tr	Tr	Tr	Tr	0.1
Marshmallows	1 oz	28	17	90	1	0	0.0	0.0	0.0
Custard, baked	1 cup	265	77	305	14	15	6.8	5.4	0.7
Gelatin dessert prepared with gelatin dessert powder and water	½ cup	120	84	70	2	0	0.0	0.0	0.0
Honey, strained or extracted	1 cup	339	17	1,030	1	0	0.0	0.0	0.0
	1 tbsp	21	17	65	Tr	0	0.0	0.0	0.0
Jams and preserves	1 tbsp	20	29	55	Tr	Tr	0.0	Tr	Tr
	1 packet	14	29	40	Tr	Tr	0.0	Tr	Tr
Jellies	1 tbsp	18	28	50	Tr	Tr	Tr	Tr	Tr
	1 packet	14	28	40	Tr	Tr	Tr	Tr	Tr
Popsicle, 3-fl-oz size	1 popsicle	95	80	70	0	0	0.0	0.0	0.0
Puddings:									
Canned:									
Chocolate	5-oz-can	142	68	205	3	11	9.5	0.5	0.1
Tapioca	5-oz-can	142	74	160	3	5	4.8	Tr	Tr
Vanilla	5-oz-can	142	69	220	2	10	9.5	0.2	0.1

[55]Made with enriched flour, margarine, and whole milk.

Cho-lesterol	Carbo-hydrate	Calcium	Phos-phorus	Iron	Potas-sium	Sodium	Vitamin A value	Thiamin	Ribo-flavin	Niacin	Ascorbic acid
Milli-grams	Grams	Milli-grams	Milli-grams	Milli-grams	Milli-grams	Milli-grams	Retinol equiv. (RE)	Milli-grams	Milli-grams	Milli-grams	Milli-grams
2	6	24	24	0.4	23	957	5	0.05	0.04	0.7	Tr
0	4	9	22	0.1	48	635	Tr	0.02	0.04	0.4	Tr
0	8	6	23	0.5	78	856	14	0.04	0.03	0.6	5
53	23	569	438	0.3	552	1,565	117	0.15	0.56	0.3	2
52	14	124	127	0.9	124	1,564	220	0.05	0.18	0.1	Tr
34	21	425	256	0.3	444	797	92	0.08	0.45	0.5	3
32	24	292	238	0.9	381	888	340	0.15	0.43	0.8	2
0	2	3	3	0.1	28	130	14	Tr	Tr	0.1	1
0	2	3	38	0.5	64	1,029	0	0.01	0.02	0.6	
7	11	14	70	1.6	189	1,305	0	0.07	0.08	1.5	0
5	13	48	69	1.1	259	1,373	264	0.04	0.10	1.1	0
0	13	17	36	1.6	252	1,357	0	0.08	0.15	1.6	0
2	14	66	47	0.2	61	1,147	0	0.04	0.09	0.9	0
3	14	39	47	0.3	62	1,134	0	0.05	0.15	0.8	3
1	22	42	35	0.4	54	64	Tr	0.01	0.05	0.1	Tr
6	16	50	61	0.4	96	23	10	0.02	0.10	0.1	Tr
5	15	65	77	0.5	125	23	8	0.02	0.12	0.2	Tr
5	13	49	83	0.4	138	19	8	0.07	0.07	1.4	Tr
6	18	48	57	0.2	100	46	8	0.01	0.08	0.1	Tr
0	97	51	178	5.8	593	24	3	0.10	0.14	0.9	Tr
0	16	7	41	0.6	86	5	1	0.01	0.04	0.1	Tr
0	27	2	Tr	0.1	1	57	0	Tr	Tr	Tr	0
1	21	22	24	0.3	42	54	Tr	0.01	0.03	0.1	Tr
0	25	2	Tr	0.1	1	10	0	0.00	Tr	Tr	0
0	28	Tr	2	0.1	1	7	0	0.10	0.00	0.0	0
0	26	1	1	0.3	11	7	0	0.00	Tr	Tr	0
0	23	1	2	0.5	2	25	0	0.00	Tr	Tr	0
278	29	297	310	1.1	387	209	146	0.11	0.50	0.3	1
0	17	2	23	Tr	Tr	55	0	0.00	0.00	0.0	0
0	279	17	20	1.7	173	17	0	0.02	0.14	1.0	3
0	17	1	1	0.1	11	1	0	Tr	0.01	0.1	Tr
0	14	4	2	0.2	18	2	Tr	Tr	0.01	Tr	Tr
0	10	3	1	0.1	12	2	Tr	Tr	Tr	Tr	1
0	13	2	Tr	0.1	16	5	Tr	Tr	0.01	Tr	1
0	10	1	Tr	Tr	13	4	Tr	Tr	Tr	Tr	0
0	18	0	0	Tr	4	11	0	0.00	0.00	0.0	0
1	30	74	117	1.2	254	285	31	0.04	0.17	0.6	Tr
Tr	28	119	113	0.3	212	252	Tr	0.03	0.14	0.4	Tr
1	33	79	94	0.2	155	305	Tr	0.03	0.12	0.6	Tr

Foods, approximate measures, units, and weight (weight of edible portion only)		Water		Food energy	Protein	Fat	FATTY ACIDS		
							Satu-rated	Mono-unsaturated	Poly-unsaturated
		Grams	Percent	Calories	Grams	Grams	Grams	Grams	Grams

SUGARS AND SWEETS—Continued

Puddings:									
Dry mix, prepared with whole milk:									
Chocolate:									
Instant	½ cup	130	71	155	4	4	2.3	1.1	0.2
Regular (cooked)	½ cup	130	73	150	4	4	2.4	1.1	0.1
Rice	½ cup	132	73	155	4	4	2.3	1.1	0.1
Tapioca	½ cup	130	75	145	4	4	2.3	1.1	0.1
Vanilla:									
Instant	½ cup	130	73	150	4	4	2.2	1.1	0.2
Regular (cooked)	½ cup	130	74	145	4	4	2.3	1.0	0.1
Sugars:									
Brown, pressed down	1 cup	220	2	820	0	0	0.0	0.0	0.0
White:									
Granulated	1 cup	200	1	770	0	0	0.0	0.0	0.0
	1 tbsp	12	1	45	0	0	0.0	0.0	0.0
	1 packet	6	1	25	0	0	0.0	0.0	0.0
Powdered, sifted, spooned into cup	1 cup	100	1	385	0	0	0.0	0.0	0.0
Syrups:									
Chocolate-flavored syrup or topping:									
Thin type	2 tbsp	38	37	85	1	Tr	0.2	0.1	0.1
Fudge type	2 tbsp	38	25	125	2	5	3.1	1.7	0.2
Molasses, cane, blackstrap	2 tbsp	40	24	85	0	0	0.0	0.0	0.0
Table syrup (corn and maple)	2 tbsp	42	25	122	0	0	0.0	0.0	0.0

VEGETABLES AND VEGETABLE PRODUCTS

Alfalfa seeds, sprouted, raw	1 cup	33	91	10	1	Tr	Tr	Tr	0.1
Artichokes, globe or French, cooked, drained	1 artichoke	120	87	55	3	Tr	Tr	Tr	0.1
Asparagus, green:									
Cooked, drained:									
From raw:									
Cuts and tips	1 cup	180	92	45	5	1	0.1	Tr	0.2
Spears, ½-in. diam. at base	4 spears	60	92	15	2	Tr	Tr	Tr	0.1
From frozen:									
Cuts and tips	1 cup	180	91	50	5	1	0.2	Tr	0.3
Spears, ½-in. diam. at base	4 spears	60	91	15	2	Tr	0.1	Tr	0.1
Canned, spears, ½-in. diam. at base	4 spears	80	95	10	1	Tr	Tr	Tr	0.1
Bamboo shoots, canned, drained	1 cup	131	94	25	2	1	0.1	Tr	0.2
Beans:									
Lima, immature seeds, frozen, cooked, drained:									
Thick-seeded types (Fordhooks)	1 cup	170	74	170	10	1	0.1	Tr	0.3
Thin-seeded types (baby limas)	1 cup	180	72	190	12	1	0.1	Tr	0.3
Snap:									
Cooked, drained:									
From raw (cut and French style)	1 cup	125	89	45	2	Tr	0.1	Tr	0.2
From frozen (cut)	1 cup	135	92	35	2	Tr	Tr	Tr	0.1
Canned, drained solids (cut)	1 cup	135	93	25	2	Tr	Tr	Tr	0.1
Beans, mature. See Beans, dry, and Black-eyed peas, dry.									
Bean sprouts (mung):									
Raw	1 cup	104	90	30	3	Tr	Tr	Tr	0.1
Cooked, drained	1 cup	124	93	25	3	Tr	Tr	Tr	Tr
Beets:									
Cooked, drained:									
Diced or sliced	1 cup	170	91	55	2	Tr	Tr	Tr	Tr
Whole beets, 2-in. diam.	2 beets	100	91	30	1	Tr	Tr	Tr	Tr
Canned, drained solids, diced or sliced	1 cup	170	91	55	2	Tr	Tr	Tr	0.1

[56]For regular pack; special dietary pack contains 3 mg sodium.

[57]For green varieties; yellow varieties contain 10 RE.

[58]For green varieties; yellow varieties contain 15 RE.

							NUTRIENTS IN INDICATED QUANTITY				
Cho-lesterol	Carbo-hydrate	Calcium	Phos-phorus	Iron	Potas-sium	Sodium	Vitamin A value	Thiamin	Ribo-flavin	Niacin	Ascorbic acid
Milli-grams	Grams	Milli-grams	Milli-grams	Milli-grams	Milli-grams	Milli-grams	Retinol equiv. (RE)	Milli-grams	Milli-grams	Milli-grams	Milli-grams
14	27	130	329	0.3	176	440	33	0.04	0.18	0.1	1
15	25	146	120	0.2	190	167	34	0.05	0.20	0.1	1
15	27	133	110	0.5	165	140	33	0.10	0.18	0.6	1
15	25	131	103	0.1	167	152	34	0.04	0.18	0.1	1
15	27	129	273	0.1	164	375	33	0.04	0.17	0.1	1
15	25	132	102	0.1	166	178	34	0.04	0.18	0.1	1
0	212	187	56	4.8	757	97	0	0.02	0.07	0.2	0
0	199	3	Tr	0.1	7	5	0	0.00	0.00	0.0	0
0	12	Tr	Tr	Tr	Tr	Tr	0	0.00	0.00	0.0	0
0	6	Tr	Tr	Tr	Tr	Tr	0	0.00	0.00	0.0	0
0	100	1	Tr	Tr	4	2	0	0.00	0.00	0.0	0
0	22	6	49	0.8	85	36	Tr	Tr	0.02	0.1	0
0	21	38	60	0.5	82	42	13	0.02	0.08	0.1	0
0	22	274	34	10.1	1,171	38	0	0.04	0.08	0.8	0
0	32	1	4	Tr	7	19	0	0.00	0.00	0.0	0
0	1	11	23	0.3	26	2	5	0.03	0.04	0.2	3
0	12	47	72	1.6	316	79	17	0.07	0.06	0.7	9
0	8	43	110	1.2	558	7	149	0.18	0.22	1.9	49
0	3	14	37	0.4	186	2	50	0.06	0.07	0.6	16
0	9	41	99	1.2	392	7	147	0.12	0.19	1.9	44
0	3	14	33	0.4	131	2	49	0.04	0.06	0.6	15
0	2	11	30	0.5	122	[56]278	38	0.04	0.07	0.7	13
0	4	10	33	0.4	105	9	1	0.03	0.03	0.2	1
0	32	37	107	2.3	694	90	32	0.13	0.10	1.8	22
0	35	50	202	3.5	740	52	30	0.13	0.10	1.4	10
0	10	58	49	1.6	374	4	[57]83	0.09	0.12	0.8	12
0	8	61	32	1.1	151	18	[58]71	0.06	0.10	0.6	11
0	6	35	26	1.2	147	[59]339	[60]47	0.02	0.08	0.3	6
0	6	14	56	0.9	155	6	2	0.09	0.13	0.8	14
0	5	15	35	0.8	125	12	2	0.06	0.13	1.0	14
0	11	19	53	1.1	530	83	2	0.05	0.02	0.5	9
0	7	11	31	0.6	312	49	1	0.03	0.01	0.3	6
0	12	26	29	3.1	252	[61]466	2	0.02	0.07	0.3	7

[59]For regular pack; special dietary pack contains 3 mg sodium.

[60]For green varieties; yellow varieties contain 14 RE.

[61]For regular pack; special dietary pack contains 78 mg sodium.

Foods, approximate measures, units, and weight (weight of edible portion only)		Water	Food energy	Protein	Fat	Satu-rated	Mono-unsaturated	Poly-unsaturated
						FATTY ACIDS		
VEGETABLES AND VEGETABLE PRODUCTS—Con.		Grams	Percent	Calories	Grams	Grams	Grams	Grams
Beet greens, leaves and stems, cooked, drained	1 cup	144	89	40	4	Tr	Tr	0.1	0.1
Black-eyed peas, immature seeds, cooked and drained:								
From raw	1 cup	165	72	180	13	1	0.3	0.1	0.6
From frozen	1 cup	170	66	225	14	1	0.3	0.1	0.5
Broccoli:								
Raw	1 spear	151	91	40	4	1	0.1	Tr	0.3
Cooked, drained:								
From raw:								
Spear, medium	1 spear	180	90	50	5	1	0.1	Tr	0.2
Spears, cut into ½-in. pieces	1 cup	155	90	45	5	Tr	0.1	Tr	0.2
From frozen:								
Piece, 4½ to 5 in. long	1 piece	30	91	10	1	Tr	Tr	Tr	Tr
Chopped	1 cup	185	91	50	6	Tr	Tr	Tr	0.1
Brussels sprouts, cooked, drained:								
From raw, 7–8 spouts, 1¼ to 1½-in. diam	1 cup	155	87	60	4	1	0.2	0.1	0.4
From frozen	1 cup	155	87	65	6	1	0.1	Tr	0.3
Cabbage, common varieties:								
Raw, coarsely shredded or sliced	1 cup	70	93	15	1	Tr	Tr	Tr	0.1
Cooked, drained	1 cup	150	94	30	1	Tr	Tr	Tr	0.2
Cabbage, Chinese:								
Pak-choi, cooked, drained	1 cup	170	96	20	3	Tr	Tr	Tr	0.1
Pe-tsai, raw, 1-in. pieces	1 cup	76	94	10	1	Tr	Tr	Tr	0.1
Cabbage, red, raw, coarsely shredded or sliced	1 cup	70	92	20	1	Tr	Tr	Tr	0.1
Cabbage, savoy, raw, coarsely shredded or sliced	1 cup	70	91	20	1	Tr	Tr	Tr	Tr
Carrots:								
Raw, without crowns and tips, scraped:								
Whole, 7½ by 1⅛ in. or strips, 2½ to 3 in. long	1 carrot or 18 strips	72	88	30	1	Tr	Tr	Tr	0.1
Grated	1 cup	110	88	45	1	Tr	Tr	Tr	0.1
Cooked, sliced, drained:								
From raw	1 cup	156	87	70	2	Tr	0.1	Tr	0.1
From frozen	1 cup	146	90	55	2	Tr	Tr	Tr	0.1
Canned, sliced, drained solids	1 cup	146	93	35	1	Tr	0.1	Tr	0.1
Cauliflower:								
Raw, (flowerets)	1 cup	100	92	25	2	Tr	Tr	Tr	0.1
Cooked, drained:								
From raw (flowerets)	1 cup	125	93	30	2	Tr	Tr	Tr	0.1
From frozen (flowerets)	1 cup	180	94	35	3	Tr	0.1	Tr	0.2
Celery, pascal type, raw:								
Stalk, large outer, 8 by 1½ in. (at root end)	1 stalk	40	95	5	Tr	Tr	Tr	Tr	Tr
Pieces, diced	1 cup	120	95	20	1	Tr	Tr	Tr	0.1
Collards, cooked, drained:								
From raw (leaves without stems)	1 cup	190	96	25	2	Tr	0.1	Tr	0.2
From frozen (chopped)	1 cup	170	88	60	5	1	0.1	0.1	0.4
Corn, sweet:								
Cooked, drained:								
From raw, ear 5 by 1¾ in.	1 ear	77	70	85	3	1	0.2	0.3	0.5
From frozen:								
Ear, about 3½ in. long	1 ear	63	73	60	2	Tr	0.1	0.1	0.2
Kernels	1 cup	165	76	135	5	Tr	Tr	Tr	0.1

Cho-lesterol	Carbo-hydrate	Calcium	Phos-phorus	Iron	Potas-sium	Sodium	Vitamin A value	Thiamin	Ribo-flavin	Niacin	Ascorbic acid
Milli-grams	Grams	Milli-grams	Milli-grams	Milli-grams	Milli-grams	Milli-grams	Retinol equiv. (RE)	Milli-grams	Milli-grams	Milli-grams	Milli-grams
0	8	164	59	2.7	1,309	347	734	0.17	0.42	0.7	36
0	30	46	196	2.4	693	7	105	0.11	0.18	1.8	3
0	40	39	207	3.6	638	9	13	0.44	0.11	1.2	4
0	8	72	100	1.3	491	41	233	0.10	0.18	1.0	141
0	10	82	86	2.1	293	20	254	0.15	0.37	1.4	113
0	9	71	74	1.8	253	17	218	0.13	0.32	1.2	97
0	2	15	17	0.2	54	7	57	0.02	0.02	0.1	12
0	10	94	102	1.1	333	44	350	0.10	0.15	0.8	74
0	13	56	87	1.9	491	33	111	0.17	0.12	0.9	96
0	13	37	84	1.1	504	36	91	0.16	0.18	0.8	71
0	4	33	16	0.4	172	13	9	0.04	0.02	0.2	33
0	7	50	38	0.6	308	29	13	0.09	0.08	0.3	36
0	3	158	49	1.8	631	58	437	0.05	0.11	0.7	44
0	2	59	22	0.2	181	7	91	0.03	0.04	0.3	21
0	4	36	29	0.3	144	8	3	0.04	0.02	0.2	40
0	4	25	29	0.3	161	20	70	0.05	0.02	0.2	22
0	7	19	32	0.4	233	25	2,025	0.07	0.04	0.7	7
0	11	30	48	0.6	355	39	3,094	0.11	0.06	1.0	10
0	16	48	47	1.0	354	103	3,830	0.05	0.09	0.8	4
0	12	41	38	0.7	231	86	2,585	0.04	0.05	0.6	4
0	8	37	35	0.9	261	[62]352	2,011	0.03	0.04	0.8	4
0	5	29	46	0.6	355	15	2	0.08	0.06	0.6	72
0	6	34	44	0.5	404	8	2	0.08	0.07	0.7	69
0	7	31	43	0.7	250	32	4	0.07	0.10	0.6	56
0	1	14	10	0.2	114	35	5	0.01	0.01	0.1	3
0	4	43	31	0.6	341	106	15	0.04	0.04	0.4	8
0	5	148	19	0.8	177	36	422	0.03	0.08	0.4	19
0	12	357	46	1.9	427	85	1,017	0.08	0.20	1.1	45
0	19	2	79	0.5	192	13	[63]17	0.17	0.06	1.2	5
0	14	2	47	0.4	158	3	[63]13	0.11	0.04	1.0	3
0	34	3	78	0.5	229	8	[63]41	0.11	0.12	2.1	4

NUTRIENTS IN INDICATED QUANTITY

[62]For regular pack; special dietary pack contains 61 mg sodium.

[63]For yellow varieties; white varieties contain only a trace of vitamin A.

Foods, approximate measures, units, and weight (weight of edible portion only)		Water		Food energy	Protein	Fat	FATTY ACIDS		
							Satu- rated	Mono- unsaturated	Poly- unsaturated
VEGETABLES AND VEGETABLE PRODUCTS—Con.		Grams	Percent	Calories	Grams	Grams	Grams	Grams	Grams
Corn, sweet:									
Canned:									
Cream style	1 cup	256	79	185	4	1	0.2	0.3	0.5
Whole kernel, vacuum pack	1 cup	210	77	165	5	1	0.2	0.3	0.5
Cowpeas. See Black-eyed peas, immature, mature.									
Cucumber, with peel, slices, ⅛-in. thick (large, 2⅛-in. diam.; small, 1¾-in. diam.)	6 large or 8 small slices	28	96	5	Tr	Tr	Tr	Tr	Tr
Dandelion greens, cooked, drained	1 cup	105	90	35	2	1	0.1	Tr	0.3
Eggplant, cooked, steamed	1 cup	96	92	25	1	Tr	Tr	Tr	0.1
Endive, curly (included escarole), raw, small pieces	1 cup	50	94	10	1	Tr	Tr	Tr	Tr
Jerusalem artichoke, raw, sliced	1 cup	150	78	115	3	Tr	0.0	Tr	Tr
Kale, cooked, drained:									
From raw, chopped	1 cup	130	91	40	2	1	0.1	Tr	0.3
From frozen, chopped	1 cup	130	91	40	4	1	0.1	Tr	0.3
Kohlrabi, thickened bulb-like stems, cooked, drained, diced	1 cup	165	90	50	3	Tr	Tr	Tr	0.1
Lettuce, raw:									
Butterhead, as Boston types:									
Head, 5-in. diam.	1 head	163	96	20	2	Tr	Tr	Tr	0.2
Leaves	1 outer or 2 inner leaves	15	96	Tr	Tr	Tr	Tr	Tr	Tr
Crisphead, as iceberg:									
Wedge, ¼ of head	1 wedge	135	96	20	1	Tr	Tr	Tr	0.1
Pieces, chopped or shredded	1 cup	55	96	5	1	Tr	Tr	Tr	0.1
Looseleaf (bunching varieties including romaine or cos), chopped or shredded pieces	1 cup	56	94	10	1	Tr	Tr	Tr	0.1
Mushrooms:									
Raw, sliced or chopped	1 cup	70	92	20	1	Tr	Tr	Tr	0.1
Cooked, drained	1 cup	156	91	40	3	1	0.1	Tr	0.3
Canned, drained solids	1 cup	156	91	35	3	Tr	0.1	Tr	0.2
Mustard greens, without stems and midribs, cooked, drained	1 cup	140	94	20	3	Tr	Tr	0.2	0.1
Okra pods, 3 by ⅝ in., cooked	8 pods	85	90	25	2	Tr	Tr	Tr	Tr
Onions:									
Raw:									
Chopped	1 cup	160	91	55	2	Tr	0.1	0.1	0.2
Sliced	1 cup	115	91	40	1	Tr	0.1	Tr	0.1
Cooked (whole or sliced), drained	1 cup	210	92	60	2	Tr	0.1	Tr	0.1
Onions, spring, raw, bulb (⅜-in. diam.) and white portion of top	6 onions	30	92	10	1	Tr	Tr	Tr	Tr
Onion rings, breaded, par-fried, frozen, prepared	2 rings	20	29	80	1	5	1.7	2.2	1.0
Parsley:									
Raw	10 sprigs	10	88	5	Tr	Tr	Tr	Tr	Tr
Freeze-dried	1 tbsp	0.4	2	Tr	Tr	Tr	Tr	Tr	Tr
Parsnips, cooked (diced or 2 in. lengths), drained	1 cup	156	78	125	2	Tr	0.1	0.2	0.1
Peas, edible pod, cooked, drained	1 cup	160	89	65	5	Tr	0.1	Tr	0.2
Peas, green:									
Canned, drained solids	1 cup	170	82	115	8	1	0.1	0.1	0.3
Frozen, cooked, drained	1 cup	160	80	125	8	Tr	0.1	Tr	0.2
Peppers:									
Hot chili, raw	1 pepper	45	88	20	1	Tr	Tr	Tr	Tr
Sweet (about 5 per lb, whole), stem and seeds removed:									
Raw	1 pepper	74	93	20	1	Tr	Tr	Tr	0.2
Cooked, drained	1 pepper	73	95	15	Tr	Tr	Tr	Tr	0.1

[64] For regular pack; special dietary pack contains 8 mg sodium.

[65] For regular pack; special dietary pack contains 6 mg sodium.

[66] For regular pack; special dietary pack contains 3 mg sodium.

[67] For red peppers; green peppers contain 35 RE.

Cho-lesterol	Carbo-hydrate	Calcium	Phos-phorus	Iron	Potas-sium	Sodium	Vitamin A value	Thiamin	Ribo-flavin	Niacin	Ascorbic acid
Milli-grams	Grams	Milli-grams	Milli-grams	Milli-grams	Milli-grams	Milli-grams	Retinol equiv. (RE)	Milli-grams	Milli-grams	Milli-grams	Milli-grams
0	46	8	131	1.0	343	[64]730	[63]25	0.06	0.14	2.5	12
0	41	11	134	0.9	391	[65]571	[63]51	0.09	0.15	2.5	17
0	1	4	5	0.1	42	1	1	0.01	0.01	0.1	1
0	7	147	44	1.9	244	46	1,229	0.14	0.18	0.5	19
0	6	6	21	0.3	238	3	6	0.07	0.02	0.6	1
0	2	26	14	0.4	157	11	103	0.04	0.04	0.2	3
0	26	21	117	5.1	644	6	3	0.30	0.09	2.0	6
0	7	94	36	1.2	296	30	962	0.07	0.09	0.7	53
0	7	179	36	1.2	417	20	826	0.06	0.15	0.9	33
0	11	41	74	0.7	561	35	6	0.07	0.03	0.6	89
0	4	52	38	0.5	419	8	158	0.10	0.10	0.5	13
0	Tr	5	3	Tr	39	1	15	0.01	0.01	Tr	1
0	3	26	27	0.7	213	12	45	0.06	0.04	0.3	5
0	1	10	11	0.3	87	5	18	0.03	0.02	0.1	2
0	2	38	14	0.8	148	5	106	0.03	0.04	0.2	10
0	3	4	73	0.9	259	3	0	0.07	0.31	2.9	2
0	8	9	136	2.7	555	3	0	0.11	0.47	7.0	6
0	8	17	103	1.2	201	663	0	0.13	0.03	2.5	0
0	3	104	57	1.0	283	22	424	0.06	0.09	0.6	35
0	6	54	48	0.4	274	4	49	0.11	0.05	0.7	14
0	12	40	46	0.6	248	3	0	0.10	0.02	0.2	13
0	8	29	33	0.4	178	2	0	0.07	0.01	0.1	10
0	13	57	48	0.4	319	17	0	0.09	0.02	0.2	12
0	2	18	10	0.6	77	1	150	0.02	0.04	0.1	14
0	8	6	16	0.3	26	75	5	0.06	0.03	0.7	Tr
0	1	13	4	0.6	54	4	52	0.01	0.01	0.1	9
0	Tr	1	2	0.2	25	2	25	Tr	0.01	Tr	1
0	30	58	108	0.9	573	16	0	0.13	0.08	1.1	20
0	11	67	88	3.2	384	6	21	0.20	0.12	0.9	77
0	21	34	114	1.6	294	[66]372	131	0.21	0.13	1.2	16
0	23	38	144	2.5	269	139	107	0.45	0.16	2.4	16
0	4	8	21	0.5	153	3	[67]484	0.04	0.04	0.4	109
0	4	4	16	0.9	144	2	[68]39	0.06	0.04	0.4	[69]95
0	3	3	11	0.6	94	1	[70]28	0.04	0.03	0.3	[71]81

[68]For green peppers; red peppers contain 422 RE.

[69]For green peppers; red peppers contain 141 mg ascorbic acid.

[70]For green peppers; red peppers contain 274 RE.

[71]For green peppers; red peppers contain 121 mg ascorbic acid.

Foods, approximate measures, units, and weight (weight of edible portion only)		Water	Food energy	Protein	Fat	FATTY ACIDS		
						Satu-rated	Mono-unsaturated	Poly-unsaturated
VEGETABLES AND VEGETABLE PRODUCTS—Con.		Grams	Percent	Calories	Grams	Grams	Grams	Grams
							Grams	Grams
Potatoes, cooked:								
Baked (about 2 per lb, raw):								
With skin	1 potato	202	71	220	5	Tr	0.1	Tr	0.1
Flesh only	1 potato	156	75	145	3	Tr	Tr	Tr	0.1
Boiled (about 3 per lb, raw):								
Peeled after boiling	1 potato	136	77	120	3	Tr	Tr	Tr	0.1
Peeled before boiling	1 potato	135	77	115	2	Tr	Tr	Tr	0.1
French fried, strip, 2 to 3½ in. long, frozen:								
Oven heated	10 strips	50	53	110	2	4	2.1	1.8	0.3
Fried in vegetable oil	10 strips	50	38	160	2	8	2.5	1.6	3.8
Potato products, prepared:								
Au gratin:								
From dry mix	1 cup	245	79	230	6	10	6.3	2.9	0.3
From home recipe	1 cup	245	74	325	12	19	11.6	5.3	0.7
Hashed brown, from frozen	1 cup	156	56	340	5	18	7.0	8.0	2.1
Mashed:								
From home recipe:								
Milk added	1 cup	210	78	160	4	1	0.7	0.3	0.1
Milk and margarine added	1 cup	210	76	225	4	9	2.2	3.7	2.5
From dehydrated flakes (without milk), water, milk, butter, and salt added	1 cup	210	76	235	4	12	7.2	3.3	0.5
Potato salad, made with mayonnaise	1 cup	250	76	360	7	21	3.6	6.2	9.3
Scalloped:								
From dry mix	1 cup	245	79	230	5	11	6.5	3.0	0.5
From home recipe	1 cup	245	81	210	7	9	5.5	2.5	0.4
Potato chips	10 chips	20	3	105	1	7	1.8	1.2	3.6
Pumpkin:								
Cooked from raw, mashed	1 cup	245	94	50	2	Tr	0.1	Tr	Tr
Canned	1 cup	245	90	85	3	1	0.4	0.1	Tr
Radishes, raw, stem ends, rootlets cut off	4 radishes	18	95	5	Tr	Tr	Tr	Tr	Tr
Sauerkraut, canned, solids and liquid	1 cup	236	93	45	2	Tr	0.1	Tr	0.1
Seaweed:								
Kelp, raw	1 oz	28	82	10	Tr	Tr	0.1	Tr	Tr
Spirulina, dried	1 oz	28	5	80	16	2	0.8	0.2	0.6
Southern peas. See Black-eyed peas, immature, mature.								
Spinach:								
Raw, chopped	1 cup	55	92	10	2	Tr	Tr	Tr	0.1
Cooked, drained:								
From raw	1 cup	180	91	40	5	Tr	0.1	Tr	0.2
From frozen (leaf)	1 cup	190	90	55	6	Tr	0.1	Tr	0.2
Canned, drained solids	1 cup	214	92	50	6	1	0.2	Tr	0.4
Spinach souffle	1 cup	136	74	220	11	18	7.1	6.8	3.1
Squash; cooked:								
Summer (all varieties), sliced, drained	1 cup	180	94	35	2	1	0.1	Tr	0.2
Winter (all varieties), baked, cubes	1 cup	205	89	80	2	1	0.3	0.1	0.5
Sunchoke. See Jerusalem artichoke.								
Sweet potatoes:								
Cooked (raw, 5 by 2 in.; about 2½ per lb):								
Baked in skin, peeled	1 potato	114	73	115	2	Tr	Tr	Tr	0.1
Boiled, without skin	1 potato	151	73	160	2	Tr	0.1	Tr	0.2

							NUTRIENTS IN INDICATED QUANTITY				
Cho-lesterol	Carbo-hydrate	Calcium	Phos-phorus	Iron	Potas-sium	Sodium	Vitamin A value	Thiamin	Ribo-flavin	Niacin	Ascorbic acid
Milli-grams	Grams	Milli-grams	Milli-grams	Milli-grams	Milli-grams	Milli-grams	Retinol equiv. (RE)	Milli-grams	Milli-grams	Milli-grams	Milli-grams
0	51	20	115	2.7	844	16	0	0.22	0.07	3.3	26
0	34	8	78	0.5	610	8	0	0.16	0.03	2.2	20
0	27	7	60	0.4	515	5	0	0.14	0.03	2.0	18
0	27	11	54	0.4	443	7	0	0.13	0.03	1.8	10
0	17	5	43	0.7	229	16	0	0.06	0.02	1.2	5
0	20	10	47	0.4	366	108	0	0.09	0.01	1.6	5
12	31	203	233	0.8	537	1,076	76	0.05	0.20	2.3	8
56	28	292	277	1.6	970	1,061	93	0.16	0.28	2.4	24
0	44	23	112	2.4	680	53	0	0.17	0.03	3.8	10
4	37	55	101	0.6	628	636	12	0.18	0.08	2.3	14
4	35	55	97	0.5	607	620	42	0.18	0.08	2.3	13
29	32	103	118	0.5	489	697	44	0.23	0.11	1.4	20
170	28	48	130	1.6	635	1,323	83	0.19	0.15	2.2	25
27	31	88	137	0.9	497	835	51	0.05	0.14	2.5	8
29	26	140	154	1.4	926	821	47	0.17	0.23	2.6	26
0	10	5	31	0.2	260	94	0	0.03	Tr	0.8	8
0	12	37	74	1.4	564	2	265	0.08	0.19	1.0	12
0	20	64	86	3.4	505	12	5,404	0.06	0.13	0.9	10
0	1	4	3	0.1	42	4	Tr	Tr	0.01	0.1	4
0	10	71	47	3.5	401	1,560	4	0.05	0.05	0.3	35
0	3	48	12	0.8	25	66	3	0.01	0.04	0.1	(1)
0	7	34	33	8.1	386	297	16	0.67	1.04	3.6	3
0	2	54	27	1.5	307	43	369	0.04	0.10	0.4	15
0	7	245	101	6.4	839	126	1,474	0.17	0.42	0.9	18
0	10	277	91	2.9	566	163	1,479	0.11	0.32	0.8	23
0	7	272	94	4.9	740	[72]683	1,878	0.03	0.30	0.8	31
184	3	230	231	1.3	201	763	675	0.09	0.30	0.5	3
0	8	49	70	0.6	346	2	52	0.08	0.07	0.9	10
0	18	29	41	0.7	896	2	729	0.17	0.05	1.4	20
0	28	32	63	0.5	397	11	2,488	0.08	0.14	0.7	28
0	37	32	41	0.8	278	20	2,575	0.08	0.21	1.0	26

[72]With added salt; if none is added, sodium content is 58 mg.

Foods, approximate measures, units, and weight (weight of edible portion only)		Water	Food energy	Protein	Fat	FATTY ACIDS Satu-rated	Mono-unsaturated	Poly-unsaturated
		Grams	Percent	Calories	Grams	Grams	Grams	Grams

VEGETABLES AND VEGETABLE PRODUCTS—Con.

Food	Measure	Grams	Percent	Calories	Grams	Grams	Grams	Grams
Sweet potatoes:								
Candied, 2½ by 2-in. piece	1 piece	105	67	145	1	3	1.4	0.7	0.2
Canned:								
Solid pack (mashed)	1 cup	255	74	260	5	1	0.1	Tr	0.2
Vacuum pack, piece 2¾ by 1 in.	1 piece	40	76	35	1	Tr	Tr	Tr	Tr
Tomatoes:								
Raw, 2⅗-in. diam. (3 per 12 oz pkg)	1 tomato	123	94	25	1	Tr	Tr	Tr	0.1
Canned, solids and liquid	1 cup	240	94	50	2	1	0.1	0.1	0.2
Tomato juice, canned	1 cup	244	94	40	2	Tr	Tr	Tr	0.1
Tomato products, canned:								
Paste	1 cup	262	74	220	10	2	0.3	0.4	0.9
Puree	1 cup	250	87	105	4	Tr	Tr	Tr	0.1
Sauce	1 cup	245	89	75	3	Tr	0.1	0.1	0.2
Turnips, cooked, diced	1 cup	156	94	30	1	Tr	Tr	Tr	0.1
Turnip greens, cooked, drained:								
From raw (leaves and stems)	1 cup	144	93	30	2	Tr	0.1	Tr	0.1
From frozen (chopped)	1 cup	164	90	50	5	1	0.2	Tr	0.3
Vegetable juice cocktail, canned	1 cup	242	94	45	2	Tr	Tr	Tr	0.1
Vegetables, mixed:								
Canned, drained solids	1 cup	163	87	75	4	Tr	0.1	Tr	0.2
Frozen, cooked, drained	1 cup	182	83	105	5	Tr	0.1	Tr	0.1
Water chestnuts, canned	1 cup	140	86	70	1	Tr	Tr	Tr	Tr

MISCELLANEOUS ITEMS

Food	Measure	Grams	Percent	Calories	Grams	Grams	Grams	Grams
Baking powders for home use:								
Sodium aluminum sulfate:								
With monocalcium phosphate monohydrate	1 tsp	3	2	5	Tr	0	0.0	0.0	0.0
With monocalcium phosphate monohydrate, calcium sulfate	1 tsp	2.9	1	5	Tr	0	0.0	0.0	0.0
Straight phosphate	1 tsp	3.8	2	5	Tr	0	0.0	0.0	0.0
Low sodium	1 tsp	4.3	1	5	Tr	0	0.0	0.0	0.0
Catsup	1 cup	273	69	290	5	1	0.2	0.2	0.4
	1 tbsp	15	69	15	Tr	Tr	Tr	Tr	Tr
Celery seed	1 tsp	2	6	10	Tr	1	Tr	0.3	0.1
Chili powder	1 tsp	2.6	8	10	Tr	Tr	0.1	0.1	0.2
Chocolate:								
Bitter or baking	1 oz	28	2	145	3	15	9.0	4.9	0.5
Semisweet, see Candy.								
Cinnamon	1 tsp	2.3	10	5	Tr	Tr	Tr	Tr	Tr
Curry powder	1 tsp	2	10	5	Tr	Tr	(¹)	(¹)	(¹)
Garlic powder	1 tsp	2.8	6	10	Tr	Tr	Tr	Tr	Tr
Gelatin, dry	1 envelope	7	13	25	6	Tr	Tr	Tr	Tr
Mustard, prepared, yellow	1 tsp or individual packet	5	80	5	Tr	Tr	Tr	0.2	Tr
Olives, canned:								
Green	4 medium or 3 ex-tra large	13	78	15	Tr	2	0.2	1.2	0.1
Ripe, Mission, pitted	3 small or 2 large	9	73	15	Tr	2	0.3	1.3	0.2
Onion powder	1 tsp	2.1	5	5	Tr	Tr	Tr	Tr	Tr
Oregano	1 tsp	1.5	7	5	Tr	Tr	Tr	Tr	0.1
Paprika	1 tsp	2.1	10	5	Tr	Tr	Tr	Tr	0.2
Pepper, black	1 tsp	2.1	11	5	Tr	Tr	Tr	Tr	Tr

[73]For regular pack; special dietary pack contains 31 mg sodium.

[74]With added salt; if none is added, sodium content is 24 mg.

[75]With no added salt; if salt is added, sodium content is 2,070 mg.

Cho-lesterol	Carbo-hydrate	Calcium	Phos-phorus	Iron	Potas-sium	Sodium	Vitamin A value	Thiamin	Ribo-flavin	Niacin	Ascorbic acid
Milli-grams	Grams	Milli-grams	Milli-grams	Milli-grams	Milli-grams	Milli-grams	Retinol equiv. (RE)	Milli-grams	Milli-grams	Milli-grams	Milli-grams
8	29	27	27	1.2	198	74	440	0.02	0.04	0.4	7
0	59	77	133	3.4	536	191	3,857	0.07	0.23	2.4	13
0	8	9	20	0.4	125	21	319	0.01	0.02	0.3	11
0	5	9	28	0.6	255	10	139	0.07	0.06	0.7	22
0	10	62	46	1.5	530	[73]391	145	0.11	0.07	1.8	36
0	10	22	46	1.4	537	[74]881	136	0.11	0.08	1.6	45
0	49	92	207	7.8	2,442	[75]170	647	0.41	0.50	8.4	111
0	25	38	100	2.3	1,050	[76]50	340	0.18	0.14	4.3	88
0	18	34	78	1.9	909	[77]1,482	240	0.16	0.14	2.8	32
0	8	34	30	0.3	211	78	0	0.04	0.04	0.5	18
0	6	197	42	1.2	292	42	792	0.06	0.10	0.6	39
0	8	249	56	3.2	367	25	1,308	0.09	0.12	0.8	36
0	11	27	41	1.0	467	883	283	0.10	0.07	1.8	67
0	15	44	68	1.7	474	243	1,899	0.08	0.08	0.9	8
0	24	46	93	1.5	308	64	778	0.13	0.22	1.5	6
0	17	6	27	1.2	165	11	1	0.02	0.03	0.5	2
0	1	58	87	0.0	5	329	0	0.00	0.00	0.0	0
0	1	183	45	0.0	4	290	0	0.00	0.00	0.0	0
0	1	239	359	0.0	6	312	0	0.00	0.00	0.0	0
0	1	207	314	0.0	891	Tr	0	0.00	0.00	0.0	0
0	69	60	137	2.2	991	2,845	382	0.25	0.19	4.4	41
0	4	3	8	0.1	54	156	21	0.01	0.01	0.2	2
0	1	35	11	0.9	28	3	Tr	0.01	0.01	0.1	Tr
0	1	7	8	0.4	50	26	91	0.01	0.02	0.2	2
0	8	22	109	1.9	235	1	1	0.01	0.07	0.4	0
0	2	28	1	0.9	12	1	1	Tr	Tr	Tr	1
0	1	10	7	0.6	31	1	2	0.01	0.01	0.1	Tr
0	2	2	12	0.1	31	1	0	0.01	Tr	Tr	Tr
0	0	1	0	0.0	2	6	0	0.00	0.00	0.0	0
0	Tr	4	4	0.1	7	63	0	Tr	0.01	Tr	Tr
0	Tr	8	2	0.2	7	312	4	Tr	Tr	Tr	0
0	Tr	10	2	0.2	2	68	1	Tr	Tr	Tr	0
0	2	8	7	0.1	20	1	Tr	0.01	Tr	Tr	Tr
0	1	24	3	0.7	25	Tr	10	0.01	Tr	0.1	1
0	1	4	7	0.5	49	1	127	0.01	0.04	0.3	1
0	1	9	4	0.6	26	1	Tr	Tr	0.01	Tr	0

[76]With no added salt; if salt is added, sodium content is 998 mg.

[77]With salt added.

Foods, approximate measures, units, and weight (weight of edible portion only)		Water	Food energy	Protein	Fat	Satu-rated	FATTY ACIDS	
							Mono-unsaturated	Poly-unsaturated
MISCELLANEOUS ITEMS—Continued		Grams	Percent	Calories	Grams	Grams	Grams	Grams	Grams

		Grams	Percent	Calories	Grams	Grams	Grams	Grams	Grams
Pickles, cucumber:									
Dill, medium, whole, 3¾ in. long, 1-¼ in. diam.	1 pickle	65	93	5	Tr	Tr	Tr	Tr	0.1
Fresh-pack, slices 1½-in. diam., ¼ in. thick	2 slices	15	79	10	Tr	Tr	Tr	Tr	Tr
Sweet, gherkin, small, whole, about 2½ in. long, ¾-in. diam.	1 pickle	15	61	20	Tr	Tr	Tr	Tr	Tr
Popcorn. **See Grain Products.**									
Relish, finely chopped, sweet	1 tbsp	15	63	20	Tr	Tr	Tr	Tr	Tr
Salt	1 tsp	5.5	0	0	0	0	0.0	0.0	0.0
Vinegar, cider	1 tbsp	15	94	Tr	Tr	0	0.0	0.0	0.0
Yeast:									
Baker's, dry, active	1 pkg	7	5	20	3	Tr	Tr	0.1	Tr
Brewer's, dry	1 tbsp	8	5	25	3	Tr	Tr	Tr	0.0

				NUTRIENTS IN INDICATED QUANTITY							
Cho-lesterol	Carbo-hydrate	Calcium	Phos-phorus	Iron	Potas-sium	Sodium	Vitamin A value	Thiamin	Ribo-flavin	Niacin	Ascorbic acid
Milli-grams	Grams	Milli-grams	Milli-grams	Milli-grams	Milli-grams	Milli-grams	Retinol equiv. (RE)	Milli-grams	Milli-grams	Milli-grams	Milli-grams
0	1	17	14	0.7	130	928	7	Tr	0.01	Tr	4
0	3	5	4	0.3	30	101	2	Tr	Tr	Tr	1
0	5	2	2	0.2	30	107	1	Tr	Tr	Tr	1
0	5	3	2	0.1	30	107	2	Tr	Tr	0.0	1
0	0	14	3	Tr	Tr	2,132	0	0.00	0.00	0.0	0
0	1	1	1	0.1	15	Tr	0	0.00	0.00	0.0	0
0	3	[78]3	90	1.1	140	4	Tr	0.16	0.38	2.6	Tr
0	3	17	140	1.4	152	10	Tr	1.25	0.34	3.0	Tr

[78]Value may vary from 6 to 60 mg.

TABLE A–2

Exchange Lists for Meal Planning

Groups/Lists	Carbohydrate (grams)	Protein (grams)	Fat (grams)	Calories
Carbohydrate Group				
Starch	15	3	1 or less	80
Fruit	15	—	—	60
Milk				
Skim	12	8	0–3	90
Low-fat	12	8	5	120
Whole	12	8	8	150
Other carbohydrates	15	varies	varies	varies
Vegetables	5	2	—	25
Meat and Meat Substitute Group				
Very lean	—	7	0–1	35
Lean	—	7	3	55
Medium-fat	—	7	5	75
High fat	—	7	8	100
Fat Group	—	—	5	45

The Exchange Lists are the basis of a meal planning system designed by a committee of the American Diabetes Association and The American Dietetic Association. While designed primarily for people with diabetes and others who must follow special diets, the Exchange Lists are based on principles of good nutrition that apply to everyone. ©1995 American Diabetes Association, Inc., The American Dietetic Association.

LIST 1

Starch List

One starch exchange equals 15 grams carbohydrate, 3 grams protein, 0–1 grams fat, and 80 calories.

Bread

Bagel	½ (1 oz)
Bread, reduced calorie	2 slices (1½ oz)
Bread, white, whole-wheat, pumpernickel, rye	1 slice (1 oz)
Bread sticks, crisp, 4 in. long × ½ in.	2 (⅔ oz)
English muffin	½
Hot dog or hamburger bun	½ (1 oz)
Pita, 6 in. across	½
Roll, plain, small	1 (1 oz)
Raisin bread, unfrosted	1 slice (1 oz)
Tortilla, corn, 6 in. across	1
Tortilla, flour, 7–8 in. across	1
Waffle, 4½ in. square, reduced-fat	1

Cereals and Grains

Bran cereals	½ cup
Bulgur	½ cup
Cereals	½ cup
Cereals, unsweetened, ready-to-eat	¾ cup
Cornmeal (dry)	3 tbsp
Couscous	⅓ cup
Flour (dry)	3 tbsp
Granola, low-fat	¼ cup
Grape-Nuts	¼ cup

Grits	½ cup
Kasha	½ cup
Millet	¼ cup
Muesli	¼ cup
Oats	½ cup
Pasta	½ cup
Puffed cereal	1½ cups
Rice milk	½ cup
Rice, white or brown	⅓ cup
Shredded Wheat	½ cup
Sugar-frosted cereal	½ cup
Wheat germ	3 tbsp

Starchy Vegetables

Baked beans	⅓ cup
Corn	½ cup
Corn on cob, medium	1 (5 oz)
Mixed vegetables with corn, peas, or pasta	1 cup
Peas, green	½ cup
Plantain	½ cup
Potato, baked or boiled	1 small (3 oz)
Potato, mashed	½ cup
Squash, winter (acorn, butternut)	1 cup
Yam, sweet potato, plain	½ cup

One starch exchange equals 15 grams carbohydrate, 3 grams protein, 0–1 grams fat, and 80 calories.

Crackers and Snacks
Animal crackers	8
Graham crackers, 2½ in. square	3
Matzoh	¾ oz
Melba toast	4 slices
Oyster crackers	24
Popcorn (popped, no fat added or low-fat microwave)	3 cups
Pretzels	¾ oz
Rice cakes, 4 in. across	2
Saltine-type crackers	6
Snack chips, fat-free (tortilla, potato)	15–20 (¾ oz)
Whole-wheat crackers, no fat added	2–5 (¾ oz)

Dried Beans, Peas, and Lentils
(Count as 1 starch exchange, plus 1 very lean meat exchange.)
Beans and peas (garbanzo, pinto, kidney, white, split, black-eyed)	½ cup
Lima beans	⅔ cup
Lentils	½ cup
Miso[S]	3 tbsp

Starchy Foods Prepared with Fat
(Count as 1 starch exchange, plus 1 fat exchange.)
Biscuit, 2½ in. across	1
Chow mein noodles	½ cup
Corn bread, 2 in. cube	1 (2 oz)
Crackers, round butter type	6
Croutons	1 cup
French-fried potatoes	16–25 (3 oz)
Granola	¼ cup
Muffin, small	1 (1½ oz)
Pancake, 4 in. across	2
Popcorn, microwave	3 cups
Sandwich crackers, cheese or peanut butter filling	3
Stuffing, bread (prepared)	⅓ cup
Taco shell, 6 in. across	2
Waffle, 4½ in. square	1
Whole-wheat crackers, fat added	4–6 (1 oz)

[S]400 mg or more of sodium per serving.

LIST 2   Fruit List

One fruit exchange equals 15 grams carbohydrate and 60 calories. The weight includes skin, core, seeds, and rind.

Fruit
Apple, unpeeled, small	1 (4 oz)
Applesauce, unsweetened	½ cup
Apples, dried	4 rings
Apricots, fresh	4 whole (5½ oz)
Apricots, dried	8 halves
Apricots, canned	½ cup
Banana, small	1 (4 oz)
Blackberries	¾ cup
Blueberries	¾ cup
Cantaloupe, small	⅓ melon (11 oz) or 1 cup cubes
Cherries, sweet, fresh	12 (3 oz)
Cherries, sweet, canned	½ cup
Dates	3
Figs, fresh	1½ large or 2 medium (3½ oz)
Figs, dried	1½
Fruit cocktail	½ cup
Grapefruit, large	½ (11 oz)
Grapefruit sections, canned	¾ cup
Grapes, small	17 (3 oz)
Honeydew melon	1 slice (10 oz) or 1 cup cubes
Kiwi	1 (3½ oz)
Mandarin oranges, canned	¾ cup
Mango, small	½ fruit (5½ oz) or ½ cup
Nectarine, small	1 (5 oz)
Orange, small	1 (6½ oz)

Papaya	½ fruit (8 oz) or 1 cup cubes
Peach, medium, fresh	1 (6 oz)
Peaches, canned	½ cup
Pear, large, fresh	½ (4 oz)
Pears, canned	½ cup
Pineapple, fresh	¾ cup
Pineapple, canned	½ cup
Plums, small	2 (5 oz)
Plums, canned	½ cup
Prunes, dried	3
Raisins	2 tbsp
Raspberries	1 cup
Strawberries	1¼ cup whole berries
Tangerines, small	2 (8 oz)
Watermelon	1 slice (13½ oz) or 1¼ cup cubes

Fruit Juice
Apple juice/cider	½ cup
Cranberry juice cocktail	⅓ cup
Cranberry juice cocktail, reduced-calorie	1 cup
Fruit juice blends, 100% juice	⅓ cup
Grape juice	⅓ cup
Grapefruit juice	½ cup
Orange juice	½ cup
Pineapple juice	½ cup
Prune juice	⅓ cup

LIST 3
Milk List

One exchange equals 12 grams carbohydrate and 8 grams protein.

Skim and Very Low-fat Milk
(0–3 grams fat per serving)

Skim milk	1 cup
1/2% milk	1 cup
1% milk	1 cup
Nonfat or low-fat buttermilk	1 cup
Evaporated skim milk	1/2 cup
Nonfat dry milk	1/3 cup dry
Plain nonfat yogurt	3/4 cup
Nonfat or low-fat fruit-flavored yogurt sweetened with aspartame or with a nonnutritive sweetener	1 cup

Low-fat Milk
(5 grams fat per serving)

2% milk	1 cup
Plain low-fat yogurt	3/4 cup
Sweet acidophilus milk	1 cup

Whole Milk
(8 grams fat per serving)

Whole milk	1 cup
Evaporated whole milk	1/2 cup
Goat's milk	1 cup
Kefir	1 cup

LIST 4
Other Carbohydrates List

One exchange equals 15 grams carbohydrate, or 1 starch, or 1 fruit, or 1 milk.

Food	Serving Size	Exchanges per Serving
Angel food cake, unfrosted	1/12th cake	2 carbohydrates
Brownie, small, unfrosted	2 in. square	1 carbohydrate, 1 fat
Cake, unfrosted	2 in. square	1 carbohydrate, 1 fat
Cake, frosted	2 in. square	2 carbohydrates, 1 fat
Cookie, fat-free	2 small	1 carbohydrate
Cookie or sandwich cookie with creme filling	2 small	1 carbohydrate, 1 fat
Cupcake, frosted	1 small	2 carbohydrates, 1 fat
Cranberry sauce, jellied	1/4 cup	2 carbohydrates
Doughnut, plain cake	1 medium (1 1/2 oz)	1 1/2 carbohydrates, 2 fats
Doughnut, glazed	3 3/4 in. across (2 oz)	2 carbohydrates, 2 fats
Fruit juice bars, frozen, 100% juice	1 bar (3 oz)	1 carbohydrate
Fruit snacks, chewy (pureed fruit concentrate)	1 roll (3/4 oz)	1 carbohydrate
Fruit spreads, 100% fruit	1 tbsp	1 carbohydrate
Gelatin, regular	1/2 cup	1 carbohydrate
Gingersnaps	3	1 carbohydrate
Granola bar	1 bar	1 carbohydrate, 1 fat
Granola bar, fat-free	1 bar	2 carbohydrates
Hummus	1/3 cup	1 carbohydrate, 1 fat
Ice cream	1/2 cup	1 carbohydrate, 2 fats
Ice cream, light	1/2 cup	1 carbohydrate, 1 fat
Ice cream, fat-free, no sugar added	1/2 cup	1 carbohydrate

LIST 4, continued

One exchange equals 15 grams carbohydrate, or 1 starch, or 1 fruit, or 1 milk.

Food	Serving Size	Exchanges per Serving
Jam or jelly, regular	1 tbsp	1 carbohydrate
Milk, chocolate, whole	1 cup	2 carbohydrates, 1 fat
Pie, fruit, 2 crusts	1/6 pie	3 carbohydrates, 2 fats
Pie, pumpkin or custard	1/8 pie	1 carbohydrate, 2 fats
Potato chips	12–18 (1 oz)	1 carbohydrate, 2 fats
Pudding, regular (made with low-fat milk)	1/2 cup	2 carbohydrates
Pudding, sugar-free (made with low-fat milk)	1/2 cup	1 carbohydrate
Salad dressing, fat-free[S]	1/4 cup	1 carbohydrate
Sherbet, sorbet	1/2 cup	2 carbohydrates
Spaghetti or pasta sauce, canned[S]	1/2 cup	1 carbohydrate, 1 fat
Sweet roll or Danish	1 (2 1/2 oz)	2 1/2 carbohydrates, 2 fats
Syrup, light	2 tbsp	1 carbohydrate
Syrup, regular	1 tbsp	1 carbohydrate
Syrup, regular	1/4 cup	4 carbohydrates
Tortilla chips	6–12 (1 oz)	1 carbohydrate, 2 fats
Yogurt, frozen, low-fat, fat-free	1/3 cup	1 carbohydrate, 0–1 fat
Yogurt, frozen, fat-free, no sugar added	1/2 cup	1 carbohydrate
Yogurt, low-fat with fruit	1 cup	3 carbohydrates, 0–1 fat
Vanilla wafers	5	1 carbohydrate, 1 fat

[S]400 mg or more per exchange.

LIST 5

Vegetable List*

One vegetable exchange equals 5 grams carbohydrate, 2 grams protein, 0 grams fat, and 25 calories.

Artichoke
Artichoke hearts
Asparagus
Beans (green, wax, Italian)
Bean sprouts
Beets
Broccoli
Brussels sprouts
Cabbage
Carrots
Cauliflower
Celery
Cucumber
Eggplant
Green onions or scallions
Greens (collard, kale, mustard, turnip)
Kohlrabi
Leeks
Mixed vegetables (without corn, peas, or pasta)

Mushrooms
Okra
Onions
Pea pods
Peppers (all varieties)
Radishes
Salad greens (endive, escarole, lettuce, romaine, spinach)
Sauerkraut[S]
Spinach
Summer squash
Tomato
Tomatoes, canned
Tomato sauce[S]
Tomato/vegetable juice[S]
Turnips
Water chestnuts
Watercress
Zucchini

*One exchange = 1/2 cup of cooked vegetable or vegetable juice
          1 cup of raw vegetables
[S]400 mg or more sodium per exchange.

Very Lean Meat and Substitutes List

One exchange equals 0 grams carbohydrate, 7 grams protein, 0–1 grams fat, and 35 calories.

One very lean meat exchange is equal to any one of the following items.

Poultry: Chicken or turkey (white meat, no skin), Cornish hen (no skin)   1 oz

Fish: Fresh or frozen cod, flounder, haddock, halibut, trout; tuna fresh or canned in water   1 oz

Shellfish: Clams, crab, lobster, scallops, shrimp, imitation shellfish   1 oz

Game: Duck or pheasant (no skin), venison, buffalo, ostrich   1 oz

Cheese with 1 gram or less fat per ounce:
   Nonfat or low-fat cottage cheese   ¼ cup
   Fat-free cheese   1 oz

Other: Processed sandwich meats with 1 gram or less fat per ounce, such as deli thin, shaved meats, chipped beef[S], turkey ham   1 oz
   Egg whites   2
   Egg substitutes, plain   ¼ cup
   Hot dogs with 1 gram or less fat per ounce[S]   1 oz
   Kidney (high in cholesterol)   1 oz
   Sausage with 1 gram or less fat per ounce   1 oz

Count as one very lean meat and one starch exchange.

Dried beans, peas, lentils (cooked)   ½ cup

Lean Meat and Substitutes List

One exchange equals 0 grams carbohydrate, 7 grams protein, 3 grams fat, and 55 calories.

One lean meat exchange is equal to any one of the following items.

Beef: USDA Select or Choice grades of lean beef trimmed of fat, such as round, sirloin, and flank steak; tenderloin; roast (rib, chuck, rump); steak (T-bone, porterhouse, cubed), ground round   1 oz

Pork: Lean pork, such as fresh ham; canned, cured, or boiled ham; Canadian bacon[S]; tenderloin, center, loin chop   1 oz

Lamb: Roast, chop, leg   1 oz

Veal: Lean chop, roast   1 oz

Poultry: Chicken, turkey (dark meat, no skin), chicken white meat (with skin), domestic duck or goose (well-drained of fat, no skin)   1 oz

Fish:
   Herring (uncreamed or smoked)   1 oz
   Oysters   6 medium
   Salmon (fresh or canned), catfish   1 oz
   Sardines (canned)   2 medium
   Tuna (canned in oil, drained)   1 oz
Game: Goose (no skin), rabbit   1 oz
Cheese:
   4.5%–fat cottage cheese   ¼ cup
   Grated Parmesan   2 tbsp
   Cheeses with 3 grams or less fat per ounce   1 oz
Other:
   Hot dogs with 3 grams or less fat per ounce[S]   1½ oz
   Processed sandwich meat with 3 grams or less fat per ounce, such as turkey pastrami or kielbasa   1 oz
   Liver, heart (high in cholesterol)   1 oz

Medium-Fat Meat and Substitutes List

One exchange equals 0 grams carbohydrate, 7 grams protein, 5 grams fat, and 75 calories.

One medium-fat meat exchange is equal to any one of the following items.

Beef: Most beef products fall into this category (ground beef, meatloaf, corned beef, short ribs, Prime grades of meat trimmed of fat, such as prime rib)   1 oz
Pork: Top loin, chop, Boston butt, cutlet   1 oz
Lamb: Rib roast, ground   1 oz
Veal: Cutlet (ground or cubed, unbreaded)   1 oz
Poultry: Chicken dark meat (with skin), ground turkey or ground chicken, fried chicken (with skin)   1 oz
Fish: Any fried fish product   1 oz
Cheese: With 5 grams or less fat per ounce
   Feta   1 oz
   Mozzarella   1 oz
   Ricotta   ¼ cup (2 oz)
Other:
   Egg (high in cholesterol, limit to 3 per week)   1
   Sausage with 5 grams or less fat per ounce   1 oz
   Soy milk   1 cup
   Tempeh   ¼ cup
   Tofu   4 oz or ½ cup

High-Fat Meat and Substitutes List

One exchange equals 0 grams carbohydrate, 7 grams protein, 8 grams fat, and 100 calories.

Remember these items are high in saturated fat, cholesterol, and calories and may raise blood cholesterol levels if eaten on a regular basis. One high-fat meat exchange is equal to any one of the following items.

Pork: Spareribs, ground pork, pork sausage	1 oz
Cheese: All regular cheeses, such as American[S], cheddar, Monterey Jack, Swiss	1 oz
Other: Processed sandwich meats with 8 grams or less fat per ounce, such as bologna, pimento loaf, salami	1 oz

Sausage, such as bratwurst, Italian, knockwurst, Polish, smoked	1 oz
Hot dog (turkey or chicken)[S]	1 (10/lb)
Bacon	3 slices (20 slices/lb)

Count as one high-fat meat plus one fat exchange.

Hot dog (beef, pork, or combination)	1 (10/lb)[S]
Peanut butter (contains unsaturated fat)	2 tbsp

[S]400 mg or more sodium per exchange.

LIST 7
Fat Lists

Monounsaturated Fats List

One fat exchange equals 5 grams fat and 45 calories.

Avocado, medium	⅛ (1 oz)
Oil (canola, olive, peanut)	1 tsp
Olives: ripe (black)	8 large
green, stuffed[S]	10 large
Nuts	
almonds, cashews	6 nuts
mixed (50% peanuts)	6 nuts
peanuts	10 nuts
pecans	4 halves
Peanut butter, smooth or crunchy	2 tsp
Sesame seeds	1 tbsp
Tahini paste	2 tsp

Polyunsaturated Fats List

One fat exchange equals 5 grams fat and 45 calories.

Margarine: stick, tub, or squeeze	1 tsp
lower-fat (30% or 50% vegetable oil)	1 tbsp
Mayonnaise: regular	1 tsp
reduced-fat	1 tbsp
Nuts, walnuts, English	4 halves

Oil (corn, safflower, soybean)	1 tsp
Salad dressing: regular[S]	1 tbsp
reduced-fat	2 tbsp
Miracle Whip Salad Dressing®: regular	2 tsp
reduced-fat	1 tbsp
Seeds: pumpkin, sunflower	1 tbsp

Saturated Fats List*

One fat exchange equals 5 grams of fat and 45 calories.

Bacon, cooked	1 slice (20 slices/lb)
Bacon, grease	1 tsp
Butter: stick	1 tsp
whipped	2 tsp
reduced-fat	1 tbsp
Chitterlings, boiled	2 tbsp (½ oz)
Coconut, sweetened, shredded	2 tbsp
Cream, half and half	2 tbsp
Cream cheese: regular	1 tbsp (½ oz)
reduced-fat	2 tbsp (1 oz)
Fatback or salt pork, see below[†]	
Shortening or lard	1 tsp
Sour cream: regular	2 tbsp
reduced-fat	3 tbsp

[S]400 mg or more sodium per exchange.

*Saturated fats can raise blood cholesterol levels.

[†]Use a piece 1 in. × 1 in. × ¼ in. if you plan to eat the fatback cooked with vegetables. Use a piece 2 in. × 1 in. × ½ in. when eating only the vegetables with the fatback removed.

LIST 8
Free Foods List

A *free food* is any food or drink that contains less than 20 calories or less than 5 grams of carbohydrate per serving. Foods with a serving size listed should be limited to three servings per day. Be sure to spread them out throughout the day. Foods listed without a serving size can be eaten as often as you like.

Fat-free or Reduced-fat Foods

Cream cheese, fat-free	1 tbsp
Creamers, nondairy, liquid	1 tbsp
Creamers, nondairy, powdered	2 tsp
Mayonnaise, fat-free	1 tbsp
Mayonnaise, reduced-fat	1 tsp
Margarine, fat-free	4 tbsp
Margarine, reduced-fat	1 tsp
Miracle Whip®, nonfat	1 tbsp
Miracle Whip®, reduced-fat	1 tsp
Nonstick cooking spray	
Salad dressing, fat-free	1 tbsp
Salad dressing, fat-free, Italian	2 tbsp
Salsa	¼ cup
Sour cream, fat-free, reduced-fat	1 tbsp
Whipped topping, regular or light	2 tbsp

Sugar-free or Low-sugar Foods

Candy, hard, sugar-free	1 candy
Gelatin dessert, sugar-free	
Gelatin, unflavored	
Gum, sugar-free	
Jam or jelly, low-sugar or light	2 tsp
Sugar substitutes*	
Syrup, sugar-free	2 tbsp

Drinks
Bouillon, broth, consommé^S

Bouillon or broth, low-sodium	
Carbonated or mineral water	
Cocoa powder, unsweetened	1 tbsp
Coffee	
Club soda	
Diet soft drinks, sugar-free	
Drink mixes, sugar-free	
Tea	
Tonic water, sugar-free	

Condiments

Catsup	1 tbsp
Horseradish	
Lemon juice	
Lime juice	
Mustard	
Pickles, dill^S	1½ large
Soy sauce, regular or light^S	
Taco sauce	1 tbsp
Vinegar	

Seasonings
Be careful with seasonings that contain sodium or are salts, such as garlic or celery salt, and lemon pepper.

- Flavoring extracts
- Garlic
- Herbs, fresh or dried
- Pimento
- Spices
- Tabasco® or hot pepper sauce
- Wine, used in cooking
- Worcestershire sauce

*Sugar substitutes, alternatives, or replacements that are approved by the Food and Drug Administration (FDA) are safe to use. Common brand names include: Equal® (aspartame), Sprinkle Sweet® (saccharin), Sweet One® (acesulfame K), Sweet-10® (saccharin), Sugar Twin (saccharin); Sweet 'n Low® (saccharin).

^{S}400 mg or more of sodium per choice.

LIST 9
Combination Foods List

Many of the foods we eat are mixed together in various combinations. These combination foods do not fit into any one exchange list. Often it is hard to tell what is in a casserole dish or prepared food item. This is a list of exchanges for some typical combination foods. This list will help you fit these foods into your meal plan. Ask your dietitian for information about any other combination foods you would like to eat.

Food	Serving Size	Exchanges per Serving
Entrees		
Tuna noodle casserole, lasagna, spaghetti with meatballs, chili with beans, macaroni and cheese^S	1 cup (8 oz)	2 carbohydrates, 2 medium-fat meats
Chow mein (without noodles or rice)	2 cups (16 oz)	1 carbohydrate, 2 lean meats
Pizza, cheese, thin crust^S	¼ of 10 in. (5 oz)	2 carbohydrates, 2 medium-fat meats, 1 fat

LIST 9, continued

Food	Serving Size	Exchanges per Serving
Pizza, meat topping, thin crust[S]	¼ of 10 in. (5 oz)	2 carbohydrates, 2 medium-fat meats, 2 fats
Pot pie[S]	1 (7 oz)	2 carbohydrates, 1 medium-fat meat, 4 fats
Frozen Entrees		
Salisbury steak with gravy, mashed potato[S]	1 (11 oz)	2 carbohydrates, 3 medium-fat meats, 3–4 fats
Turkey with gravy, mashed potato, dressing[S]	1 (11 oz)	2 carbohydrates, 2 medium-fat meats, 2 fats
Entree with less than 300 calories[S]	1 (8 oz)	2 carbohydrates, 3 lean meats
Soups		
Bean[S]	1 cup	1 carbohydrate, 1 very lean meat
Cream (made with water)[S]	1 cup (8 oz)	1 carbohydrate, 1 fat
Split pea (made with water)[S]	½ cup (4 oz)	1 carbohydrate
Tomato (made with water)[S]	1 cup (8 oz)	1 carbohydrate
Vegetable beef, chicken noodle, or other broth-type[S]	1 cup (8 oz)	1 carbohydrate

[S]400 mg or more sodium per exchange.

LIST 10

Fast Foods*

Food	Serving Size	Exchanges per Serving
Burritos with beef[S]	2	4 carbohydrates, 2 medium-fat meats, 2 fats
Chicken nuggets[S]	6	1 carbohydrate, 2 medium-fat meats, 1 fat
Chicken breast and wing, breaded and fried[S]	1 each	1 carbohydrate, 4 medium-fat meats, 2 fats
Fish sandwich/tartar sauce[S]	1	3 carbohydrates, 1 medium-fat meat, 3 fats
French fries, thin	20–25	2 carbohydrates, 2 fats
Hamburger, regular	1	2 carbohydrates, 2 medium-fat meats
Hamburger, large[S]	1	2 carbohydrates, 3 medium-fat meats, 1 fat
Hot dog with bun[S]	1	1 carbohydrate, 1 high-fat meat, 1 fat
Individual pan pizza[S]	1	5 carbohydrates, 3 medium-fat meats, 3 fats
Soft-serve cone	1 medium	2 carbohydrates, 1 fat
Submarine sandwich[S]	1 sub (6 in.)	3 carbohydrates, 1 vegetable, 2 medium-fat meats, 1 fat
Taco, hard shell[S]	1 (6 oz)	2 carbohydrates, 2 medium-fat meats, 2 fats
Taco, soft shell[S]	1 (3 oz)	1 carbohydrate, 1 medium-fat meat, 1 fat

[S]400 mg or more of sodium per serving.

*Ask at your fast-food restaurant for nutrition information about your favorite fast foods.

TABLE B–1

Approximate Conversions to and from Metric Measures

If Measure Is in	Multiply by	To Find
Length		
Inches	2.5	Centimeters
Feet	30	Centimeters
Centimeters	0.4	Inches
Meters	3.3	Feet
Weight		
Ounces	28	Grams
Pounds	0.45	Kilograms
Grams	0.035	Ounces
Kilograms	2.2	Pounds
Volume		
Teaspoon	5	Milliliters
Tablespoons	15	Milliliters
Fluid ounces	30	Milliliters
Cups (8 ounce)	0.24	Liters
Pints	0.47	Liters
Quarts	0.95	Liters
Milliliters	0.03	Fluid ounces
Liters	2.1	Pints
Liters	1.06	Quarts
Energy		
Kilocalories	4.184	Kilojoules
Kilojoules	0.239	Kilocalories
Temperature		
Fahrenheit	Subtract 32; then multiply by $\frac{5}{9}$	Celsius
Celsius	Multiply by $\frac{9}{5}$; then add 32	Fahrenheit

TABLE B–2

Conversion from Selected Common Units to Système International (SI) Units

Component	Common Unit	SI Unit	Conversion Factor, Common to SI Unit
Hematology			
erythrocyte count	mm^{-3}	$10^6/l$	1
hematocrit	%	l	0.01
hemoglobin	g/dl	g/l	10
leukocyte count	mm^{-3}	$10^9/l$	0.001
Other			
albumin, serum	g/dl	g/l	10
cholesterol, total, HDL, LDL	mg/dl	mmol/l	0.026
glucose	mg/dl	mmol/l	0.056
protein, total serum	g/dl	g/l	10
transferrin, serum	mg/dl	g/l	0.01
triglycerides, serum	mg/dl	mmol/l	0.0113
urea nitrogen, serum	mg/dl	mmol/l	0.357

TABLE B–3
Nutrient Arithmetic Conversions

Vitamins	

Vitamin A [Retinol Equivalents (RE)]
  1 RE = 1 μg retinol
  1 RE = 6 μg β-carotene
  1 RE = 12 μg other provitamin
      A carotenoids

Vitamin D [Cholecalciferol]
  10 ug cholecalciferol = 400 IU of vitamin D

Vitamin E [α-Tocopherol Equivalents (α-TE)]
  1 α-TE = 1 mg δ-α tocopherol

Niacin [Niacin Equivalents (NE)]
  1 NE = 1 mg niacin
  1 NE = 60 mg dietary tryptophan

Minerals	

Sodium
Sodium Chloride (NaCl) = 40% sodium; 60% chloride

Example 1: 1,000 mg salt (NaCl) = 400 mg sodium [1,000 × .40 = 400]

Example 2: Given: 1 teaspoon table salt weighs 5 grams
      5 gm NaCl = 5,000 mg NaCl = 2,000 mg sodium

ANTHROPOMETRIC MEASUREMENTS

Body Frame: See Figure D–5
Weight for Height and Body Frame: See Table D–1
Healthy Weight: See Figure D–6

$$\text{Percent of Usual Body Weight} = \frac{\text{Present Weight}}{\text{Usual Weight}} \times 100$$

$$\text{Percent of Healthy Body Weight} = \frac{\text{Present Weight}}{\text{Healthy Weight}} \times 100$$

$$\text{Relative Weight} = \frac{\text{Measured Body Weight}}{\text{Midpoint of Medium Frame Weight in Table D–1}}$$

$$\text{Body Mass Index (BMI)} = \frac{\text{Weight, kg}}{\text{Height, m}^2}$$

Standards for BMI

	Females	Males
Underweight	<19	<20
Desirable BMI	19–24	20–25
Overweight	25–30	25–30
Obese	>30	>30

LABORATORY MEASUREMENTS

Total Lymphocyte Count (TLC) = Percent Lymphocytes × White Blood Cells (WBC)

$$\text{Nitrogen Balance} = \frac{\text{Protein Intake, g}}{6.25} - \text{Urinary Urea Nitrogen (UUN)} + 4$$

FIGURE B–1
Calculations for Nutrition Assessment

The Warning Signs of poor nutritional health are often overlooked. Use this checklist to find out if you or someone you know is at nutritional risk.

Read the statements below. Circle the number in the yes column for those that apply to you or someone you know. For each yes answer, score the number in the box. Total your nutritional score.

DETERMINE YOUR NUTRITIONAL HEALTH

	YES
I have an illness or condition that made me change the kind and/or amount of food I eat.	2
I eat fewer than 2 meals per day.	3
I eat few fruits or vegetables, or milk products.	2
I have 3 or more drinks of beer, liquor or wine almost every day.	2
I have tooth or mouth problems that make it hard for me to eat.	2
I don't always have enough money to buy the food I need.	4
I eat alone most of the time.	1
I take 3 or more different prescribed or over-the-counter drugs a day.	1
Without wanting to, I have lost or gained 10 pounds in the last 6 months.	2
I am not always physically able to shop, cook and/or feed myself.	2
TOTAL	

Total Your Nutritional Score. If it's —

0-2 **Good!** Recheck your nutritional score in 6 months.

3-5 **You are at moderate nutritional risk.** See what can be done to improve your eating habits and lifestyle. Your office on aging, senior nutrition program, senior citizens center or health department can help. Recheck your nutritional score in 3 months.

6 or more **You are at high nutritional risk.** Bring this checklist the next time you see your doctor, dietitian or other qualified health or social service professional. Talk with them about any problems you may have. Ask for help to improve your nutritional health.

These materials developed and distributed by the Nutrition Screening Initiative, a project of:

AMERICAN ACADEMY
OF FAMILY PHYSICIANS

THE AMERICAN
DIETETIC ASSOCIATION

NATIONAL COUNCIL
ON THE AGING, INC.

Remember that warning signs suggest risk, but do not represent diagnosis of any condition. Turn the page to learn more about the Warning Signs of poor nutritional health.

FIGURE C–1
Nutritional Health Checklist. (*Source:* Reprinted with permission of the Nutrition Screening Initiative, 1010 Wisconsin Avenue, Suite 800, Washington, DC 20007-3603.)

**The Nutrition Checklist is based on the Warning Signs described below.
Use the word <u>DETERMINE</u> to remind you of the Warning Signs.**

Disease

Any disease, illness or chronic condition which causes you to change the way you eat, or makes it hard for you to eat, puts your nutritional health at risk. Four out of five adults have chronic diseases that are affected by diet. Confusion or memory loss that keeps getting worse is estimated to affect one out of five or more of older adults. This can make it hard to remember what, when or if you've eaten. Feeling sad or depressed, which happens to about one in eight older adults, can cause big changes in appetite, digestion, energy level, weight and well-being.

Eating Poorly

Eating too little and eating too much both lead to poor health. Eating the same foods day after day or not eating fruit, vegetables, and milk products daily will also cause poor nutritional health. One in five adults skip meals daily. Only 13% of adults eat the minimum amount of fruit and vegetables needed. One in four older adults drink too much alcohol. Many health problems become worse if you drink more than one or two alcoholic beverages per day.

Tooth Loss/ Mouth Pain

A healthy mouth, teeth and gums are needed to eat. Missing, loose or rotten teeth or dentures which don't fit well or cause mouth sores make it hard to eat.

Economic Hardship

As many as 40% of older Americans have incomes of less than $6,000 per year. Having less--or choosing to spend less--than $25-30 per week for food makes it very hard to get the foods you need to stay healthy.

Reduced Social Contact

One-third of all older people live alone. Being with people daily has a positive effect on morale, well-being and eating.

Multiple Medicines

Many older Americans must take medicines for health problems. Almost half of older Americans take multiple medicines daily. Growing old may change the way we respond to drugs. The more medicines you take, the greater the chance for side effects such as increased or decreased appetite, change in taste, constipation, weakness, drowsiness, diarrhea, nausea, and others. Vitamins or minerals when taken in large doses act like drugs and can cause harm. Alert your doctor to everything you take.

Involuntary Weight Loss/Gain

Losing or gaining a lot of weight when you are not trying to do so is an important warning sign that must not be ignored. Being overweight or underweight also increases your chance of poor health.

Needs Assistance in Self Care

Although most older people are able to eat, one of every five have trouble walking, shopping, buying and cooking food, especially as they get older.

Elder Years Above Age 80

Most older people lead full and productive lives. But as age increases, risk of frailty and health problems increase. Checking your nutritional health regularly makes good sense.

FIGURE C–1, continued

A. Personal Data (for a patient, most, if not all, of this information is usually available from the medical record)

Name _____ Identification No._____

Address _____ Telephone_____

_____ Gender_____

Age _____ Marital status_____

Race _____Ethnic origin _____Religion_____

Education: elementary school _____high school _____college_____

Occupation: _____Where employed_____

    Hours of work _____Travel time to work_____

    homemaker _____school _____unemployed _____retired_____

Family (list members of immediate family and their ages)

_____ _____

_____ _____

_____ _____

Housing: room _____apartment _____single family dwelling_____

    lives alone _____lives with family _____other_____

Income: employment _____public assistance _____pension _____savings_____

    Social Security _____other_____

Do you participate in any of the following programs? WIC _____ School lunch _____ Food stamps _____

    Congregate feeding for older persons _____ Home delivered meals _____

    Do you have health insurance? _____ Does it cover medical nutrition therapy? _____

B. Anthropometric Data

Height: _____inches _____cm   Weight: _____pounds _____kg

Usual weight: _____pounds _____kg

Healthy weight: _____pounds _____kg

Has weight changed recently? _____if so, how much?_____

Was this weight change planned?_____

% of usual weight (see Table D–2)

$$\frac{\text{Present weight}}{\text{Usual weight}} \times 100 = \underline{\hspace{2cm}} \quad \text{Standard} \underline{\hspace{2cm}}$$

% of healthy body weight (see Figure D–6)

$$\frac{\text{Present weight}}{\text{Healthy weight}} \times 100 = \underline{\hspace{2cm}} \quad \text{Standard} \underline{\hspace{2cm}}$$

Body Mass Index (see Figure B–1)

$$\text{Body Mass Index} = \frac{\text{Weight, kg}}{\text{Height, m}^2} = \underline{\hspace{2cm}} \quad \text{Standard} \underline{\hspace{2cm}}$$

FIGURE C–2
Sample Form for Diet History and Nutritional Assessment

C. **Clinical Findings** (For a patient these would include the medical history and the physical examination.)

Physical Signs:

Hair:_____

Skin:_____

Eyes:_____

Ears:_____

Mouth:_____

      Lips:_____

      Teeth: good repair _____missing, no dentures _____dentures_____

      chews well _____chews with difficulty_____

      Gums:_____

      Tongue:_____

Nails:_____

Posture:_____

Daily Routines:

Time rising _____Time leave for work/school _____Return_____

Meal times (include snack times):

Alcoholic beverages (state frequency and amount):
    beer _____wine _____hard liquors_____

Smoking: _____how much_____

Exercise: kind _____how often _____how long_____

Recreational activities and hobbies:_____

Medications: prescribed_____

_____

    over the counter_____

Gastrointestinal Function:

appetite: good _____fair _____poor_____

sense of taste _____sense of smell_____

digestive upsets: seldom _____frequent_____

nature of upsets: nausea _____vomiting _____heartburn_____

    distention _____cramping _____other (specify)_____

elimination, bowel: frequency _____character_____

      constipation _____ diarrhea_____

FIGURE C–2, continued

D. Laboratory Studies (The following basic studies are widely used in screening for patients who may be at risk. In this exercise some students may be able to supply data, while others will skip this section.)

Hemoglobin: _____g/dl   Standard: _____g/dl
Hematocrit: _____percent   Standard: _____percent
Total serum protein: _____g/dl   Standard: _____g/dl
Serum albumin: _____g/dl   Standard: _____g/dl
Serum transferrin: _____mg/dl   Standard: _____mg/dl
Total lymphocytes: _____mm^3 Standard: _____mm^3
Glucose: _____mg/dl   Standard: _____mg/dl
Total cholesterol: _____mg/dl   Standard: _____mg/dl
HDL: _____mg/dl   Standard: _____mg/dl
LDL: _____mg/dl   Standard: _____mg/dl
Triglycerides: _____mg/dl   Standard: _____mg/dl
Other:

E. Dietary History

Name some of your favorite foods. How often do you eat them?

Name some foods that you never eat.

Name some foods that you seldom eat.

How would you describe your feelings about food? (Let the individual express feelings; do not suggest possibilities.)

Are you allergic to any foods? _____If so, which food(s)?

How was diagnosis of allergy made:
by physician _____by self _____by other (specify)_____

Have you ever been on a special diet? _____Why?_____

If yes, what kind of diet?_____

FIGURE C–2, continued

How long were you on the diet? _____ How well did you stick to the diet? very closely? _____

most of the time _____ about half of the time _____ not at all _____

Is any member of your family on a modified diet? _____

If yes, what kind of diet? _____

Are you taking any nutritional supplements (vitamins, minerals, protein)? _____

If yes, what kind? _____

How long have you taken them? _____

Why are you taking them? _____

Indicate where you eat each of your meals/snacks: at home _____ school cafeteria _____

work cafeteria _____ fast-food restaurant _____

carried from home _____ other (specify) _____

Who prepares the meals at home? _____

What facilities are available for meal preparation? none _____ eats all meals away from home _____

hot plate only _____ full kitchen facilities _____ refrigerator _____ microwave oven _____

What transportation is available for food shopping? walk _____ public transportation _____ drive _____

need transportation _____

Food Intake for 24-hour Period

Time of Day and Where Eaten	Kind of Food	Amounts

Is this intake record typical of your usual pattern?
    If no, how does it differ?
How does weekend food intake compare with weekdays?

FIGURE C–2, continued

FIGURE D–1
Physical Growth NCHS Percentiles: Girls, Birth to 36 Months (Length/Age and Weight/Age) (*Source:* Adapted from Hamill PVV, Drizd TA, Johnson CL, Reed RB, Roche AF, Moore WM: *"Physical Growth: National Center for Health Statistics percentiles." Am J Clin Nutr.* 32:607–629, 1979. Data from the Fels Longitudinal Study, Wright State University School of Medicine, Yellow Springs, Ohio.)

©1982 Ross Laboratories.

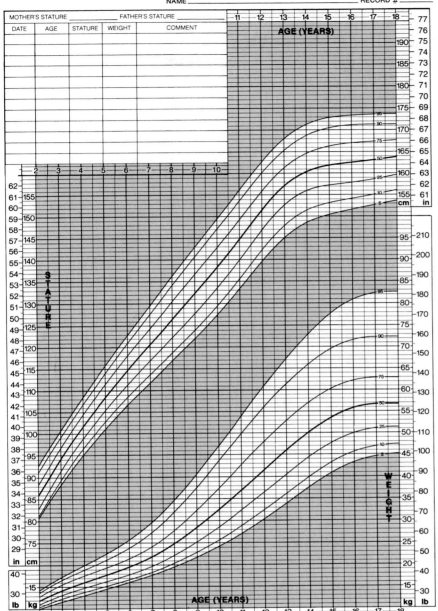

FIGURE D–2
Physical Growth NCHS Percentiles: Girls, 2 to 18 Years (Stature/Age and Weight/Age)
Ross Laboratories, Columbus, Ohio. (*Source:* Adapted from Hamill, PVV, et al.:
"Physical Growth: National Center for Health Statistics," *Am J Clin Nutr.* 32:607–629,
1979. Data from National Center for Health Statistics (NCHS), Hyattsville, Maryland.)

FIGURE D–3
Physical Growth NCHS Percentiles: Boys, Birth to 36 Months (Length/Age and Weight/Age) (*Source:* Adapted from Hamill PVV, Drizd TA, Johnson CL, Reed RB, Roche AF, Moore WM: *"Physical Growth: National Center for Health Statistics percentiles." Am J Clin Nutr.* 32:607–629, 1979. Data from the Fels Longitudinal Study, Wright State University School of Medicine, Yellow Springs, Ohio.)

©1982 Ross Laboratories.

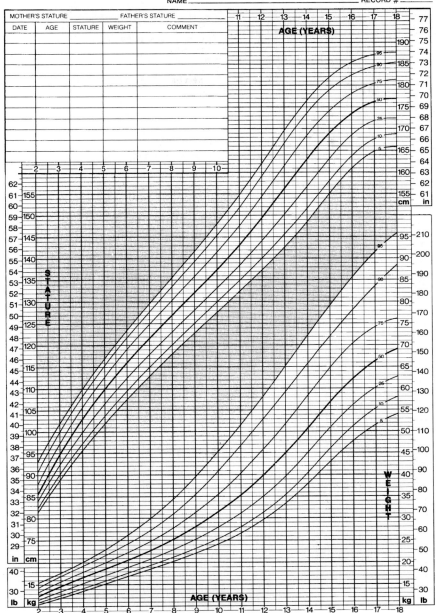

FIGURE D–4

Physical Growth NCHS Percentiles: Boys, 2 to 18 Years (Stature/Age and Weight/Age)
Ross Laboratories, Columbus, Ohio. (*Source:* Adapted from Hamill, PVV, et al.:
"Physical Growth: National Center for Health Statistics," *Am J Clin Nutr.* 32:607–629,
1979. Data from National Center for Health Statistics (NCHS), Hyattsville, Maryland.)

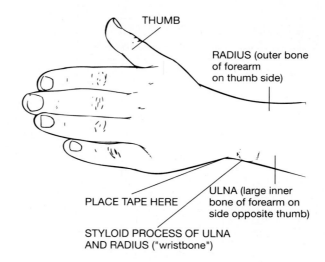

THUMB

RADIUS (outer bone
of forearm
on thumb side)

ULNA (large inner
bone of forearm on
side opposite thumb)

PLACE TAPE HERE

STYLOID PROCESS OF ULNA
AND RADIUS ("wristbone")

The wrist is measured distal to styloid process of radius and
ulna at smallest circumference.

Use height without shoes and inches for wrist size to
determine frame type from this chart.

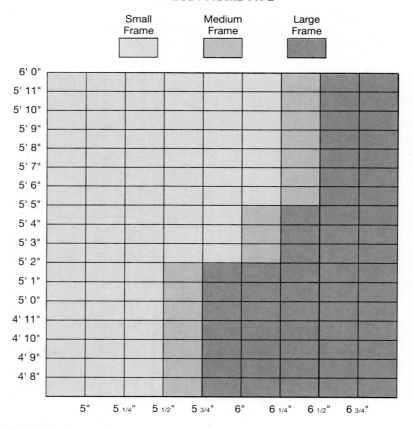

FIGURE D–5

Determination of Body Frame According to Wrist Measurement. (*Source:* Copyright ©
1973, Peter G. Lindnec, M.D. All rights reserved. Reproduced with permission.)

TABLE D–1

Heights and Weights for Men and Women*

	Men				Women		
	Small Frame	Medium Frame	Large Frame		Small Frame	Medium Frame	Large Frame
Height		(pounds)†		Height		(pounds)**	
5 ft. 2 in.	128–134	131–141	138–150	4 ft. 10 in.	102–111	109–121	118–131
5 ft. 3 in.	130–136	133–143	140–153	4 ft. 11 in.	103–113	111–123	120–134
5 ft. 4 in.	132–138	135–145	142–156	5 ft. 0 in.	104–115	113–126	122–137
5 ft. 5 in.	134–140	137–148	144–160	5 ft. 1 in.	106–118	115–129	125–140
5 ft. 6 in.	136–142	139–151	146–164	5 ft. 2 in.	108–121	118–132	128–143
5 ft. 7 in.	138–145	142–154	149–168	5 ft. 3 in.	111–124	121–135	131–147
5 ft. 8 in.	140–148	145–157	152–172	5 ft. 4 in.	114–127	124–138	134–151
5 ft. 9 in.	142–151	148–160	155–176	5 ft. 5 in.	117–130	127–141	137–155
5 ft. 10 in.	144–154	151–163	158–180	5 ft. 6 in.	120–133	130–144	140–159
5 ft. 11 in.	146–157	154–166	161–184	5 ft. 7 in.	123–136	133–147	143–163
6 ft. 0 in.	149–160	157–170	164–188	5 ft. 8 in.	126–139	136–150	146–167
6 ft. 1 in.	152–164	160–174	168–192	5 ft. 9 in.	129–142	139–153	149–170
6 ft. 2 in.	155–168	164–178	172–197	5 ft. 10 in.	132–145	142–156	152–173
6 ft. 3 in.	158–172	167–182	176–202	5 ft. 11 in.	135–148	145–159	155–176
6 ft. 4 in.	162–176	171–187	181–207	6 ft. 0 in.	138–151	148–162	158–179

Source: Metropolitan Life Insurance Company, New York, 1983. Data from 1979 Build Study. Society of Actuaries and Associates of Life Insurance Medical Directors of America, 1980.

*Weight at ages 25 to 59 based on lowest mortality.

†Weight in indoor clothing weighing 5 pounds, shoes with 1-inch heel.

**Weight in indoor clothing weighing 3 pounds, shoes with 1-inch heel.

FIGURE D–6

Are You Overweight? Suggested Weights for Adults (*Source:* Report of the Dietary Guidelines Advisory Committee on the Dietary Guidelines for Americans, 1995, pages 23–24.)

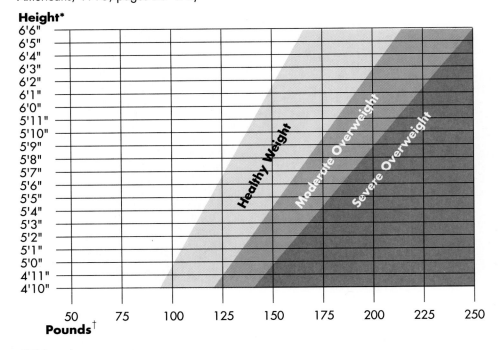

*Without shoes.

†Without clothes. The higher weights apply to people with more muscle and bone, such as many men.

TABLE D–2
Some Standards for Nutritional Assessment

		Normal	Mild to Moderate Depletion	Severe Depletion
Anthropometric				
Weight loss, unplanned, 1 month			5%	>5%
Weight loss, unplanned, 6 months			10%	>10%
Percent of usual weight		90–110	75–89	<75
Percent of healthy weight		90–110	75–89	<75
Triceps skinfold, percent of standard		90	60–89	<60
Laboratory				
Serum albumin, g/dl		3.5–4.5	2.1–2.7	<2.1
Prealbumin, g/dl		>17	10–17	<10
Transferrin, mg/dl		250–300	100–149	<100
Total lymphocyte count, mm³		1,500	800–1,500	<800
Red blood cells, m/cu mm, Males		4.2–6.3		
Females		4.2–5.5		
Hemoglobin, g/dl, Males		14–18		
Females		12–16		
Hematocrit, %, Males		41–51		
Females		37–47		

		Desirable	Increased Risk	High Risk
Total Cholesterol, mg/dl	Adults	<200	200–239	≥240
	Children	<170	170–199	≥200
HDL Cholesterol, mg/dl	Adults	≥35	<35	
LDL Cholesterol, mg/dl	Adults	<130	130–159	≥160
	Children	<110	110–129	≥130
Triglycerides, mg/dl	Adults	<250	250–500	>500

ACE: angiotensin-converting enzyme
ACTH: adrenocorticotropic hormone
ADH: antidiuretic hormone
ADL: activities of daily living
ADP: adenosine diphosphate
AHCPR: Agency for Health Care Policy and Research
AIDS: Acquired Immunodeficiency Syndrome
ASHD: arteriosclerotic heart disease
ATP: adenosine triphosphate
BE: barium enema
BEE: basal energy expenditure
BHA: butylated hydroxyanisole (an antioxidant)
BHT: butylated hydroxytoluene (an antioxidant)
bid: twice daily
BMI: body mass index
BMR: basal metabolic rate
BUN: blood urea nitrogen
BV: biologic value
C: Celsius; centigrade
CAD: coronary artery disease
CAPD: continuous ambulatory peritoneal dialysis
CBE: charting by exception
CCPD: continuous cyclic peritoneal dialysis
CDC: Centers for Disease Control and Prevention
CHD: coronary heart disease
CHF: congestive heart failure
cm: centimeter
cm^3 : cubic centimeter
CMV: *cytomegalovirus*
CoA: coenzyme A
COPD: chronic obstructive pulmonary disease
CVA: cerebrovascular accident
CVD: cardiovascular disease
DBFC: double blind food challenge
DCCT: Diabetes Control and Complications Trial
DHHS: Department of Health and Human Services
dl: deciliter
DNA: deoxyribonucleic acid
DRV: Daily Reference Value
DTR: Dietetic Technician Registered
EAA: essential amino acid
EF: extrinsic factor
EFA: essential fatty acid
EPA: Environmental Protection Agency
ESRD: end-stage renal disease
ESWL: extracorporeal shock wave lithotripsy
F: Fahrenheit
FAO: Food and Agriculture Organization of the United Nations
FAS: fetal alcohol syndrome
FDA: U.S. Food and Drug Administration
FSIS: Food Safety Inspection Service
FTC: Federal Trade Commission

g: gram
GFR: glomerular filtration rate
GI: gastrointestinal
GRAS: generally regarded as safe (FDA list of additives)
GTF: glucose tolerance factor
GTT: glucose tolerance test
HAACP: Hazard Analysis Critical Control Point
Hb: hemoglobin
Hb A_{1c}: glycosylated hemoglobin
Hct: hematocrit
HDL: high-density lipoprotein
HIV: human immunodeficiency virus
HMO: health maintenance organization
hs: at bedtime
I and O: intake and output
IDDM: insulin-dependent diabetes mellitus
IDL: intermediate density lipoprotein
IF: intrinsic factor
IHD: ischemic heart disease
INQ: index of nutritional quality
IPD: intermittent peritoneal dialysis
IU: international unit
IV: intravenous
JCAHO: Joint Commission on Accreditation of Healthcare Organizations
JNC: Joint National Committee on Detection, Evaluation, and Treatment of High Blood Pressure
kcal: kilocalorie
kg: kilogram
L: liter
lb: pound
LBW: low birth weight
LCT: long-chain triglyceride
LDL: low-density lipoprotein
LDOPA: levodopa
LES: lower esophageal sphincter
LTT: lactose tolerance test
MAC: *mycobacterium avium* complex
MAC: midarm circumference
MAMC: midarm muscle circumference
MAO: monoamine oxidase
mcg: microgram; also μg
MCT: medium-chain triglyceride
MDRD: Modification of Diet in Renal Disease
mEq: milliequivalent
mg: milligram
MI: myocardial infarction
ml: milliliter
mm: millimeter
MSG: monosodium glutamate
MSUD: maple syrup urine disease
NCEP: National Cholesterol Education Program
NE: niacin equivalent

NFCS: National Food Consumption Survey
NG: nasogastric
NHANES: National Health and Nutrition Examination Survey
NIDDM: non-insulin-dependent diabetes mellitus
NLEA: Nutrition Labeling and Education Act
NMF: National Marine Fisheries
NPN: nonprotein nitrogen
NPO: nothing by mouth
NRC: National Research Council
NSAID: nonsteroidal anti-inflammatory drug
OBRA: Omnibus Budget Reconciliation Act
OGTT: oral glucose tolerance test
OTC: over the counter
oz: ounce
PABA: para-aminobenzoic acid
PBI: protein bound iodine
pc: after meals
PCP: *pneumocystis carinii* pneumonia
PEG: percutaneous endoscopic gastrosomy
PEM: protein-energy malnutrition
PER: protein efficiency ratio
pH: hydrogen ion concentration
PICC: peripherally inserted central catheter
PIH: pregnancy-induced hypertension
PKU: phenylketonuria
PMS: premenstrual syndrome
POMR: problem-oriented medical record
ppm: parts per million
PPN: peripheral parenteral nutrition
PTH: parathyroid hormone
PUFA: polyunsaturated fatty acid
qid: four times daily
RBC: red blood cells
RD: Registered Dietitian

RDA: Recommended Dietary Allowance
RDI: Reference Daily Intake
RE: retinol equivalent
REE: resting energy expenditure
RME: resting metabolic expenditure
RMR: resting metabolic rate
RNA: ribonucleic acid
RQ: respiratory quotient
SDA: specific dynamic action
SGA: small for gestational age
SGOT: serum glutamate oxalacetate transaminase
SI: Système International
SMA: sequential multiple analyzer
SOAP: subjective, objective, assessment, plan
TBSA: total body surface area
tbsp: tablespoon
TCA: tricarboxylic acid cycle
α-TE: alpha-tocopherol equivalent
TEE: total energy expenditure
TG: triglyceride
TIBC: total iron-binding capacity
tid: three times daily
TLC: total lymphocyte count
TPN: total parenteral nutrition
TSF: triceps skinfold
TSH: thyroid-stimulating hormone
tsp: teaspoon
UNICEF: United Nations Childrens Fund
USDA: United States Department of Agriculture
VLCD: very low calorie diets
VLDL: very-low-density lipoprotein
WBC: white blood cells
WHO: World Health Organization
WIC: Women, Infants, and Children

Glossary

A

Acetylcholine: a body substance that transmits nerve impulses

Acid-producing foods: foods in which the anions remaining in the body after metabolism exceed the cations

Acrolein: an irritating substance that results from the breakdown of glycerol by high heat

Adipocytes: fat cells

Adipose: fat

Aerobic: requiring oxygen

Aerophagia: swallowing air

Aging: the continuum of change that occurs from birth to death

Albumin: protein in blood for maintenance of fluid and acid-base balance

Alkali-producing foods: foods in which the cations remaining in the body after metabolism exceed the anions

Alkaloid: a substance in leaves, seeds, or bark of plants; usually bitter and alkaline in reaction

Ambient temperature: temperature in one's immediate location

Amenorrhea: absence of menstruation

Amino acids: organic acids that contain an amino group (NH_2) and an acid or carboxyl group (COOH) attached to the same carbon

Amphetamine: central nervous system stimulant

Amylophagia: starch eating

Anabolic steroids: derivatives of the male sex hormone testosterone

Anaerobic: without oxygen

Analgesic: pain reliever

Anaphylaxis: severe allergic reaction

Android: manlike

Anergy: failure of the immune system to recognize and respond to foreign proteins

Angiotensin II: a vasopressor

Anion: an ion carrying a negative charge of electricity

Anorexia: lack of appetite

Anorexia nervosa: self-imposed starvation

Anthropometry: science of measuring the human body

Anticholinergic: reducing acid secretion

Antigen: a substance, usually foreign to the organism, that causes formation of antibodies specific to it

Antioxidant: substance added to foods or present in the body, to prevent or delay interaction of another substance with oxygen

Antipyretic: fever-reducing agent

Antispasmodic: preventing or relieving spasm

Antitussive: agent that prevents or relieves coughing

Anuria: lack of urine or < 100 ml urine per day

Appetite: pleasurable sensations provided by food

Areola: the pigmented area surrounding the nipple

Ascites: fluid accumulation in the abdominal cavity

Ascorbic: without scurvy

Aspiration: drawing in or out as by suction

Asterixis: flapping tremor of the arms and legs

Atheroma: soft, mushy deposits of fatty materials in the lining of blood vessels

Autoimmune disease: disease in which the body produces immune response against itself

B

Bacterial infection: illness produced by rapid bacterial growth

Bacterial intoxication: illness resulting from toxin produced by bacterial growth

Bariatric: pertaining to prevention, control, or treatment of obesity

Basal metabolism: energy needs of the body while at rest

Behavior modification: manipulating environmental conditions to bring about a desired behavioral change

Beriberi: disease of thiamin deficiency

Bifidus factor: a factor in human milk that promotes growth of desirable intestinal bacteria in the infant

Biliary stasis: stagnation of bile flow

Biliary tract: organs and ducts involved in the secretion, storage, and delivery of bile to the duodenum

Bioavailability: the amount of nutrients absorbed from the intestinal tract

Biologic value: the percentage of absorbed nitrogen retained by the body

Biotechnology: application of living organisms and systems to technical, industrial, and medical processes

Blood pressure: pressure exerted by blood on artery walls

Body mass index: a measure of body mass or fat; weight (kg) ÷ height $(m)^2$

Bonding: the close relationship that is established when the baby is cuddled while being fed

Bradycardia: lowered heart rate

Bronchi: branches from trachea to lungs, providing passage for air movement

Buffer: a substance that can react with an acid or alkali without much change occurring in the pH

Bulimia nervosa: bingeing and inappropriate compensatory behavior

C

Cachexia: malnutrition and wasting that responds poorly to nutritional therapy

Calcitonin: hormone secreted by the thyroid gland that lowers blood calcium

Carbonic anhydrase: zinc-containing enzyme that releases carbon dioxide from the red blood cells

Cardiovascular: pertaining to the heart and blood vessels

Carotene: pigment in yellow and deep-green vegetables and fruits

Carotenoids: compounds in yellow and green vegetables and fruits; related to vitamin A

Catalyst: a substance that increases a chemical reaction without becoming a part of the compound that is formed

Cation: an ion carrying a positive charge of electricity

Cell color: normochromic; hypochromic (low color); hyperchromic (high color)

Cell size: normocytic; microcytic (small); macrocytic (large)

Challenge test: giving an amount of a substance sufficient to test the body's metabolic ability to respond

Cheilosis: cracking of skin at the corners of the lips

Chelate: bind with a metal

Cholecystectomy: removal of the gallbladder

Cholecystitis: inflammation of the gallbladder

Cholelithiasis: formation or presence of stones in the gallbladder or common duct

Chronic disease: any impairment or variation from normal with one or more of the following: it is permanent, leaves residual disability, is caused by nonreversible pathological alterations, requires special training for rehabilitation, or is expected to require a long period of supervision or observation

Chylomicrons: microscopic fat particles that give milky appearance to the lymph; consist chiefly of triglycerides

Cilia: hairlike projections from epithelial cells

Claudication: ischemia caused by narrowing of the arteries

Cofactor: a mineral element that helps to bring about the activity of an enzyme

Collagen: protein matrix of cartilage, connective tissue, and bone

Colostomy: opening of colon onto abdominal wall

Colostrum: thin, yellowish fluid first secreted from the breast after delivery

Complete protein: contains the kinds and amounts of essential amino acids to support maintenance and growth

Cretinism: severe thyroid deficiency in prenatal life leading to marked physical and mental retardation

Cross reactivity: allergy to foods of the same botanical group

Cruciferous: vegetables whose flowers have the shape of a cross

Cyanosis: bluish, purplish color of skin or mucous membranes due to lack of oxygen

Cyclamate: an artificial sweetener not currently allowed in the United States

Cytoprotective: protecting the cell

D

Daily Reference Value: nutrients not included in RDA: used in labeling

Deamination: removal of the amino group from an amino acid

Decompensation: inability of the heart to maintain adequate circulation

Decarboxylase: thiamin-containing enzyme that removes carbon dioxide

Dehydration: reduced body water

Dental caries: tooth decay

Diastolic: when the heart is relaxing

Diet: the kinds of foods and beverages consumed

Dietary fibers: indigestible plant matter

Diglyceride: two fatty acids linked to a glycerol molecule

Distress: negative stress

Diverticula: tiny sacs or pouches in the intestinal wall

Diverticulitis: inflamed diverticula

Diverticulosis: presence of diverticula

dl: deciliter; 100 ml

Double blind: investigation method in which neither the subject nor the researcher knows what treatment, if any, the subject is receiving

Dysgeusia: altered, unpleasant, or uncharacteristic taste sensation

Dyspepsia: indigestion

Dysphagia: difficulty in swallowing, or transfer of food from mouth to stomach

Dyspneic: labored breathing due to air hunger

E

Eclampsia: convulsions or coma that are the end result of untreated preeclampsia

Ectomorph: thin, angular body type

Eczema: inflammation of the skin

Edema: accumulation of excess fluid in the tissues

Electrolyte: any compound that in solution breaks up into its constituent ions

Embolism: obstruction of a blood vessel by a foreign substance or blood clot

Emetic: substance that induces vomiting

Emulsion: liquid in which particles of one substance are suspended in another

Endemic: peculiar to a particular people or locality

Endocrine: internal secretion of a gland

Endogenous: from within the body

Endomorph: round, soft body type

Energy: the power that keeps the body functioning

Enrichment: the addition of thiamin, riboflavin, niacin, and iron (in 1998 folate will be included) to grain foods; state, not federal, law

Enteral: involving the intestines

Enteric coated: tablets or capsules coated with a substance that will not dissolve until reaching the small intestine

Enteritis: inflammation of the mucous tissue of the small intestine

Enteropathy: any intestinal disease

Enzyme: a living catalyst

Epidemiology: study of the patterns of occurrence of disease

Epigastric: over the pit of the stomach

Equivalent weight: atomic weight ÷ valence

Ergogenic: work improving

Erythema: redness

Erythroblasts: large nucleated red cells produced in the bone marrow

Erythropoietin: a hormone that stimulates the bone marrow to produce red blood cells

Esophageal varices: twisted, engorged veins in the lower esophagus

Essential amino acid: an amino acid that cannot be synthesized in the body and must be supplied by the diet

Essential hypertension: elevated blood pressure of unknown cause

Estrogen: female sex hormone

Etiology: pertaining to the cause of disease

Eustress: positive stress

Exchange (food): an amount of food that has approximately the same nutritive values as other foods in the lists into which they are grouped

Exocrine: external secretion of a gland

Exogenous: from outside the body

Extracellular: outside the cells

Extrusion reflex: tongue movements in the infant that push food out from the mouth

Exudate: fluid leaking from capillaries

F

Fad: a style or custom that many people adopt for a short time

Familial: common in the same family

Fatty acid: compound in foods and the body that contains carbon, hydrogen, and oxygen (hydrocarbon); component of other compounds, e.g., fats and cell membranes

Febrile: pertaining to a fever

Ferritin: storage form of iron in the liver, spleen, and bone marrow

Fetor hepaticus: fecal odor on the breath

Fever: elevation of temperature above normal

Fluorosis: discoloration or mottling of the teeth caused by excessive fluoride intake

Food: anything that nourishes the body

Food additives: substances present in food as a result of any aspect of agriculture, processing, and commercial or household handling

Food lore: teaching and stories about food, passed from generation to generation of people

Fortification: the addition of one or more nutrients that may or may not have been present in the original food

G

Gastrectomy: removal of all or part of the stomach

Gastroparesis: delayed emptying of the stomach

Gastroplasty: plastic surgery on the stomach

Generic: general; not protected by trademark registration

Genetic engineering: repair or modification of hereditary cellular material by technical methods

Geophagia: clay or dirt eating

Geriatrics: health science dealing with diseases, debilities, and care of aging persons

Gerontology: the study of aging

Gestational diabetes: glucose intolerance first occurring during pregnancy

Gliadin: protein fraction of gluten

Gluconeogensis: formation of glucose from noncarbohydrate sources

Glucose polymer: substance with repeating glucose units

Glucose tolerance factor: a chromium-containing organic compound that improves the action of insulin on the uptake of glucose by the cells

Glucosuria: glucose in the urine

Gluteal: pertaining to the buttocks

Gluten: a protein in some grains

Glycemic response: the rise in blood glucose produced by the carbohydrate of a given food

Glycogenesis: formation of glycogen from glucose

Glycogenolysis: hydrolysis of glycogen to glucose

Glycolipids: fats in which glucose or galactose replaces one of the fatty acids

Glycosylated hemoglobin: hemoglobin to which glucose is attached

Goiter: enlarged thyroid caused by iodine deficiency

Goitrogen: an antithyroid compound found in brussels sprouts, cabbage, cauliflower, radishes, rutabagas, and turnips

Gynoid: womanlike

H

HDL: high-density lipoproteins; high in protein; appear to reduce the risk of coronary heart disease

Health: state of complete physical, mental, and social well-being

Heartburn: burning sensation caused by reflux of gastric contents into esophagus

Hematocrit: packed red cell volume

Hematuria: blood in the urine

Heme iron: iron associated with heme, the red coloring matter in hemoglobin

Hemochromatosis: excessive iron deposits leading to organ damage

Hemoglobin: iron-containing pigment in red blood cells

Hemolysis: destruction of red blood cells

Hemosiderosis: excessive iron stores without tissue damage

Hepatic: relating to the liver

Hepatic encephalopathy: neurological and neuromuscular abnormalities in patients with severe liver dysfunction

Hepatomegaly: enlarged liver

Hiatus hernia: abnormal gap in the diaphragm through which part of the stomach protrudes

Hormone: a substance produced by one organ and carried in the blood to stimulate another organ or tissue

Hunger: controls within the body that stimulate the consumption of food

Hydrogenation: addition of hydrogen atoms to double-bond carbons in the fatty acid molecule

Hydrolysis: the splitting of a compound by the addition of water; for example, starch to glucose

Hydroxylation: the introduction of hydroxyl (–OH) groups into various compounds

Hyperemesis gravidarum: pernicious vomiting of pregnancy

Hyperesthesia: acute sense of pain, heat, cold, and touch

Hyperglycemia: elevated blood sugar

Hyperkalemia: abnormal elevation of blood potassium level

Hyperlipidemia: elevation of any blood lipids

Hypermetabolic: increased or excessive metabolic rate

Hyperosmolar: having osmolality greater than body fluids

Hyperoxaluria: increased oxalate in the urine

Hyperplasia: excessive number of normal cells

Hypertension: elevated blood pressure

Hypervitaminosis: excessive accumulation of vitamin in body stores leading to toxic symptoms

Hypoallergenic: lacking potential to cause an allergic reaction

Hypogeusia: diminished taste sensation

Hypoglycemia: low blood sugar

Hyposmia: diminished sense of smell

Hypotension: low blood pressure

Hypothermia: temperature below a set point

Hypothesis: a theory; a guide to experimental investigation

Hypovolemia: diminished blood supply

Hypoxia: diminished oxygen supply

I

Iatrogenic: doctor or hospital-induced

Idiopathic: disease without known cause

IDL: intermediate-density lipoproteins; contain residues remaining after removal of triglycerides from VLDL

Ileocecal valve: valve between ileum and colon

Ileostomy: opening of ileum onto the abdominal wall

Ileus: cessation of peristalsis

Immunocompetence: ability to develop an immune response

Inborn error of metabolism: genetic enzyme deficiency causing metabolic dysfunction

Incomplete protein: protein in which one or more essential amino acids are absent

Indole: nitrogenous compound responsible for the strong odors associated with foods of the cabbage family

Induration: hardening

Infarction: death of tissue due to inadequate blood supply

Infection: invasion of the body by a pathogenic agent that multiplies and produces harmful effects

Initiator (carcinogen): a substance or event that changes a normal cell into a potentially cancerous cell

Insensible perspiration: water evaporated from skin as fast as it is excreted

Intentional additive: any substance of known composition that is added to food to serve some useful purpose

Interstitial: situated between cells or tissues

Intracellular: within the cell or cells

Intrinsic factor: a mucoprotein produced in the stomach

Iodized salt: salt to which potassium iodide is added; 1 part potassium iodide to 10,000 parts salt

Ion: an atom or group of atoms carrying a charge of electricity

Irradiation: technique using energy, rays, or ions; used in commercial synthesis of vitamin D, or on food to kill microorganisms causing foodborne illness

Ischemia: deficiency of blood supply to an area of the body

Isomer: any compound containing the same chemical composition, but able to exist in more than one configuration

Isotonic: having the same osmolality as body fluids

J

Jaundice: yellow color of the skin

K

Kwashiorkor: a protein-deficiency disease

L

Lacto-ovo-vegetarian diet: diet including plant foods, milk, milk products, and eggs

Lacto-vegetarian diet: diet including plant foods, milk, and milk products

Lactulose: synthetic compound of lactose and fructose

Lanugo: growth of fine hair on the skin

LDL: low density lipoproteins; synthesized by liver from IDL residues, chief carrier of cholesterol

Lignin: woody part of plants; not a carbohydrate

Lipogenesis: the synthesis of new fats

Lipolysis: breakdown of fats

Liposuction: suctioning of body fat

Lithotripsy: shattering of stones by ultrasonic shock waves

Lower esophageal sphincter: band of muscle between the esophagus and stomach

Lumen: passage inside a tubular organ

M

Macrobiotic diet: strict vegetarian diet consisting of several stages

Macronutrients: nutrients occurring in largest amounts in foods and in the body

Malnutrition: impairment of health resulting from a deficiency, excess, or imbalance of nutrient intake or utilization

Marasmus: a protein energy deficiency disease

Medical nutrition therapy: use of specific nutrition services to treat an illness, injury, or condition

Medium chain triglycerides: fats with 8–12 carbon chains that do not exist in nature and must be manufactured

Megadose: an amount exceeding the RDA by 10 times or more

Megaloblasts: larger nucleated cells that are precursors of mature red cells

mEq/liter: mg/liter ÷ equivalent weight

Mesomorph: muscular, athletic body type

Metastasis: appearance of a neoplasm in parts of the body away from the primary tumor

Micelles: aggregation of tiny molecules held in suspension, e.g., clumps of fatty acids held within the watery contents of the small intestine

Micronutrients (trace nutrients): nutrients occurring in very small amounts in foods and in the body

Milliequivalent (mEq): weight in milligrams of an element that combines with or replaces 1 mg hydrogen

Miscible: capable of being mixed

Monoglyceride: a single fatty acid linked to a glycerol molecule

Myocardium: heart muscle

Myoglobin: iron-containing protein in muscle

Myxedema: severe hypothyroidism

N

Necrosis: death of living tissue

Neonatal: period immediately after birth through the first month of life

Neoplasm: new uncontrolled growth of cells; tumor

Nephropathy: disease of the kidney

Neuroleptic: producing symptoms similar to diseases of the nervous system

Neuropathy: disease of the nerves

Niacin equivalent: 1 mg niacin or 60 mg tryptophan

Nitrogen equilibrium: balance of the intake and excretion of protein (nitrogen)

Nocturia: urination, especially excessive, during the night

Norepinephrine: hormone produced by the adrenal; a vasoconstrictor

Nosocomial: pertaining to a hospital

Nursing diagnosis: statement of a health-related problem or potential problem that can be treated legally and independently by nurses

Nutrient density: the quantity of protein, vitamin(s), and/or mineral(s) supplied by a food in relation to its caloric content

Nutrients: the 50 or more substances needed by the body

Nutrition: processes by which the body utilizes food

Nutrition support: enteral and/or parenteral feedings to prevent or reverse malnutrition

Nutritional status: health as it is related to use of food by the body

Nystagmus: constant, involuntary movement of the eyeball

O

Obesity: excessive body fat

Obligatory excretion: the amount of fluid required to hold nitrogenous and mineral wastes in solution

Occlusion: closing or shutting off

Omega fatty acid: naming system applied to unsaturated fatty acids, which identifies the first double bond, counting from the methyl (CH_3) end of the chain

Opportunistic infection: infection that exploits the opportunity of a compromised immune system

Osmolality: number and size of particles per kg of water

Ossification: to change into bone

Osteodystrophy: defective bone development

Osteomalacia: softening of bones

Osteoporosis: reduction in quantity of bone

Overweight: body weight in excess of some standard

Oxalic acid: an organic acid found in rhubarb leaves, cocoa, and some greens

P

Parathyroid hormone: hormone secreted by the parathyroid gland when blood calcium is low

Parenteral: route other than the alimentary canal; usually intravenous or subcutaneous

Pellagra: disease of niacin deficiency

Peptic ulcer: erosion of the gastrointestinal mucosa

Peptide linkage: linkage of the amino group of one amino acid to the carboxyl group of another amino acid

Periodontal disease: disease of the gums, bone, and other supporting structures of the teeth

Peristalsis: progressive involuntary wavelike movement occurring in hollow tubes in the body, especially the intestines

Peritoneum: membrane lining the abdominal cavity and surrounding the abdominal organs

pH: symbol used to express concentration of hydrogen ions in a solution

Phenylketonuria: inborn error of metabolism with inability to metabolize phenylalanine

Phospholipids: fats in which a phosphate and a nitrogen group have been substituted for one of the fatty acids

Photosynthesis: process by which plants manufacture carbohydrates from carbon dioxide and water

Phytic acid: an organic acid found in the outer layers of grains

Phytochemicals: compounds naturally present in plants

Pica: craving for and consumption of substances generally considered not edible

Placebo: an inactive substance that looks and tastes like the compound being tested

Platelet aggregation: clumping of platelets at the site of an injury; can cause arterial blockage

Polyneuritis: inflammation of two or more nerves

Prealbumin: serum protein with short half-life

Precursor (vitamin): compound that can be changed into the active vitamin

Preeclampsia: appearance of hypertension, edema, and proteinuria around the twentieth week of pregnancy

Preformed vitamin: active form of a vitamin

Premenstrual syndrome (PMS): group of physical and emotional changes preceding the onset of menstruation

Prevention, primary: activities to reduce the risk or prevent the occurrence of disease or disability

Prevention, secondary: early diagnosis and prompt intervention to halt the progress or reduce the severity of a disease

Prevention, tertiary: therapies to limit sequelae of diagnosed chronic disease and to promote rehabilitation

Promoter (cancer): any factor that causes cancerous cells to grow and multiply

Prostaglandins: a group of hormone-like substances

Proteases: protein splitting enzymes

Protein-sparing: body's use of carbohydrate and fat for energy, rather than protein

Pulmonary: relating to the lungs

Pulmonary edema: fluid accumulation in the lungs

Q

Quackery (foods): pertaining to false claims for the health virtues or curative properties of foods

R

Rancidity: the change in flavor and odor of fats when they are oxidized

Recumbent: lying down

Reference Daily Intakes: protein, vitamin, and mineral values used in labeling

Reflux: backward flow

Remodeling (bone): continuous process of maintenance and repair

Renal: relating to the kidney

Renal threshold: point at which a substance in the blood such as glucose spills into the urine

Resorption: removal of bone mineral

Respiratory quotient: ratio of CO_2 produced to O_2 consumed

Restored: addition of nutrients of a processed food so it has the same value as the original food

Retinol-binding protein: linkage of vitamin A with protein for transport by the blood to the tissues

Retinopathy: disease of the retina

Retrovirus: virus with reverse coding

Rhodopsin (visual purple): pigment in rods of retina sensitive to light

Rickets: disease of vitamin D deficiency

Rooting reflex: response to touch that causes the infant to turn toward the nipple

S

Saponification: combination of a fatty acid with a cation such as calcium to form a soap

Satiety: state of being satisfied; internal controls that shut off food consumption

Sclerosis: hardening

Scurvy: disease of vitamin C deficiency

Sepsis: disease-producing organisms in the blood

Seroconversion: development of evidence of antibody response to a disease or vaccine

Serotonin: a compound that transmits nerve impulses; a vasoconstrictor

Somatotype: body type

Stomatitis: sore mouth

Stress: any event that threatens the body's steady state or homeostasis

Stress ulcers: superficial erosions of the gastric mucosa without previous erosion

Subscapular: below the shoulder

Sympathometric: sympathetic nervous system stimulant

Synergism: interaction yielding a total effect greater than the sum of its parts

Systemic: pertaining to the whole body rather than one of its parts

Systemic anaphylaxis: severe allergic reaction

Systolic: when the heart is contracting

T

Thrombus: blood clot that obstructs a blood vessel

Tophi: deposits of uric acid crystals in the joints

Trabeculae: lacelike deposits of calcium and phosphorus in the ends of the bones

Transferrin: protein for transporting iron in the blood

Trauma: injury caused by external force or violence, e.g., multiple fractures, gunshot wounds, stab wounds, crushing injuries

Triceps: muscle in the upper arm

Triglyceride: three fatty acids linked to a glycerol molecule

U

Ulcerative colitis: chronic inflammation of the mucosa of the colon

Uremia: an excess of urea and other wastes in the blood

V

Vagotomy: severing of the vagus nerve

Vegan diet: includes only plant foods

Ventromedial center: satiety center in the hypothalamus

Viscera: internal body organs

Vitamin: an organic chemical compound that occurs in minute quantities and is necessary for life and growth

VLDL: very-low-density-lipoproteins; consist primarily of triglycerides

W

Wilson's disease: inborn error of metabolism involving excess deposit of copper in the liver and other organs

X

Xerophthalmia: dryness of the eye caused by vitamin A deficiency

Xerostomia: dry mouth

Index

Information provided in tables is indicated by the letter t following the page number, boldface type is used to indicate items appearing in margins; illustrations are identified by italic type

Median Heights and Weights and Recommended Energy Intake

Category	Age (years) or Condition	WEIGHT (kg)	WEIGHT (lb)	HEIGHT (cm)	HEIGHT (in)	REE (kcal/day)	AVERAGE ENERGY ALLOWANCE (kcal)[a] Multiples of REE	AVERAGE ENERGY ALLOWANCE (kcal)[a] Per kg	AVERAGE ENERGY ALLOWANCE (kcal)[a] Per day[b]
Infants	0.0–0.5	6	13	60	24	320		108	650
	0.5–1.0	9	20	71	28	500		98	850
Children	1–3	13	29	90	35	740		102	1,300
	4–6	20	44	112	44	950		90	1,800
	7–10	28	62	132	52	1,130		70	2,000
Males	11–14	45	99	157	62	1,440	1.70	55	2,500
	15–18	66	145	176	69	1,760	1.67	45	3,000
	19–24	72	160	177	70	1,780	1.67	40	2,900
	25–50	79	174	176	70	1,800	1.60	37	2,900
	51+	77	170	173	68	1,530	1.50	30	2,300
Females	11–14	46	101	157	62	1,310	1.67	47	2,200
	15–18	55	120	163	64	1,370	1.60	40	2,200
	19–24	58	128	164	65	1,350	1.60	38	2,200
	25–50	63	138	163	64	1,380	1.55	36	2,200
	51+	65	143	160	63	1,280	1.50	30	1,900
Pregnant	1st trimester								+0
	2nd trimester								+300
	3rd trimester								+300
Lactating	1st 6 months								+500
	2nd 6 months								+500

Source: National Academy of Sciences, *Recommended Dietary Allowances,* 10th ed., National Academy Press, Washington, DC, 1989.

[a]In the range of light to moderate activity, the coefficient of variation is ±20%.

[b]Figure is rounded.

Approximate Energy and Protein Requirements During Stress (based on healthy weight)

	kcal per kilogram	kcal per pound
Energy Requirements		
Bed rest	23–25	11–12
Infection, mild to severe	28–40	13–18
Surgery, major	28–30	13–14
Injury	31–40	14–18
Burns	35–48	16–22

Protein Requirements (calories to nitrogen ratio)
Normal: 225:1 to 175:1
Stress: 150:1
Severe stress: 100:1–80:1
Nitrogen, g × 6.25 = Protein, g